The Routledge Companion to Environmental Planning

This Companion presents a distinctive approach to environmental planning by: situating the debate in its social, cultural, political and institutional context; being attentive to depth and breadth of discussions; providing up-to-date accounts of the contemporary practices in environmental planning and their changes over time; adopting multiple theoretical and analytical lenses and different disciplinary approaches; and drawing on knowledge and expertise of a wide range of leading international scholars from across the social science disciplines and beyond.

It aims to provide critical reviews of the state-of-the-art theoretical and practical approaches as well as empirical knowledge and understandings of environmental planning; encourage dialogue across disciplines and national policy contexts about a wide range of environmental planning themes; and, engage with and reflect on politics, policies, practices and decision-making tools in environmental planning. The Companion provides a deeper understanding of the interdependencies between the themes in the four parts of the book (Understanding 'the environment', Environmental governance, Critical environmental pressures and responses, and Methods and approaches to environmental planning) and its 37 chapters. It presents critical perspectives on the role of meanings, values, governance, approaches and participations in environmental planning. Situating environmental planning debates in the wider ecological, political, ethical, institutional, social and cultural debates, it aims to shine light on some of the critical journeys that we have traversed and those that we are yet to navigate and their implications for environmental planning research and practice.

The Companion provides a reference point mapping out the terrain of environmental planning in an international and multidisciplinary context. The depth and breadth of discussions by leading international scholars make it relevant to and useful for those who are curious about, wish to learn more, want to make sense of, and care for the environment within the field of environmental planning and beyond.

Simin Davoudi is Professor of Environmental Policy and Planning and Director of the Global Urban Research Unit (GURU) at the School of Architecture, Planning and Landscape, Newcastle University, UK.

Richard Cowell is Professor of Environmental Planning at the School of Geography and Planning, Cardiff University, UK.

Iain White is Professor of Environmental Planning, the University of Waikato, New Zealand.

Hilda Blanco is Emeritus Professor and Project Director at the Centre for Sustainable Cities at Sol Price School of Public Policy, the University of Southern California, USA.

The Routledge Companion to Environmental Planning

The Routledge Companion to Environmental Planning

Edited by Simin Davoudi, Richard Cowell, Iain White and Hilda Blanco

LONDON AND NEW YORK

First published 2020
by Routledge
4 Park Square, Milton Park, Abingdon, Oxon OX14 4RN

and by Routledge
605 Third Avenue, New York, NY 10017

First issued in paperback 2023

Routledge is an imprint of the Taylor & Francis Group, an informa business

British Library Cataloguing-in-Publication Data
A catalogue record for this book is available from the British Library

Library of Congress Cataloging-in-Publication Data
Names: Davoudi, Simin, editor. | Cowell, Richard, editor. | White, Iain, editor. | Blanco, Hilda, editor.
Title: The Routledge companion to environmental planning / edited by Simin Davoudi, Richard Cowell, Iain White and Hilda Blanco.
Description: Abingdon, Oxon ; New York, NY : Routledge, 2020. | Includes bibliographical references and index.
Subjects: LCSH: Environmental management—Planning. | Environmental protection—Planning. | Environmental policy. | Sustainable development.
Classification: LCC GE300 .R68 2020 | DDC 333.7—dc23
LC record available at https://lccn.loc.gov/2019018523

ISBN: 978-1-03-257000-6 (pbk)
ISBN: 978-1-138-89480-8 (hbk)
ISBN: 978-1-315-17978-0 (ebk)

DOI: 10.4324/9781315179780

Typeset in Bembo
by Apex CoVantage, LLC

Publisher's Note
The publisher has gone to great lengths to ensure the quality of this reprint but points out that some imperfections in the original copies may be apparent.

Contents

Figures

Tables

Boxes

Contributors

Saad Almutairi is a PhD student at the School of Engineering, Newcastle University, UK. His PhD research is investigating the impacts of increased uptake of electric vehicles on air quality and health.

Tomas Badura is a Senior Research Associate at the Centre for Social and Economic Research on the Global Environment (CSERGE) at the University of East Anglia. Tomas is an economist focusing on valuing preferences for ecosystem related goods and services and natural capital accounting. His interest lies in how this information can support policy and decision making across spatial, temporal and decision scales. Tomas took part in a number of leading initiatives in this area, including TEEB, UKNEA-FO and KIP-INCA. Previous to his PhD research at UEA, Tomas worked at the Institute for European Environmental Policy.

John Barry is Professor of Green Political Economy in the School of History, Anthropology, Philosophy and Politics at Queens University Belfast. His areas of research include green moral and political theory; green and heterodox political economy; the social economy and normative aspects of environmental and sustainable development politics and policy; action and engaged research; the greening of citizenship and civic republicanism. His latest book is *The politics of actually existing unsustainability: human flourishing in a climate-changed, carbon-constrained world* (2012, Oxford University Press). He is currently working on a book provisionally entitled *The story of unsustainable growth: understanding economic growth as ideology, myth and religion.*

Riccardo Beltramo is Full Professor at the University of Torino, Department of Management, Area of Commodity Sciences. Since 1985, he has been involved in the field of Industrial Ecology and researches on integrated management systems, applied to manufacturing and service sectors as well as industrial areas. He currently teaches Operations Management, Industrial Ecology and Eco-management of Tourism. In the field of Internet of Things, he has invented the Scatol8® System, a remote sensing network to monitor, display and elaborate environmental and management variables, giving rise to an academic spin-off company 'The Scatol8 for Sustainability srl'.

Nathalie Berny is Senior Lecturer in Political Science in Sciences Po Bordeaux and researcher at Centre Émile Durkheim. She has been visiting fellow at the University of Oxford, Nuffield College, and the Graduate Institute in Geneva. Her research interests revolve around the logics of collective action (interest groups politics and social movements), participatory/deliberative democracy as well as policy change in the field of the environment. Comparison of organisations or institutional arrangements is the common thread of her different works on France, the

United Kingdom and the European Union. She has recently published a monography on the French environmental movement.

Ross Beveridge is an Urban Studies Foundation Senior Research Fellow at the University of Glasgow. His main research interests are urban political activism, infrastructure politics and the spatialities of anti-politics.

Hilda Blanco (PhD, City and Regional Planning, UC Berkeley, 1989) has held tenured appointments at Hunter College Department of Urban Affairs (1988–96) and the University of Washington (1996–2009), where she chaired the Department of Urban Design and Planning (2000–7) and is currently an Emeritus Professor. From 2010–16, she was a research professor and Interim Director of the Center for Sustainable Cities at the University of Southern California and is currently the Project Director of the Center. Her research areas include sustainable cities and cities and climate change. She is an associate editor of the *Journal of Environmental Planning and Management.*

Jonathan Boston is Professor of Public Policy in the School of Government at Victoria University of Wellington. He has served at various times as the Director of the Institute of Policy Studies and the Director of the Institute for Governance and Policy Studies. His research interests include child poverty, climate change, tertiary education funding, public management and intergenerational governance. His recent books include *Governing for the future: designing democratic institutions for a better tomorrow* (2017, Emerald) and *Social investment: a New Zealand policy experiment* (co-edited with Derek Gill, 2017, Bridget Williams Books).

Maxwell T. Boykoff is Director of the Center for Science and Technology Policy, which is part of the Cooperative Institute for Research in Environmental Sciences at the University of Colorado Boulder. He is also an Associate Professor in the Environmental Studies programme. He earned a PhD in Environmental Studies from the University of California-Santa Cruz and Bachelor of Sciences in Psychology from The Ohio State University. Max has ongoing interests in cultural politics and environmental governance, science and environmental communications, science-policy interactions, political economy and the environment, and climate adaptation. He has consequently produced many peer reviewed journal articles, book chapters and books in these subjects.

John R. Campbell has been researching population and environment issues in Pacific Island countries since the 1970s. He is currently working on the human dimensions of climate change adaptation and disaster risk reduction including environmental migration. He obtained a PhD at the University of Hawaii where his thesis was on population and environment interrelations on a small island in northern Vanuatu. He has written a book on development and disasters in Fiji, co-authored a book on climate change in Pacific Islands and a number of book chapters and articles on disasters, environmental management and global change, especially in Oceania.

Noel Castree is a Professor of Geography at Manchester University, UK and a Professorial Research Fellow at the University of Wollongong, Australia. He has written about representations of nature, 'race' and environment, most notably in the book *Making sense of nature* (2014, Routledge). He is editor of the journal *Progress in Human Geography* and a co-editor of the *International encyclopedia of geography* (2017, Wiley-Blackwell).

Ellen Clarke graduated from Oxford University and went on to complete an MA in War Studies at King's College London where her research interests included the connection between conflict and environmental risk. She has worked as a research assistant at Sciences-Po Bordeaux and for BirdLife International, a global nature conservation NGO, based in their regional Middle East office in Jordan. She is pursuing a professional career in political risk analysis, building on a previous role which included reporting on environmental activist groups across the globe.

Jon Coaffee is Professor in Urban Geography at the University of Warwick, based in the School of Politics and International Studies. His research has been supported by a significant number of UK and EU Research Council grants linked to building resilience across different socio-technical systems in response to a range of shock and stress events. This work has been published in multiple disciplinary areas and most notably includes: *The everyday resilience of the city* (2008, Palgrave Macmillan), *Urban resilience: planning for risk, crisis and uncertainty* (2016, Macmillan International) and the *The handbook of international resilience* (2016, Routledge).

Matthew Cotton is a Senior Lecturer in Human Geography in the Department of Environment and Geography at the University of York. His research interests lie in the social and ethical dimensions of environmental planning and policy processes, specifically focusing upon the environmental justice dimensions of shale gas development, electricity transmission and distribution systems, and the nuclear fuel cycle. His recent published work includes the monographs *Nuclear waste politics* (2016, Routledge), *Ethics and technology assessment* (2014, Springer) and a co-edited volume *Governing shale gas* (2018, Routledge). His contribution to this volume is supported by the Euratom research and training programme 2014–18 under grant agreement No. 662268.

Richard Cowell is Professor of Environmental Planning at the School of Geography and Planning, Cardiff University. His research examines the interface between planning systems and environmental sustainability, especially governance re-scaling, policy change and sectoral integration, and the relationship between knowledge and decision making. His most recent work focuses on energy transitions, especially land use planning dimensions, and he is currently analysing the environmental implications of Brexit (www.brexitenvironment.co.uk). In addition to the book, *Land and limits: interpreting sustainability in the planning process* (with Susan Owens, 2010, Routledge), Richard has authored numerous academic journal papers and delivered evidence to government, Parliamentary Committees and environmental NGOs.

Robin Curry is an environmental scientist and engineer whose work focuses on evidence-based approaches to sustainability. His research centres on the application and development of a range of metrics including Material Flow Analysis, the Ecological Footprint and Life Cycle Analysis. He has a particular interest in the evaluation and optimisation of bioenergy systems. Robin teaches in the School of Chemistry and Chemical Engineering (SCCE) at Queen's University Belfast.

Simin Davoudi is Professor of Environmental Policy and Planning and Director of the Global Urban Research Unit (GURU) at the School of Architecture, Planning and Landscape at Newcastle University. She is past President of the Association of the European Schools of Planning (AESOP) and Fellow of the: Royal Town Planning Institute; Academy of Social Sciences; and Royal Society of Arts. She has advised various UK government departments and EU Directorate Generals and held visiting professorships in the universities of: Amsterdam; Nijmegen; BTH (Sweden); Virginia Tech; RMIT; and Tampere. Her research and publications cover various aspects of urban planning, environmental governance and resilience. Her latest book is 'The resilience machine' (2018, Routledge).

Mina Di Marino is Associate Professor at the Department of Urban and Regional Planning, Norwegian University of Life Sciences. She holds a Master's degree in Architecture and Doctoral degree in Urban, Regional and Environmental Planning from the Polytechnic of Milan, Italy. Since the 2000s, her research has focused on sustainability and ecology, and their integration into regional and urban planning. She is currently working on green infrastructure and ecosystem services in the contexts of urban growth. She has conducted her research in Italy, Canada, Finland, and more recently, Norway. Recent studies appeared in *Landscape Research* and *Urban Forestry and Greening*.

Geraint Ellis is Professor of Environmental Planning at Queen's University Belfast, an Adjunct Professor at the University of Western Ontario and Editor of the *Journal of Environmental Policy and Planning*. He researches and teaches a variety of issues related to planning and sustainability, particularly energy transition and health in the built environment, and has published widely on these topics. He holds a number of advisory and voluntary positions, including being a Director of Belfast Healthy Cities and an independent member of the Irish Republic's National Economic and Social Council, which advises the Taoiseach on strategic matters. Twitter @ gellis23

Ruben Fernandes is part of the research team of CITTA – Research Centre for Territory, Transports and Environment. To date, he has been involved in projects covering topics related to spatial planning policies, urban metabolism, governance and urban regeneration, transport planning and urban mobility. At the moment, he works as a public officer at Norte Regional Coordination and Development Commission – a decentralised body of the Portuguese Government tasked with implementing regional development policies in the Northern Region of Portugal. He got his Master's degree in Economics and his Doctorate degree in Civil Engineering, both at the University of Porto.

Silvia Ferrini joined The Centre for Social and Economic Research on the Global Environment (CSERGE) in 2007 and is a Research Fellow in Applied Economics. She has a long experience in environmental research projects on water management, environment and finance (Aquamoney (EU-F6), ChREAM 2005–9 and Fessud (EU-F7) 2011–16) and ecosystem services approaches (contributing author of UKNEA-FO). She is specialised in modelling economic and environmental data to inform policy decision making. She has been involved in the KIP-INCA European project and is now leading the international project GROW COLOMBIA to preserve biodiversity in Latin America.

Michael K. Goodman is Professor of Geography in the Department of Geography and Environmental Science at the University of Reading. He is currently a visiting professor at the Centre for Research in Spatial, Environmental and Cultural Politics (SECP), University of Brighton and also the Centre for Place, Space and Society, Wageningen University, Netherlands. His research interests include the cultural politics of food, humanitarianism and climate change, with his most recent work focused on digital foodscapes and the celebritisation of food and the environment. He is currently editing a special issue of *Climatic Change* on 'everyday climate cultures' and one in *The European Journal of Cultural Studies* on 'digital food cultures'.

Michael Gunder is an Associate Professor of Urban Planning at the School of Architecture and Planning, University of Auckland, New Zealand. He is a co-editor of *The handbook of planning theory* (2017, Routledge) and from 2011–15, he was the Managing Editor of the journal *Planning Theory*. His research predominately draws on Lacanian derived poststructuralism to analyse the ideological dimensions of built environment public policies and their related supportive narratives, such as sustainability, resilience, liveability or smart growth and city.

Nicole Gurran is Professor of Urban and Regional Planning at the University of Sydney, where she leads Urban Housing Lab@Sydney and directs the University's AHURI (Australian Housing and Urban Research Institute) research centre. Her research focuses on intersections between urban planning and the housing system and she has led and collaborated on a series of studies on aspects of urban policy, housing, sustainability and planning. She has authored and co-authored numerous publications and books including *Australian urban land use planning: principles, policy and practice* (2011, Sydney University Press) and *Urban planning and the housing market* (with Glen Bramley, 2017, Palgrave).

Nick Hacking has researched waste issues amongst communities and other waste and resources actors since 2009. Most recently he has examined the importance of standards to the future circular economy for the Economic and Social Research Council (ESRC). The likely impact of Brexit on the waste and resources sector has also been an important topic of analysis with a co-authored report produced for the ESRC in 2017. Nick obtained his PhD at Cardiff University with a thesis about using notions of space and place to improve theoretical approaches to innovation linked to hydrogen fuel cells (used to cleanly and efficiently store renewable energy).

Carlo Ingrao obtained a five-year (MSc) Degree in Environment and Land Management Engineering at the University of Catania (Italy), in 2007. Later, in 2012, he was titled as PhD in Geotechnical Engineering at the University of Catania. His research activities and interests are mainly focused upon the development of energy, environmental and economic assessments in the fields of: industrial and environmental engineering; commodity sciences; buildings; biomass and bioenergy; agriculture and food production; and food packaging. Within those themes, since 2013 he has authored and co-authored around fifty publications in international journals, book chapters and conference proceedings: most of them are indexed by Scopus and/or WoS.

Giuseppe Ioppolo graduated in Civil Engineering for Environment, is Associate Professor in Commodity Science (Environmental Management) at the University of Messina (Italy). His main research topics are: sustainability, environmental governance, urban metabolism and environmental resource management tools at local and urban scale. He is Editor-in-chief of the journal *Sustainability*. He was visiting researcher at CML-IE Leiden University on 'Urban metabolism and sustainability', Netherlands; and he spent time as visiting professor at Tsinghua University (China), Tokyo University (Japan) and UCSB California (USA). He is author of about 100 publications in books and international refereed journals.

Kirsi Pauliina Kallio is Senior Researcher at the University of Tampere, RELATE Centre of Excellence, Space and Political Agency Research Group. Her research focuses on political agency from a critical geographical perspective, including theorisation of political subjectivity and lived citizenship, and developing topological and geosocial approaches in spatial theory. She has introduced the idea of political agency as a developing 'human condition', drawing attention to the largely neglected political agencies of children and youth. Her publications include 'Tracing children's politics' in *Political Geography* (2011), 'Shaping subjects in everyday encounters' in *Society and Space* (2016), and the volume *The beginning of politics* (2014, Routledge).

Sue Kidd is an academic and chartered town planner from the University of Liverpool's Department of Geography and Planning. She has acted as an advisor to the EU, government departments, regional and local authorities and NGOs. Sue has an interest in integrated approaches to planning and much of her work has focused on sustainable development in coastal and marine areas. Sue has been at the forefront of the theory and practice of Marine Spatial Planning (MSP) and has been engaged in a variety of ways in assisting the roll out of MSP in European Seas.

Kimmo Lapintie is Professor of Urban and Regional Planning at the Department of Architecture, Aalto University School of Arts, Design and Architecture. He holds a Master's Degree in Architecture and PhD from Tampere University of Technology and a Licentiate Degree from the University of Turku. He has done research on argumentation and participation in planning, theory of space, green concepts and ecosystem services, and the emerging phenomenon of multilocality of living and working. Recent studies have been published by Routledge, International Planning Studies, European Planning Studies, and Library and Information Science Research.

Mick Lennon is a Lecturer in Planning and Environmental Policy in University College Dublin, Ireland. A central strand of his research focuses on how the interpretation of nature influences the formulation and implementation of planning and environmental policy.

Timothy W. Luke is University Distinguished Professor in the Department of Political Science as well as the School of Public and International Affairs at Virginia Polytechnic Institute and State University, Blacksburg, Virginia. He writes about urban political economy and ecology from a research perspective grounded in critical theory, environmental politics, social and political theory in relation to issues in global governance, political economy, and cultural conflict, particularly with regard to the theories and practices of environmental management, resilience, and sustainability.

Juliana Maantay (MUP, PhD) is Professor of Urban Environmental Geography at City University of New York (Lehman College, CUNY Graduate Center, and CUNY School of Public Health) since 1998. She directs the graduate programme in Geographic Information Science as well as the Urban GISc Lab and has edited several compendia and written two widely used textbooks and numerous other publications on the urban environment and geospatial analysis. Her main research foci are environmental justice, health disparities, and exposure/risk assessment, specifically in urban areas. For 25 years prior to her academic career, she was an urban planner, environmental analyst and architect.

Patricia McCarney is Professor of Political Science and the Director of the Global Cities Institute at the University of Toronto, Canada and is President and CEO of the World Council on City Data (WCCD). She has served as Associate Vice President, International Research and Development at the University of Toronto. Professor Patricia McCarney received her PhD from MIT in 1987. Before joining the University of Toronto, between 1983 and 1994, Professor McCarney worked as a professional staff member in a number of international agencies, including the World Bank in Washington, and the UN-Habitat in Nairobi.

Darryn McEvoy is a Research Professor in Urban Resilience and Climate Change Adaptation at the School of Engineering, RMIT University. Employed as an Innovation Professor in 2009, he has extensive experience of leading multi-partner, multi-disciplinary, climate change research projects in Australia and the wider Asia Pacific region. He acts as a scientific adviser to UN-Habitat in the Asia-Pacific region and is currently the scientific lead for the new UNFCCC Adaptation Fund project 'Climate resilient Honiara' in the Solomon Islands (2019–22).

Marisa B. McNatt completed a PhD in Environmental Studies in 2018 from the University of Colorado, Boulder and also earned an MA in Environmental Journalism from CU Boulder. Her dissertation looks at how factors beyond NIMBY (not in my back yard), like political power, narratives, and culture affect outcomes for proposed US offshore wind farms. She is a member of the Media and Climate Change Observatory (MeCCO) that monitors international media coverage of climate change. Marisa has also worked for the Pacific Ocean Energy Trust (POET), an organisation that is committed to the responsible development of marine renewable energy in the Pacific Region.

Antonio Messineo is Full Professor in Applied Physics and Director of the Energy and Environment Laboratory at the University of Enna 'Kore'. He is author of more than 100 scientific papers with more than 50 papers in international journals. His main research topics are related to energy and environmental planning, refrigerating engineering, renewable energies, energy and building. At the moment, his research interests are mainly related to the energetic valorization of biomass. He has been responsible for numerous research projects at regional, national and European level mainly related to energy and environmental planning, and renewable energies.

Jonathan Metzger is an Associate Professor at the Department of Urban Planning and Environment at KTH Royal Institute of Technology in Stockholm. He has a broad social scientific background and concrete experiences from working as a planning practitioner on the regional and transnational levels. Most of his research deals with decision making concerning complex environmental issues – often (but not exclusively) with a focus on urban and regional policy and politics.

David Mitchell is an Associate Professor with the School of Science, RMIT University. He holds a PhD from RMIT University, a degree in Surveying, and postgraduate studies in Urban Development. His research focus is on the impact of climate change and natural disasters on land tenure, and he explores how safeguarding land tenure rights and effective land use planning can enhance climate resilient pathways. David has extensive experience in research and international consulting in Asia and the Pacific; including Indonesia, Cambodia, Lao, Philippines, Fiji, and the Solomon Islands.

Petter Næss is Professor in Planning in Urban Regions at the Norwegian University of Life Sciences. He has contributed theoretically on urban sustainability, philosophy of science within the field of spatial planning and methodology within infrastructure planning. Næss has published extensively on urban sustainability issues covering a wide, interdisciplinary perspective, including sustainability-relevant impacts of land use and the built environment as well as societal conditions and processes influencing urban development (including the role of public planning). A special interest in his research is the relationship between urban structures and transport.

Anil Namdeo is a Chartered Environmentalist, a Chartered Scientist and a Reader in Air Quality Management at Newcastle University, UK. He has special research interests on traffic emissions, air quality and health. He is a member of the Steering Committee, Institute of Air Quality Management, UK. He was a member of the Transportation Research Board (TRB) Committee on Transportation and Air Quality (2014–18) and a topic expert for NICE (National Institute for Health and Care Excellence) Public Health Advisory Committee on Air Pollution (Outdoor Air Quality and Health). He has been involved in several research projects on environmental and sustainability assessment of land use and transport policies. He is actively engaged in international research collaborations.

Eoin O'Neill is Associate Professor of Environmental Policy at University College Dublin (Earth Institute, Geary Institute, School of Architecture, Planning and Environmental Policy) and a Head of School. His research interests include risk perception and behaviour, and planning and environmental policy design and he has published widely on these in international journals. Prior to being appointed at UCD he worked as a Technical Specialist in Flood Risk Management at the UK Environment Agency. He is a chartered planner with the Royal Town Planning Institute (MRTPI).

James Palmer is a Lecturer in Environmental Governance at the School of Geographical Sciences, University of Bristol. His research examines the interactions between knowledge, expertise and environmental policy making, and has focused in particular on controversies surrounding bioenergy production, climate geoengineering and road vehicle emissions. He obtained his PhD in Geography from the University of Cambridge in 2013.

Robert Paterson is Associate Professor of Community and Regional Planning and Co-Director of the Urban Information Lab at the University of Texas at Austin. He has served as an Associate Dean for Research and Facilities and co-founded the School of Architecture's Center for Sustainable Development with Dean Frederick Steiner and Dr Steven Moore. His current research is focused on disaster resilient community planning, sustainable brownfield redevelopment and the use of scenario-based planning tools and processes to promote sustainable urban forms.

Paulo Pinho is Full Professor of Spatial and Environmental Planning, at the Faculty of Engineering, University of Porto and, at present, the Secretary General of AESOP, the Association of European Schools of Planning. He was the founder and has been the Director of CITTA, the Research Centre for Territory, Transports and Environment. He graduated in Civil Engineering with a major in Territorial Planning from the University of Porto and has a postgraduate diploma in Urban and Regional Planning and a PhD in Environmental Planning from Strathclyde University, Glasgow. His recent research and publications focus on urban and metropolitan morphologies and dynamics, urban metabolism and planning policies for low carbon cities.

Michael Redclift is a British Environmental Sociologist with an interest in sustainable development and Latin American rural societies. From 1973 until 1987 he worked at the Institute of Latin American Studies, University of London and at Wye College, where he became Professor of Environmental Sociology in 1991. Currently he is Emeritus Professor of International Environmental Policy in the Department of Geography, at King's College, University of London. His research interests include sustainable development, global environmental change, environmental security and the modern food system. He is currently working on austerity and sufficiency in relation to past and future economic and environmental crises.

Erin Roberts is a Research Associate working within an interdisciplinary space in the School of Social Sciences at Cardiff University. Her disciplinary background is in Human Geography – specialising in cultural, rural and energy geographies. Her research interests focus on the relationship between people, place, energy and society. Specifically, she is interested in how identity, place and practice combine to create unique cultural landscapes with their own set of opportunities and barriers to change. At present, she is working on projects relating to lived experiences of energy system change and valuing natural environments for human health and wellbeing.

Inger-Lise Saglie is Professor of Urban and Regional Planning at the Norwegian University of Life Sciences. Her work has centred on environmental concerns in planning but also social aspects. A particular theme is urban development and the contradictory concerns that need to be balanced, such as in the compact city policies. Urban quality, green structure and public space are areas of interest. She has also been interested in planning theory and planning processes, including power relations, the role of private market actors in urban development and the possibilities for public participation.

Alister Scott is a Geographer, Chartered Planner (MRTPI) and 'Pracademic' who works at disciplinary and professional boundaries and edges in dealing with interdisciplinary problems. His research addresses 'messy' problems concerning policy and decision making across both built and natural environments. He has published in peer review papers, professional and popular journals; produced policy briefs, videos, web portals, plays and even game boards and has written regularly for national and regional newspapers. Current research and knowledge exchange work is focused on improving the way nature is mainstreamed in policy and decision making which is reflected in his contribution to this book.

Delyse Springett has a long-term interest in education for sustainability for all. She was co-founder of the New Zealand Natural Heritage Foundation at Massey University. She developed and taught the University's first Master's course on sustainability for business students and taught a similar Master's course for The University of Hong Kong. At the corporate level, she developed and administered a national annual survey of Corporate Social and Environmental Responsiveness for New Zealand companies. Her doctoral research 'Corporate perspectives on sustainable development in New Zealand: a critical analysis' was completed at Durham University. Employing a critical theorisation, her papers have been regularly published in international journals and she has contributed chapters to edited publications.

Frederick Steiner is Dean and Paley Professor at the University of Pennsylvania School of Design. He served for 15 years as Dean of the School of Architecture at The University of Texas at Austin, having taught at Arizona State University, Washington State University, the University of Colorado at Denver, and Tsinghua University. Dean Steiner was a Fulbright-Hays scholar at Wageningen University and a Rome Prize Fellow at the American Academy in Rome. A Fellow of the American Society of Landscape Architects, he has written, edited, or co-edited 18 books, including *Making plans* (2018, UT Press).

Kerry Turner has been at the forefront of the interface between ecological and environmental economics and the natural sciences in a wide range of environmental management situations since the 1970s, covering ecosystem conservation and management and land and coastal/catchment management. He has helped establish and has worked in two ESRC funded research centres as Research Fellow, Executive Director and Director (The Public Sector Economics

Research Centre, Leicester University and East Anglia 1976–79; The Centre for Social and Economic Research on the Global Environment (CSERGE) 1991–2010) and led on policy-relevant projects such as the UKNEA and UKNEA-FO, and in VNN projects.

Iain White is Professor of Environmental Planning at the University of Waikato, New Zealand. He is committed to engaging beyond the discipline to researchers, practitioners and communities to generate real world impact. This focus typically investigates critically the nature of the science-policy-practice interface, tracing theory and science through to policy and outcomes. This approach has been applied most notably with regard to natural hazards, risk and resilience. In addition to academic articles, Iain has authored or co-authored: *Water and the city* (2010, Routledge), *Environmental planning in context* (2015, Palgrave) and *Why plan? Planning theory for practitioners* (2019, Lund Humphries).

Mark Whitehead is a Professor of Human Geography at Aberystwyth University. His early research focused on the changing forms of urban policy under the New Labour government in the UK. His subsequent work has spanned various aspects of political and environmental studies with a particular concern for the changing nature of state power. Mark has written widely on issues of urban sustainability and was an Associate and Managing Editor of the journal *Environmental Values*. Mark's most recent work has been exploring the nature of neoliberal oriented urban adaptation strategies. In 2014 Mark published his latest book, which explored the geographical implications of living in the Anthropocene.

Ken Willis is Emeritus Professor of Environmental Economics at Newcastle University. His research concentrates on environmental valuation (using stated preference-contingent valuation and choice experiments; and revealed preference travel-cost and hedonic price models) and cost-benefit analysis. He has researched the benefits of agri-environment schemes, biodiversity, cultural heritage, energy, forests, landscape, mineral extraction, recreation, renewable energy, transport, visual amenity, waste disposal, and water quality and supply. He has written a number of books on environmental valuation, and numerous articles in journals. As well as the UK, he has worked in Ghana, Iran, Italy, Malaysia, and South Africa, amongst others, on valuation issues.

Cecilia Wong is Professor of Spatial Planning, Fellow of the UK Academy of Social Sciences and the Royal Town Planning Institute. She has research expertise on strategic spatial planning, policy monitoring and analysis, urban and regional development, and housing and infrastructure planning. She has conducted major research projects for the UK government, Economic and Social Research Council, Joseph Rowntree Foundation, Royal Town Planning Institute, and regional and local bodies. She is currently carrying out a three-year research project, funded by the Economic and Social Research Council's Newton Fund, on 'Eco-urbanisation: promoting sustainable development in metropolitan regions of China'.

Bronwyn E. Wood is a Senior Lecturer in Education at Victoria University of Wellington in New Zealand's capital city. Her research interests lie at the intersection of sociology, geography and education and centre on issues relating to youth participation, citizenship and education. Her environmental research has theorised aspects of young people's ecological wellbeing and strategies for more effective education for sustainability in schools and higher education. She is an editor of the journal *Theory, Research and Social Education*.

Tan Yigitcanlar is an Associate Professor at the School of Civil Engineering and Built Environment, Queensland University of Technology, Brisbane, Australia. He has been responsible for research, teaching, training and capacity building programmes in the fields of urban and regional planning, development and management in esteemed Australian, Brazilian, Korean, Finnish, Japanese and Turkish universities. The main foci of his research interests are clustered around the following three interrelated and interdisciplinary themes: 'Knowledge-based urban development and knowledge cities', 'Sustainable urban development and sustainable cities', 'Intelligent urban technologies and smart cities'. He has published over 120 high-impact journal articles and 11 books.

Wei Zheng is a Post-doc Research Associate at the University of Manchester. She obtained her Bachelor of Science degree from Nanjing University, China and her PhD from the Hong Kong Polytechnic University. Her research interests include urban renewal, urbanisation, GIS application in spatial planning, and green infrastructure. She has participated in several research projects, such as 'Developing an integrated collaborative platform for sustainable urban renewal: a pilot study', 'Investigating the green building development in China' and 'Carbon footprint of citizens and the driving mechanism: a case of Nanjing, China'.

Acknowledgements

We are very grateful to Dr Faith Goodfellow for her brilliant editorial assistance and tireless support throughout the preparation of this Companion. It would have been impossible for us to undertake this major and long-term book project without her generous, highly diligent and thoughtful contributions. Over the last three years, Faith helped us generously with: liaising with 59 contributors, keeping track of the progress, ensuring that various drafts of the chapters are submitted on time and according to guidelines, formatting and copy editing final draft chapters, putting together the prelims and making the manuscript submission-ready. Drawing on her academic knowledge and expertise in environmental sciences and her interest in philosophy, Faith also contributed with her insightful comments and observations to address the intellectual and practical challenges that a major Companion such as this often faces. We cannot thank her enough!

The cover image: Kiki Smith, *Spinners (moths and spiders webs)*, 2014
Cotton Jacquard tapestry, hand-painting and gold leaf, edition 4/10; 294.6 x 193 cm
© Kiki Smith. Courtesy Pace Gallery. Photo: Tom Barratt

Chapter 4: Sustainable development: history and evolution of the concept
Delyse Springett and Michael Redclift
This chapter is abstracted by Simin Davoudi from Springett D. and Redclift M. (2015) 'Sustainable development: History and evolution of the concept' in, Redclift M. and Springett, D. (eds.) *Routledge International Handbook of Sustainable Development*, London: Routledge, pp. 3–38

Chapter 19: Anthropocene communications: cultural politics and media representations of climate change
Marisa B. McNatt, Michael K. Goodman and Maxwell T. Boykoff
This chapter is adapted from Boykoff M. T., McNatt M. B. and Goodman M. K. (2015) 'Communicating in the Anthropocene: the cultural politics of climate change news coverage around the world', in Hansen A. and Cox R. (eds.) *The Routledge Handbook of Environment and Communication*, London: Routledge, pp. 221–31

Chapter 24: Biodiversity, ecosystem services and environmental planning
Kimmo Lapintie and Mina di Marino
The main research findings of this study were supported by the Academy of Finland, Strategic Research Council (SRC) on Urbanizing Society (Beyond MALPE-coordination: Integrative Envisioning – BeMInE), grant number 13303549 STN

Chapter 29: Indicator-based approaches to environmental planning
Cecilia Wong and Wei Zheng
The material in this chapter is based in part on research funded by the UK Economic and Social Research Council (ES/N010698/1) on 'Eco-urbanisation: promoting sustainable development in metropolitan regions of China'

Chapter 36: Mainstreaming the environment in planning policy and decision making
Alister Scott
This chapter is based on work carried out as part of the following grant: Mainstreaming Green Infrastructure in Planning Policy and Decision Making NE/R00398X/1

Introduction

Environmental planning: meanings, governance, pressures, responses and approaches

Simin Davoudi, Richard Cowell, Iain White and Hilda Blanco

Introduction

> Right now, we are facing a man-made disaster of global scale. Our greatest threat in thousands of years. Climate change. If we don't take action, the collapse of our civilisations and the extinction of much of the natural world is on the horizon. . . . The world's people have spoken. Their message is clear. Time is running out. They want you, the decision-makers, to act now.
>
> *(Attenborough, 3 December 2018, Katowice)*[1]

The above statement by Sir David Attenborough speaking at the opening ceremony of the United Nations' climate talks (COP24) captures the importance and the essence of environmental planning which, put simply, is about making purposeful planning decisions to protect and enhance the environment now and in the future. It also highlights that what are commonly known as *environmental* problems are in fact social and political problems; they are human-made. Climate breakdown, for example, is not just an environmental issue; it is crucially a social, cultural and political phenomenon, one that demands a rethinking of who we are, how we relate to each other in society, and how we understand 'humanity's place on Earth' (Hulme, 2009, p. v). It is a powerful reminder of our intimate and inescapable interdependencies with nonhuman nature. This, however, is not to suggest that climate breakdown has led to a clearer understanding of our relationship with nonhuman nature. If anything, it has added to the complexity of interpreting it, adding new layers to myriad definitions that have been created since the origin of the term 'nature' in Ancient Greece and its multiple mutations and Darwinian interpretation.

The ambiguities involved in making sense of the relationship between human and nonhuman nature are reflected in the title of this Companion. It includes two words, environment and planning, neither of which are easy to define. Both have been subject to intensive debate and contestation that defy commonly agreed definitions. The challenge becomes greater when the two terms are combined into 'environmental planning'. Hiding behind the seemingly straightforward definition mentioned in our opening paragraph is not only competing interpretations but also ongoing tensions and contradictions about questions such as, what is to be protected and enhanced, how, by whom, for whom, and for what purpose, as well as what it means to plan.

A key motivation for this Companion is to bring the human-nonhuman interrelationship and the tensions that arise from it to the fore and explore its implications for various aspects of environmental planning. While the field of environmental planning is served by a diverse range of rich and valuable contributions, we believe that there is still a need for situating the subject in wider debates and its broader social, political, ecological, ethical and institutional context. Such contextual matters are no mere background; as the chapter contributions make clear, social, political and ethical concerns can strongly configure the form, scope and outcomes of environmental planning.

A second rationale for the Companion is the complexity and dynamism of the debate which are reflected in the ontological, epistemological and methodological changes over the past decades, especially since the concept of sustainable development drove a steep upswing in environmental planning in the 1990s. *Ontologically*, the singular, unproblematic understanding of 'the environment' as an object of inquiry, or as a problem waiting to be solved has been challenged. Indeed, it has been suggested that we should 'move beyond a definition of nature that relies on the absence of any trace of human influence' and instead consider 'history, rather than ontology' as 'the proper guide for understanding nature today' (Arias-Maldonado, 2015, p. 19). *Epistemologically*, novel and diverse ways of knowing the environment and mobilising political action have emerged. Practices such as citizen science are reconfiguring the boundaries between epistemology and ontology. *Methodologically*, innovative rules, approaches and tools have replaced or complemented the traditional techniques in environmental planning. There have also been changes in scale (for example, from local to global), focus (for example, from ozone to carbon) and discourse (for example, from sustainability to resilience) in environmental planning and research. While these changes bear witness to the dynamic and evolving nature of the field, they also risk masking what have not changed of which a striking example is the continuing dominance of anthropocentric views of nature.

Reflecting upon the changing frameworks and uneven progress bring us to a third rationale for this Companion: rendering visible the journeys travelled and untravelled.[2] The discipline has come a long way since its initial development in response to the pressures of the industrial age, but, while some pathways to change are productive and well-trodden, others seem not only stubbornly persistent but also actively resisting exploration and action. The reasons for this disparity are not necessarily connected to criticality or gaps in science or policy, rather they may be related to fundamental issues of framing, economics, or politics. For example, one of the most high-profile environmental successes of the previous century was the gradual closing of the hole in the ozone layer. Within the space of a few years, scientists discovered the problem and linked it to the chlorofluorocarbons (CFCs), commonly used in aerosols and refrigeration. Policy makers across the globe acted swiftly to ban their use. It is illustrative to consider why this journey appears so straightforward in comparison to the arch 21st-century problem of climate change. In hindsight, we can appreciate the importance of strong political leadership to gain consensus, but it is also important to note that because only a few countries produced CFCs, there was funding to aid transition and an alternative technical replacement was almost readily available. Put differently, while it was a critical global environmental problem, it did not appear to represent an intractable economic and political one. The roads we have yet to travel may lack these characteristics.

As such, the four parts of the book are designed to reflect the complex and challenging contexts in which environmental planning operates, emphasising the interdisciplinary nature of many contemporary issues. Each part aims to provide an ever deeper understanding that weaves together our definition and understanding of: the environment; the pivotal influence of governance, power and politics; the always emergent knowledge of pressures and impacts; and the highly political nature of decision making and planning approaches. In doing so, it helps us better appreciate not only that many of the most critical journeys lie ahead but also those as yet untravelled might be the most challenging to navigate. Just as environmental pressures change over time, so too does the discipline need to continuously reflect on and adjust its practices and research agendas.

The Companion presents a distinctive approach to environmental planning by: situating the debate in its social, cultural, political and institutional context; being attentive to depth and breadth of discussions; providing up-to-date accounts of the contemporary practices in environmental planning and their changes over time; adopting multiple theoretical and analytical lenses and different disciplinary approaches; and drawing on knowledge and expertise of a wide range of leading international scholars. Its aims are:

- To provide critical reviews of the state of the art theoretical and practical approaches as well as empirical knowledge and understandings of environmental planning;
- To encourage dialogue across disciplines and national policy contexts about a wide range of environmental planning themes;
- To engage with and reflect on politics, policies, practices and decision-making tools in environmental planning.

These aims underpin the logic for the structure of the Companion which consists of four parts. **Part 1 Understanding 'the environment'** focuses on different and evolving perspectives on, the relationship between human and nonhuman nature and the ways in which these have informed influential concepts in environmental planning. It contributes to the theoretical and conceptual context of the Companion and enables the reader to situate environmental controversies in policy and practice within critical social science perspectives. **Part 2 Environmental governance** discusses the governing processes that shape how: environment is problematised and dealt with; knowledge is produced and utilised; and the roles played by key actors change across time and place. **Part 3 Critical environmental pressures and responses** focuses on understanding the complexity of environmental challenges and how they are addressed and contested within environmental planning. **Part 4 Methods and approaches to environmental planning** discusses the changing modes of regulation and the methods and tools used in environmental planning to aid decision making. In the following account, we provide a brief overview of the themes covered in each part of the Companion.

Understanding 'the environment'

A key rationale for Part 1 of the Companion is to discuss the multiple ways in which the environment is framed and nature is imagined in environmental planning. Even the Dictionary definition of the environment is split between two meanings: one refers to 'the surroundings or conditions in which a person, animal, or plant lives or operates'; and the other, to 'the natural world, as a whole or in a particular geographical area, especially as affected by human activity' (Oxford English Dictionary).[3] The first definition resembles that of the French word 'environ', which indeed is the root of the term environment. It also comes close to the meanings of habitus and milieu. However, as Cooper (1992, p. 169, original emphasis) suggests, 'an environment as milieu is not something a creature is merely *in*, but something it *has*'. This implies that our environment (understood as our milieu) is part of who we are. As Barry (1999, p. 13) puts it, 'the environment is a *relational* concept' which 'in social theory is defined in relation to ourselves and particular human social relations, and particular historical and cultural contexts'.

The second definition equates the environment with 'nature' or 'the natural world' with the added qualification that the only nature that counts as the environment is that which is affected by human activity. However, neither this qualification, nor the use of the word nature leads to a clear and singular definition of the environment. If anything, it makes the matter more complicated because as Williams (1980, p. 219) suggests, nature is 'perhaps the most complex word in the language', whose wide semantic scope defies any attempts to arrive at an unequivocal meaning. Added to the complexity of its mystical and metaphysical meanings is the dynamic and evolving phenomena that the concept of nature tries to grasp. It is frequently used to invoke binaries such as: nature versus nurture; natural versus artificial; nature versus culture; and nonhuman nature versus human.

These definitional difficulties are not merely semantic problems; they are indicative of different worldviews and ways of thinking. However, as Foster (1997, p. 10) suggests, '"the environment" . . . is something upon which very many frames of references converge. But there is no frame of reference which is as it were "naturally given", and which does not have to be contended for in environmental debate'. Environmental planning is situated at the heart of such debate. It is both producer and product of the discursive processes through which a particular meaning of the environment is, temporarily, fixed. A notable example is the role played by the discipline in perpetuating the divide between society and nature by persisting on a dualistic distinction between the built and the natural environment.

The institutions and practices of environmental planning provide key arenas in which various conceptions of the environment are called into being to legitimise certain courses of action. There have been several attempts within, and outside, the environmental planning scholarship to identify multiple and changing environmental discourses in planning (Whatmore and Boucher, 1993; Healey and Shaw, 1994). For example, it is argued that in the English planning system the environment has been defined as: local amenity, heritage landscape, nature reserve, storehouse of resources, tradable commodity, problem, sustainability, and risk (Davoudi, 2012, 2014). These and other ways of seeing the environment (or nature) sometimes compete with one another for a position at the top table of political action and other times reinforce each other for particular effects.

Seeing the environment as 'risk' is a particular concern because inherent in that discourse is a tendency to shift the focus from what we do to the environment to what the environment does to us (Davoudi, 2014). This language of risk is fed by and feeds into the narratives of security, leading to a recasting of social and environmental problems as security problems. For example, there has been a subtle shift of emphasis from renewable energy to energy security or from biodiversity to food security. Parallel to this securitisation trend has been the rising popularity of resilience. On the one hand, resilience, and its epistemological origin in complexity theory, has changed our thinking about uncertainty by highlighting the nonlinearity, discontinuity and self-organising capacity of complex social and ecological systems. On the other hand, its translation into the social domains and its instrumental application in policy and practice across a wide range of areas – from psychology to disaster risk management and environmental planning – has been considered as conceptually problematic and normatively contested with negative implications for just and democratic processes and outcomes (Bohland et al., 2018), issues that are discussed next under the theme of environmental governance.

Environmental governance

The complex and contested nature of 'the environment' as a societal concern has its reflection in the challenges of governance. Governance, too, is a term subject to diverse and fluid meanings (Jordan, 2008). It can be summarised, broadly, as the patterns that emerge from the diverse ways in which governing takes place, i.e. controlling, steering or guiding aspects of society (Kooiman, 1993). Research into governance around environmental challenges has contributed to this sense of complexity and diversity, not least because environmental governance has been characterised by considerable flux, with pressures for change coming from multiple directions. There are changes driven by the purposive desire for better environmental governance, as governments and other actors re-think the scale, the actors that must be engaged and the modes of governing to better 'fit' the tasks (OECD, 2002; Meadowcroft et al., 2005; Moss and Newig, 2010). However, environmental governance is also profoundly affected by governance changes that seem to have little direct relevance to the environment, such as wider shifts in political ideology, state re-scaling (devolution, (de)centralisation (Cowell et al., 2017)), and technology. A prominent example at the time of writing is the UK's impending exit from the European Union (EU), a decision propelled by a range of political, social and economic concerns but which risks creating an enormous 'environmental governance gap' for the United Kingdom (Burns et al., 2016, 2018), as it disarticulates from the relatively powerful environmental governance machinery of the EU.

Research into environmental planning spans these different drivers of change, advancing arguments about how better governance can be achieved, offering better ways of conceptualising the governance challenges in hand, and critiquing contextual conditions that are unhelpful. This is reflected in the chapters brought together in Part 2 of the Companion. Together they provide valuable new perspectives on governance processes, the actors involved, how knowledge is utilised and how these vary across time and space. In some cases, the focus is on long-standing, systemic problems within governance systems, such as the obsessions with economic growth as the dominant objective of government or the struggles to ensure that the way we address environmental problems accommodates the demands of justice. Whilst the controversies surrounding climate change provide a high-profile illustration of how environmental governance is permeated with justice concerns, engagement with justice has been an expanding feature of research across many spheres of environmental planning.

Another thread that characterises much environmental planning scholarship since the 1990s is the recognition that the elements that constitute the world of environmental planning are more numerous and less likely to have simple essential properties than was widely believed or asserted. Thus, when it comes to the arenas and scales of environmental governance, successive waves of research into 'local action' on environmental problems has highlighted how the capacity of 'local actors' to achieve change is interconnected with rules, flows of resources, and networks with other bodies that reach beyond the locality (Marvin and Guy, 1997). Actors may not always behave according to neat, structural categories: just as sections of business have positioned themselves as pro-environmental agents, so environmental NGOs display diverse connections to civil society, closer relations with government (Lowe and Goyder, 1983) and help operate market-style systems of regulation, such as systems of sustainability standards. Careful inspection of the growth of network-based modes of governance in the environmental sphere, where power appears to be dispersed across numerous actors, has also revealed a whole host of intermediaries that are brokering connections and shaping governance effects. Analysts are also increasingly recognising the agency of nonhuman elements in helping hold together relations between human actors – be they documents, technologies, other species or ecological processes (Callon, 1986; Beauregard, 2012). In all of these spheres, close analysis of the actual practices of environmental governance has raised challenges for environmental concepts and questioned simple normative prescriptions.

One path of environmental governance that has become increasingly well-travelled in many countries is the desire to enrol individuals and publics in 'co-producing' environmental outcomes. As with shifts in governance generally, this has diverse drivers including not only recognition of the centrality of consumption behaviours in many environmental problems but also ideologically driven governance shifts away from (direct) government responsibility towards individuals, communities and markets. Well-travelled also is the pursuit of 'better communication' with publics as a governance strategy for shifting behaviour, which remains a remarkably persistent government reflex despite long-standing criticisms (Blake, 1999). Any such reliance on information is also profoundly affected by wider shifts in the technological and cultural context. In particular, the expansion of mass media to include the internet and social media has created arenas for new sets of actors (tech firms, internet influencers, climate change denialists) as well as the old ones, fostered new modes of public engagement and enabled new linkages to be formed.

What has also become increasingly apparent, since the 1990s, is that for all the unifying rhetoric of sustainable development (Myerson and Rydin, 1996), the governance of environmental problems is characterised by fragmentations and tensions. This means more than enduring tensions between economic growth and environmental protection, or between the interests of nation states. Tensions also pervade governance processes themselves, arising from the fact that prospective governance principles – reliance on markets, delivery against long-term targets, democratic legitimacy and public engagement – are difficult to reconcile without trade-offs or omissions, and the losers from any compromises may then form the basis of resistance (Boltanski, 2011). A good example would be transitions towards more sustainable forms of energy, which require balances to be struck between open deliberation of possible pathways and the creation of institutional structures sufficiently stable to enable new infrastructures to be put in place (Smith and Stirling, 2007). Another tension lies between the construction of governance arrangements able to exercise reach across time and space, which tends to favour abstraction and simplification, and the need for sensitivity to contextual differences and unpredictable events (Scott, 1998). Key ingredients in any governance arrangements are knowledge and expertise but – far from floating above this world of tensions and compromises – they are very much enmeshed within them.

Such tensions help us to grasp a central paradox of environmental governance: 'good governance' for environmental problems is widely heralded as essential in steering society towards long-term sustainability (Jordan, 2008), yet governance – like government – is also a 'congenitally failing operation' (Rose and Miller, 1992, p. 190), always struggling to align the heterogeneous elements involved. It is always tempting – and sometimes perfectly justifiable – to place the blame for failures in environmental governance at the hands of human agents, whether it be for poor choices, poor information, insufficient resourcing or deliberate obstruction. However, to do so underplays the importance of the environmental problems themselves, as intertwining ecological and human processes that together generate complexity (and therefore significant risk and uncertainty), and in forever escaping the temporal and spatial horizons of human organisation. Moreover, Andrew Jordan neatly articulates how, if we are to understand why sustainable development continually frustrates the pursuit of governance coherence, '(t)his messiness has a lot to do with the fact that sustainability concerns nothing less than the future direction of human civilisation' (Jordan, 2008, p. 28). Dissent and difficulty is precisely what one should expect.

This leads us to an important aspect of theorisation about governance and the challenges of drawing general conclusions. As Jessop (1997, p. 105) explains, it is difficult to theorise about tendencies in governance since there is no governance 'in general'; there are only 'definite objects of governance that are shaped in and through definite modes of governance'. As objects, the contours of environmental problems are themselves complicit in the frailties of our responses, and it is to this issue that we turn next.

Critical environmental pressures and responses

The third part of the Companion aims to provide a thorough and wide-ranging understanding of the criticality and complexity of environmental problems and appraise how key issues are responded to within environmental planning. In many respects the subject matter here is at the very heart of the discipline. Without an appreciation of how, and to what extent, human activities create environmental pressures, there is not only no mandate to act but also no data to consider and weigh the inevitable social, economic, and environmental trade-offs which lie at the centre of much of what environmental planners do. All responses, whether concerning regulation, incentives or monitoring, influence individual or market behaviours in different ways. Therefore, in contrast to much political debate which simply posit environmental planning interventions as a tension between a strong state and a free market, it is worth noting that there is no such thing as a 'free market'. All markets are a product of the historical assemblage of laws, regulations, culture, policy, and behavioural logics that serve to create the various frameworks of decision making relating to land and resources. So, our current issues – and market failures – do, in part, stem from previous environmental planning decisions and approaches (often called government failures) that have played a role in constituting what is now business-as-usual. This is readily apparent within auto-dependent landscapes that rely heavily on fossil fuels, or the varied ways by which nature is valued within assessments and decision making.

Consequently, the spatiality of pressures in a place links to the spatiality of cultural, political, or economic ideologies that shape the balance between the state, market, and citizens. This means that both acceptance of impacts and the powers and scope of environmental planning are uneven and socially constructed. For example, in broad terms, the EU Water Framework Directive advocates that water bodies should achieve a 'good' quality status, a proactive aim which is deemed to reflect the state of the system before anthropogenic pressures. Whereas in other parts of the world the onus may be much more permissive and reactive, using terms like thresholds or carrying capacity, which may lead to approaches that seek to understand how much pressure an ecosystem can take before significant, and possibly irreversible, effects are experienced.

While there may be stark national differences between environmental planning regimes, there are also common connecting threads. The spatiality of environmental pressures ranges from the local scale to encompass the entire biosphere, while the temporal remit balances the immediate with considerations of intergenerational equity and the possible effects on people who are not yet born. Likewise, international trends in globalisation and neoliberalisation emphasise how capital is increasingly mobile, with competition apparent both within and between countries. A strong contemporary narrative in this regard is that an increase in environmental 'red-tape' can serve to depress investment as businesses shift to locations with lighter regulation. Yet, the notion of planning mediating between environmental pressures and economic success is relatively simplistic. In reality, good quality environments attract skilled labour and mobile multinationals, too. Perhaps that is why the environment is often wrapped up in the place marketing packages. Given the intrinsic nature of politics, competition is also apparent between differing environmental issues seeking political attention. It is common to see polls of the most pressing environmental concerns, which may influence what attention they garner within public imaginaries or potential policy cycles. These may be typically split into thematic areas like air quality, climate change, or biodiversity loss, but the environment is not so easily siloed. These boundaries are artificial and are related to the constraints and historical development of scientific disciplines or managerial institutions. In reality, ecosystems are integrated and the ways that systemic changes provide feedbacks, may be uncertain. Reconciling the actuality of this uncertainty and complexity with demands for more simple, efficient, and market-oriented planning is a key contemporary challenge.

Similarly, an emerging research agenda relates to why progress is not made; again seeking to shine the spotlight on the boundaries and limits of environmental planning. In a polarised, post-truth world the regulation of industrial activities at the site, city or national scale may be easier than harmonising international taxes or laws, or developing new global agreements. So, success in one area, such as the huge advances during the 20th century relating to the reduction of pollution from the ends of pipes in many countries, may be as much a function of governance or scientific fit as the urgency of the problem. Indeed, the continued existence of well-established and critical environmental pressures may indicate that they map poorly onto the ways and scales by which we traditionally manage land and resources. For instance, it is a feature of environmental planning that evidence of effects is often retrospective, and more sustainable, precautionary or long-term approaches struggle to gain political attention in contrast to the loud voices of the present (White and Haughton, 2017). Equally, the siren calls of current growth, human ingenuity and the potential for a technical 'fix' cast a perennial shadow over the urgency to act. Therefore, evolving scientific understanding of the criticality and persistence of not just pressures but also institutions and governance helps drive the development of new responses with the potential to facilitate change on a more systemic level. Indeed, the rise in environmental consciousness and scientific knowledge during the latter half of the 20th century was a factor in the shift towards assessment methodologies, while the next few decades promise to offer ever more sophisticated means to recognise, record and mitigate environmental impacts.

This perspective serves to illustrate not just the changing nature and complexity of environmental planning but also the challenges of balancing public and private interests, and delivering what are sometimes competing economic, environmental and social objectives. The term 'balance' however, while frequently used in environmental planning, may not adequately acknowledge the way that the scales can be tipped towards certain powerful interests or the status quo. Beyond the more obvious and visible impacts of population growth and economic behaviour, or effectiveness of diverse systems of policy and regulation, it is possible to appreciate how environmental pressures and responses are intertwined with the governance issues discussed in the previous section and the approaches and methods discussed in the next.

Methods and approaches to environmental planning

Part 4 of the Companion focuses on methods and approaches used to aid decision making related to the interaction between human and nonhuman nature. How do human actions impact on nonhuman nature? How do we measure such impacts? How could we reduce such impacts? Over the past half century, approaches and methods have been developed, expanded, and increasingly applied to respond to these questions. Environmental Impact Assessment (EIA) is a key example. It was established by the new United States (US) Environmental Protection Agency in 1970 to assess the potential impacts (on both human and nonhuman nature) of major new developments and consider ways to reduce such impacts. The environmental impact review practice was expanded beyond the individual project scale to apply to programmes and plans in the 1990s, through the development of Strategic Environmental Assessments (SEAs), which the EU institutionalised through its 2001 SEA Directive, and the Protocol for the Directive was ratified by the EU at the end of 2008 (Tetlow and Hanusch, 2012). Impact assessments are oriented to assess the future potential impacts of development at different scales, EIA at project scales, SEAs at program and plan scales. They are both intended to aid the decision-making process by identifying potential negative consequences of proposed actions to the environment, both human and nonhuman, with the intent of reducing such consequences. They remain major environmental methods.

Another set of approaches, developed and institutionalised since then, focus on analysing both the environmental inputs and outputs of decisions. Metabolic and life cycle analyses focus on inputs and outputs of development decisions or processes, and they can be applied at different scales: to countries, such as the energy metabolism of Scotland (Scottish Government, 2006); to cities (Kennedy et al., 2007), such as the urban metabolism of Hong Kong (Warren-Rhodes and Koenig, 2001); to the use of certain products, such as the use of cloth or disposable nappies in the United Kingdom (UK Environment Agency, 2008); to sectors, such as the built environment (Lotteau et al., 2015); and to individuals, such as ecological footprints (Global Footprint Network, 2019). Such methods have a dual directional focus, analysing inputs as well as outputs, aiming to reduce natural resource use and impacts, and not just the impacts of outputs. Many such studies are voluntary or products of scholarly research, but some methods are recognised internationally, such as life cycle assessment for which an international standard was published in 2006 (ISO, 2006).

Indicator systems have become a popular way to track progress on environmental goals or policies. At the international level, organisations such as the UN and the EU (especially to track progress towards the UN Sustainable Development Goals at the national scale) and professional associations, such as the International Council for Local Environmental Initiatives' indicator system for sustainable cities (ICLEI, 2019), have developed various sets of environmental or sustainability indicators. Indicator systems enable the comparison of outcomes over time for individual countries or cities, as well as comparisons among countries and cities. Whether this type of analysis provides useful information on progress depends on the extent to which the indicators selected are appropriate to measure the environmental goals contained in associated policies or programs, as well as the causal linkages among policies and outcomes.

Other methods have focused on how to incorporate the value of nature in market economies. Public finance still relies on a cost-benefit analysis framework, but the development of environmental economics since the 1990s has introduced various ways to value 'environmental externalities', i.e. to value nature. Costanza and colleagues (1997, 2014) have developed new ways to put a monetary price on ecosystems. More recently, another environmental economic approach seeks to keep track of natural assets and their importance for national wealth (Lange et al., 2018; UN and EU, 2014). But the underlying issue, even in these approaches, is whether human societies have the capacity now or in the near future to substitute human-dominated processes for natural processes, that is, whether we can assume that economic capital can be substituted for natural 'capital'. The present-oriented perspective of the economic system prevailing in the world today assumes this, and it systematically discounts the future benefits of ecosystems or natural processes within a 50- to 100-year horizon to almost nothing. These efforts to adopt economic language and analysis to support environmental quality are likely to be motivated by the belief that only such economic analyses of environmental quality will be taken seriously by decision makers. But such beliefs are more often asserted than tested, and it remains an important challenge for environmental planners to assess whether presenting environmental qualities as capital, services or infrastructure actually does shift the dial towards greater environmental protection.

Environmental methods are important in helping the profession of environmental specialists who are looking for systematic and replicable processes such as urban metabolism analysis or life cycle analysis. However, methods in environmental planning need to serve other major objectives, namely, to enable publics and decision makers to better visualise or conceptualise nature and the environment and better address the negative environmental consequences of human activities. Suitability analysis is a good example of a method developed by landscape architects that can incorporate multiple features of a place to arrive at development/preservation decisions. This method, enhanced by GIS technology, enables both decision makers and publics to visualise and take into account multiple layers of the environment in making decisions about new development or areas for environmental protection or preservation. While the thrust of major environmental methods, such as impact statements or metabolic assessments, is to reduce the environmental impacts of human actions, the green infrastructure approach aims to proactively incorporate nature or nature's services in cities. The history of 'greening cities' dates back to the design of large urban parks, both within and around cities, such as New York City's Central Park or London's Green Belt that have served as green lungs and/or natural drainage areas for cities. Today, the green infrastructure approach can provide positive ways to enhance urban environments by aiding in the regulation of air quality, temperature, or flooding through natural processes (or rather human processes that incorporate knowledge of natural processes into our cities). Based on up-to-date knowledge of environmental processes, the approach also has high visibility and promise to engage publics and decision makers in generating models that integrate urban and natural features and processes.

One interpretation of the constant development and invention of analytical techniques to support environmental planning is that it is an outcome of learning. As societies strive for better ways of capturing – more accurately, more comprehensively, more ethically – a slippery subject matter, new methods and techniques are developed. Another interpretation is that it is a sign of environmental planning's continuing systemic weakness which drive those concerned with environmental protection to constantly seek out arguments with greater traction on policy processes and hence using economic languages and measures that already have such traction.

Conclusion

As we mentioned at the beginning of this introduction, the field of environmental planning is rich with a wide range of scholarship as well as practical guides covering a variety of environmental concerns. However, the dynamic and evolving nature of environmental planning demands periodic revisiting of the field. A notable example of this is Anthropocene, a term describing the humans' influence on the planet's ecosystems (Crutzen and Stoermer, 2000) at a scale that has previously been unimaginable. Anthropocene has unsettled our ways of knowing and imagining our relationship with nature and demanded a revisiting of our actions. While a decade ago, O'Riordan (2009, p. 313) suggested that, 'Environmentalism is morphing into sustainability', the question that is now being posed is 'whether sustainability is morphing into something that does not fit into the traditional green expectations about what a sustainable society ought to be' (Arias-Maldonado, 2013, p. 429). Questions such as these are indicative of the complex and challenging contexts in which environmental planning operate and to which it contributes.

By bringing together a wide range of contributors from different disciplines and different parts of the world, we aim to provide a deeper understanding of the interdependencies between the themes in the four parts of the book and its 36 chapters and to present critical perspectives on the role of meanings, values, governance, approaches and participations in environmental planning. Situating environmental planning debates in the wider ecological, political, ethical, institutional, social and cultural debates has helped us shine light on some of the critical journeys that we have traversed and those that we are yet to navigate and their implications for environmental planning research and practice. Perhaps the most challenging of all is coming to terms with the understanding of nature, not as an entity affected by the harmful or helpful actions of an external agent, humans, standing outside the realm of nature but rather as a dynamic and evolving assemblage made of human and nonhuman influences; a socio-natural hybrid that White and Wilbert (2009, p. 6) call 'technonature'.

Our aspiration for this Companion is to provide a reference point mapping out the terrain of environmental planning in an international and multidisciplinary context. We believe that the depth and breadth of discussions by leading international scholars from across the social science disciplines and beyond make this volume relevant to and useful for those who are curious about, wish to learn more, want to make sense of, and care for the environment within the field of environmental planning and beyond. While the coverage is relatively comprehensive, it could never be exhaustive; thus, the chapters should be read as entry points into the themes they raise. For in-depth understandings of the issues discussed, we would urge the readers to consult with the cited references in the chapters.

Notes

1 www.bbc.co.uk/news/science-environment-46398057.
2 These points were highlighted by Dr Ruth Machen in the opening of an event organised by her in November 2018 at Newcastle University.
3 https://en.oxforddictionaries.com/definition/environment.

References

Arias-Maldonado, M. (2013). 'Rethinking sustainability in the Anthropocene'. *Environmental Politics*, 22(3): 428–46.

Arias-Maldonado, M. (2015). 'What is nature?' in M. Arias-Maldonado (ed.) *Environment and society* (Springer briefs in political science). New York: Springer, pp 17–32.

Barry, J. (1999). *Environment and social theory*. London: Routledge.

Beauregard, R. (2012). 'Planning with things'. *Journal of Planning Education and Research*, 32(2): 182–90.

Blake, J. (1999). 'Overcoming the "value-action gap" in environmental policy: tensions between national policy and local experience'. *Local Environment*, 4(3): 257–78.

Bohland, J., Davoudi, S. and Lawrence, J. (eds.) (2018). *The resilience machine*. London: Routledge.

Boltanski, L. (2011). *On critique: a sociology of emancipation*. Translated by Gregory Elliot. Cambridge: Polity Press.

Burns, C., Jordan, A., Gravey, V., Berny, N., Bulmer, S., Carter, N., Cowell, R., Dutton, J., Moore, B., Oberthür, S., Owens, S., Rayner, T., Scott, J. and Stewart, B. (2016). *The EU referendum and the UK environment: an expert review*. Available at: www.brexitenvironment.co.uk/wp-content/uploads/dlm_uploads/2017/07/Expert-Review_EU-referendum-UK-environment.pdf

Burns, C., Carter, N., Cowell, R., Eckersley, P., Farstad, F., Gravey, V., Jordan, A., Moore, B. and Reid, C. (2018). *Environmental policy in a devolved United Kingdom: challenges and opportunities after Brexit*. ESRC Changing Europe Programme. Available at: www.brexitenvironment.co.uk/wp-content/uploads/2018/10/BrexitEnvUKReport.pdf

Callon, M. (1986). 'Some elements of a sociology of translation: domestication of the scallops and fishermen of St Brieuc Bay', in J. Law (ed.) *Power, action and belief. A new sociology of knowledge?* London: Routledge and Kegan Paul, pp 196–233.

Cooper, D. (1992). 'The idea of environment', in D. Cooper and J. Palmer (eds.) *The environment in question: ethics and global issues*. London: Routledge, pp 165–80.

Costanza, R., d'Arge, R., de Groot, R., Farber, S., Grasso, M., Hannon, B., Limburg, K., Naeem, S., O'Neil, R. V., Paruelo, J., Raskin, R. G., Sutton, P. and van den Belt, M. (1997). 'The value of the world's ecosystem services and natural capital'. *Nature*, 387: 253–60.

Costanza, R., de Groot, R., Sutton, P., van der Ploeg, S., Anderson, S. J., Kubiszewski, I., Farber, S. and Turner, R. K. (2014). 'Changes in the global value of ecosystem services'. *Global Environmental Change*, 26(1): 152–8.

Cowell, R., Ellis, G., Sherry-Brennan, F., Strachan, P. and Toke, D. (2017). 'Re-scaling the governance of renewable energy: lessons from the UK devolution experience'. *Journal of Environmental Policy and Planning*, 19(5): 480–502.

Crutzen, P. J. and Stoermer, E. F. (2000). 'The "Anthropocene"'. *Global Change Newsletter*, 41: 17–18.

Davoudi, S. (2012). 'Climate risk and security: new meanings of "the environment" in the English planning system'. *European Planning Studies*, 20(1): 49–69.

Davoudi, S. (2014). 'Climate change, securitisation of nature, and resilient urbanism'. *Environment and Planning C: Government and Policy*, 32(2): 360–75.

Foster, J. (1997). 'Introduction', in J. Foster (ed.) *Valuing nature? Economics, ethics and environment*. London: Routledge, pp 1–20.

Global Footprint Network (2019). *What is your ecological footprint*. Available at: www.footprintcalculator.org

Healey, P. and Shaw, T. (1994). 'Changing meanings of "environment" in the British planning system'. *Transactions of the Institute of British Geographers* [New series], 9: 425–38.

Hulme, M. (2009). *Why we disagree about climate change? Understanding controversy, inaction and opportunity*. Cambridge: Cambridge University Press.

ICLEI (2019). *Star communities*. Available at: www.starcommunities.org/about/framework/

ISO (2006). *ISO 14040:2006: Environmental management – life cycle assessment – principles and framework*. Geneva: International Organization for Standardization.

Jessop, B. (1997). 'The complexity of governance and the governance of complexity: preliminary remarks on some problems and limits of economic guidance', in A. Amin and J. Hausner (eds.) *Beyond market and hierarchy: interactive governance and social complexity*. Cheltenham: Edward Elgar, pp 111–47.

Jordan, A. (2008). 'The governance of sustainable development: taking stock and looking forwards'. *Environment and Planning C: Government and Policy*, 26: 17–33.

Kennedy, C., Cuddihy, J. and Engel-Yan, J. (2007). 'The changing metabolism of cities'. *Journal of Industrial Ecology*, 11(2): 43–59.

Kooiman, J. (1993). *Modern governance*. Newbury Park, CA: Sage.

Lange, G. M., Wodon, Q. and Carey, K. (2018). *The changing wealth of nations 2018: building a sustainable future*. Washington, DC: The World Bank.

Lotteau, M., Loubet, P., Pousse, M., Dufrasnes, E. and Sonnemann, G. (2015). 'Critical review of life cycle assessment (LCA) for the built environment at the neighborhood scale'. *Building and Environment*, 93: 165–78.

Lowe, P. and Goyder, J. (1983). *Environmental groups in politics*. London: Allen and Unwin.

Marvin, S. and Guy, S. (1997). 'Creating myths rather than sustainability: the transition fallacies of the new localism'. *Local Environment*, 2(3): 311–18.

Meadowcroft, J., Farrell, K. N. and Spangenberg, J. (2005). 'Developing a framework for sustainability governance in the EU'. *International Journal of Sustainable Development*, 8: 3–11.

Moss, T. and Newig, J. (2010). 'Multilevel water governance and problems of scale: setting the stage for a broader debate'. *Environmental Management*, 46(1): 1–6.

Myerson, G. and Rydin, Y. (1996). *The language of environment. a new rhetoric*. London: UCL Press.

O'Riordan, T. (2009). 'Reflection on the pathways to sustainability', in W. N. Adler and A. Jordan (eds.) *Governing sustainability*. Cambridge: Cambridge University Press, pp 307–28.

OECD (2002). *Improving policy coherence and integration for sustainable development: a checklist*. Paris: OECD.

Rose, N. and Miller, P. (1992). 'Political power beyond the state: problematics of government'. *The British Journal of Sociology*, 43(2): 173–205.

Scott, J. (1998). *Seeing like a State. How certain schemes to improve the human condition have failed*. New Haven: Yale University Press.

Scottish Government (2006). *Scottish energy study – volume 1 – energy in Scotland, supply and demand, section 6.1: energy flow in Scotland*. AEA Technology for the Scottish Executive [Online]. Available at: www.scotland.gov.uk/Publications/2006/01/19092748/8

Smith, A. and Stirling, A. (2007). 'Moving outside or inside? Objectification and reflexivity in the governance of socio-technical systems'. *Journal of Environmental Policy and Planning*, 9(3/4): 351–73.

Tetlow, M. F. and Hanusch, M. (2012). 'Strategic environmental assessment: the state of the art'. *Impact Assessment and Project Appraisal*, 30(1): 15–24.

UK Environment Agency (2008). *An updated lifecycle assessment study of disposable and reusable nappies*. Science Report-SC010018/SR2. Bristol: Environment Agency. Available at: https://assets.publishing.service.gov.uk/government/uploads/system/uploads/attachment_data/file/291130/scho0808boir-e-e.pdf

UN and EC (2014). *System of environmental-economic accounting 2012 – experimental ecosystem accounting*. New York: UN. Available at: https://seea.un.org/sites/seea.un.org/files/seea_eea_final_en_1.pdf

Warren-Rhodes, K. and Koenig, A. (2001). 'Escalating trends in the urban metabolism of Hong Kong: 1971–1997'. *Ambio*, 30(7): 429–38.

Whatmore, S. and Boucher, S. (1993). 'Bargaining with nature: the discourse and practice of environmental planning gain'. *Transactions of the Institute of British Geographers*, 18(2): 166–78.

White, D. F. and Wilbert, C. (2009). 'Introduction: inhabiting technonatural time/spaces', in D. F. White and C. Wilbert (eds.) *Technonatures: environments, technologies, spaces, and places in the twenty-first century*. Waterloo: Wilfred Laurier University Press, pp 1–31.

White, I. and Haughton, G. (2017). 'Risky times: hazard management and the tyranny of the present'. *International Journal of Disaster Risk Reduction*, 22: 412–19.

Williams, R. (1980). *Problems in materialism and culture*. London: Verso.

Part 1

Understanding 'the environment'

Simin Davoudi, Richard Cowell, Iain White and Hilda Blanco

> In this actual world there is then not much point in counterposing or restating the great abstraction of Man and Nature . . . if we go on with the singular abstractions, we are spared the effort of looking . . . at the whole complex of social and natural relationships which is at once our product and our activity.
>
> *(Williams, 1980, p. 83)*

A key rationale for the opening part of this Companion is to take Raymond Williams' provocation seriously and explore the complexity and contingencies of social and natural relationships as they are manifested in and performed through discourses and practices of environmental planning. How controversies in environmental planning are dealt with often depends on whose 'nature' or 'environment' is called into being by planners and other actors and whose is enacted. Environmental planning is a key site of discursive contestation over questions such as: what is nature? and what is the environment? Conceptual clusters that structure environmental discourses influence the processes and outcomes. It is, therefore, crucial that we start this Companion with contributions that focus primarily on unpicking some of the key concepts used in environmental planning and unravel the taken-for-granted assumptions that lie behind them.

The first two chapters engage with two of the most elemental concepts: 'nature' and 'environment'. In their different ways, the chapters explore their multiple meanings and practical applications and warn against naturalising and reifying nature and environment as singular and universal concepts. Instead, **Noel Castree**, in **Chapter 1**, urges greater reflexivity and awareness about the basic concepts that structure the discourses *of* and *about* environmental planning. He highlights the need for and the value of interrogating the languages used to invoke 'the nature of environment' by different actors in the planning process. Revisiting some of the early works by geographers and anthropologists, he shows how in planning, concepts of nature and environment are normalised as the mirror images of a physical reality beyond the human realm. The chapter includes a synthesis of different meanings (the signifieds) of nature (the signifier) under four categories: universal, external, superordinate and intrinsic natures. Castree argues that in the latter three categories 'the environment' can be used as a synonym for nature. He also highlights the challenge of deep multiculturalism in planning and the incommensurability of languages used which makes it difficult to give voice to, for example, indigenous peoples in planning arenas. The chapter presents a double critique of: misrecognition – concealing the social character of claims about 'the nature of environment' – and hypostatisation – fixing of the meaning of mutable and dynamic social constructs (such as the environment) as a given.

This theme is further explored by **Timothy W. Luke** in **Chapter 2.** He flags conceptual and practical dilemmas in scanning the complex and conflictual discourses about the environment and their implication for environmental planning. The chapter presents a critique of 'who speaks for nature' and crucially 'who speaks through nature' and for what purposes. Which nature, or the environment, is brought into being in discourses and practices of environmental planning and with what consequences is a key theme of this chapter. Luke identifies and critiques four currents of discussion through which 'Nature', understood as being shaped discursively and materially by human, is made into 'the environment' for human sake. These are: reification and revitalisation, resourcification and regulation, rationalisation and restoration, and ruination and resilience. A striking example of the first current is how environmental planning conceives of whole regions of the Earth in development related terms of: built environment, yet-to-be-built environment or never-to be-built environment. A recurring theme is the critique of modernity and its techno-scientifically driven command and control mode of environmental planning, one that is rooted in an anthropocentric view of nature. For Luke, discourses embedded in such mentality are fraud and lead to degrading of 'Nature'. To oppose them requires environmental discourses that invoke workable alternatives for implementing fair and democratic politics for their realisation.

In planning, alternative future imaginaries have often been presented in idealised visions of the environment and the place of humans in it. The history of environmental planning, as a future oriented activity, is inseparable from the history of visionary idealism, and none can be fully understood without interrogating the role played by ideology, as suggested by **Michael Gunder** in **Chapter 3.** He discusses three pathways through which visionary environmental idealism is materialised. The first and most directly aligned with environmental planning is the creation of eco-cities (environmentally desirable human settlements). The second is through post-political ecological modernisation which seeks to make the current neoliberal capitalist world more environmentally friendly (see also Chapter 5). The third is through radical 'deep green' movements that seek fundamental changes to current social and economic systems such as, sustainable de-growth or deep ecology, which are also advocated as the way forward by Naess and Saglie in Chapter 5. Gunder presents a critique of visionary environmental idealism as it materialises through these three pathways. For him, 'sustainability', which is a common goal of all three, is an ideological 'master signifier' (a signifier without signified) whose meaning (signification) can be understood clearly only when it is embroidered with more explicit signifiers (see also Chapter 4). Drawing on Jacques Lacan (2006), Gunder suggests that sustainability is also presented as a transcendental ideal external to human experience and knowledge, yet one that is nevertheless perceived to be an ultimate desired goal currently lacking in the world. Like several other contributors to this Companion, Gunder emphasises the need for unpicking the specific meaning of and implicit values that underlie key concepts used in environmental planning discourses and the knowledges pertaining to its master signifiers such as sustainability.

Sustainability is indeed one of the most readily recognised and frequently used concepts in environmental planning. Despite or because of this, its meanings have been subject to much contestations among political actors at all levels from the local to international bodies. To shed light on this highly influential concept, in **Chapter 4**, **Delyse Springett** and **Michael Redclift** trace its genealogy, problematise its meanings and provide a detailed account of the international contestations of sustainable development. The chapter focuses on the political economy of environmental conditions and highlights the connection between capitalist liberal markets and cultures and social injustices and environmental crisis. Emphasising the inextricable link between social and environmental responsibilities, the authors show how power and knowledge are used to exploit both nature and people. A key theme is the problematisation of the discursive dominance of Global North over Global South. They argue that the root causes of environmental as well as social problems are economic and cultural globalisation of neoliberal conceptual forces. The overall message, as in several other chapters in this part, is that concepts such as sustainable development should not be taken-for-grated and reified in and through environmental planning processes and practices. Environmental planning concepts are carriers of power and politics; they make certain ideas and interests visible and dominant while masquerading and undermining others.

Chapter 5 by **Peter Næss** and **Inger-Lise Saglie** focuses on the concept of ecological modernisation which emerged in environmental sociology in the 1990s in relation to the implementation of sustainable development (see also Chapter 4). The authors argue that contemporary environmental planning is grounded in ecological modernisation and its particular rationality of development. They unpack the core premises of ecological modernisation and use them to critically assess the compact city ideal which has gained hegemonic status as a sustainable urban form internationally. Næss and Saglie draw on their work on Oslo, a city that has rigorously pursued compact city policy in recent decades, to examine whether the policy has fulfilled the ambitions of ecological modernisation. Central to these ambitions is the claim that the paths to sustainability do not require radical changes to current economic and social processes, that economic growth and the use of environmental resources can be decoupled. The Oslo case shows the limitations and weaknesses of eco-modernisation and reveals the inescapable tensions that exist between economic growth and environmental sustainability. While acknowledging the merits of densification strategies versus urban sprawl, the authors argue that environmental planning benefits by looking beyond eco-modernisation and engaging with alternative discourses and perspectives such as 'risk society', 'de-growth' and 'eco-socialism' (see also Chapter 3).

Chapter 6 by **Mark Whitehead** focuses on one of the most recent concepts that have entered the conceptual repertoire of environmental planners, the Anthropocene. The chapter considers whether and how the Anthropocene is changing our understandings of and approaches to human-nature relations. It also discusses the particular challenges that this new conceptualisation of human-induced changes to global environmental change present to the practices of environmental planning. While acknowledging the multifaceted nature of the concept of the Anthropocene (with ontological, epistemological, practical and moral dimensions), Whitehead focuses on the ontological and moral forms of the Anthropocene which he considers to have most valence for environmental planning. He explores the former in relation to the concept of planetary boundaries and the nine critical areas analysed by scientists to identify the upper limits of human interventions in them. Whitehead argues that the significance of planetary boundaries framework for environmental planning is its representation of the complex and interconnected forms of environmental change and its definition of safe and unsafe operating spaces as the context for establishing policy targets. For him, environmental planning as a facilitator of collective action has a significant role to play in responding to challenges of the Anthropocene. Such a role is likely to be performed through a mix of technocratic (technological and managerial solutions) and eco-centric (nature-based solutions) strategies. Either way, there is a need for major extension of the temporal and spatial scope of environmental planning beyond its current form. There is also a need for deeper understanding of risk and uncertainty in the complex and ever-changing world in which environmental planning operates and by which its scope and principles are defined. The final two chapters of this part of the Companion focus on this theme. Both emphasise the importance of risk analyses as an integral part of environmental planning.

The meanings of risk and risk management are explored in **Chapter 7** by **Eoin O'Neill.** He shows how the concept of risk frequently includes references to other concepts such as danger, hazard or harm and how risk is defined as a combination of hazard, vulnerability and exposure or as the product of probability and consequences. The latter is argued to be the one used frequently in environmental planning practices. O'Neill discusses the different perspectives on risk in relation to people's different perceptions of risk. Drawing on landmark analysis by the psychologist Paul Slovic (1987), he summarises various factors that influence risk perception such as prior experience, attitude and perceived control. O'Neill flags real and perceived geographical proximity as an influential factor which is directly relevant to environmental planning practices. The chapter highlights some of the key challenges to the implementation of risk-informed environmental planning and advocates a greater stakeholder engagement in decision-making processes.

In **Chapter 8, Jon Coaffee** draws on Ulrich Beck's 'Risk society' thesis to discuss how the magnitude and boundless nature of the global risks is transforming how risks are imagined, managed and governed and how the concept of resilience has emerged as a key response to risks and uncertainty. Its rapid and extensive inclusion in discourses of environmental planning has led to the suggestion that resilience has replaced sustainability as the master signifier, representing a shift of emphasis from putting 'the world back into balance' to looking 'for ways to manage in an imbalanced world' (Zolli, 2012, no page). The chapter discusses how resilience processes and practices relate to traditional notions of risk and risk management. Resilience studies reflect the new understanding of risk but interpret resilience in two distinct ways: one is based on an equilibristic view of the world (the socio-ecological model) and the other on an evolutionary perspective. The chapter discusses the implications of these two interpretations for environmental planning and argues that despite the rise of evolutionary resilience thinking, there remains a considerable implementation gap which can only be bridged in environmental planning if resilience is seen as a set of transformative learning processes, and not merely as outputs or outcomes. While there is a growing critique of resilience as a carrier of neoliberal strategies (see Bohland et al., 2018), Coaffee highlights its transformative potentials not least in 'breaking planning out of its obsession with order, certainty and stasis' (Porter and Davoudi, 2012, p. 330).

If there is one theme which binds the contributions to this part of the Companion together, it is their emphasis on the importance of concepts in environmental planning discourses. Although concepts are not ontologically fixed, they can be hegemonic in the sense that they can temporarily stabilise meaning through nomination, and anchor and structure discourses which in turn shape the nature and scope of environmental planning and influence its outcomes. The chapters echo the view that 'environmental narratives in planning are selective abstractions which amplify one meaning of the environment and marginalize others'; that 'their formation is a contested political act, infused with power relations' (Davoudi, 2012, p. 50).

Two other key themes emerge from the contributions to this part. The first one is a reminder that the dichotomy between human and nature is no longer tenable and neither is the belief in a pristine nature, untouched by human activities. The second theme is the fallacy of the modernist assumptions about the ability of humans to tame and order nature with techno-managerial and rational environmental planning for their own exploitation with little or no consequences. In addition, the contributors to this part provide thought provoking ideas about the role of environmental planning in imagining and delivering environmentally and socially better futures.

References

Bohland, J., Davoudi, S. and Lawrence, J. (eds.) (2018). *The resilience machine*. New York: Routledge.

Davoudi, S. (2012). 'Climate risk and security: new meanings of "the environment" in the English planning system'. *European Planning Studies*, 20(1): 49–69.

Lacan, J. (2006 [1996]). *Ecrits*. London: Norton.

Porter, L. and Davoudi, S. (2012). 'The politics of resilience for planning: a cautionary note'. *Planning Theory and Practice*, 13(2): 329–33.

Slovic, P. (1987). 'Perception of risk'. *Science*, 236(17 April): 280–5.

Williams, R. (1980). *Problems in materialism and culture*. London: Verso.

Zolli, A. (2012). 'Learning to bounce back'. *New York Times*, 2 November. Available at: www.nytimes.com/2012/11/03/opinion/forget-sustainability-its-about-resilience.html

1

Perspectives on the nature of environmental planning

Noel Castree

> . . . the environment seems the most natural of all topics: it is 'out there', a physical entity . . . [presenting] . . . real problems with real solutions.
>
> *(Myerson and Rydin, 1996, p. 2)*

Environmental planning is a very practical affair. In the rather formal words used in the current 'Wikipedia' entry on the topic:

> Environmental planning concerns itself with the decision making processes . . . required for managing relationships that exist within and between natural systems and human systems. [It] . . . endeavours to manage these processes in an effective, orderly, transparent and equitable manner for the benefit of all constituents within such systems for the present and for the future.
>
> *(Wikipedia, 2018)*

We should add that environmental planning involves more than 'managing decision making processes', important though this is; implementation of decisions and monitoring outcomes are also key aspects of the process. Everything from regulating flood plain use along the Yangtze River to introducing 'green infrastructure' in Toronto to conserving wildlife in the Galapagos Islands involves environmental planning. It is typically a long and complex process in which planners are central actors but hardly the only ones. Planning regimes – that is, relatively stable ensembles of laws, rules, institutions and personnel prevailing in any one place or region – tend to have long term impacts. For instance, a national park created five years ago is likely to benefit future generations five or more decades down the line.

In countries that are democratically governed, such planning is ostensibly in the public interest – though because democracies encourage a range of political viewpoints, it's no surprise that planning can become controversial at times (i.e. the question 'exactly whose interests are being served?' arises). Controversies can manifest 'upstream', 'mid-stream' or 'downstream' of the planning process – that is, before, during and after the setting of planning guidelines, rules and procedures. As planning regimes have become less top-down and more participatory in nations like Holland, the United States and New Zealand, the room for disagreement and contention has grown.

Typically, disputes arise because one or more persons, groups or organisations oppose a key element of the process – such as the location or character of the environmental phenomena to be planned for (for example, a wetland used by migratory birds that some want in-filled for house building purposes), the specific norms and regulations used in planning (for example, that only certified university researchers will be asked to help design 'compensatory habitat' for the birds) or the eventual impacts of planning decisions (for example, the loss of lake amenity by local fisher people, bird watchers and dog walkers). Controversies speak to the very 'real' nature of environmental planning – it's focus on physical phenomena like forests, watercourses, grasslands and so on that, in various ways, *matter* to people (for example, for economic, recreational or spiritual reasons). They matter materially *and* symbolically by virtue of how they are used and perceived by different constituencies.

This chapter aims to demonstrate that environmental planning has important conceptual underpinnings that are essential to its practical expression in particular times and places. The way that planners, and the many stakeholders they need to engage with, conceive of the world has a significant bearing on the answer to the key practical question: 'What should we plan for, why, when and how?' A lack of awareness of, and reflexivity about, the basic conceptual vocabulary structuring discourses of (or about) environmental planning can constitute a significant blind spot for planning professionals and those that their decisions impact. Environmental planning can become controversial – or not – depending on (i) which ensemble of concepts structure discourse about everything from urban green belt plans to plans for new wildlife corridors in the countryside and (ii) whether these concepts' meanings are taken for granted or not.

Fortunately, over the last two decades or more, a number of academic researchers in planning and the wider social sciences have paid close attention to such concepts and how they get variously 'normalised' and challenged in different situations. Their key point is that there's nothing 'natural' about how certain concepts come to animate discourse during the planning process, however elemental they may seem. This may seem an obvious point to make about concepts that are, in historical terms, still quite young – such as 'sustainability' and 'biodiversity' (both emergent in the 1980s) or 'natural capital' (emergent from around 2000). But it's less so for concepts that are so established as to seem entirely commonplace and thus uncontentious. Key among them are the concepts of 'nature' and 'environment'. Both are often used unproblematically to denote an objectively existing world of insects, river catchments, ecosystems, mountains and so on. But what happens when their 'realist' credentials are called into question and we regard them as categories that reflect particular, culturally specific ways of viewing the world?

In the pages to follow some of the ways that environmental planning discourse has been 'denaturalised' are illustrated and some of the benefits of scrutinising its fundamental concepts itemised. The arguments and examples shared are hardly new – as previously said, they have been presented over 20 years or more. However, they continue to have relevance to planning theory and practice in the present and are thus worthy of serious consideration. This chapter begins by defining the two key concepts whose 'denaturalisation' is the focus of the subsequent two sections. It can usefully be read together with the chapter by Timothy W. Luke in this book.

The nature of the environment

Definitions: fundamental concepts as semantic 'mirrors'?

Environmental planners focus on biophysical phenomena that are part of our 'environment' at local, regional, national and even international scales. In English, an environment is 'that which surrounds us' or 'The physical . . . conditions in which a person or other organism lives'

(Oxford English Dictionary, 2018). As societies have grown in scale, complexity and impact, it has become ever more important to plan how various terrestrial and aquatic environments are utilised. Planning can facilitate *exploitation* (for example, mining); *conservation* (for example, of rare ecosystems); *restoration* (for example, of diverted rivers); or, in a blend of the first and third, *compensation* (or 'off-setting', where exploitation is permitted in one location so long as efforts at protecting the environment occur elsewhere within the same plan). Aside from utilising or protecting the nonhuman world, environmental planning can also promote human welfare (for example, by reducing environmental risks to people). It can be animated by an 'anthropocentric' ethos, a more 'ecocentric' one or some blend of the two. In a strict sense, 'environmental planning' is that part of planning where biophysical phenomena are a *central* focus – as in measures to prevent damage to Australia's Great Barrier Reef, say. But, in a broader sense, pretty much *all* planning is environmental planning insofar as, these days, the environmental impact of everything from new housing estates and railway lines to new palm oil plantations must be carefully considered by those proposing them.

In order to work, all environmental planning must presume to know two things about the world: first, the particular elements of biophysical reality in question (a forest, a cave, a section of beach, a small upland catchment, an atoll, an industrial site to be remediated, etc.) and second, the character or quality of these elements (for example, a wetland of average ecological value, a stretch of flood plain where the risk of severe inundation is low, an alpine lake containing unique amphibians, a block of badly polluted soil). In both cases (delimiting and describing), a 'common sense' belief that the world exists 'out there' is in play, a belief finding linguistic expression in the seemingly uncontentious words 'environment' and 'nature'.

The former is a special case of the latter. Nature is a very old term in the English language, and its meanings have varied through time. It refers to a wide range of phenomena in a plurality of different ways – this is why it's an unusually complicated word; some would argue the most complex of all. As a linguistic philosopher might say, it's a 'signifier' (word or sound) that has more than one 'signified' (a specific meaning) and these signifieds attach themselves to an astonishing number and range of material things ('referents'). Today, the term has four principal meanings. First, it means the nonhuman world, especially those parts untouched or barely affected by humans ('the natural environment'); second, it signifies the entire physical world, including humans as biological entities and products of evolutionary history; third, it means the power or force governing some or all living things (such as gravity or the conservation of energy); finally, it means the essential quality or defining property of something (for example, it is natural for birds to fly and fish to swim).

As a shorthand, we can (respectively) call these meanings 'external nature', 'universal nature', 'superordinate nature' and 'intrinsic nature' (see Figure 1.1). In this context, 'the environment' can readily be a synonym for nature in three of the four cases. Any given environment can be characterised on a spectrum from wholly 'natural' to entirely artificial, but even in the latter case planners and their stakeholders will presume to know the 'nature' of the place or elements in question (in the third or fourth sense of the term). To 'get at' this nature they may commission scientists and other experts to analyse the environment using various observational techniques. Specific empirical representations of environmental phenomena are, as it were, 'situated' in the broader descriptive category of 'nature'. Even if these phenomena are not wholly 'natural' in degree or kind, their (partly artificial) 'nature' is presumed to exist independently of the planners and others who must manage their proper use. Likewise, any planned changes to their use can be imagined to have real effects whose nature can be accurately represented and projected – as, for instance, when a 'brownfield' site in London is 'rewilded' so as to become more natural over time.

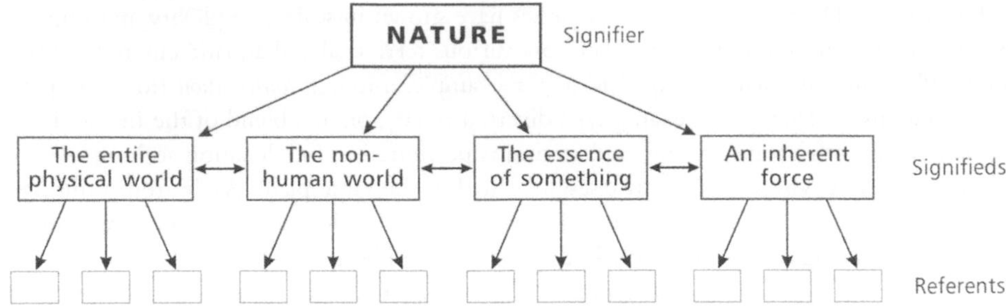

Figure 1.1 The concept of nature: its meanings and referents

The two-way horizontal arrows indicate that more than one of the four meanings can be in play in any given reference to 'nature'. The meanings can be attached to the totality of what we call 'nature' and, equally, to any of the myriad of different phenomena we consider to be natural in kind or degree. The second meaning is otherwise signified as 'the environment'.

Source: Author's own, previously published in Castree (2014, p. 10)

In sum, for many people when they use the words nature and environment, they typically think they are making *ontological* statements – that is to say, statements about a biophysical reality that exists beyond the words used to make linguistic sense of it. In this respect, we are apt to assume that nature and environment are 'mimetic concepts' whose meanings capture in words actually existing phenomena that exist outside discourse. This assumption is a prior necessity for the practical business of environmental planning.

The conceptual mediation of the material world

But is the assumption of mimesis valid and uncontentious? Some believe not, even as others would find it difficult to question. Beginning in the mid-1980s, a number of scholars in the critical social sciences and humanities began to question whether 'the environment' and 'nature' could be referred to 'objectively' – be it by planners, citizens, governments, business leaders or even scientists. In disciplines like anthropology, human geography, and sociology, and interdisciplinary fields like cultural studies and STS (Science and Technology Studies), there was a 'discursive turn'. That is, there was a focus on how language – and other 'representational media' such as maps, films, documentaries and the like – *constituted* reference and meaning rather than simply 'revealing' it. As philosopher Kate Soper expressed it, in what became an influential book (entitled *What Is Nature?*), 'there can be no adequate attempt to explore "what nature is" that's not centrally concerned with what it has been *said* to be' (Soper, 1995, p. 21). Soper, and like-minded others, were in part inspired by the philology of English cultural historian Raymond Williams. Williams' book *Keywords* (1976) inquired not into the 'proper' meanings of words – as if material reality somehow dictates this – but their social history of invention and usage. According to him, the meanings of 'nature' and 'environment' are as much a reflection of those societies in which the terms have currency, as they are the properties of the phenomena to which the terms by convention refer. When we talk about nature and environment, Williams famously opined, we are talking about *ourselves* without necessarily knowing it or admitting it.

Through the 1990s, these claims about 'discursive construction' were controversial in Western academia. Critics decried the apparent 'anti-realism' (ontologically) and perspectival 'relativism' (epistemologically) of these claims. For instance, an American scientist, Alan Sokal (1996), (in)famously played a hoax on STS in the academic journal *Social Text*. His aim was to show that there *is* an objectively knowable world existing outside of various discourses about it. A quarter century on, these controversies have died down and the debates evolved. Critical social science and the humanities have, in many ways, now adopted a post-constructivist perspective where the world's 'agency' and 'materiality' are being acknowledged and the 'power' of discourse given less prominence. Given the eminently 'practical' orientation of environmental planning – indeed, planning of whatever stripe – this may seem like a welcome return to the real world in the eyes of some.

However, the insights of the 'discursive turn' are not passé. Looking carefully and critically at the origin and content of claims made about 'the nature of environment' remains as relevant to planning practice as the finer details of tools like cost-benefit analysis or land use zoning rules. Indeed, it is central to a truly democratic process of environmental planning. The next two sections illustrate how and why using extended examples, before the argument is then summarised.

Investing the environment with 'unnatural' meaning

It has been at least 25 years since the first studies of how 'the nature of environment' is invoked by different actors in the planning process were published. For instance, geographers Carolyn Harrison and Jacquelin Burgess (1994) analysed a dispute over a site of special scientific interest in southeast England, Rainham Marshes. The Marshes were a natural place on the edge of London. Harrison and Burgess demonstrated the mixture of 'strategic framing' and unconscious bias that led a developer (Music Corporation of America) and a group of nature conservationists to represent the 'reality' of Rainham in very different ways. Their point was not that the latter's discourse was more scientific and thus 'realistic' – while the former's was commercially self-interested. Instead, it was that *both* discourses were shot-through with particular metaphors, values and norms that reflected the rival interests in play. Yet both the developer and conservationists pretended otherwise, making 'realist' claims about the character and potential of the marshes (thus disavowing the social nature of their representations).

Though, as noted in the previous section, research like this is considered passé by some; others continue to see the value in interrogating the basic language of environmental planning. To illustrate this value, let us consider the future of a fascinating place in the US. We live on an increasingly human planet, in an epoch some geoscientists are calling 'the Anthropocene'. A lot of environmental planning now focuses on places that, for some people, could or should become – or remain – natural (for example, biodiverse, off-limits to road building, etc.). The Salton Sea, located in southern California, is a particular example of a wider issue for environmental planning today: namely, what to do about 'anthropogenic environments' in which an original natural area of land or water has been massively altered by human activities. Such areas are, unlike the Rainham Marshes, manifestly 'unnatural'. The question is: should they be restored, allowed to 'run wild' or be further developed?

The Salton Sea is a shallow, highly saline inland water body in southern California (see Figure 1.2). Historically, it was variously a freshwater reservoir, a saline lake or a desert flat depending on the natural movements of the Colorado River. By the time Europeans colonised North America, it was the latter (a playa). But in an accident of hydraulic engineering, the area was inundated by the California Development Company in 1905. Lacking a natural outlet, the water created a large lake (nearly 22,000 km^2) over a two-year period. During the 20th century, the lake developed a peculiar ecology. Fish were introduced for angling and an ecosystem of birds, aquatic creatures and underwater biomass evolved. Meanwhile, farming of surrounding lands made it a sink for pesticides and herbicides. Evaporation of the lake surface in the hot sun, along with insufficient supply of new water, turned it into a euthrophied lake replete with

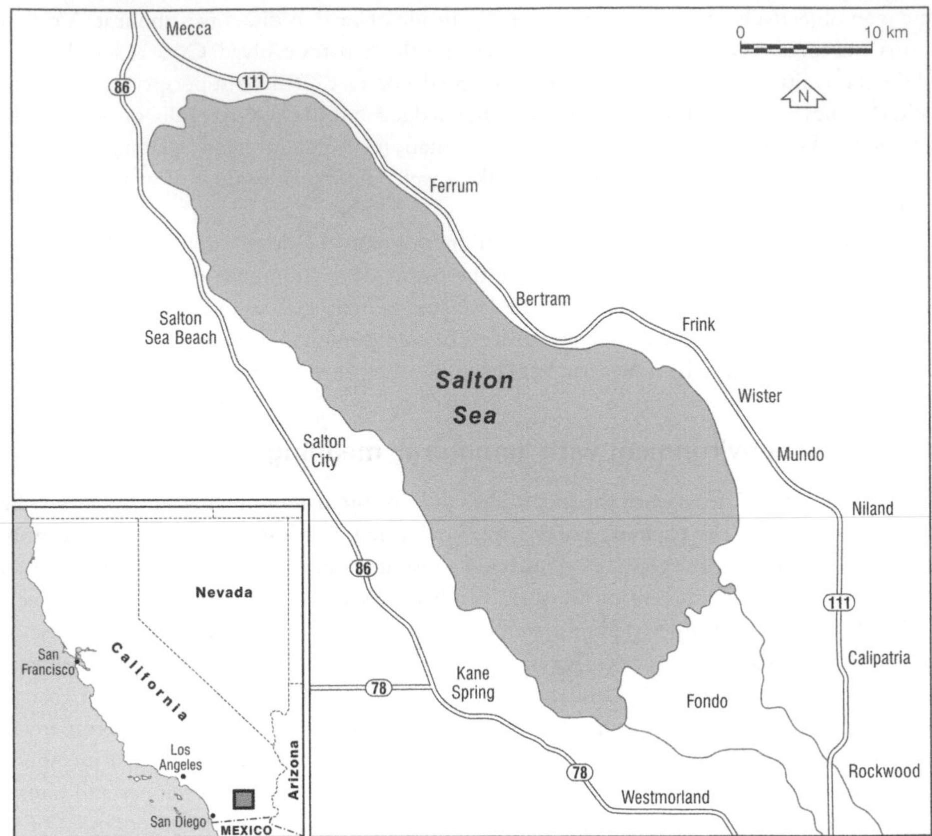

Figure 1.2 Salton Sea map
Source: Author's own

algal blooms and a frequent rotten egg stench. The knock-on effects on birds, fish and marine ecology have been severe since the 1970s. As the water area has shrunk, a dessicated lake bed has been exposed – full of agricultural pollutants and salt. When liberated as dust by strong winds, this threatens to become a health hazard for people living or working around the Sea's long shoreline. In sum, the Salton went from being an ecologically functional – albeit artificial – water body by the 1950s to something of a 'monstrous' water body by 2000. Recognising the need to do something about its condition, the State of California passed the Salton Sea Restoration Act (SSRA) in 2003. Though little in practice had happened by 2018, the Act did inspire a number of plans for restoration.

While these plans varied considerably in their detail, what's interesting here is how each tried to 'fix' the Sea's current and future potential character conceptually. The Act called for 'restoration of long-term stable aquatic and shoreline habitat for the historic levels and diversity of fish and wildlife . . .' (State of California, 2003). But this not only opens the question of which historic 'baseline' over the last century the Lake should be restored to. In light of economic constraints, it also opens the question of which exact plan will be perceived to 'deliver' a suitable kind of restoration. This last question is important because any 'restoration' involves all sorts of manifestly *human* interventions, such as massive pipes, rock berms or artificial lakes created in the desert playa around the Sea. For instance, the Imperial Irrigation District (IID) – a company with rights to Colorado River water – proposed to open geothermal power plants as the Sea recedes, utilising a natural underground energy source. This 'green energy' system would then fund restoration of a reduced size Sea and of smaller lakes created around its shore. While still clearly an anthropogenic sea, the IID plan would nonetheless 'restore' Salton in ways that appear more 'natural' than the present-day toxic water body in question. Yet, paradoxically, if the geothermal plan were implemented in adjacent desert areas protected by the US federal government they would be perceived as *degrading* a 'pristine' dryland environment. This illustrates how the definition of 'environmental restoration' depends fundamentally on the way the nonhuman world is categorised by those looking at it (for more on the Salton case see Cantor and Knuth, 2018).

The key point here is that artificial environments like the Salton Sea do not allow those with plans to change them to point to clear, contention-free evidence of what *counts* as a more natural alternative. Instead, evidence gets freighted with conceptual meanings to suit the wishes of those advocating specific restoration plans. Yet, as with the IID plan, the cultural constructedness of these meanings is backgrounded for fear that advocates will lose credibility in the planning process. Even in our 'post-natural' world, the desire to make 'real' claims about the nature of environments dies hard. Indeed, the desire is arguably *accentuated* because hybrid places like Salton raise questions about the 'right' natural baseline (and its achievability, practically speaking). Planning needs to be alert to the social origin and content of these claims lest it unthinkingly naturalise some of them.

Planning beyond the nature-society dualism: multi-cultural environments

The example presented indicates that 'Nature is . . . not simply a physical entity but a repository of meaning' (Wapner, 2014, p. 41), whether it be pristine nature or a more artificial one we are dealing with. Does this suggest that 'nature' does not exist, only the concept and its proxy terms like 'environment'? In one sense, yes. This is not the same as arguing that the material things to which the concepts refer do not exist; they assuredly do. But it is purely a matter of convention to call them 'natural' or 'environmental' in the various ways and contexts that we choose to. For many (perhaps including some readers of this chapter), this is no doubt a difficult argument to accept. It accords a lot of importance to language, its social origins and its practical effects, and it challenges the conventional idea that many or most concepts are semantic 'mirrors' that faithfully represent the material world. Nature, *by definition*, seems to be that which lies outside and is irreducible to linguistic conventions. But appearances are deceptive. In the end, there is nothing 'natural' about our habit of designating certain things as belonging to 'nature' and 'environment', nor about using those things as a reference point to anchor our wishes and desires for stability or for change.

Is the nature-society dualism an ontological given?

Over the years, some analysts have pushed this argument even further. They have claimed that not only are claims *about* 'the environment' saturated with social content; more than this, they have suggested that the reality-representation distinction that *underpins* these claims is *itself* socially constructed and 'unnatural'. Among the first to do this in the context of a planning dispute was Canadian geographer Bruce Braun. Let us revisit his arguments and case study

(Braun, 1997), which appeared in print around the same time as Harrison and Burgess published their Rainham Marsh study.

At first sight, his approach to the case seems just the same as Harrison and Burgess'. Braun examined a very heated dispute that erupted in the Canadian province of British Columbia in the early 1990s. The dispute was about government plans to allow logging to occur in an area of 'old growth' temperate rainforest on Vancouver Island. It involved a stand-off between a major forest products company (MacMillan Bloedel) and environmentalists determined to preserve an increasingly rare natural landscape (much of British Columbia's native forest had been cleared through the 19th and 20th centuries). The latter saw Clayoquot as one of the last remaining spaces of 'pristine nature', while the former regarded it as a valuable economic resource that should be logged in a responsible way for the good of the Canadian economy and those communities dependent on forestry jobs. In a detailed analysis of both sides' representations of Clayoquot, Braun shows how the region's 'realities' were made to appear quite different depending who was doing the looking. There was a clash between representations that made Clayoquot appear wild, intricate, threatened and special on the one side and those (on the other side) that made it appear as one more 'resource zone' to be rationally harvested by logging firms using cutting-edge technology. In both cases, the authors of the representations claimed to be depicting Clayoquot as it actually was. But Braun's point was that these representations – which comprised books, pamphlets, newsletters and media-worthy events reported on television – reflected the specific agendas of those promoting them. In other words, one could not adjudicate between them by testing their veracity against the non-representational actualities of Clayoquot's old growth trees.

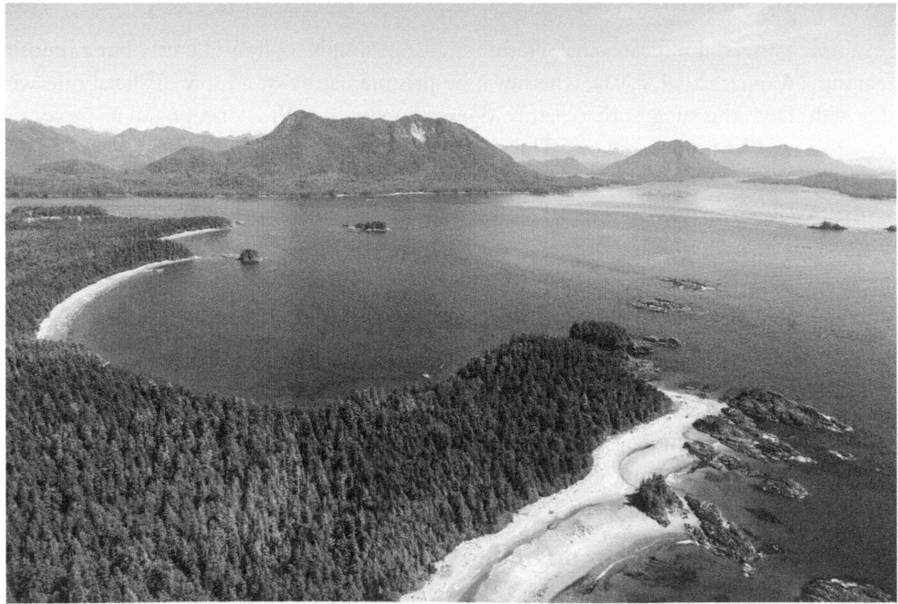

Figure 1.3 Clayoquot Sound, British Columbia

Source: Josh Lewis (permission obtained)

The Sound was declared a UNESCO Biosphere Reserve in 2000. As with the Rainham Marshes in England, its defenders have sought to depict it as a thoroughly natural region.

Arresting though this insight was at the time of publication, Braun's research contained a further surprise. Notwithstanding the *differences* in content and message between the environmentalists' and forest company's representations, Braun argued that they ultimately shared the *same* symbolic universe: a specifically Anglo-North American one that reflects the linguistic conventions and cultural suppositions of those colonists who spread through the United States and Canada from the 17th century onwards. What's more – these conventions and suppositions functioned as a former of cultural power. He makes this point with reference to Clayoquot Sound's small groups of remaining 'native' (or indigenous) people, the Nu-chah-nulth. These groups had, historically, lived a peripatetic existence and had used the forest for generations to meet their material and symbolic needs. Yet both the environmentalists and the logging company fighting over Clayoquot's future assumed that the region was largely *empty*. This, Braun argued, constituted a geographical expression of a specifically Western belief that nature and society are two separate things. Clayoquot's indigenous peoples, he concluded, were thus victims of symbolic violence, even in the supposedly *post*-colonial conditions of modern Canada. Their history, and present day claims to regain control of Clayoquot, simply did not register in the unthinking assumptions made by the descendants of the original European colonisers. As such, these assumptions functioned as discursive, and all too real, forms of exclusion and oppression.

'Deep multi-culturalism': a challenge to environmental planning

Braun's study was an early illustration of why environmental planning needs to be acutely, and deeply, aware of what 'multi-culturalism' means. In the contemporary world, planning routinely operates in contexts where different groups of people (according to religion, culture, ethnicity, etc.) come into contact. But they do so unequally. In such contexts, as Braun illustrates, *incommensurability* is an issue that can too easily be overlooked. It is an issue of, metaphorically, Martians trying to communicate with Earthlings, the languages are not compatible. The British anthropologist Tim Ingold noted this in his studies of non-Western hunter-gather peoples. As he put it, they:

> [D]o *not*, as a rule, approach their environment as an external world of nature that has to be 'grasped' conceptually . . . [Unlike Westerners] . . . [t]hey do not see themselves as mindful subjects having to contend with a . . . world of physical objects; indeed, the separation of mind and nature has no place in their thought and practice.
>
> *(Ingold, 1986, p. 120)*

The challenge for environmental planning implied by Ingold's insight is being taken more seriously than when Braun published his research about Clayoquot Sound. In former settler-colonies like Australia, New Zealand and Canada this is especially true. There, indigenous cosmologies are no longer forced to defer to Western perspectives on reality. Moves towards 'participatory' and inclusive planning are now having to reckon with the challenge of worldviews such as that signified by the word Country in Australia.

Classically, 'giving voice' to indigenous peoples in the planning arena has involved researchers undertaking sustained fieldwork, or more rapid 'consultation', in order to understand non-Western life-worlds from the inside. Planners can then take the resulting knowledge into account. However, this is arguably inadequate. For instance, Australian geographers Kate Lloyd, Sandie Suchet-Pearson and Sarah Wright have pushed far beyond this classic practice in their path-breaking research on, with and for Bawaka Country in north east Arnhem Land (part of Australia's Northern Territory). Over a period of years, the trio has developed very close relationships with individuals and families who inhabit Bawaka Country. These relationships have led them not merely to understand indigenous cosmology from the outside-looking-in. More profoundly, they have endeavoured to *embrace* this cosmology in the conduct and dissemination of their research. This has involved seeing themselves not as *analysts* of Country but as *elements* of Country. Bawaka, as Country et al. report:

> [I]ncorporates people, animals, plants, water and land. But Country is more than just people and things; it is also what connects them to each other and the multiple spiritual and symbolic realms. It relates to laws, custom, movement, song, knowledges . . . histories, presents, futures . . . Country can be talked to, it can be known, it can itself communicate, feel and take action.
>
> *(Country et al., 2016, p. 458)*

In this light, the trio have fundamentally questioned the standard protocols of academic research into people and nonhumans. By implication, their research questions the way indigenous knowledge is treated in the environmental planning process. In fact, it poses fundamental questions *about* the process in countries like Australia. What would it mean to 'plan for Country', rather than plan, say, for a lake restoration by merely *taking account* of aboriginal views *of* Country? (cf. Mould et al., 2018). This important question becomes legitimate to ask once one questions the idea that nature-society, environment-people, object-subject divisions are a given. To continue the metaphor, it forces the Earthlings to speak Martian not vice versa, challenging the neutrality of planning's most basic categories of thought.

Conclusion: denaturalising environmental planning

As noted at the start of this chapter, humans relate to the environment in ways that are variously classified as harmful or careful, exploitative or restorative. Planning seeks to shape these relations, responding to governmental laws and the views of affected parties. Today, environmental planning occurs, increasingly, in a world where what's considered 'nature' is in decline, where 'artificial environments' are growing, and where multi-culturalism is writ large. The discourses of planning – that is, the words used, and the claims made, by all those involved in the process – are what enable dialogue and decision making, prior to any practical actions. This chapter seeks to illustrate the value of a deep examination of planning discourse, focussing on the elemental concepts of 'nature' and 'environment'.

Lest this value not be entirely clear by this point, let it be spelled out by way of a conclusion. Interrogating the social content of ostensibly 'mimetic' concepts allows us to reap the rewards of a double critique: namely, a critique of *misrecognition* and *hypostatisation*. As the examples presented in this chapter show, claims about the nature of environment, so common in the planning process, routinely conceal their social character. As such, they can be misconstrued as mimetic – that is, as mirrors of an external world. Secondly, the exposure of routine acts of misrecognition is linked to critique of hypostatisation. To hypostatise something is to ossify and freeze it when, in fact, the thing so conceived is part of a changeable and dynamic process. To show that apparently natural (and thus 'given') things are, in fact, mutable social constructs varying between location and actors is useful. It allows analysts to contest acts justified in the name of nature and environment. It allows them to highlight the social relations animating apparently non-social ideas or entities. It enables them to show that 'nature' and 'environment' can be mediums for the expression of power relations. And it offers them the chance to imagine alternative social and ecological arrangements that honestly confront the non-naturalness not just of ideas of nature and environment but also, in many cases, the phenomena those ideas referred to. In short, the constructionist ethos remains very empowering for those who want to ensure that environmental planning is democratic and fair. Though the arguments presented in this chapter are now well established, even old, they remain relevant and useful. This is not to say they are problem free (for instance, see Proctor (1998) for an early, constructive critique of 'discursive idealism'). But used with care, they can enrich environmental planning and make the planning process more open and inclusive.

References

Braun, B. (1997). 'Buried epistemologies: the politics of nature in (post)colonial British Columbia'. *Annals of the Association of American Geographers*, 87(1): 3–32.

Cantor, A. and Knuth, S. (2018). 'Speculations on the post-natural: restoration, accumulation and sacrifice at the Salton Sea'. *Environment & Planning A: Economy and Space*, 31 August [Online].

Castree, N. (2014). *Making sense of nature*. London/New York: Routledge.

Country, B., Wright, S., Suchet-Pearson, S., Lloyd, K., Burrarwanga, L., Ganambarr, R., Ganambarr-Stubbs, M., Ganambarr, B., Maymuru, D. and Sweeney, J. (2016). 'Co-becoming Bawaka: towards a relational understanding of place/space'. *Progress in Human Geography*, 40(4): 455–75.

Harrison, C. and Burgess, J. (1994). 'Social constructions of nature: a case study of conflicts over the development of Rainham Marshes'. *Transactions of the Institute of British Geographers*, 19(3): 291–310.

Ingold, T. (1986). 'Hunting and gathering as ways of perceiving the environment', in R. Ellen and K. Fukui (eds.) *Redefining nature*. Oxford: Berg, pp 117–55.

Mould, S., Fryirs, K. and Howitt, R. (2018). 'Practicing socio-geomorphology'. *Society and Natural Resources*, 31(1): 106–20.

Myerson, G. and Rydin, Y. (1996). *The language of environment*. London: UCL Press.

Oxford English Dictionary (2018). Available at: www.oed.com/view/Entry/63089?redirectedFrom=environment#eid

Proctor, J. (1998). 'The social construction of nature: relativist accusations, pragmatist and critical realist responses'. *Annals of the Association of American Geographers*, 88(3): 352–76.

Sokal, A. D. (1996). 'Transgressing the boundaries: toward a transformative hermeneutics of quantum gravity'. *Social Text*, 46–7: 217–52.

Soper, K. (1995). *What is nature?* Oxford: Blackwell.

State of California (2003). *Salton Sea Restoration Act (SSRA)*. Available at: www.leginfo.ca.gov/pub/03-04/bill/sen/sb_0251-0300/sb_277_bill_20030911_enrolled.html

Wapner, P. (2014). 'The changing nature of nature'. *Global Environmental Politics*, 14(4): 36–54.

Wikipedia (2018). *The free encylopedia. Environmental planning entry*. Available at: https://en.wikipedia.org/wiki/Environmental_planning

Williams, R. (1976). *Keywords*. London: Fontana.

2

Discourses and meanings of 'the environment'

Timothy W. Luke

Introduction

As daily updates in the media confirm, the most prominent landmarks on terrain of 'the environment' in the 21st century are discourses and their meanings, but this presence is much older. From the Bible's account of God enjoining Adam and Eve, 'Be fruitful and multiply and fill the earth and subdue it and have dominion over the fish of the sea and over the birds of the heavens and over every living thing that moves on the earth' (Genesis 1:28) to contemporary apocalyptic tales about 'learning to die in the Anthropocene' (Scranton, 2015; Chin et al., 2013; McNeill and Engelke, 2014; Zizek, 2010), discourses and their conflicting meanings frame how humans cope with their built, never to be built or unbuilt environments as they confront the demands of environmental planning and sustainability.

The symbolic and practical powers of modern science and technology discursively are described either as gaining the high Enlightenment goals of human power and knowledge to secure control over nature or as gambling with all earthly life on an out-of-control processed world of chaotic climate degradation. On the one hand, 'enlightenment is totalitarian', its message directs human beings 'to learn from nature', especially 'how to use it to dominate wholly both it and human beings. Nothing else counts' (Horkheimer and Adorno, 2002, pp. 2 and 4). On the other hand, economic change at this juncture is said only to produce 'a risk society', as Beck asserts, in which 'what is important is to exploit and develop the superiority of doubt against industrial dogmatism' (Beck, 1992, p. 17). Tracing environmental discourse and its meanings for planning applications, then, requires linking closely the redoubts of the policy, engineering, life, physical, and social sciences (Commoner, 1990) that often are alien to each other.

In fact, under conditions of modern industrial life, as Beck suggests, '"progress" can be understood as legitimate social change without democratic political legitimation' (Beck, 1992, p. 214). Here those who would enjoy 'progress' and its benefits accord its goals, practices and outcomes a great degree of legitimacy, but they do this with hardly any discussion, participation or reflection. Progress is poorly understood because it essentially is an inhumanly destructive process tied to the permanent revolution of unending commodification of everyday life through technics and economics (Jameson, 1992; Lyotard, 1991).

Nearly 50 years after the first Earth Day during 1970, and following decades of expansive environmental legislation, guided by intensive scientific investigation since the 19th century, hesitation is spreading about the efficacy of any environmental planning and regulation along with greater resistance to green governance. This is leading to 'a proliferation of discourses about the environment from most quarters of society' (Darier, 1999, p. 2). It is important, then, to reconsider how complex and conflicted these discourses are, while probing the meaning of 'the environment' since doubt, denial and defiance are increasing across a diverse array of cultural, political, and social movements that are focused on 'ecology' (Soper, 1997; Meyer, 2001; Morton, 2013; Davies, 2016).

Save for the name itself, 'virtually nothing unites the bioregionalists, Gaians, eco-feminists, eco-Marxists, biocentricists, eco-anarchists, deep ecologist, and social ecologists who pursue their idea and actions in its name' (Ross, 1994, p. 5). What then are the meanings of 'the environment' and the referents of those discourses circulating meanings among these and other movements? By asking, 'Who speaks for nature? Who speaks for the environment?' discourses of 'the environment' come into play as meaning-makers of the environment for policy and planning as they affect 'who is in position to shape what is and what is not being said and what is being and not being done in relation to government policy' (Wander, 1996, p. ix).

With regard to discourse and the meaning of 'the environment,' is the question, as Ross (1994) intimates, not 'who speaks for Nature' but rather who, like planners, speaks through nature for themselves? In turn, is who, like planners' clients, hears, listens and responds to those discursive developments, free to do what they will with what becomes understood as 'the environment'? Whether or not environmental discourses are local, intense and well-intentioned or global, expansive and ill-considered (Young, 1990), how are their discursive understandings put to work? Environmental discourses are, on the one hand, productive codes of power/knowledge exerting cognitive and emotional effects in/through technoscience within an economy and society rooted in the postmodern conditions of accelerating performativity (Lyotard, 1984). These discourses, on the other hand, are also practices of 'green governmentality' within grounded styles for acting how, knowing what, acting where, and knowing why it must be done (Burchell et al., 1991; Luke, 1997). Seen in this way, practices of environmental discourse add up to 'a dynamic process that is situated and provisional, collective and distributed, purposive and pragmatic, and mediated and contested' (Davoudi, 2015, p. 316).

Arguably, it is impossible to untangle discourses and their meanings from 'the environment' as the interwoven discursive narratives of 'reification' and 'revitalization', 'resourcification' and 'regulation', 'rationalization' and 'restoration' or 'ruination' and 'resilience' all suggest. To whom these environmental discourses speak, from where they write, and for what they work, however, are quite problematic. This chapter, therefore, flags conceptual and practical dilemmas in scanning these discursive streams about 'the environment', and what they mean for, among other things, environmental planning.

Working up 'the environment'

To profile environmental discourse and its meaning reticulates the institutional and intellectual pressures rolling through contemporary life. Knowledge must be delimited, defined and delivered by trained specialists to fellow specialists, competing cohorts and interested laymen through the many conduits of discursive engagement. Thinking about 'the environment', then, parses out the ever-shifting orders and organisations of 'real knowledge' as formulae for ecological, environmental or green modes of action and thought via multiple discourses.

Nature, as it is shaped discursively and materially by human efforts to preserve and protect it, essentially is made into 'the environment' only to the extent it becomes suited to support human beings (Luke, 1999a; Davoudi, 2014). For Foucault, this space of movement mastered by science and engineering requires 'broadening and organizing that space, methods of power and knowledge assumed responsibility for the life processes and undertook to control and modify them' (Foucault, 1980, p. 142). Four currents of discussion suggest where these spaces are organised and broaden as elaborated below.

'Reification' and 'revitalization'

First, life scientists and physical scientists often subscribe to discourses with shared epistemic prejudices, entailing the 'reification' of Nature and Society. These critical, ethical or normative expert discourses resist opposing discourses from green radical groups objectifying, estranging and alienating natural and social life as they tout the imperatives of 'revitalization' for the natural and social life-worlds together (Devall and Sessions, 1985; Eckersley, 1992; Bennett, 2009). Similarly, green-leaning policy makers and social thinkers, who worry about the liberty, freedom, and equality of human subjects, find themselves tied down by reified epistemic constraints as they aspire to realise urban sustainability, ecological carrying capacity or environmental justice more inclusive, resilient or sustainable through environmental planning (Sandilands, 1999; Vogel, 1996; Rothenberg, 1993). Ironically, the narratives of reification and revitalization in environmental discourses show how their terms of thinking both have failed, and sustained, many efforts to stage environmental defence to limit ecological damage due to deeply rooted reasons (Meyer, 2001; Torgerson, 1999).

Discourses of reification pulse continuously within the policy capillaries shaped for modern markets, mechanistic methods and materialist masses. As Husserl argues, mathematics and mathematical science 'as a garb of ideas, or the garb of symbols of the symbolic mathematical theories, encompasses everything which, for scientists and the educated generally, *represents* the life-world, *dresses it up* as 'objectively actual and true' nature. It is through the garb of ideas that we take for *true being* what is actually a *method*' (Husserl, 1970, p. 51).

Anthropocentric discourse methodically does presume that the forms of human experience are uniform, constant and inalienable. Whether represented in paints or by equations, human rationality shapes systems and structures around this inherent methodological consistency. Nature and its environments always are apprehended by environmental planners in anthropocentric terms because they are already appraised in anthropomorphic and anthropometric categories (Deleuze and Guattari, 1994).

> Nature, though explicitly nonhuman, is *ours*: we do not so much read the "book of nature", as Galileo desired, as write it. It is a human artefact, and like the traffic light, its only purpose is provided by the use we give it.
>
> *(Evernden, 1992, p. 60)*

Hence, 'the environment', as it is discursively understood in modernised economies and societies, displays two paradoxical principles which express the ellipses of modernity, namely (i) 'even though we construct Nature, Nature is as if we did not construct it' and (ii) 'even though we do not construct Society, Society is as if we did construct it' (Latour, 1993, p. 32). Modern individuals inhabit a civic social order that may protect some freedoms once held in the state of nature, and it might alleviate some liabilities raised by living in civil society. To assay the merits of this trade-off of freedom, a domain of pristine natural givenness must be constituted by society as 'wilderness', or 'nature', or 'the environment' to define where and how such progressive social freedom might be secured. Nature and Society, or the biological and the historical, are kept conceptually pure and distinct by these worldviews, although their continuous operational remediation in commerce, industry, and science is the daily work of experts in modernity (Latour, 1993, pp. 30–3). After 250 years of industrial revolution (Mitchell, 2013) Nature – as vast expanses around the planet of untamed wildness – largely has vanished into modern markets. Enmeshed in complex networks of scientific rationalization and commercial exploitation, 'the environment' obviously becomes then an even more contingent cluster of constructs. Strangely, whole regions of the planet – from megapolitan sprawl to state protected wilderness – are increasingly now either a 'built environment' or a 'planned habitat' thanks to multiple waves and different layers of planning. Experts and managers use wilderness to define the spaces of 'yet-to-be-built' or 'never-to-be-built' environments, but abstract reified form delineates these regions, which await further resourcification to serve as a productive 'built environment' of commodity production and consumption (Cox, 2012).

'Resourcification' and 'regulation'

Second, many discourses on the environment are actively engaged in concretely working out liberal capitalist values as resource-oriented narratives (Radkau, 2008). The role assigned to environments in these scripts is one of rough and ready 'resourcification' for the global economy and national society (Luke, 2009). The environment at minimum functions as a standing reserve, a resource supply depot and a waste reception centre, while providing human markets with multiple sites for these resourcified flows of energy, information, and matter. Still, its fungibilization, liquidification and capitalisation cannot occur efficiently without expert intellectual labour to prep it, produce it and then provide it in properly regulated ways to the global marketplace (Bauman, 2007).

Artists, decision makers, planners and scientists shape 'environments', which humans are enjoined to protect with environmental discourses, in anthropocentric terms around mathematical formalisation (Ihde, 1990). Within these boundaries of action, Nature is 'resourcified' from its inception as an ensemble of multipurpose environments, brimming with primary resources awaiting their secondary extraction and tertiary use.

Fascination with abstract forms, intrinsic mechanisms and mathematical mobilisation discursively reify energy and matter as 'the environment'. When it comes to the modernised nature of Nature, one finds from the Renaissance 'that the logic of mathematics goes hand in hand with the *theory of art*. Only out of this union, out of this alliance, does the concept of "necessity" of nature emerge. Mathematics and art now agree upon the same fundamental requirement: the requirement of "form"' (Cassirer, 1972, p. 152). The arts and sciences all presumed Nature unfolds 'full of "rational principles" that have not yet been part of experience' which the human mind can articulate instrumentally and rationally, if it will 'only create true, necessary knowledge by its own principles' (Cassirer, 1972, pp. 57–8). Rather than seeking divine wisdom, cosmic meaning or inner truths from Nature, anthropocentric analysis leverages environmental regularities to seek 'reasons – or rather, *provided* the reasons of Nature through their creative endeavor. Nature was thus subsumed to the human enterprise, a puppet of the rational forces discernible to the human mind' (Evernden, 1992, p. 59). As the anthropocentric growth imperatives behind ecology-degrading science and technology are released, the ellipses of modernity hide another difficult regulatory truth: 'for the humanist concept of "Human" to exist, we must first invent Nature: our freedom rests on the bondage of nature to the "Laws" which we prescribe' (Evernden, 1992, p. 60).

Initiatives to affirm ecocentrism, on the other hand, try to co-opt 'new scientific discoveries' for their own discursive purposes. Many ecocentrists assert they are not 'anti-science', and the premises of ecocentrism 'are actually *more* consistent with modern science than the premises of anthropocentrism' (Eckersley, 1992, pp. 50–1). Unfortunately, as Evernden notes, environmental discourses of many stripes – ecocentric or not – basically lose everything before they begin. 'By basing all arguments on enlightened self-interest the environmentalists have ensured their own failure whenever self-interest can be perceived as lying elsewhere' (Evernden, 1993, p. 10). Fundamental attitudes about the nonhuman might well be open to otherness. However, 'at bottom, nothing has really changed', and the environment remains vulnerable 'whenever there are short-term benefits to be had by sacrificing environmental protection' (Evernden, 1993, p. 10). Implicitly, many environmental discourses accept 'the industrialist and the environmentalist are brothers under the skin; they differ merely as to the best use the natural world ought to be put to' (Brandt, 1977, p. 49).

'Rationalization' and 'restoration'

Third, environmental discourses express the normalisation projects that liberal schools of ethics and economics have sustained along with the rise of modernity. As green discourses play into the ongoing reification and resourcification of Nature, they only drape yet another green wrapper around the market's industrial rationalization of life and its agro-industrial infrastructuralisation of the planet as they unfold with environmental design and planning. Both moves forestall the prospects for ecological restoration, once the battles against scarcity in nature are won. Nevertheless, how clear is it discursively when and where this outcome might be avoided or simply just managed?

Critical environmental discourses acquire new importance when cultural, economic and political debates 'include the question of our relationship to, and impact upon, the nonhuman world' (Eckersley, 1992, p. 2). On the one hand, such declarations of intent reveal cultural, political and social thinkers turning to nature from society for conceptual energies and ethical resources to fuel the unending debates about restoring nature's original conditions. Ultimately, 'humans are a part of nature; we are, to recall Aristotle's words, political *animals* . . . it is actually in human's *interest* to integrate ecological concerns fully into their political institutions and policies' (Hayward, 1998, p. 1). These reminders count because, on the other hand, they are new comparative conceptual inconsistencies found during Enlightenment-based rationalization to be hidden in pristine green niches until natural restoration efforts can rise against rationalization for those willing to take the risks (Commoner, 1990; Cantrill and Oravec, 1996).

Discourses of 'the environment' are central to new rationalizing modes of command, control planning and communication beneath modern modes of mediating 'between history and life: in this dual position of life that placed it at the same time outside history, in its biological environment, and inside human historicity, penetrated by the latter's techniques of knowledge and power' (Foucault, 1980, p. 143). That is, environmental discourses rationalize environmental planning and economic practices for managing living human and nonhuman beings in 'the domain of value and utility' (Foucault, 1980, p. 144) by qualifying, measuring, appraising and hierarchising the ranges of nonhuman life in, around and with the commercialised requirements of human life.

These epistemic conventions behind Nature/Society relations count among the key energising factors occluded by the ellipsis of modernity. Evernden argues that these precepts discount 'the possibility of there being anything "human" in nature, including purpose and meaning, but then we proceed to use nature as a refugium for social ideals' (Evernden, 1992, p. 24).

The modern struggle of human freedom against nonhuman necessity puts the natural before, beneath or behind the historical to securitise human control (Marzec, 2012).

Under this reifying regimen, power/knowledge systems of modernity brought 'life and its mechanisms into the realm of explicit calculations', making the manifold disciplines of scientific knowledge and discourses of state power into a new type of productive agency that led to the 'transformation of human life' (Foucault, 1980, p. 145) with the industrial revolution. Once this control threshold is crossed, experts speculate how the environmental interactions of human economics, politics, and technologies continually place all human beings' existence at risk as the survival of nonhuman living beings comes under deep doubt (Luke, 1999b).

Once the planet becomes infrastructure, science allegedly shows how it has 'with only occasional localized failures' provided 'services upon which human society depends consistently and without charge' (Cairns, 1995, p. 3) because 'the environment' is reified and reduced to its 'ecosystem services' (Westmen, 1978, pp. 960–4). Human life will continue, however, only if these survival-promoting services continue. As telluric infrastructure, planetary ecologies require planners, ethicists and builders to monitor, measure and manage the system of systems that sustain the flow of all these robust services (Smil, 2013).

Environmental discourse in this register is another fragmented shard of modernity shooting from 'an explosion of numerous and adverse techniques for achieving the subjugation of bodies and the control of populations', which as Foucault (1980, p. 140) claims, mark 'the beginning of an era of "bio-power"'. With rationalizing institutions, the individual bodies of particular life forms, and massed populations of specific living species, can collectively be inserted rationally by planning agencies into the machineries of production to fit these biotic populations to the regulated economic processes of capitalism. Various schools of environmentalism today, and their production-driven agendas, simply remediate another side of the institutions of power, which were created in the 18th century 'to have methods of power capable of optimizing forces, attitudes, and life in general without at the same time making them more difficult to govern' (Foucault, 1980, p. 141).

These discursive inventions mobilised by different political interests and social forces for the resolution of environmental problems also resourcify applied ethics, discursive deliberation, and participatory politics in a manner not unlike Dryzek's efforts to propound a definitive 'ecological rationality' by fusing human 'practical reason' with ecosystemic properties. Because Dryzek holds that 'the nature of the collective choice mechanisms in place will largely determine the kind of world that ensures' (Dryzek, 1987, p. 9), his solution for the environmental crisis leads him to tout new institutional mega-designs. Since 'practical reason both allows and requires individuals to reconstruct their relationships with one another', Dryzek asserts ecological rationality also will be served by regulative innovations in economic and social institutions: 'for in remaking our institutions, we also remake ourselves: who we are, what we value, how we interact, and what we can accomplish' (Dryzek, 1987, p. 247).

Nature, once again, becomes infrastructural in being viewed simultaneously as a life support system losing 'carrying capacity', while always already having a regulatory potential for making its own restorative recovery. Its embedded systemic qualities must be tuned with corrective rational interventions to attain greater ecological rationality, which Dryzek sees 'as the capacity of human and natural systems in combination to cope with problems of this sort' (Dryzek, 1987, p. 11).

'Ruination' and 'resilience'

Fourth, questions tied to the environment, nonhuman being, and nature's ruination are not always soundly assayed. Ecocentric thinkers claim that they have 'rarely been aired explicitly by political theorists, let alone given prominence' (Eckersley, 1992, p. 2). With the climate change challenges of the 21st century, there is new ground to be broken by building up resilience as a crucial discourse in the ruins of modernity. The bearers of this resilient message might reveal 'the outline of a more encompassing, non-anthropocentric political framework with which to approach social and ecological problems' (Eckersley, 1992, p. 2) as well as demonstrate 'how ecological values can be presented in terms that require a reasoned response' (Hayward, 1998, p. 2) in discursive engagements.

Before defining and developing 'the environment,' decision makers, planners and theorists must be mindful of how processed, produced and preserved 'the environment' already is. Resilience thinking in these spaces might aid rescuers from the thick tangle of ontic compromise and epistemic convention which often make tangible improvements for nature and society against the economy almost impossible (Walker and Salt, 2012; Gunderson and Holling, 2002) and resilience discourse, for example, can map out defences for the Earth against the depredations of modern technology, global exchange and anthropocentric thinking (Hamilton, 2013; Evans and Reid, 2014). Yet, the Nature they seek to tie human interests might never restore a resilient biocentric primacy for the Earth because contemporary 'risk society' only bringing tighter security to its nonhuman assets to mitigate the ruination running out of control (Chakrabartty, 2009; Davoudi, 2014).

Efforts to advance ethical briefs on Nature's behalf against its ruination too often read uncomfortably like the prologue to new green papers for ecological modernisation. Finding, and then defending, human and nonhuman interests establish a zone of material interest for a resilient existence whose material success demands continuous human guardianship. Revealing and aggregating nonhuman interests create a sphere for moral agency and existence, which also necessitate finding trustworthy human spokespersons. Efforts to find something to preserve and protect beyond humanity must enlist humans to care about, or even for, the nonhuman. What some human beings are doing to the nonhuman will not stop; instead, resilience work points to how it might be moderated, regulated, controlled by other humans to advance more ecological modes of life (Bohland et al., 2018)

Discourse as development debate

Environmental discourses arise in struggles to dominate Nature, and one must recognise they become part of that planning project and sustainability struggle. As the conquest of Nature advances by fits and starts, it also thoroughly tests the nature of discursive thinking about Nature (Radkau, 2008). To answer Evernden's question, 'to whom are we to entrust the engineering of the concept of nature' (Evernden, 1992, p. 28), environmental discourses respond: 'us'.

This claim poses a big question for discourses rooted in reification and revitalization, resourcification and regulation, rationalization and restoration or ruination and resilience: who is 'us'? Who is speaking for whom, to whom, and about what in environmental discourse? By and large, those who are speaking are typically mixed lots of more elite political players, ranging from academics and activists to designers, planners and developers. While the common conceit of environmental discourses is that they always address concerned publics, worried governments or cautious managers. In fact, they usually circulate between each other and a few worried elite audiences rather than facing the environment per se.

'The big question' for environmental discourse and its meanings, and everyone who might be taken by their suggestions for change, 'is whether it is possible or desirable to speak of the green movement as a *we* – and, if so, in what sense' (Torgerson, 1999, p. 19).

Recognising that any 'we-ness' rests upon organising instrumentally defined collective action and at other times requires hearing all of these discursive chords in an open *green public sphere* in which '*we* would not be an instrument but a space of appearance, a common world' (Torgerson, 1999, p. 20). At the end of the day, however, few environmental discourses credit environmental activists with much more than developing a new 'language of environment'. Dryzek claims, discursive developments are '*the* enduring legacy' (Dryzek, 1997, p. 121). As Torgerson concludes, 'the emergence, expansion, and enhancement of a green public sphere denies industrialism its unquestioned supremacy', but 'if one wishes to assess the potential of the green public sphere to contribute to historical change, that potential is no doubt modest and cannot develop on its own' (Torgerson, 1999, p. 168).

Conclusion: contesting the natures of Nature

Clearly, Nature is an essentially contested concept, and worried analyses of environmental discourses express another side of its continuously contested essence (Luke, 1997; Soper, 1997; Meyer, 2001; Morton, 2013). Because of Nature's resourcification in today's global economy and society, the centrality of modernising formalisations, which presume knowledge of a pure, objective, unmediated Nature brought under the power of humans accumulating modern scientific knowledge, remain decisively useful ideas. And while they also are heavily contested, these precepts are still dying very hard. The exploitative qualities of liberal economic and social relations frequently go unquestioned. Indeed, it is made fundamental by claiming it is right and works well when rational self-interest in a serviceable Nature is made clear, consistent and cautious. To whom this sort of environmental discourse speaks, from where it writes and for what it works, however, are very problematic, not least when it is adopted in policies and practices of environmental planning.

Yet, as the work in some environmental discourses suggest, there also will be more disquiet about the methodological and technological articles of faith behind modernity (Eder, 1996). While their celebrants praise this system of science, and its derivative technologies, for their demonstrated ability to raise industrial output, overcome deadly diseases, speed methods of travel, and enhance a longer, richer human lifespan (McNeill and Engelke, 2014), few of them discuss how these same mechanisms of scientific knowledge and technical action also generate noxious by-products, cause new afflictions, create frustrations from mobility and detract from the quality of life (Ihde, 1990).

Whole movements of people – scientists and laypersons alike – have increasingly led to doubt, or even openly to protest these modernist formulas for legitimating scientific authority, technical power, and command and control style of environmental planning. Sometimes they are derogated for being mere ideologists; other times, it is clear elements among the resistance are themselves deeply embedded in the modes of scientific production (Ehlers and Krafft, 2010). Nonetheless, more resistances are emerging, and environmental radicals are facilitating their spread during each successive new generation.

Whether they seek deeper meanings in environmental discourse or not, many streams in the environmental movement have proven to be among the most ardent opponents of the orthodox readings given to science and technology. Feminists, minority peoples and working-class groups, who rarely benefit from having scientific authority or technical power, also join environmentalists in questioning the allegedly neutral knowledge that science provides about Nature (Eder, 1996). In its emergent days, science put forth foundational epistemologies for dividing facts and values, theory and observation, experiments and explanations, or truth and opinion to challenge religious-feudal authority, whose place, power and property in early modern society rested upon the legitimating principles of the divine right of royal dynasties (Cassirer, 1972). Today, countervailing forces against scientific planning often are recruiting allies for their resistance from competing groups of expert planners with different discursive grounding.

To conclude, discourses of 'the environment', to the degree they vigilantly preserve and protect Nature serviceable to humanity, are fraud. Instead of seeking to define and defend the last pristine tracts of an embattled Nature, they often take as much, or even more, effort to assess and attack those technical and cultural systems in the economy and society whose impact protects nature. Nature is actually more resistant and survivable than human beings admit. The end of Humanity would not be the end of Nature, only Nature as human beings know, need and expect it. Yet, this historicised Nature is what the productive forces of today's economy and society are degrading, and those sources of the degradation need to be more limited.

Truly effective environmental discourse must interrogate every aspect of all environments – built and unbuilt, human and nonhuman, ecological and economic – to find what is systematically wrong with what 'is' as well as what is structurally preventing what 'ought to be' from being realised in both environmental planning and policy. In identifying the normative imperatives, however, environmental discourse must push forward workable alternatives to implement fair, just and democratic politics for their realisation. This vision of politics, in turn, must address, if not answer, the same big theoretical questions: Who, whom? Why, when, where and how? Here, there or everywhere?

References

Bauman, Z. (2007). *Liquid times: living in an age of uncertainty*. Cambridge: Polity Press.

Beck, U. (1992). *Risk society: towards a new modernity*. London: Sage.

Bennett, J. (2009). *Vibrant matter: a political ecology of things*. Durham, NC: Duke University Press.

Bohland, J., Davoudi, S. and Lawrence, J. (eds.) (2018). *The Resilience Machine*. New York: Routledge.

Brandt, A. (1977). 'Views'. *The Atlantic Monthly*, July: 49–51.

Burchell, G., Gordon, C. and Miller, P. (eds.) (1991). *The Foucault effect: studies in governmentality*. Chicago: University of Chicago Press.

Cairns, J. (1995). 'Achieving sustainable use of the planet in the next century: what should Virginians do?' *Virginia Issues and Answers*, 2(Summer): 3–6.

Cantrill, J. G. and Oravec, C. L. (eds.) (1996). *The symbolic Earth: discourse and our creation of the environment*. Lexington: University of Kentucky Press.

Cassirer, E. (1972). *The individual and the cosmos in Renaissance philosophy*. Philadelphia: University of Pennsylvania Press.

Chakrabartty, D. (2009). 'The climate of history: four theses'. *Critical Enquiry*, 35(2): 197–222.

Chin, A., Fu, R., Harbor, J., Taylor, M. P. and Vanacker, V. (2013). 'Anthropocene: human interactions with earth systems' [Editorial]. *Anthropocene*, 1: 1–2.

Commoner, B. (1990). *Making peace with the planet*. New York: Pantheon.

Cox, R. (2012). *Environmental communication and the public sphere*. New York: Sage.

Darier, E. (ed.) (1999). *Discourses of the environment*. Oxford: Wiley-Blackwell.

Davies, J. (2016). *The birth of the Anthropocene*. Berkeley: University of California Press.

Davoudi, S. (2014). 'Climate change, securitisation of nature, and resilient urbanism'. *Environment and Planning C: Government and Policy*, 3(2): 360–75.

Davoudi, S. (2015). 'Planning as practice of knowing'. *Planning Theory*, 14(3): 316–31.

Deleuze, G. and Guattari, F. (1994). *What is philosophy?* New York: Columbia University Press.

Devall, B. and Sessions, G. (1985). *Deep ecology*. Layton, UT: Peregrine Smith Books.

Dryzek, J. (1987). *Rational ecology: environment and political economy*. London: Blackwell.

Dryzek, J. (1997). *The politics of the earth: environmental discourses*. Oxford: Oxford University Press.

Eckersley, R. (1992). *Environmentalism and political theory: toward an ecocentric approach.* Albany: State University of New York Press.

Eder, K. (1996). *The social construction of nature: a sociology of ecological enlightenment.* London: Sage.

Ehlers, E. and Krafft, T. (eds.) (2010). *Earth system science in the Anthropocene: emerging issues and problems.* Berlin: Springer Verlag.

Evans, B. and Reid, J. (2014). *Resilient life: the art of living dangerously.* Cambridge: Polity Press.

Evernden, N. (1992). *The social creation of nature.* Baltimore: Johns Hopkins University Press.

Evernden, N. (1993). *The natural alien: humankind and environment* (2nd edn). Toronto: University of Toronto Press.

Foucault, M. (1980). *The history of sexuality, volume I: an introduction.* New York: Vintage.

Gunderson, L. H. and Holling, C. S. (2002). *Panarchy: understanding transformation in human and natural systems.* Washington, DC: Island Press.

Hamilton, C. (2013). *Earth masters: the dawn of the age of climate engineering.* New Haven: Yale University Press.

Hayward, T. (1998). *Political theory and ecological values.* New York: St Martin's Press.

Horkheimer, M. and Adorno, T. W. (2002). *Dialectic of enlightenment.* Stanford: Stanford University Press.

Husserl, E. (1970). *The crisis of European science and transcendental phenomenology.* Evanston: Northwestern University Press.

Ihde, D. (1990). *Technology and the lifeworld: from garden to earth.* Bloomington: Indiana University Press.

Jameson, F. (1992). *Postmodernism, or, The cultural logic of late capitalism.* Durham: Duke University Press.

Latour, B. (1993). *We have never been modern.* London: Harvester Wheatsleaf.

Luke, T. W. (1997). *Ecocritique: contesting the politics of nature, economy and culture.* Minneapolis: University of Minnesota Press.

Luke, T. W. (1999a). 'Training eco-managerialists: academic environmental studies as a power/knowledge formation', in F. Fischer and M. Hajer (eds.) *Living with nature: environmental discourse as cultural politics.* Oxford: Oxford University Press, pp 103–20.

Luke, T. W. (1999b). 'Environmentality as green governmentality', in E. Darier (ed.) *Discourses of the environment.* Oxford: Blackwell, pp 121–51.

Luke, T. W. (2009). 'Developing planetarian accountancy: fabricating nature as stock, service, and system for green governmentality'. *Current Perspectives in Social Theory*, 26: 129–59.

Lyotard, J. F. (1984). *The postmodern condition: a report on knowledge.* Minneapolis: University of Minnesota Press.

Lyotard, J. F. (1991). *The inhuman: reflections on time.* Stanford: Stanford University Press.

Marzec, R. P. (2012). *Militarizing the environment: climate and the security state.* Minneapolis: University of Minnesota Press.

McNeill, J. R. and Engelke, P. (2014). *The great acceleration: an environmental history of the Anthropocene since 1945.* Cambridge, MA: Harvard University Press.

Meyer, J. (2001). *Political nature.* Cambridge, MA: MIT Press.

Mitchell, T. J. (2013). *Carbon democracy: political power in the age of oil.* London: Verso.

Morton, T. (2013). *Hyperobjects: philosophy and ecology after the end of the world.* Minneapolis: University of Minnesota Press.

Radkau, J. (2008). *Nature and power: a global history of the environment.* New York: Cambridge University Press/German Historical Institute.

Ross, A. (1994). *The Chicago gangster theory of life: nature's debt to society.* London: Verso.

Rothenberg, D. (1993). *Hand's end: technology and the limits of nature.* Berkeley: University of California Press.

Sandilands, C. (1999). *The good-natured feminist: ecofeminism and the quest for democracy.* Minneapolis: University of Minnesota Press.

Scranton, R. (2015). *Learning to die in the Anthropocene.* San Francisco: City Lights Publishers.

Smil, V. (2013). *Harvesting the biosphere: what we have taken from nature.* Cambridge, MA: MIT Press.

Soper, K. (1997). *What is nature?* Oxford: Blackwell.

Torgerson, D. (1999). *The promise of green politics: environmentalism and the public sphere.* Durham, NC: Duke University Press.

Vogel, S. (1996). *Against nature: the concept of nature in critical theory.* Albany: State University of New York Press.

Walker, B. and Salt, D. (2012). *Resilience practice: building capacity to absorb disturbance and maintain function.* Washington, DC: Island Press.

Wander, P. C. (1996). 'Foreword', in R. D. Bessel and B. K. Duffy (eds.) *Defending nature in American civic discourse*. Albany, NY: State University of New York Press, pp ix–xlii.
Westmen, W. E. (1978). 'How much are nature's services worth?' *Science*, 197: 960–4.
Young, J. (1990). *Sustaining the earth*. Cambridge, MA: Harvard University Press.
Zizek, S. (2010). *Living in the end times*. London: Verso.

3

Visionary idealism in environmental planning

Michael Gunder

Most people would consider it only rational to act, when it is possible to do so, in a manner that will create a better future, first for themselves, then for their family and friends, and then for their community and perhaps, if they are somewhat altruistic and/or if it does not take much effort, even for their wider society or the world itself. Of course, the devil is always in the details, and one person's better future may be considerably at odds with those of others – and that, of course, is the problem! This is particularly an issue if a person's better future is concerned with, say, the materialist things constituting collectively a successful economy versus another person's better future who is concerned about sustaining their natural environment. The tensions between these types of diverse positions is what traditionally constitutes politics and visionary idealism can often lie at the heart of this process.

The adjective 'visionary' according to the online Cambridge Dictionary (no date, no page) means 'with the ability to imagine how a country, society, industry, etc. will develop in the future'. 'Idealism' has a formal meaning derived from philosophy in which things, concepts and ideas that we sense in and about the world can only exist in the minds of the those that mediate (think) about them as mental representations (Kant, 1934). Idealism also has a more informal everyday meaning: 'the belief that your ideals can be achieved, often when this does not seem likely to others' (Cambridge Dictionary, no date, no page). Accordingly, this chapter will use the term 'visionary idealism' to mean: a belief about, and in, a mental idea (a virtual concept) that has become an ideal (a lofty aim) of what should become real (actual) in the world. Of course, visionary idealism is inherently a core dimension of all forms of spatial planning as the discipline strives to first imagine and then plan for a better future.

A term especially applicable to visionary idealism and environmental planning is ideology. An ideology is a person's discursive beliefs, views and values as to how one thinks about the world and how one wants it to become (Eagleton, 2013 [1994]). But this is seldom, if ever, purely self-determined. Rather, this is largely constructed and shaped by an individual's society via their socialisation, education and wider media influences (Gunder, 2011). Indeed, environmental planning can only operate within the scope of dominant global, national and even local ideologies that largely define planning's specific territorial purpose and powers of agency (Gunder, 2010, 2016). Consequently, both planning for a better future and the concept of visionary idealism are often entwined and entangled with ideology, especially when considering and/or implementing a desired ideal that many people share among themselves about the future (Gunder and Hillier, 2009), as will be discussed extensively in this chapter.

For reasons of logical organisation, this chapter suggests that visionary environmental idealism is materialised in the world in three distinct but often overlapping modes, where the ideological concept of sustainability, or similar ecological concepts aligned with sustainability, are important factors underlying each of these three pathways. The first and most directly aligned with environmental planning practice is materially creating an environmentally desirable human settlement, the more sustainable of these habitations are often called 'eco-cities'. This is hardly a new task for planning; think of Ebenezer Howard's (1965 [1902]) famous *Garden cities of tomorrow* first published at the start of the last century. Current examples of eco-cities include Masdar City, Abu Dhabi, UAE or that proposed for Dongtan Eco-city, Shanghai, China (Cugurullo, 2016). Or consider this built environment materialisation occurring in more moderate form where there is contemporary planning for existing cities but in a sustainable manner so that they may achieve a reduced carbon footprint and/or enhanced urban resilience. Well known examples globally are Curitiba, Brazil (Macedo, 2013), or Freiburg, Germany (Kronsell, 2013).

Second, these ideals may be materialised through popular movements, or political parties, that attempt to democratically modify the existing largely neo-liberal capitalist world to one which has a greater environmental responsibility for biodiversity and carbon neutrality (Swyngedouw, 2010). This is perhaps most strongly personified by the popular face of many Green Parties currently contesting democratic election around the world, who advocate moderate, rather than radical, built-form practices and behaviours for carbon neutrality, climate change adaptation or urban resilience within capitalist democracies (Blühdorn, 2013). These types of moderate environmental policies have also, of course, been deployed by other elected political parties in government (Allmendinger, 2016). Further, in this regard, Swyngedouw (2010, 2015), as well as others including Allmendinger (2016), warn how sustainability has often been deployed in contemporary politics so as to de-politicise democratic practices of the traditional state and replace them with a consensual techno-rational overarching form of governance. This is a form of governance which incorporates spatial and environmental planning as one of its functions, yet in a manner that tends to predominantly facilitate market interests, while often only providing lip service to the environment (Allmendinger, 2016). This second trend will be referred to in this chapter as 'post-political ecological modernisation'.

Third, and perhaps most profoundly, these ideals are materialised as radical or 'deep green movements' (Dodson, 2007 [1990]) that attempt to profoundly change the world's current dominant pro-market ideology of neoliberalism (Gunder, 2010, 2016). These are radical movements that strive to overturn the current global anthropocentric capitalist imperative for material consumption and wealth creation with its resultant adverse environmental and social consequences. While their specific goals may differ, all of these radical movements give an overarching precedent to sustaining the global environment in a manner that is well within the world's steady state ecological carrying capacity and often in a way that reduces the need for traditional government and planning, as least as we currently know them (Dodson, 2007 [1990]; Martinez-Alier et al., 2010; Rees, 2018).

The following sections will consider each of these three perspectives in more detail. But to do so, the chapter will first consider what constitutes ideology. It will then consider the concept of sustainability from this viewpoint. The chapter will contend that sustainability is largely an ideological concept that underlies all three perspectives materialising environmental visionary idealism, but it is not the only ideological dimension influencing and impacting on these perspectives. The chapter will conclude that most moderate environmental visionary idealism is largely dominated and over shadowed by pro-market ideology. Accordingly, possibly the only type of environmental visionary idealism able to displace this pro-market domination is via the implementation of a radical deep green alternative, however unlikely this may seem today.

Ideology's universal promise: the illusion of safety in an uncertain world

Ideology's initial meaning was that of the 'science of ideas' as developed by Destutt de Tracy in 1797, but, with the rise of Marxian thought, this meaning largely disappeared by the end of the 19th century (Fine and Sandstrom, 1993, p. 22). For Marx, ideology was about the creation in a society of a false consciousness about material life through 'various guises such as morality, religion, and metaphysics' so that the 'ideas of the ruling class [became] the ruling ideas' (Freeden, 2003, pp. 5–6). Of course, if one is aware of false consciousness, it becomes rather difficult to be readily duped by it, so any contemporary understanding of ideology and how it may now work must be considerably more nuanced. The psychoanalytical perspective of Lacan and his follower Žižek provides this necessary sophistication (Gunder and Hillier, 2009). Indeed, Frederic Jameson (2003 [1977], p. 37), writing over 40 years ago, observed that we can ascribe to Lacan 'the first new and as yet insufficiently developed concept of the nature of ideology since Marx'.

The concept of lack is core to Lacan (2006 [1966]), for it is the lack of something that we want that constitutes desire and this fundamentally provides 'the ontological underpinnings of human existence' (Ruti, 2008, p. 485). Gunder and Hillier (2009, p. 24, emphasis in original) observed 'that the political or technical deployment of a "lack" or "deficiency" is a powerful planning and political trope for response and action, [as] who would wish to live in a "deficient" city lacking in *safety, competitiveness, sustainability* or some other shortfall?' Indeed, these authors contend that one of the roles of planning is to provide a fantasy of how the planned future city will resolve its identified problems – lacks – and be enjoyably complete and fulfilling, free of any future fear or worry.

In this regard, for Žižek (1989, p. 45), '[i]deology . . . is a fantasy-construction which serves as a support for our "reality" itself: an "illusion" which structures our effective, real social relations and thereby masks some insupportable, real, impossible kernel', which underlies our insecurities in an uncertain world, especially when engaging with the future. In this context, Žižek contends that it is 'the horror of contemplating the unknowable' that 'leads people to weave imaginary webs, or fantasies, of what they claim can be known, and to fabricate harmonies where antagonisms reign' (Freeden, 2003, p. 111). Here, importantly, planning facilitates 'this ideological task by harmoniously articulating how populations should enjoyably use their settlements, spaces and environments when seeking a better future', or, alternatively, act in a more sustainable and resilient manner to stave off the fear of future environmental catastrophe (Gunder, 2010, p. 306).

Importantly, planning is considered by many to be a 'magpie discipline', which deploys numerous important concepts drawn from the social and natural sciences to structure the discipline's understandings of the natural and built environment. These concepts often inherently have, or often gain in their public policy acceptance, an ideological dimension when they are deployed in planning for a community's perceived security and hence desired future (Gunder and Hillier, 2009). The following section considers one of these important planning concepts: 'sustainability', as an illustrative example.

Sustainability as an ideological concept

> In search of a new 'vision' for planning . . . one which can 'reach out to society as a whole, addressing its wants, needs and insecurities' . . . a 'vision to rank with those of Ebenezer Howard a century ago' . . . There is a consensus that such a vision can now emerge . . . sustainability
>
> *(Davoudi, 2001, p. 86).*

The word 'sustainable' has a fuzzy meaning that is impossible to clearly or concisely define, so that it can mean many things to different people. Hence business people, socialists and ecologists may desire a sustainable future, especially under the rubric of 'sustainable development' (WCED, 1987), with one group thinking that such a future is about sustainable economic growth, another group about having a socially just society and a third group about a future of sustained biodiversity and climate stability. Sustainable development supposedly promises 'it all: economic growth, environmental conservation, social justice; and not just for the moment but in perpetuity' and in implementing it '[n]o painful changes are necessary' to our current way of life (Dryzek, 2005, p. 157)! Indeed, for this very reason, ideals like sustainability are called empty signifiers, as they can make diverse promises to many different perspectives (Davidson, 2010; Gunder, 2006; Swyngedouw, 2010, 2015).

Further, sustainability is a transcendental ideal, which means that it is a lofty desired societal goal that resides external to existing human experience or established knowledge (Gunder, 2006, p. 212; Žižek, 1993, p. 16). Accordingly, no one really knows what a transcendental ideal ultimately is or will become. Sustainability is a signifier constituting what we fundamentally perceive the world is lacking, even if we cannot fully describe what this lack actually is or what its fulfilment will actually mean. But we believe it is "'the real thing", the unattainable X, the object-cause of desire' (Žižek, 1989, p. 96). In this regard, Žižek (1989) referred to such ideals in the title of his book as 'sublime objects of ideology' for even if they are unclear and ambiguous with diverse meanings, most people are still attracted to them as highly important concepts of desirous aspiration and identification.

Lacan (2007) calls these special words: master signifiers. Indeed, these master signifiers of aspiration and identification have 'to remain empty in order to serve as the underlying organizing principle of a series' – the central organising element – of ordinary signifiers comprising groupings of related narratives and discourses (Žižek, 2000, p. 52). The master signifier gives the illusion of unanimity in the discursive field being engaged with by these diverse signifiers, even if the field is actually 'riddled with differences, antagonisms, and contradictions' (Bryant, 2008, p. 18). A master signifier, such as sustainability, is not a guarantor of specific meaning, or even the meaning itself. Rather it is metaphorically rather more like the light which attracts all the moths to the flame, like the flag we all pledge allegiance to, something unquestionably good like motherhood, the *Thing* that we profoundly believe in and defend with our honour and sometimes even our life. Indeed:

> [I]ts role is purely structural, its nature is pure performative . . . it is an element which represents the agency of the signifier within the field of the signified' – in short: it is 'the dazzling splendor of the element which holds [everything else] together.
>
> *(Žižek, 1989, p. 99)*

Each master signifier of ideology has a similar function. That is to structure a discourse in a specific way so that the other words in that discourse have specific meanings aligned with that of the master signifier (Žižek, 1989, p. 102). When an environmental planner considers a discourse through the prism of sustainability, or through the ideological lenses of another master signifier, such as neoliberalism, the same signified word in these two different discourses will have different specific meanings and connotations. Consider how a 'sustainable city' conjures a very different vision of a desirable future than that of a 'neoliberal city'! Indeed, it can be argued that much that constitutes an environmental planning education is learning the specific meanings and often implicit values that underlie the words used in the different planning discourses constituting the knowledges pertaining to planning master signifiers like sustainability, resilience, social justice, governance, globalisation and even neoliberalism (Gunder, 2004).

Sustainability is an empty ideological element, a master signifier par excellence, whose signification – its own meaning – can only be partially gathered through embroidering it extensively with more explicit and more clearly understandable signifiers. Sustainability, similar to other inter-related environmental master signifiers, such as 'nature' (Swyngedouw, 2010), 'green' (Dodson, 2007 [1990]; Stavrakakis, 1997) or 'ecology' itself (Žižek, 1996, p. 131) are each constituted by a complex symbolic ideological montage of supporting signifiers constituting a discourse or knowledge about them. This is a tapestry comprised of a bricolage of concepts/terms/signifiers tied together with what Lacan (2006) calls 'points de capiton' – quilting or nodal points (Laclau and Mouffe, 1985, p. 112) – which anchor and stop their master signifier from sliding around too much in its overarching, but still fuzzy, meaning. For sustainability, these include words/concepts such as: 'carbon-neutral' or 'carbon-footprint', 'environmental wellbeing', 'biodiversity', 'sea level rise', 'de-growth', 'global warming', 'Anthropocene', 'resilience' and so on. All act as quilting points through which an always fluid and unbounded matrix of general representations and narratives constituting the master signifier itself – sustainability – are voiced and communicated. All these quilting terms in aggregate constitute a discursive mosaic that is 'an articulation of various separate moments around [the] master signifier' comprised of 'a family of nodal points that bind them together' (Stavrakakis, 1997, p. 266).

Importantly, before the emergence of 'this articulation – before the intervention of the nodal points . . . these moments pre-existed as floating signifiers, as proto-ideological elements with no particular . . . connotation' attributable to their yet to emerge master signifier: sustainability (Stavrakakis, 1997, p. 266). Indeed, much debate emerged about protecting the environment during the 1970s, if not before, and the Brundtland Report constituting sustainable development was published in 1987 (WCED, 1987). But it was only when these nodal points began to universally coalesce around sustainability in the latter half of the 1990s that sustainability emerged as a populist movement and key environmental planning issue, even though to today's practitioners, sustainability may now seem as if it has always existed (Gunder, 2006, p. 213).

Moreover, this master signifier and its nodal points are more than just mere anchor points; 'they refer to a "beyond" of meaning, a certain enjoyment expressed as fantasy – notably, the desire for an environmentally balanced and socially harmonious order' (Swyngedouw, 2015, p. 133). Indeed, it is via this kernel of enjoyment created by this illusion/promise of future completeness/wholeness that the master signifier will provide once it is somehow being achieved which allows this fantasy element of 'sustainability' to act as a sublime object of ideology. Further, in doing so 'sustainability' is able to grip and bind itself to the subject at the unconscious level of desire and hence of personal belief, identification and aspiration (Gunder, 2016, p. 23).

The range of environmental visionary ideals

The Eco-City as the materialisation of economic sustainable development

Davoudi (2013, pp. 254–5) considers the two most significant drivers of contemporary visionary idealism within planning and its related built environment fields to be rapid global urbanisation and climate change. She contends that this is culminating in numerous utopian visions of carbon-neutral future communities ranging from high-tech de-carbonating smart cities to low-tech communal settlements attempting to create sustainable closed-cycle spatial metabolisms.

But Davoudi (2013, p. 257) cautions in this regard that 'the story of planning is littered with the ruins of utopias which were demolished by the vested interest of powerful players'. Indeed, what goes unsaid by Davoudi is a third driver of utopian visionary idealism and most everything else globally, that is the ideological dominance of neoliberalism and its privileging of the market and capital accumulation over all other societal 'goods' (Gunder, 2010, 2016).

An excellent exemplar of this is Masdar City. In 2007, the Abu Dhabi government launched its six square kilometre area eco-city development of Masdar City. Globally, it immediately 'became a relentlessly repeated reference and virtual blueprint for future sustainable development' (Jensen, 2016, p. 45). The proposed totally renewable energy-based, zero waste and zero carbon city was to have 50,000 residents and an additional daily 50,000 mass transit commuting workers when finally completed in 2016. However, with the world financial crisis of 2008 and the more recent collapse of global oil prices impacting on Gulf State incomes, the project has largely been downgraded and diminished to a university campus with the majority of its environmentally innovated initiatives sidelined (Jensen, 2016, p. 50).

Critics have subsequently referred to the project as 'greenwashing' and accused the Abu Dhabi government 'of using Masdar as a façade for the world while avoiding clamping down on unsustainable domestic lifestyles and public infrastructure' (Crot, 2013, p. 2814). Indeed, Crot (2013, p. 2810) contends that the government's 'interpretation of sustainability is clearly skewed towards the combination of economic success and environmental progress'. With the latter being achieved via technological solutions and the use of renewable resources without regard to the protection of wildlife and its habitats, or with any regard for social sustainability. Cugurullo (2016, p. 2430) observes that what underlay the design of Masdar City was not ecological research on its surrounding biophysical environment, 'but rather market analyses studying the economic environment that surrounds the clean technology market'. Further, Cugurullo (2016, p. 2421) declares that the eco-city of Masdar 'reflects one of the most international manifestations of the ideology of sustainability: ecological modernisation', which 'rejects environmental concerns as antithetical to economic priorities, and advances technological innovation as the equaliser of economic growth and environmental preservation'.

Post-political ecological modernisation

Similarly, numerous studies globally of planning and public environmental policy for sustainable development in more conventional cities and at both the local and higher levels of government have found a similar privileging of the economy over that of the environment, with at best only token regard to social considerations (for example: Allmendinger, 2016; Coffey and Marston, 2013; Gunder, 2006; Rico and Lin, 2012). These manifestations of ecological modernisation, as identified by these works along with the research of Swyngedouw (2010, 2015), have also manifested a post-political dimension. This is a dimension that displaces traditional modes of local political democratic decision making with regimes of techno-scientific governance predicated on a pro-market sustainable development consensus supported globally by the United Nations and World Bank as the best way that the world's cities should develop (Rico and Lin, 2012, p. 192).

Much contemporary visionary idealism about the environment contains a significant component of fear intertwined within it. The diverse narratives through which the contemporary understanding of our environmental status is constructed is one systemically entwined with invocations of fear for forthcoming catastrophes and apocalypses, of significant human distress, if not outright ecological annihilation, in the not so distant future (Swyngedouw, 2010, 2013). A wide range of imaginary fantasies of catastrophe are proposed as a consequence of non-adaption of particular saving prescriptions. While some are based on valid scientific modelling, such as sea level rise, others are even more apocalyptic and often without significant clear scientific evidence, be it endemic drought with wild fires and resultant eventual desertification, widespread ecological collapse, tsunamis, global epidemics, the supposed consequences of post-peak-oil (not withstanding now market driven ongoing transitions to alternative energy and transportation technologies), an extensive period of large volume volcanism and the like. 'In sum, our ecological predicament is sutured by millennialism fears sustained by an apocalyptic rhetoric and representational tactics, and by a series of performative gestures signalling an overwhelming, mind-boggling danger, one that threatens to undermine the very co-ordinates of our everyday lives and routines' (Swyngedouw, 2010, p. 308).

In this regard, Swyngedouw (2013, p. 10) contends that this catastrophic language primarily serves a useful ideological empowerment function in our post-political contemporary world of governance. It acts in a manner so as 'to turn nightmare into crisis management' so that this 'nurturing of fear, which is invariably followed by a set of techno-managerial fixes . . . serves precisely to depolitize'. Accordingly, while our dominant elites and leaders can admit that while the situation may now be grave, they can still insist that through the application of correct techno-managerial responses our 'homeland security (ecological, economic, or otherwise) is in good hands' (Swyngedouw, 2013, p. 10). Of course, environmental planning is a central technological process in this post-political ideological pro-market process (Allmendinger, 2016; Swyngedouw, 2015).

Radical deep green movements

Beyond an acceptance of the global neoliberal status quo are radical environmental movements with a goal to change the world and its fundamental ideological values away from those of ever-increasing consumer material consumption and the market. These include ecofeminism (Mies and Shiva, 2014 [1993]), sustainable de-growth (Martinez-Alier et al., 2010; Trainer, 2015), social ecology (Bookchin, 1988), and deep ecology (Naess, 1984). While all have strong elements of sustainability embedded within them, they go beyond mere 'market' sustainability. Moreover, these deep green movements do propose a fundamental requirement for a radically changing, or replacing, of capitalist growth so as to live within the Earth's carrying capacity.

Any such prescription for such radical action will take tremendous political will. Indeed, 'neither *sustainability* nor *justice* can be achieved unless global resource use, consumption, "living standards" and GDP within rich countries are reduced to a small fraction' of their current levels – 'in the region of 10 percent' – 'and are kept there' (Trainer, 2015, p. 59). Further, many may question if the necessary popular will exists to so radically change our world from one of considering economic growth as a positive to living in a world of de-growth so as to maintain humankind within the Earth's truly sustainable carrying capacity. Undoubted, to do so will require a very strong new 'imaginary' of the future with an appropriate de-growth ideology that the vast majority of the global population can readily agree with and be willing to live within. This is a world which will require a profound change of ideological spirit, one that contradicts the core elements that have propelled most of global 'society for 200 years – above all, the quest for material wealth' (Trainer, 2015, p. 67).

Conclusion

This chapter has discussed visionary idealism and environmental planning in their entanglement with ideology. It has then discussed the concept of sustainability as one example of an ideological master signifier. The chapter then drew on this ideological analysis to critique visionary environmental idealism as it is materialised in the world via three diverse ways: planning for contemporary environmental built-forms; contemporary post-political ecological modernisation and governance; and as movements for radical ideological change. In doing so, it examined the domination of market ideology in our neoliberal world and the consequences that this may have for effective visioning within environmental planning. The chapter suggests that perhaps only radical green movements may provide effective means to curtail the worst attributes of the Anthropocene. But this would only occur if sufficient people, including those in environmental planning, are willing to adopt the necessary ideological perspective and implement the necessary profound action to bring this radical visionary idealism into being.

Many members of the public, as well as environmental planners, believe strongly in their responsibility to sustain the environment. But as this chapter has illustrated, this belief is open to capture by our dominant neoliberal ideology so as to channel these beliefs in a manner that ultimately sits at odds with the attainment of an ecologically sustainable future. Perhaps a radical engagement with visionary environmental idealism is the only way to overcome this pernicious domination. Of course, it is the responsibility of each reader to decide for him or herself what future world that he or she may wish to help create.

References

Allmendinger, P. (2016). *Neoliberal spatial governance*. London: Routledge.

Blühdorn, I. (2013). 'The governance of unsustainability: ecology and democracy after the post-democratic turn'. *Environmental Politics*, 22(1): 16–36.

Bookchin, M. (1988). *Toward an ecological society*. Montreal: Black Rose Books.

Bryant, L. (2008). 'Žižek's new universe of discourse: politics and the discourse of the capitalist'. *International Journal of Zizek Studies*, 2(4): 1–48.

Cambridge Dictionary [Online]. Available at: http://dictionary.cambridge.org/dictionary/english/

Coffey, B. and Marston, G. (2013). 'How neoliberalism and ecological modernization shaped environmental policy in Australia'. *Journal of Environmental Policy and Planning*, 15(2): 179–99.

Crot, L. (2013). 'Planning for sustainability in non-democratic polities: the case of Masdar City'. *Urban Studies*, 50(14): 2809–25.

Cugurullo, F. (2016). 'Urban eco-modernisation and the policy context of new eco-city projects: where Masdar City fails and why'. *Urban Studies*, 53(11): 2417–33.

Davidson, M. (2010). 'Sustainability as ideological praxis: the acting out of planning's master-signifier'. *City*, 14(4): 390–405.

Davoudi, S. (2001). 'Planning and the twin discourses of sustainability', in A. Layard, S. Davoudi and S. Batty (eds.) *Planning for a sustainable future*. London: Spon, pp 81–99.

Davoudi, S. (2013). 'Urban futures', in N. Phelps, R. Freestone and M. Tewdr-Jones (eds.) *The planning imagination*. London: Routledge, pp 252–66.

Dodson, A. (2007 [1990]). *Green political thought*. New York: Routledge.

Dryzek, J. (2005). *The politics of the earth*. Oxford: Oxford University Press.

Eagleton, T. (2013 [1994]). 'Introduction', in T. Eagleton (ed.) *Ideology*. Abingdon: Routledge, pp 1–20.

Fine, G. and Sandstrom, K. (1993). 'Ideology in action: a pragmatic approach to a contested concept'. *Sociological Theory*, 11(1): 21–38.

Freeden, M. (2003). *Ideology*. Oxford: Oxford University Press.

Gunder, M. (2004). 'Shaping the planner's ego-ideal: a Lacanian interpretation of planning education'. *Journal of Planning Education and Research*, 23(3): 299–311.

Gunder, M. (2006). 'Sustainability: planning's saving grace or road to perdition?' *Journal of Planning Education and Research*, 26(2): 208–21.

Gunder, M. (2010). 'Planning as the ideology of (neo-liberal) space'. *Planning Theory*, 9(4): 298–314.

Gunder, M. (2011). 'A metapsychological exploration of the role of popular media in engineering public belief on planning issues'. *Planning Theory*, 10(4): 325–43.

Gunder, M. (2016). 'Planning's "failure" to ensure efficient market delivery: a Lacanian deconstruction of this neoliberal scapegoating fantasy'. *European Planning Studies*, 1(24): 21–38.

Gunder, M. and Hillier, J. (2009). *Planning in ten words or less*. Farnham: Ashgate.

Howard, E. (1965 [1902]). *Garden cities of tomorrow*. Cambridge, MA: MIT Press.

Jameson, F. (2003 [1977]). 'Imaginary and symbolic in Lacan: marxism, psychoanalytic criticism, and the problem of the subject', in S. Žižek (ed.) *Jacques Lacan: critical evaluations in cultural theory, volume III.* London: Routledge, pp 3–43.

Jensen, B. (2016). 'Masdar City: a critical retrospective', in S. Wippel, K. Bromber, C. Steiner and B. Krawietz (eds.) *Under construction: logics of urbanism in the Gulf Region*. London: Routledge, pp 45–54.

Kant, I. (1934). *Critique of pure reason*. London: Dent.

Kronsell, A. (2013). 'Legitimacy for climate policies: politics and participation in the Green City of Freiburg'. *Local Economy*, 18(8): 965–82.

Lacan, J. (2006 [1966]). *Ecrits*. London: Norton.

Lacan, J. (2007). *The seminar of Jacque Lacan: the other side of psychoanalysis, book XVII.* London: Norton.

Laclau, E. and Mouffe, C. (1985). *Hegemony and socialist strategy*. London: Verso.

Macedo, J. (2013). 'Planning a sustainable city: the making of Curitiba, Brazil'. *Journal of Planning History*, 12(4): 334–53.

Martinez-Alier, J., Pascual, U., Vivien, F.-D. and Zaccai, E. (2010). 'Sustainable de-growth: mapping the context, criticisms and future prospects of an emergent paradigm'. *Ecological Economics*, 69(9): 1741–7.

Mies, M. and Shiva, V. (2014 [1993]). *Ecofeminism*. London: Zed Books.

Naess, A. (1984). 'A defence of the deep ecology movement'. *Environmental Ethics*, 6(3): 265–70.

Rees, W. (2018). 'Planning in the Anthropocene', in M. Gunder, A. Madanipour and V. Watson (eds.) *The Routledge handbook of planning theory*. London: Routledge, pp 53–66.

Rico, M. and Lin, W. (2012). 'Urban sustainability, conflict management, and the geographies of postpoliticism: a case study of Taipei'. *Environment and Planning C*, 30(2): 191–208.

Ruti, M. (2008). 'The fall of fantasies: a Lacanian reading of lack'. *Journal of the American Psychoanalytic Association*, 56(2): 483–508.

Stavrakakis, Y. (1997). 'Green ideology: a discursive reading'. *Journal of Political Ideologies*, 2(3): 259–79.

Swyngedouw, E. (2010). 'Trouble with nature: "ecology as the new opium for the masses"', in J. Hillier and P. Healey (eds.) *The Ashgate research companion to planning theory*. Farnham: Ashgate, pp 299–318.

Swyngedouw, E. (2013). 'Apocalypse now! Fear and doomsday pleasures'. *Capitalism Nature Socialism*, 24(1): 9–18.

Swyngedouw, E. (2015). 'Depoliticized environments and the promises of the Anthropocene', in R. Bryant (ed.) *The international handbook of political ecology*. Cheltenham: Edward Elgar Publishing, pp 131–45.

Trainer, T. (2015). 'The degrowth movement from the perspective of the simpler way'. *Capitalism Nature Socialism*, 26(2): 58–75.

WCED (World Commission on Environment and Development) (1987). *Our common future*. Oxford: Oxford University Press.

Žižek, S. (1989). *The sublime object of ideology*. London: Verso.

Žižek, S. (1993). *Tarrying with the negative*. Durham: Duke University Press.

Žižek, S. (1996). *The indivisible remainder*. London: Verso.

Žižek, S. (2000). *The fragile absolute*. London: Verso.

4

Sustainable development

History and evolution of the concept[1]

Delyse Springett and Michael Redclift

When the Club of Rome coined the term, 'The Global Problèmatique' for the environmental crisis of the early 1970s, it was intended to capture the connections and dynamic interactions between the various aspects of the problem – those linkages and knock-on effects that reverberate throughout the world (Reid, 1995; Rockström et al., 2009). The institutional roots of the crisis, with its social, political and economic dimensions and the associated cultural, spiritual and intellectual implications, can be traced back to the emergence of the capitalist economy from the scientific and industrial revolutions in England (Merchant, 1980; Spretnak and Capra, 1985; Carley and Christie, 1992). Central to the changing worldview was the shift in attitudes towards nature wrought by the ideology of the Enlightenment, leading to nature's 'disenchantment' and the dissipating of its power over physical and spiritual aspects of human life (Merchant, 1980; Eckersley, 1992). The new scientific paradigm at the core of the Enlightenment that transformed the human-nature relationship, combined with the capitalist model of production and consumption, produced a *degree* of change and *scale* of degradation not previously possible (Merchant, 1980). Along with this, the Northern process of domination, effected through colonisation in pursuit of resources, markets and land – and later extended through the globalisation of trade, technological expertise, the money market and communications (The Ecologist, 1993) – eventually resulted in global impacts on nature and the lives of people. Two decades ago, Vitousek et al. (1986, p. 1861) stated: 'any clear dichotomy between pristine ecosystems and human-altered areas that may have existed in the past has vanished'. Today, the Earth is beyond the point where boundaries can be ascribed to environmental problems and the associated social impacts. However, the sharing of the impacts is not equitable, as the eco-justice movement underlines: the poor disproportionately shoulder the consequences of environmental degradation (Dobson, 1998; Agyeman et al., 2003; Martínez-Alier, 2003). These social and environmental impacts and the struggle to deal with them led to the coining of the concept of 'sustainable development' and its appearance on the international agenda in the 1970s (Carley and Christie, 1992).

The international contestation of sustainable development

The environmental movement of the 1960s was based largely upon a concept of nature that was scientifically constructed by the North (Hays, 1959; Evernden, 1992; Eder, 1996a), chiefly rooted in the earlier American 'conservation' movement and perceived by O'Riordan (1981) as organised resource exploitation and regional economic planning. As the debate became affected by ideas and concepts from the field of development (Redclift, 1987; Adams, 1990; Goulet, 1995a, 1995b), the dialectics of 'environment and development' produced a new discourse, though the North continued to identify the problems and solutions, chiefly from a 'conservation' perspective. The adoption of the term, 'sustainable development', brings with it epistemological and practical problems that have led to strong contestation, but it signifies a transformation being made in the environmental discourse. The contestation – even repudiation – of the term has not excluded its capture by some groups to become a key concept in the rhetoric of 'green' business. Against negative perceptions, some authors always understood the concept as capable of emancipating more democratic and inclusive approaches to living with nature and each other (O'Connor, J., 1998), while others saw it as legitimating perspectives from the South (Redclift, 1987; Jacobs, 1991).

The power of Northern hegemony met some resistance from the World Commission on Environment and Development (WCED), which included a large number of Commissioners from the South. The Brundtland Report (1987) placed the discourse much more firmly in the economic and political context of international development. Efforts to limit the agenda to 'environmental' matters and a critique of conventional environmental management as practised in developed countries were resisted (Redclift, 1987). The preliminary consultative process itself provided something of a model of democratic participation (Redclift, 1987), and the Report was altogether more 'political' and radical than the Stockholm Declaration (1972) or the World Conservation Strategy (1980).

Despite the criticisms, the Commission presented a political vision of sustainable development: it called for institutional restructuring of national politics, economics, bureaucracy, social systems of production and technologies, requiring a new system of international trade and finance. It was, perhaps, the neo-Marxist movement, newly taking the environment into its consideration in the late 1980s, that best perceived the potential the Report brought for significantly new ways of doing things within a revised capitalist framework.

The Report did, however, offer a challenge to traditional sources of power, of whatever hue, by lifting the debate from a focus on scarcity and counteracting 'the sectoral bias and compartmentalism' that had marked much of the work on the environment (Redclift, 1992, p. 33). The United Nations Conference on Environment and Development (UNCED, 1992), the agenda of which arose largely from the Brundtland Report, demonstrated what may happen to any serious challenge to traditional forms of power. The Conference potentially represented a 'turning point' (Gore, 1992; Frankel, 1998) and the opportunity to address the worsening socio-economic disparities between North and South along with the environmental degradation associated with these. Opinions on the achievements of UNCED are divided between confidence in significant progress being made and the belief that the Conference was a failure, even a charade stage-managed by business. The UNCED process revealed that it served powerful interests. The critique of the process and the Alternative Treaties produced by an international consortium of NGOs reveal the key 'silences' and 'non-decision making' that characterised the formal agenda.

However, since UNCED, the balance of power has shifted. While the struggle at that and earlier fora can be seen as being between 'North' and 'South', the gap today is also between the poorest countries, with no resources to attract investment, the developed countries and the new 'rapidly developing' economies.

The discourse of sustainable development: problematising the concept

This brief genealogy of sustainable development, the contestation for the concept at international level and the changing realities that the progress of globalisation brings with it explain the power and hegemony exercised in the struggle for 'ownership' and definition of the concept. It discloses why the discourse has been narrowly controlled and why a dialectical, relational approach is needed to open up the still evolving process (Harvey, 1996). A more dialectical approach might produce not a two-dimensional, undialectic 'map' but something more discursive, akin to multi-dimensional 'cognitive mapping' of the many discourses of sustainable development. The importance of maintaining discursivity is that it is the discourse that is 'creating' sustainable development (Foucault, 1972); the process is a dynamic one, where the concept should not be allowed to become a naturalised, 'reified' thing (Foucault, 1972). It comes down to a struggle between discursivity and control, an inherently ideological process (Redclift, 1996), which is witnessed at the international level. The international literature reflects the 'stakes in the ground' of specific groups: economics, ecology, environmental management, environmental philosophy, the claims and contestations of academic disciplines; views from the South and political and corporate positions all reveal the political, ideological, epistemological, discipline-based and philosophical approaches that compete for legitimacy. Broadly speaking, these fall into three major camps: ecology-centred, market-based and neo-Marxist approaches. From a critical perspective, sustainable development is perceived not only as a social construct but a multi-constructed and strongly contested concept (Eder, 1996b; Dobson, 1996) that is political and radical (Jacobs, 1991). The dismissive charge of 'vacuousness' that has been made needs to be explored to discover whether such 'vacuity' is used as an obfuscatory gag on the radical aspects of the concept – a way of excluding competing views in the struggle for ownership – or whether the concept is, indeed, vapid jargon.

'Sustainable development' or 'sustainability'?

The contestation for the definition of sustainable development is made additionally problematic by the ways in which the terms 'sustainable development' and 'sustainability' have been counterposed (Dobson, 1998). For purists, the terms are almost diametrically opposed; sustainable *development* represents a threat to sustainability on account of its 'dangerous liaison', particularly since the Brundtland Report, with economic growth. This liaison smacks of positivism and modernism since the concept is seen as emanating from the very cultural and economic sources that gave rise to 'unsustainability'. Much of the concern focuses upon Northern domination and the assumption that (Northern) 'management' can solve the sustainable development dilemma. The increasing domination and 'eco-cracy' (Gudynas, 1993) stem from the fact that, institutionally, we have bought into an all-engulfing management paradigm (Redclift, 1996) that introduces new institutional structures for *environmental management* that give scant attention to the actual processes through which the environment has been transformed and commodified. Against this is the body of opinion that believes that sustainable development encapsulates the understanding of the need for radical change to a different way of life – what has been characterised as a 'painfully difficult turn towards material simplicity and spiritual richness' (Worster, 1993, p. 132). In this sense, it is a strongly normative goal imbued with values and implying that value judgements need to be made (Redclift, 1996): a social goal for guiding behaviour at the individual, institutional, national and global levels. This shifts sustainable development out of the paradigm of management where business locates the concept (Springett, 2003, 2006). It also confirms it as a political concept. It is not surprising, then, that discussions of sustainable development generally ignore the epistemological dimension of the construct, the assumption being that Northern knowledge and expertise have developed a 'universal epistemology', whereas, in reality, the ubiquity of Northern science succeeds in fragmenting the knowledge of the South (Redclift, 1991), even though this knowledge may be increasingly important in terms of sustainable development.

Some argue that the ambiguous theoretical basis of sustainable development and the lack of consensus about its meaning make its implementation almost impossible: there are conceptual, political and ethical dilemmas in recasting 'development' activities as 'sustainable', and then declaring this a new paradigm for human interaction with the environment (Sneddon, 2000). In its mainstream guise, sustainable development is in danger of privileging *global* environmental problems and global (i.e.'powerful local', Shiva, 1993) institutions which are largely the province of the North and which choose to focus, for example, on the *problem* of poverty rather than the origins of poverty-*production*. This curtails the ability of the concept to act as an instrument for a 'transformative politics', whereas the concept of 'sustainability' is seen as not having been co-opted into the unilinear, mainstream hegemony to the same degree (Adams, 1995; Sunderlin, 1995; Sneddon, 2000). It 'carries less political baggage' (Paehlke, 1999), sparing us some of the problems associated with sustainable development. It is seen as having a 'multiplicity' of meanings, for example, leaving open the question of GNP (Paehlke, 1999, p. 243), whereas sustainable development assumes that growth is possible and desirable. Both terms view the economy, the environment and society as inevitably bound up with each other, but sustainability does not assume that economic growth is essential, nor that economic growth will inevitably result in net environmental harm (Paehlke, 1999).

However, like sustainable development, *sustainability* has a 'complex conceptual structure' (Paehlke, 1999, p. 246), and is also deplored for its 'vague, ill-defined character' (Becker et al., 1999). It is also seen as introducing 'normative commitments to the development problematic', calling for justice for future generations and implying that the economic process should be 'subordinated to social and ecological constraints' (Becker et al., 1999, p. 5). This strongly accords with the conception of sustainable development propounded by Redclift and others. Despite the calls for sustainability to be extricated from the sustainable development discourse – or to replace it – there is also evidence that a number of writers have in mind an all-embracing concept that eschews neo-classical economics, calls for better understanding and treatment of nature, demands social equity and eco-justice based on a less instrumental understanding of democracy, and that this overall conception of 'the good life' is sometimes referred to as 'sustainability' and sometimes as 'sustainable development'.

A question of definition: competing certainties versus discourse

Part of the 'problem' of sustainable development is the contestation for its definition: so intrinsically political is the concept that it elicits attempts by widely disparate vested interests to frame its meaning. The power of definition, and of determining the language that characterises a concept, are seminal ways of staking and holding claims to domination (Beder, 1996; Livesey, 2001; Ralston Saul, 2001), while dismissing that concept on account of its *lack* of clear definition also restricts any inherent potential for change from being liberated. The debate on sustainable development has ranged from a call for consensus on a definition that can lead to action (Carpenter, 1994) to proposals that the term be abandoned on account of its 'vacuity' and 'malleability' (Lélé, 1991; Sneddon, 2000) and its lack of 'objective analysis' (Reboratti, 1999). Redclift notes that it is 'about meeting human needs, *or* maintaining economic growth, *or* conserving natural capital, *or* all three' (Redclift, 1999, p. 37, emphasis added). The alleged vagueness and ill-defined character of the concept (Becker and Jahn, 1999) have been attributed both to a lack of theoretical underpinning and to the ways in which the concept itself was constructed and framed (Sneddon, 2000). Built upon the dual and opposing concepts of ecological sustainability and development/growth, the complexity of the construct promulgates not only different and conflicting theoretical perspectives but also the ensuing 'semantic confusion' that arises from these (Sachs, 1999). Its conceptual capacity and the normative and political dimensions of the concept only increase the ambiguity; it has come to be used *as though* it has 'universal and temporal validity' and general acceptance (Reboratti, 1999, p. 209; see also Smith and Warr, 1991), while, at the same time, its lack of objective analysis has led to its being dismissed as a cliché.

Some perceive the ideological repackaging of the discourse of development planning in the 1980s as a cynical attempt to construct a 'green cover' for business-as-usual and the ongoing exploitation of people and resources (Willers, 1994; Adams, 1995; Escobar, 1995): a political cover for otherwise unacceptable corporate practices (Paehlke, 1999) and an attempt at 'semantic reconciliation' of the irreconcilable ideologies of ecological transformation and economic growth. The lack of clear definition of sustainable development – its 'opaqueness' – is also seen as symptomatic of this underlying ideological struggle. However, it might also be argued that the failure to deliver a tight definition reflects the futility – even the danger – of trying to capture a complex construct in simplistic terms. Perhaps the most serious aspect of the problematic for 'sustainable development' is that the ambiguous theoretical basis and lack of context-specificity and clarity (Sneddon, 2000) disable *implementation* of a concept that does not have time on its side (Redclift, 1987; Lélé, 1991; Frazier, 1997). The dismissal of the concept as a force for power has been widespread: its 'populism' is seen as resulting in confusion and ambiguity (Lélé, 1991; Redclift, 1991; Reboratti, 1999), reducing it to a 'quasi-rhetorical term' and a 'must word' (Reboratti, 1999). Lack of academic rigour in the initial formulation of the term has relegated it to the popular status of a 'catch-phrase' (Lélé, 1991), with an accompanying 'fuzziness' surrounding its definition and interpretation. Indiscriminate use of the term disguises the fact that it is 'hard to pin down and convert into a useful methodological tool' (Reboratti, 1999); even the 'relatively acceptable' WCED needs-based definition focusing on inter- and intra-generational equity is dismissed as 'wishful thinking rather than conceptual framework' (Reboratti, 1999, p. 213). It has lost further credibility and meaning on account of the ease with which it has 'passed into the everyday language of politicians' (O'Brien, 1991) with the consequent danger of losing all meaning, though it has not impacted substantially on the *platforms* of political parties (Reboratti, 1999). The other cause of scepticism is the ease with which the construct has been colonised by business and become part of its own rhetoric.

The debate reflects the contestation by those who aim to neutralise the potentially political role that lies at the heart of the concept. This prevents serious change from taking place (Lélé, 1991) and disempowers its radical core of meaning. The general use of the concept indicates a poor understanding of the institutional causes of poverty and environmental degradation, confusion about the role of economic growth, lack of clarity about the concepts of sustainability and *participation*, with all of this constraining the democratic force of the concept (Lélé, 1991). It has also been argued that the vagueness surrounding the concept forms part of its 'appeal' (Redclift, 1991).

Such 'vagueness' may be a politically expedient aspect of the concept not only to play down its potential power but also to emancipate that power (Lélé, 1991): a more specific definition might represent a reactionary force, a means of control that restricts discourse (Ralston Saul, 2001). In other words, the 'ambiguity' of the concept may be its central virtue and strength, inviting discourse (Redclift, 1987; O'Riordan, 1993; Wilbanks, 1994).

Dryzek (2000) advocates not a definition but a discourse about sustainable development that is shaped by a shared set assumptions and capabilities and embedded in enabling language. Discourses are social and act as sources of order by coordinating the behaviour of individuals who subscribe to them. At the heart of the debate over sustainable development lies the question of power and, specifically, the potential for political and structural change that is central to a radical interpretation of the concept (Springett, 2005). Its political significance is underlined in part by the fact that it has been generated through the power of Northern institutions, as well as academic debate (Reboratti, 1999). At the same time, the lack of specificity clouds its *normative* role as a social goal which can only be achieved through examination of our own behaviour (Redclift, 1996), not 'fixed' by management and technology. For Redclift, it is a policy objective rather than a methodology – an overarching concept and 'unapologetically normative' (Redclift, 1996, p. 37), calling for a more 'human-focused' approach. The discourse is full of contradictions. Borrowing from the natural and social sciences, the concept is seen as a major constraint on human 'progress' – the price the conventional growth model must pay if the 'biospheric imperative' is ignored, calling for different technologies and more realistic assessment of environmental losses. Another contradiction concerns the implications of 'human progress' for nature, with people from different ideological persuasions calling for an examination of the 'ends' as well as the 'means' of development. Central to the problem are the unanswered questions about recovery of our control over consumption (Redclift, 1996). The Brundtland Report's focus on 'needs' still left unanswered questions about the needs of future generations, the changes in needs, the ways in which development contributes to or creates needs, and how needs are defined in different cultures. No answer has been found to the question of *what* is to be sustained (Redclift, 1999, p. 60). Redclift defines the key question as being distributive, calling for a redefinition that would incorporate future population growth and the ensuing demands on the environment, as well as necessary changes in individual consumption patterns. The discourse rarely stops to examine those real needs (largely of the South and the poor of the North) that are consistently not being met (Durning, 1992; Elkington, 1995), and this brings the heart of the problem back to the materiality of the environmental experience without which culture itself cannot exist (Ingold, 1992). Concepts of nature are always cultural statements (Beinart and Coates, 1995; Redclift, 1999), and the 'environment' is the creation of human activity, socially constructed like all discourses and based upon ecological principles that are themselves constructs of a science that is part of human culture (Redclift, 1999, p. 67).

One danger of the contestation over definition is that it will deflect attention from these unanswered questions that signify the need for an essentially political project to bring about changes in human behaviour. Competition over definition helps to obscure the more basic need to redefine the roles and functions of public and private institutions that support unsustainable behaviour – not only business but political and administrative institutions. It is a political act to contest the definition of sustainable development, and the endless contestation may cover up embarrassing questions such as government unwillingness to promote, for example, major fiscal or financial reforms, to significantly decentralise power or to recognise that scientific knowledge as a basis for 'rational' decision making has limitations. In a sense, the debate about definition can be seen as a displacement activity or a deliberate barrier to the recognition of the sustainable development imperative. Contemporary market economies have ideological mechanisms for silencing opposition (O'Connor, J., 1994), one being the act of 'semiotic conquest' of language and agenda. Endless contestation deflects the radical core of sustainable development into a confusing, de-energising struggle for 'meaning' rather than action. In terms of business, the capitalist appropriation of nature and communities is seen by O'Connor as attempting to find its own legitimation through the 'sinister double play' of the rhetoric of 'greened growth' as opposed to a focus on sustainable development. Radical constructions of sustainable development view it as a potentially energising force in its own right (Redclift, 1987; Dovers, 1989; O'Connor, M., 1994; O'Riordan and Voisey, 1997), with the potential to create important social change, but calling for a myriad of institutional changes that are not necessarily promoted by the sustainable development agenda. This radical view suggests that many strategies will be employed to obscure or dilute that power, not least by capitalist business itself.

For social change to take place, there needs to be, not a 'definition', but some consensus about the *core meaning* of the term and the moral imperative it offers for 'the good life'. This is not easy when the concept is viewed as propping up the fundamental processes of capitalist exploitation (Jacobs, 1999, p. 22). The demand for a cut-and-dried – and, therefore, almost inevitably 'technological' – definition raises the spectre of 'reason' metamorphosing into 'technology' (Horkheimer, 1947), already seen in the domination and instrumentalisation of nature. A dialectical approach to sustainable development, not pinned to a specific definition, would be more likely to question the instrumentalist epistemic shift of science in the 1920s, the rapid growth of big bureaucracies in public administration, humanity's colonisation of nature through technology and the capitalist management of the administrative apparatus of the state that worked together to create the need for the construct. Such dialectical discourse would be more likely to unearth the origins of the term, and the archaeology of the institutional infrastructure that supports these systems. Shifting from 'definition' to 'discourse' might elevate the power of sustainable development as a 'site of political contest', the source of a new political worldview that contests the status quo (Jacobs, 1999). It would suggest that sustainable development may become part of the deliberative turn to a more discursive theory of democracy (Dryzek, 2000), whereby, through a process of dialectical discourse, sustainable development could contribute to a new, more inclusive theory of 'the good life'. Inherent in such a theory would be considerations of environment, equity and ethical issues – factors it is difficult to 'value'.

The areas of core meaning that characterise the belief in the political power of sustainable development, as identified by Jacobs (1991), are:

- The entrenchment of environmental considerations in economic policy making;
- A commitment to equity;
- An appreciation that 'development' is wider than growth.

Based on this, any interpretation implies change for economic policy and exposes the additional conflict that sustainable development is the beginning, not the end, of the debate: it provides a 'common currency', bringing together conflicting vocabularies to a common, though contested, one (Jacobs, 1999). The focus on social equity, global justice and human rights presents a constructivist interpretation based on human relations, culture and politics (Lash et al., 1996). This moves away from the major response since Brundtland focused on 'managing' the Earth through technological expertise, and the framing of the concept by powerful groups of the North (Becker, 1999). Nevertheless, much of the debate has continued to focus on 'definition' rather than imperatives, and the business incursion into the debate has increased the focus on both definition and 'management'.

Sustainable development: an oxymoron?

Polanyi (1967) stressed that economic rationalism, in the strict sense, does not answer questions of motivations and valuations of a moral and practical order. Yet the compromise constructed between sustainable development and economic growth suggests that equity, conservation and economic growth, while uncomfortable companions, are not incompatible (Jacobs, 1991). Opponents view this as 'a fatal co-option' into technocentric management designed not to disturb the power processes of the growth economy and capitalist exploitation (Reboratti, 1999, p. 22). Sustainable development has become part of the historical process linked to economics and political structures, transformed both existentially and by economic growth but inextricably linked with the expansion and contraction of the world economic system (Redclift, 1987). However, it calls for a competing paradigm that breaks with the linear model of growth and accumulation. This would be more inclusive, with economic forces seen as related to the behaviour of social classes and the role of the state in accumulation. The social and environmental impacts of capitalist development would not be regarded as beyond the aegis of market economics; they would no longer be permitted as 'externalities borne chiefly by those without power, and which now need to be internalized within the economic model' (Redclift, 1987, p. 13). By strengthening the emphasis upon human need, the Brundtland Report itself provided an opportunity for a radical shift away from an economics epistemologically predisposed to a modernist, reductionist view of resources and exchange value (Norgaard, 1985). Nevertheless, it is a 'dangerous liaison' (Sachs, 1991), an attempt to reconcile the irreconcilable (Benton, 1999). It can be read as appropriation of the agenda of environmental responsibility and social justice by economists, still reliant upon economic instruments for environmental protection; and no more than a vehicle for 'free market environmentalism' dominated by neo-classical concepts for allocating resources (Beder, 1996, p. 89). International agencies such as the OECD and fora such as UNCED have favoured such ideologically-based market solutions; but others see it as resulting in economic valuation that is another kind of 'semiotic conquest' (O'Connor, J., 1994), converting ecological entity to 'natural capital' and placing it on a par with other forms of capital.

It seems improbable that any agreement about sustainable development that adheres to the core themes identified in this chapter can be based on current global, cultural and political tradition (Reboratti, 1999). Rather, it needs a new social covenant and a new set of 'rules', including economic rules and ways of thinking about growth. For example, instead of following the neo-liberal theory of the free play of markets as the system of economic regulation, economic activity would be re-located within society (Gowdy, 1999). An emancipatory shift of this kind might mean learning from the complex social systems that have been sustained for long periods of time by people in developing nations, requiring a powerfully different conception of the role of economics in creating the 'good life'.

Note

1 This chapter is abstracted from Springett, D. and Redclift, M. (2015) 'Sustainable development: history and evolution of the concept', in M. Redclift and D. Springett, D. (eds.) *Routledge international handbook of sustainable development*. London: Routledge, pp 3–38.

References

Adams, W. M. (1990). *Green development: environment and sustainability in the third world*. London: Routledge.

Adams, W. M. (1995). 'Green development theory? Environmentalism and sustainable development', in J. Crush (ed.) *Power of development*. London: Routledge, pp 87–99.

Agyeman, J., Bullard, R. D. and Evans, B. (2003). *Just sustainabilities: development in an unequal world*. Cambridge, MA: MIT Press.

Becker, E. (1999). 'Fostering transdisciplinary research into sustainability in an age of globalization: a short political epilogue', in E. Becker and T. Jahn (eds.) *Sustainability and the social sciences*. London: Zed Books, pp 284–9.

Becker, E. and Jahn, T (eds.) (1999). *Sustainability and the social sciences. A cross-disciplinary approach to integrating environmental considerations into theoretical reorientation*. London. Zed Books.

Becker, E., Jahn, T. and Stiess, I. (1999). 'Exploring uncommon ground: sustainability and the social sciences', in E. Becker and T. Jahn (eds.) *Sustainability and the social sciences*. London: Zed Books, pp 1–22.

Beder, S. (1996). *The nature of sustainable development* (2nd edn). Melbourne: Scribe Publications.

Beinart, W. and Coates, P. (1995). *Environment and history*. London: Routledge.

Benton, T. (1999). 'Sustainable development and accumulation of capital: reconciling the irreconcilable?' in A. Dobson (ed.) *Fairness and futurity: essays on environmental sustainability and social justice*. Oxford: Oxford University Press, pp 199–229.

Carley, M. and Christie, I. (1992). *Managing sustainable development*. London: Earthscan.

Carpenter, R. (1994). 'Can sustainability be measured?' *Ecology International Bulletin*, 21: 7–36.

Dobson, A. (1996). 'Environment sustainabilities: an analysis and a typology'. *Environmental Politics*, 5(3): 401–28.

Dobson, A. (1998). *Dimensions of social justice: conceptions of environmental sustainability*. Oxford: Oxford University Press.

Dovers, S. (1989). 'Sustainability: definitions, clarifications and contexts'. *Development*, 2/3: 33–6.

Dryzek, J. (2000). *Deliberative democracy and beyond: liberals, critics and contestations*. Oxford: Oxford University Press.

Durning, A. T. (1992). *How much is enough? The consumer society and the future of the earth*. New York: W. W. Norton.

Eckersley, R. (1992). *Environmentalism and political theory: toward an ecocentric approach*. London: UCL Press.

Eder, K. (1996a). *The social construction of nature: a sociology of ecological enlightenment*. London: Sage.

Eder, K. (1996b). 'The institutionalisation of environmentalism: ecological discourse and the second transformation of the public sphere', in S. Lash, B. Szerszynski and B. Wynne (eds.) *Risk, environment and modernity: towards a new ecology*. London: Sage, pp 203–21.

Elkington, J. (1995). *Who needs it? Market implications of sustainable life-styles*. London: SustainAbility Ltd and ERP.

Escobar, A. (1995). 'Imagining a post-development era', in J. Crush (ed.) *Power of development*. London: Routledge, pp 211–27.

Evernden, N. (1992). *The social creation of Nature*. Baltimore, MD: John Hopkins University Press.

Foucault, M. (1972). *The archaeology of knowledge*. London: Tavistock.

Frankel, C. (1998). *In earth's company: business, environment and the challenge of sustainability*. Gabriola Island, BC, Canada: New Society Publishers.

Frazier, J. (1997). 'Sustainable development: modern elixir or sack dress?' *Environmental Conservation*, 24: 182–93.

Gore, A. (1992). *Earth in the balance: forging a new common purpose*. London: Earthscan.

Goulet, D. (1995a). *Development ethics: a guide to theory and practice*. London: Zed Books.

Goulet, D. (1995b). 'Authentic development: is it sustainable?' in T. C. Trzyna (ed.) *A sustainable world*. London: IUCN/Earthscan, pp 44–59.

Gowdy, J. (1999). 'Economic concepts of sustainability: relocating economic action within society and the environment', in E. Becker and T. Jahn (eds.) *Sustainability and the social sciences*. London: Zed Books, pp 162–81.

Gudynas, E. (1993). 'The fallacy of eco-messianism: observations from Latin America', in W. Sachs (ed.) *Global ecology: a new arena of political conflict*. London: Zed Books, pp 170–8.

Harvey, D. (1996). *Justice, nature and the geography of difference*. Cambridge, MA: Blackwell.

Hays, S. (1959). *Conservation and the gospel of efficiency: the progressive conservation movement 1890–1920* (1979 edn). New York: Atheneum.

Horkheimer, M. (1947). *Eclipse of reason*. New York: Oxford University Press.

Ingold, T. (1992). 'Globes and spheres: the topology of environmentalism', in K. Milton (ed.) *Environmentalism: the view from anthropology*. London: Routledge, pp 31–42.

Jacobs, M. (1991). *The green economy: environment, sustainable development and the politics of the future*. London: Pluto Press.

Jacobs, M. (1999). 'Sustainable development as a contested concept', in A. Dobson (ed.) *Fairness and futurity: essays on environmental sustainability and social justice*. Oxford: Oxford University Press, pp 21–45.

Lash, S., Szerszynski, B. and Wynne, B. (1996). *Risk, environment and modernity: towards a new ecology*. London: Sage.

Lélé, S. M. (1991). 'Sustainable development: a critical review'. *World Development*, 19(6): 607–21.

Livesey, S. (2001). 'Eco-identity as discursive struggle: Royal Dutch/Shell, Brent Spar and Nigeria'. *The Journal of Business Communication*, 8(1): 58–91.

Martínez-Alier, J. (2003). *The environmentalism of the poor: a study of ecological conflicts and valuation*. Cheltenham: Edward Elgar Publishing.

Merchant, C. (1980). *The death of Nature: women, ecology and the scientific revolution*. London: Wildwood House.

Norgaard, R. (1985). 'Environmental economics: an evolutionary critique and a plea for pluralism'. *Journal of Environmental Economics and Management*, 12: 382–94.

O'Brien, P. (1991). 'Debt and sustainable development in Latin America', in D. Goodman and M. Redclift (eds.) *Environment and development in Latin America*. Manchester: Manchester University Press, pp 24–47.

O'Connor, J. (1994). 'Is capitalism sustainable?' in M. O'Connor (ed.) *Is capitalism sustainable? Political economy and the politics of ecology*. New York: Guilford, pp 152–75.

O'Connor, J. (1998). *Natural causes: essays on ecological Marxism*. New York: Guilford.

O'Connor, M. (1994). 'Introduction: liberate, accumulate – and bust?' in M. O'Connor (ed.) *Is capitalism sustainable? Political economy and the politics of ecology*. New York: Guilford, pp 1–21.

O'Riordan, T. (1981). *Environmentalism* (2nd edn). London: Pion.

O'Riordan, T. (1993). 'The politics of sustainability', in R. K. Turner (ed.) *Sustainable environmental management: principles and practice*. London: Belhaven and ESRC, pp 37–69.

O'Riordan, T. and Voisey, H. (1997). 'The political economy of sustainable development'. *Environmental Politics*, 6: 1–23.

Paehlke, R. (1999). 'Towards defining, measuring and achieving sustainability: tools and strategies for environmental evaluation', in J. Becker and T. Jahn (eds.) *Sustainability and the social sciences*. London: Zed Books, pp 243–64.

Polanyi, K. (1967). *The great transformation*. Boston: Beacon.

Ralston Saul, J. (2001). *On equilibrium*. Toronto: Penguin.

Reboratti, C. E. (1999). 'Territory, scale and sustainable development', in J. Becker and T. Jahn (eds.) *Sustainability and the social sciences*. London: Zed Books, pp 207–22.

Redclift, M. (1987). *Sustainable development: exploring the contradictions*. London: Routledge.

Redclift, M. (1991). 'The multiple dimensions of sustainable development'. *Geography*, 76(1): 36–42.

Redclift, M. (1992). 'Sustainable development and global environmental change: implications of a changing agenda'. *Global Environmental Change*, 2(1): 32–42.

Redclift, M. (1996). *Wasted: counting the costs of global consumption*. London: Earthscan.

Redclift, M. (1999). 'Sustainability and sociology: northern preoccupations', in E. T. Becker and T. Jahn (eds.) *Sustainability and the social sciences*. London: Zed Books, pp 59–73.

Reid, D. (1995). *Sustainable development: an introductory guide*. London: Earthscan.

Rockström, J., Steffen, W., Noone, K., Persson, A., Chapin III, F. S., Lambin, E. F., Lenton, T. M., Scheffer, M., Folke, C., Schellnhuber, H. J., Nykvist, B., De Wit, C. A., Hughes, T., van der Leeuw, S., Rodhe, H., Sörlin, S., Snyder, P. K., Costanza, R., Svedin, U., Falkenmark, M., Karlberg, L., Corell, R. W., Fabry, V. J., Hansen, J., Walker, B., Liverman, D., Richardson, K., Crutzen, P. and Foley, J. A. (2009). 'A safe operating space for humanity'. *Nature*, 461(7263): 472–5.

Sachs, W. (1991). 'Environment and development: the story of a dangerous liaison'. *The Ecologist*, 21(6): 252–7.

Sachs, I. (1999). 'Social sustainability and whole development: exploring the dimensions of sustainable development', in J. Becker and T. Jahn (eds.) *Sustainability and the social sciences. A cross-disciplinary approach to integrating environmental considerations into theoretical reorientation*. London: Zed Books, pp 25–36.

Shiva, V. (1993). 'The greening of the global reach', in W. Sachs (ed.) *Global ecology: a new arena of political conflict*. London: Zed Books, pp 149–56.

Smith, P. M. and Warr, K. (1991). *Global environmental issues*. London: Hodder and Stoughton.

Sneddon, C. S. (2000). '"Sustainability" in ecological economics, ecology and livelihoods: a review'. *Progress in Human Geography*, 24(4): 521–49.

Spretnak, C. and Capra, F. (1985). *Green politics: the global promise*. London: Paladin.

Springett, D. V. (2003). *Corporate conceptions of sustainable development in New Zealand: a critical analysis*. PhD thesis, Durham University.

Springett, D. V. (2005). 'Structural limits to sustainable development: managers and progressive agency'. *International Journal of Innovation and Sustainable Development*, 1(1–2): 127–48.

Springett, D. V. (2006). 'Managing the narrative of sustainable development: "discipline" of an "inefficient" concept'. *International Journal of Green Economics*, 1(1–2): 50–67.

Sunderlin, W. D. (1995). 'Managerialism and the conceptual limits of sustainable development'. *Society and Natural Resources*, 8: 481–92.

The Ecologist (1993). *Whose common future?* London: Earthscan.

UNCED (United Nations Conference on Environment and Development) (1992). *Agenda 21: programme of action for sustainable development*. New York: United Nations Publications.

Vitousek, P. M., Erlich, P. R., Erlich, A. H. and Matson, P. A. (1986). 'Human appropriation of the products of photosynthesis'. *Bioscience*, 34: 368–73.

Wilbanks, T. J. (1994). '"Sustainable development" in geographic perspective'. *Annals of the Association of American Geographers*, 84(4): 541–56.

Willers, B. (1994). 'A new world deception'. *Conservation Biology*, 8: 1146–8.

Worster, D. (1993). 'The shaky ground of sustainability', in W. Sachs (ed.) *Global ecology: a new arena of political conflict*. London: Zed Books, pp 36–48.

5

Ecological modernisation

Achievements and limitations of densification

Petter Næss and Inger-Lise Saglie

Introduction

Contemporary environmental planning is grounded in a particular rationality of development – ecological modernisation – which contains unquestioned assumptions about sustainable urban development. This chapter takes one of the most mainstreamed urban policies as the case, densification or the compact city, and develops a critique of this policy that examines it in terms of the core premises of ecological modernisation. By using Oslo as an example, it is explored whether compact city policy achieves the ambitions of ecological modernisation, notably the decoupling or partly decoupling of environmental impacts from economic growth. The chapter then questions whether urban sustainability can be reasonably achieved by policies grounded in ecological modernisation. First, current debates about ecological modernisation and their consequent rationality of development are reviewed, and then compact city policy is situated within this rationality. Based on these, a critique of urban containment (compact city) is presented which considers it as a relevant and useful strategy for reduced pressure on environmental resources but insufficient for achieving environmental sustainability.

Eco-modernisation is a term that emerged in environmental sociology in the 1990s as a part of the discussion on the implementation of the principles of sustainable development. A core topic was whether sustainability would require a complete reorganisation of present wealthy societies. The eco-modernist theorists argued that this was not necessary. Ecological modernisation acknowledges that societal institutions need to change but not necessarily in a fundamental way. They believe that rationality, science, technological innovations, new governance forms, efficiency of market forces as well as institutional learning (Mol and Spaargaren, 2000) will be sufficient tools to meet the challenges. It is argued that it will be possible to accommodate further growth without overburdening the natural resources because we will be able to produce with much less input of natural resources, thus breaking the hitherto close connection between economic growth and environmental degradation. New technological innovation will ensure higher degree of eco-efficiency. In addition, new government forms and economic incentives will be implemented to stimulate and secure these improvements in institutional arrangements. There are also expectations that we may turn our consumption away from material products to the consumption of culture and experiences, the so-called post-material turn (Inglehart, 1995).

Thus, we will see a decoupling from economic growth and use of resources, as hypothesised in the so-called Environmental Kuznets Curve. According to this hypothesis, the link between growth and environmental degradation will be broken once a certain affluence level has been reached, beyond which further growth will lead to reduced environmental load (Stern, 2004). The eco-modernists point out that the hitherto strong growth in the world's population may peak and possibly decline (Asafu-Adjaye et al., 2015). This assumption is based on the observation that in richer countries, there are fewer children in families than before. As income levels increase in less affluent countries, a similar population decline may take place in those countries, too. Taken together, both eco-efficiency and post-material turn, as well as a possible peak in the global population, will make it probable that our imprint on the world's natural resources will peak and then decline, in line with the Environmental Kuznets Curve hypothesis. Ecological modernisation proponents acknowledge that the planet has a finite resource base, with the exception of solar energy, which is abundant if we find good ways to utilise it. However, they concur that the planetary limits are more theoretical than practical because of the expected decoupling and peak in the world's population. The kinds of environmental impacts and the time horizon for which this is supposed to happen is left unstated.

Environmental impacts of urban development

In many parts of the world, contemporary urban development implies a high conversion of natural areas and farmland into urbanised land, with serious negative effects on food security as well as on ecosystems and biodiversity. According to Beatley (2000), loss of habitats is a main cause of extinction of species, and habitat loss and fragmentation are to an increasing extent caused by urban development. Most of this has resulted from outward urban expansion, but areas within the urban area demarcation may also comprise localities of high importance for biodiversity.

Although some sprawl protagonists like Gordon and Richardson (1997) have claimed the opposite, a number of studies have shown empirically that low-density suburban development increases the need for motorised travel (Newman and Kenworthy, 1999; Zegras, 2010; Ewing and Cervero, 2010; Næss et al., 2019). It is difficult and expensive to provide high quality public transport services in such urban districts. Combined with road construction in a response to future demands, urban sprawl has contributed to the creation of highly car dependent cities.

The development of more environmentally friendly cars (for example, electrical vehicles) can reduce some local as well as global impacts but cannot make urban sprawl and increased car mobility environmentally sustainable. While electric cars reduce local emissions, they will depend in most countries on electricity produced at least partly from fossil fuels. Moreover, lifecycle studies of alternative vehicle and fuel technologies have shown that technologies which, seen in isolation, appear to offer reduced greenhouse gas emissions are not necessarily climate and environmentally friendly when all chains in the production and consumption cycle are taken into consideration (Holden, 2007). Furthermore, urban car traffic is land consuming and causes a number of social and environmental problems in addition to energy use and emissions, such as traffic accidents, barrier effects, congestion, noise and the encroachments of transport infrastructure on green areas and existing built environments.

In the context of urban development, economic growth is usually associated with increased demand for floor space and technical performance of buildings, increased availability of facilities and increased accessibility to these facilities. However, as mentioned earlier, growth in these parameters tends to leave its undesirable ecological traces (Høyer and Næss, 2001). The environmental impacts of the building sector include *construction impacts* as well as *operational impacts*. For urban development, a main challenge of decoupling economic growth from negative environmental impacts is thus to find ways of accommodating growth in the existing building stock and ensuring accessibility to facilities while reducing negative environmental impacts of the construction and use of buildings and infrastructure.

The compact city ideal

Fighting against urban sprawl has for a long time been high on the agenda of environmental planning. Today, the compact city has attained hegemonic status as a sustainable urban form internationally, particularly in Norway (Jenks et al., 1996; Andersen and Røe, 2016). In a European context, the Norwegian pursuit of more compact urban development over the last couple of decades could be considered as one of the most pronounced examples of ecological modernisation as a strategy for urban spatial development (Næss et al., 2011), this being promoted also with regard to rather small towns in an international context. Compared to urban sprawl, densification can bring important benefits in terms of the protection of natural landscapes, arable land and biodiversity, particularly if the densification can be channelled to 'brownfield' sites such as derelict or under-utilised industrial areas, obsolete harbour areas and parking space incompatible with an aim of reducing car traffic in the urban centre. Multifamily houses require, other things being equal, less energy for space heating and cooling per square metre than detached single family houses (Høyer and Holden, 2001) and less material for the construction of the buildings themselves, sewers, cables and access roads (Burchell et al., 1998). Admittedly, alternative energy solutions such as 'plus energy houses' (Voss and Musall, 2012) can level out the difference between single family houses and apartments in energy requirement. However, such technological improvements do not do away with the differences in land consumption and material consumption for infrastructure and buildings. A compact urban development can provide accessibility to facilities through proximity instead of by means of a high mobility by car, thus combining important environmental and social aspects of sustainability. Suburbanites tend to travel more often by car and drive longer distances than inner-city dwellers do, with pronounced differences in energy use for transportation and related emissions as a result (see references in the previous section). Moreover, the proportion of travel to and from the workplace carried out by car is considerably lower among employees at workplaces located close to the city centres in cities such as Oslo and Copenhagen than among employees at suburban workplaces (Næss and Sandberg, 1996; Hartoft-Nielsen, 2001; Næss et al., 2019). The inner-city is the part of the metropolitan area where accessibility by public transport is usually the highest while accessibility by car is at its lowest.

Some environmental planners argue that the lower use of car as travel mode for journeys to central workplaces would be compensated by shorter commuting distances to decentralised jobs interspersed with suburban residential areas. However, in contemporary specialised and high mobility societies, people do not choose jobs (or recruit workers) mainly from within their local neighbourhood. In the largest Norwegian cities, employees at suburban workplaces tend to commute longer, not shorter distances than employees at inner-city workplaces do (Næss and Sandberg, 1996; Næss et al., 2019).

Oslo as an ecological modernisation case: partial decoupling

The urban development of the Norwegian capital Oslo may serve as an illustration of the possibilities and limitations of obtaining a decoupling between growth in the building stock and negative environmental consequences (Næss et al., 2011). After several decades of urban sprawl and reduced population densities, the outward urban expansion was significantly reduced in

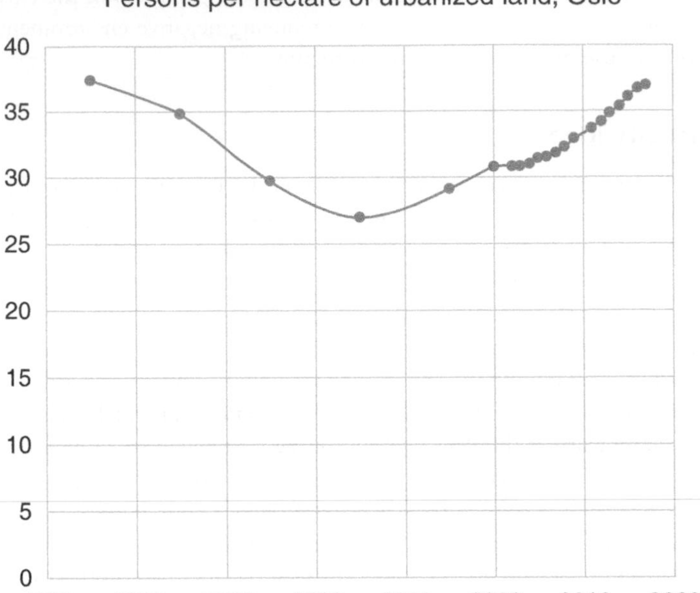

Figure 5.1 Changes in population density within the continuous urban area of Oslo (the morphological city) between 1955 and 2017

Source: Author drawn from data in Engebretsen (1993), Riksrevisjonen (2007) and Statistics Norway (2018)

the mid-1980s (Figure 5.1). Since the turn of the millennium, population density within Oslo's continuous urban area (i.e. the morphological city) has increased by 20% (Statistics Norway, 2018). The density increase was particularly high within the inner-city of Oslo. The average population density for the urban settlements in the functional urban region outside the morphological city of Oslo and the neighbouring city of Drammen also increased somewhat.

Compared to current urban development in most European cities and to its own development in the post war period until the mid-1980s, Oslo metropolitan area has managed to combine high growth in population and building stock and low encroachments on natural and cultivated areas. While the annual consumption of land for urban expansion of the morphological city of Oslo was on average 3.8 square kilometres during the period 1955–85, it was only 0.7 square kilometres annually during the period 1985–2015, despite population growth that was more than twice as high during the latter period (Engebretsen, 1993; Riksrevisjonen, 2007; Statistics Norway, 2018).

Partly as a result of densification policies increasing the population base for public transport, the network of metro, tram and bus lines has been expanded, and the number of departures per hour has increased. The number of trips by public transport increased per inhabitant by more than 20% between 1990 and 2012; from 2000 to 2012, the per capita car traffic decreased by 8%.

In absolute numbers, there was still traffic growth both in the 1990s and over the years 2000–12.

Slowing down the growth in car traffic is not sufficient to bring about an environmentally sustainable mobility. In Oslo, some increase in road capacity has worked against the fulfilment of the sustainability targets. The trends for intra-metropolitan traffic have also been accompanied with rapid growth in air travel.

The densification policy has no doubt made an important contribution to the relative stagnation in car traffic despite high increase in purchasing power over the period. Oslo's road tolls have also played a role, although their traffic reducing effect has been estimated to be relatively moderate (Lian, 2004). However, some of the stagnation may also reflect that the car has lost much of its fascination as a symbol of status and freedom. There is, for example, an international tendency among young people to drop or postpone acquiring a driver's licence. Instead, international travel has taken over much of the symbolic role that driving used to have.

Parts of the growth in flights may even be an indirect effect of urban densification since this has reduced the need for car ownership. People who do not need to spend money on owning and driving a car can spend their money on other things instead, for example, leisure flights. This is an example of a *rebound effect*, where increased resource efficiency (here: less car dependent ways of providing accessibility to daily destinations within a metropolitan area) leads to increased consumption of other services and commodities and to related increase in the environmental damages caused by this consumption (Næss, 2016; Holden and Norland, 2005).

Negative side effects of densification

Despite its low overall conversion of natural areas into building sites, Oslo's urban densification has had its negative impacts on intra-urban vegetation and ecosystems. Over the five-year period 1999–2004, there was a 5% reduction of the open access areas (defined as areas not including buildings, roads, railroads, quays, farmland, church yards, sea or major rivers) within the continuous urban area of Greater Oslo (Engelien et al., 2005). Partly, this was a result of transport infrastructure construction, but green areas also had to yield to new kindergartens or schools in districts where densification caused population growth beyond the capacity of existing social infrastructure. Together with the rapid inner-city population growth, this has diminished the availability of open access land per resident in these districts.

Protecting the city's green infrastructure is also important in order to counteract negative environmental and health impacts of climate change. Unless measures to retain a higher proportion of the precipitation in the soil are implemented, densification may increase the risk of overload of sewage systems resulting in contamination of water. Urban densification in old harbour or industrial waterfront areas may also conflict with the need to avoid flood prone building sites – at least if the densification is not accompanied with measures to protect the new buildings from floods.

Inner-city residential densification will, other things being equal, increase the number of inhabitants exposed to the higher levels of noise and concentrations of air pollution typically found in the inner parts of metropolitan areas.

Limits to densification

As the densification process continues, it will lead to a diminishing amount of land available for the construction of new buildings with small negative environmental impacts. Subsequent growth in the building stock must then be located to areas where the construction requires the conversion of natural areas or agricultural land.

A high rate of growth in the building stock and technical infrastructure may also make it an even more challenging task to combine climate change mitigation and adaptation strategies in urban land use. It will be more difficult to combine urban containment with the provision of sufficient pervious surfaces within the city if the amount of new floor space constructed is high, compared to a situation where the amount of construction is moderate.

Limits to the growth of building stock

Buildings are seldom, if ever, constructed with environmental protection as a main purpose. Instead, construction takes place to accommodate growth in the number of households, jobs, etc. and in the floor space per resident or employee.

According to ecological modernisation, we will be able to accommodate this growth with higher eco-efficiency. There are signs that such efficiency has increased. For example, energy efficiency in newly constructed dwellings has increased due to public regulation demanding higher standards in newly built and renovated buildings. Buildings may even be producers of valuable electricity, and it is expected that building regulations may set tougher targets for maximum energy consumption. However, increase in the building stock is at best environmentally friendly in a relative sense, not in absolute terms (Høyer and Næss, 2001).

In the Nordic as well as other European countries, studies have been conducted to investigate the potential for reducing the energy consumption and environmental load per unit produced by 'factor four', 'factor ten' and even 'factor twenty', i.e. down to 25%, 10% and 5%, respectively, of present levels (Nordic Council of Ministers, 1999). When undeveloped land as a resource is to be economised, such rates of reduction will only be possible if it is accepted that important characteristics of the 'products' are changed. Among other things, it may require a complete halt in the construction of detached single-family houses in urban areas. Replacing single-family house neighbourhoods with the construction of apartment buildings would, however, be highly controversial. In the long run, it is also dubious whether this would be sufficient to make continual growth in the housing stock environmentally sustainable (Xue, 2014).

Similar lines of argument can be made regarding the construction of other elements of the physical urban development, such as the construction of transport infrastructure, energy supply systems and sewage systems. Investments in transport infrastructure usually lead to major intrusions into the natural environment, and infrastructure cannot be recycled in the same way as the products we throw away as consumers.

In recent years, the consumption of residential floor area per capita in Norway has levelled out at a national scale after a steady growth until the turn of the millennium (Hille, 2013). In the municipality of Oslo, it has probably even decreased over the last decades, partly as a result of rapidly increasing housing prices. However, whether this will be a lasting trend is an open question. Moreover, if people reduce their consumption on housing while the purchasing power increases, they are likely to spend their money on other items instead, and who knows how environmentally friendly that consumption will be?

Limits to eco-modernisation

Decoupling

Oslo's densification and public transport improvements are examples of successful ecological modernisation strategies in the field of environmental planning. Nevertheless, the Oslo case shows that economic growth has only been partly decoupled from traffic growth and land take for urban development. If economic growth without negative environmental consequences were to be possible anywhere, this would most likely be in societies with a high level of prosperity, a high degree of economic freedom of action, as well as a high level of knowledge among the citizens. Oslo metropolitan area has a higher score than most other metropolitan areas worldwide on all these criteria. In this respect, Oslo might be considered a 'critical case' where the thesis that a non-environmentally-harmful economic growth is feasible could be tested. So far, however, it looks as if no city region – neither in Scandinavia nor elsewhere in the world – wishes, or is able, to implement more than a partial decoupling between growth and negative environmental impacts.

Apart from more eco-efficient technology, a change of consumption patterns where consumption of material commodities is replaced by consumption of culture, experiences, etc. is often referred to as a part of decoupling strategies. As mentioned earlier, Inglehart (1995) predicted such changes as part of an overall change towards 'post-material' values, including higher environmental awareness, as countries get richer. However, in Norway no such change can be seen over the period 1985–2013. Although there have been several fluctuations, the proportion of young people (15–26 years) in Norway who are highly worried about environmental problems was lower in 2013 than in 1985. On the other hand, the proportion who consider that they are missing a number of material goods in order to live the life they want was lower in 2013 than in 1985, particularly among those 15–20 years of age (Dalen et al., 2014). The latter is in line with Inglehart's (1995) hypothesised trend of replacing consumption of material goods with experiences. But the latter kinds of consumption too need a material base in the form of buildings and equipment. In Oslo, this is evident in the boom of new cultural buildings along the harbour front. Search for new experiences in exotic places also leads to increase in air travel.

Rebound effects

Some of the increase in air travel may be a rebound effect of densification resulting in lower daily life travel expenses, as discussed previously. The decoupling strategy of substituting consumption of material goods with experiences thus does not necessarily lead to higher environmental sustainability.

Rebound effects appear very difficult to avoid unless the purchasing power among the population decreases enough to compensate for the money saved from the resource efficient solutions. The situation is similar to squeezing a balloon: if you squeeze it at one place, it just bulges out elsewhere. If there is also economic growth and people get more affluent, the squeezing of the metaphoric balloon takes place while it is being pumped up with more and more gas.

Decoupling can in the end only be partial, not absolute. Environmental sustainability therefore requires that growth must sooner or later come to a halt. While economic growth is still needed in many low-income countries in order to combat poverty, the authors think a transition to non-growth or de-growth in already rich countries should happen sooner rather than later. This calls for urban environmental planning policies that do not take growth for granted as an uncritical 'good' (Fairclough, 2006) but instead focus on qualitative improvement of the built environment.

Social justice

The narrow focus of eco-modernisation on environmental technology and its lack of focus on social justice (Seippel, 2000) implies a risk that the implementation of eco-efficiency measures may worsen the living conditions for vulnerable groups and increase social segregation and inequalities. A contraction of the per capita consumption must be accompanied with redistribution of wealth if it is to be socially sustainable (Martinez-Alier et al., 2010; see Næss and Xue, 2016, on implications for housing policies). Obviously, strong social institutions and policies must be in place to manage the double challenge of reversing the growth trajectory of the economy while securing the quality of life for all members of society.

Beyond eco-modernisation

The dominant rationality of environmental planning has been ecological modernisation. This theory, or ideology, contains serious weaknesses that tend to go unquestioned. The inherent tension between environmental sustainability and economic growth is downplayed and neglected in this discourse. The prevailing assumption, backed by most politicians and by neoclassical environmental economists, is that this tension can be overcome through technological improvements and institutional reforms within the confines of the existing political-economic structure.

Despite its considerable environmental advantages compared to the land consuming and car dependent urban development characterising most of the post-World War II period, the compact city strategy, and ecological modernisation approaches generally, can be challenged in several ways (Davoudi, 2000; York and Rosa, 2003; Latour, 2015). For example, the reluctance to challenge consumer sovereignty implies that more eco-efficient and environmentally friendly solutions tend to be implemented alongside environmentally harmful solutions demanded by market segments less aware about environmental issues (Næss and Vogel, 2012). Moreover, environmental benefits from 'smart' solutions such as urban densification, improved public transport and road pricing tend to be counteracted by growth in the building stock, infrastructure, mobility and consumption levels in general.

Within the field of urban environmental planning, densification is increasingly considered as the most relevant eco-modernisation strategy. This strategy has several environmental merits compared to urban sprawl (outward urban expansion), such as reduced loss of natural areas and farmland, less car driving and lower material and energy consumption. However, even a strongly pursued eco-modernisation strategy of densification cannot entirely decouple growth in buildings and infrastructure from its negative environmental impacts. It is very difficult to entirely avoid conversion of unbuilt land if the volume of the building stock is to increase. The easiest opportunities for densification are gradually being used up. In addition, rebound effects from efficiency gains are likely to shift impacts to other places or sectors. Furthermore, full realisation of the technically possible potentials for eco-modernisation in environmental planning would require a degree of public coordination and control that may be difficult to implement.

The situation therefore calls for approaches to environmental planning beyond the eco-modernisation perspective. Several counter discourses attacking the prevailing eco-modernisation paradigm from different angles have emerged over the last decades, including the 'risk society' discourse (Beck, 1998), degrowth (Latouche, 2009; Kallis et al., 2015) and eco-socialism (Kovel, 2007; Xue, 2016). The authors think environmental planning would benefit from engaging more with each of these perspectives. It is, of course, better to apply environmentally friendly than environmentally harmful solutions, but there is a need to go beyond the eco-modernist focus on technological improvements if real sustainability is to be achieved. While eco-friendly technological innovation should be welcomed, we think eco-friendly and people-friendly social innovation is even more crucial.

Efforts to develop more environment and human-friendly cities should therefore not be limited to the creation of smarter building designs and urban planning solutions for a growing building stock. We need a new paradigm for environmental planning and urban sustainability that is not committed to (per capita) growth in production and consumption. From a sustainability perspective, we should give priority to qualitative improvements of the built environment instead of quantitative growth. Such limits to consumption must be accompanied with radical redistribution of wealth if they are to be socially sustainable.

References

Andersen, B. and Røe, P. G. (2016). 'The social context and politics of large scale urban architecture: investigating the design of Barcode, Oslo'. *European Urban and Regional Studies*, 24(3): 304–17.

Asafu-Adjaye, J., Blomqvist, L., Brand, S., Brook, B., DeFries, R., Ellis, E., Foreman, C., Keith, D., Lewis, M., Lynas, M., Nordhaus, T., Pielke, R., Pritzker, R., Roy, J., Sagoff, M., Shellenberger, M., Stone, R. and Teague, P. (2015). *An ecomodernist manifesto*. Available at: www.ecomodernism.org

Beatley, T. (2000). 'Preserving biodiversity: challenges for planners'. *Journal of the American Planning Association*, 66(1): 5–20.

Beck, U. (1998). *Democracy without enemies*. Cambridge: Polity Press.

Burchell, R. W., Shad, N. A., Listokin, D., Phillips, H., Downs, A., Seskin, S., Davis, J. S., Moore, T., Helton, D. and Gall, M. (1998). *The costs of sprawl – revisited*. Washington, DC: National Academy Press.

Dalen, E., Brekke, J. P. and Bakke, I. H. (2014). *Norsk monitor 2013–2014: Slik er ungdommen*. Oslo: Ipsos MMI.

Davoudi, S. (2000). 'Sustainability: a new vision for the British planning system'. *Planning Perspectives*, 15(2): 123–37.

Engebretsen, Ø. (1993). *Arealbruk i tettsteder 1955–1992*. TØI Report 177/1993. Oslo: Institute of Transport Economics.

Engelien, E., Steinnes, M. and Holst Bloch, V. V. (2005). *Tilgang til friluftsområder. Metode og resultater* (Notater 2005/15). Oslo: Statistics Norway.

Ewing, R. and Cervero, R. (2010). 'Travel and the built environment'. *Journal of the American Planning Association*, 76(3): 1–30.

Fairclough, N. (2006). *Language and globalization*. London/New York: Routledge.

Gordon, P. and Richardson, H. W. (1997). 'Are compact cities a desirable planning goal?' *Journal of the American Planning Association*, 63(1): 95–105.

Hartoft-Nielsen, P. (2001). *Arbejdspladslokalisering og transportadfærd*. Hørsholm: Forskningscenteret for skov og landskab.

Hille, J. (2013). *Norsk forbruk og miljøet*. Framtiden i våre hender rapport 5/2013. Oslo: Framtiden i våre hender.

Holden, E. (2007). *Achieving sustainable mobility: everyday and leisure-time travel in the EU*. Aldershot: Ashgate.

Holden, E. and Norland, I. T. (2005). 'Three challenges for the compact city as a sustainable urban form: household consumption of energy and transport in eight residential areas in the Greater Oslo region'. *Urban Studies*, 42(12): 2145–66.

Høyer, K. G. and Holden, E. (2001). 'Housing as basis for sustainable consumption'. *International Journal of Sustainable Development*, 4(1): 48–58.

Høyer, K. G. and Næss, P. (2001). 'The ecological traces of growth'. *Journal of Environmental Policy and Planning*, 3(3): 177–92.

Inglehart, R. (1995). 'Public support for environmental protection: objective problems and subjective values in 43 societies'. *Political Science and Politics*, 28(1): 57–71.

Jenks, M., Burton, E. and Williams, K. (eds.) (1996). *The compact city: a sustainable urban form?* London: E. and F. N. Spon.

Kallis, G., Demaria, F. and D'Alisa, G. (2015). 'Introduction: degrowth', in G. D'Alisa, F. Demaira and G. Kallis (eds.) *Degrowth. a vocabulary for a new era*. Oxon: Routledge, pp 1–17.

Kovel, J. (2007). *The enemy of nature: the end of capitalism or the end of the world?* London: Zed Books.

Latouche, S. (2009). *Farewell to growth*. Cambridge: Polity Press.

Latour, B. (2015). 'Fifty shades of green'. *Environmental Humanities*, 7(1): 219–25.

Lian, J. I. (2004). *Delvis brukerfinansiert utbygging av transportsystemet i Oslo og Akershus. Evaluering av Oslopakke 1 og 2*. TØI rapport 714. Oslo: Institute of Transport Economics.

Martinez-Alier, J., Pasqual, U., Vivien, F.-D. and Zaccai, E. (2010). 'Sustainable de-growth: mapping the context, criticisms and future prospects of an emergent paradigm'. *Ecological Economics*, 69: 1741–7.

Mol, A. P. J. and Spaargaren, G. (2000). 'Ecological modernization theory in debate: a review', in A. P. J Mol and D. Sonnenfeld (eds.) *Ecological modernization around the world*. London: Routledge, pp 17–49.

Næss, P. (2016). 'Urban planning: residential location and compensatory behaviour in three Scandinavian cities', in T. Santarius, H. J. Walnum and C. Aall (eds.) *Rethinking climate and energy policies: new perspectives on the rebound phenomenon*. Switzerland: Springer, pp 181–207.

Næss, P., Næss, T. and Strand, A. (2011). 'Oslo's farewell to urban sprawl'. *European Planning Studies*, 19(1): 113–37.

Næss, P. and Sandberg, S. L. (1996). 'Workplace location, modal split and energy use for commuting trips'. *Urban Studies*, 33(3): 557–80.

Næss, P., Strand, A., Wolday, F. and Stefansdottir, H. (2019). 'Residential location, commuting and non-work travel in two urban areas of different size and with different center structures'. *Progress in Planning*, 128, pp. 1–36.

Næss, P. and Vogel, N. (2012). 'Sustainable urban development and the multi-level transition perspective'. *Environmental Innovation and Societal Transitions*, 4: 36–50.

Næss, P. and Xue, J. (2016). 'Housing standards, environmental sustainability and social welfare', in P. Næss and L. Price (eds.) *Crisis system: a critical realist and environmental critique of economics and the economy*. London/New York: Routledge, pp 130–48.

Newman, P. W. G. and Kenworthy, J. R. (1999). *Sustainability and cities. Overcoming automobile dependence*. Washington, DC/Covelo, California: Island Press.

Nordic Council of Ministers (1999). *Factors 4 and 10 in the Nordic countries: the transport sector, the forest sector, the building and real estate sector, the food supply chain*. Copenhagen: Nordic Council of Ministers.

Riksrevisjonen (2007). *Riksrevisjonens undersøkelse av bærekraftig arealplanlegging og arealdisponering I Norge* (Dokument nr. 3: 11 (2006–7)). Oslo: Norwegian National Auditing Office.

Seippel, Ø. (2000). 'Ecological modernization as a theoretical device: strengths and weaknesses'. *Journal of Environmental Policy and Planning*, 2(4): 287–302.

Statistics Norway (2018). *Areal og befolkning i tettsteder, etter tettsted, statistikkvariabel og år*. Available at: www.ssb.no/statbank/table/04859/?rxid=5a5561ea-67b0-4bed-937b-fbb177ba3b82

Stern, D. I. (2004). 'The rise and fall of the environmental Kuznets curve'. *World Development*, 32(8): 1419–39.

Voss, K. and Musall, E. (2012). *Net zero energy buildings – international projects of carbon neutrality in buildings* (2nd edn). Munich: Institut für internationale Architektur-Dokumentation GmbH & Co.

Xue, J. (2014). *Economic growth and sustainable housing: an uneasy relationship*. London/New York: Routledge.

Xue, J. (2016). 'China at the crossroad: ecological modernization or ecosocialism?' in P. Næss and L. Price (eds.) *Crisis system: a critical realist and environmental critique of economics and the economy*. London/New York: Routledge, pp 192–208.

York, R. and Rosa, E. (2003). 'Key challenges to ecological modernization theory'. *Organization and Environment*, 16(3): 273–88.

Zegras, C. (2010). 'The built environment and motor vehicle ownership and use: evidence from Santiago de Chile'. *Urban Studies*, 47(8): 1793–1817.

6

Anthropocene

The challenge for environmental planning

Mark Whitehead

Introduction: planning in the 'humanised Earth'

The Anthropocene is one of the key 'buzz concepts' of the early 21st century, except that its status as a *concept* is somewhat uncertain. Strictly speaking, the Anthropocene is a formally proposed but, as yet, unformalized geological epoch. The concept was famously proposed by Paul Crutzen and Eugene Stoermer in 2000 (Crutzen and Stoermer, 2000). As a kind of fin de siècle statement on the state of human-environmental relations, the notion of the Anthropocene suggests that our collective impact on the global environment necessitates a new designation in geological time. The bases on which the Anthropocene could be formally designated are currently being investigated by the International Commission on Stratigraphy (ICS). It is already clear that this designation process is going to be hotly contested as it raises significant questions for the scientific assumptions and working practices of stratigraphy. But beyond its geological foundations, as a concept, the Anthropocene is already beginning to catch the intellectual and, to a lesser extent, public imagination. The Anthropocene signals the significant problems that human-induced environmental change may generate in the near-future. It also suggests new levels of responsibility among the human race for managing environmental processes and systems.

This chapter has two primary goals. The first is to consider the varied implications of the Anthropocene for how we think about, and approach, late-modern human-environmental relations. The second is to consider the particular challenges that the Anthropocene presents the discipline and profession of environmental planning with. In exploring these issues, it is important to note the high levels of uncertainty that surround the Anthropocene. Precisely how one approaches these two issues may thus depend on the particular way in which one understands the Anthropocene. Given these uncertainties, we must also acknowledge the possibility that the Anthropocene may, in the long-term, only have very limited implications for environmental planning. With these caveats in mind, let us begin our exploration of the geological epoch of humankind.

What is the Anthropocene: planning for planetary boundaries

Put most simply, the Anthropocene suggests that we may have entered a new period in the history of the Earth. The geological epoch that we are currently in at the time of writing this chapter (at least in official geological terms) is known as the Holocene. The Holocene is a so-called interglacial epoch that incorporates, approximately, the last 11,700 years of history – which includes all written history and the emergence of all of the world's major civilizations. To suggest that this geological epoch may be at an end is thus no insignificant matter. The Holocene has been characterised by a relatively stable (in geological contexts) set of environmental conditions, which have been conducive to the emergence of modern agriculture and the subsequent onset of urbanisation. The scientific case for the end of the Holocene and the commencement of the Anthropocene is made by Zalasiewicz et al. (2008, p. 4) in the following statement:

> [S]ince the start of the Industrial Revolution, Earth has endured changes sufficient to leave a global stratigraphic signature distinct from that of the Holocene or of previous Pleistocene interglacial phases, encompassing novel biotic, sedimentary, and geochemical changes. These changes, although likely only in their initial phases, are sufficiently distinct and robustly established for suggestions of the Holocene-Anthropocene boundary in the recent historical past to be geologically reasonable.

The idea of the Anthropocene is thus predicated on the notion that human interventions in the biological, geophysical and chemical composition of the Earth, and the environmental systems with which it is associated, reflect a geologically significant destabilisation of the Holocene.

The forms of human impacts on the global environmental systems that are associated with the Anthropocene take different forms. They can be seen, for example, in *changes to physical sedimentation* (related to agriculture and dam building projects); *carbon cycle perturbation and temperature* (expressed most dramatically in relation to climate change); *biotic change* (evidenced in the assertion that humans are currently driving the Earth's sixth mass extinction event); *ocean changes* (particularly sea level change and acidification); and the agricultural forcing of the nitrogen cycle (Zalasiewicz et al., 2008, pp. 5–6). Calls for the designation of the Anthropocene epoch are thus based upon the significance of cumulative human transformations of Earth and the impacts that these cumulative forms of change are having on the operation of planetary systems (Whitehead, 2014).

Despite the accumulation of evidence supporting the human transformation of the global environment, controversies remain about the designation of a formal Anthropocene era. These controversies take different forms. Some are *technical* and relate to the difficulties of effectively demarcating a clear temporal boundary line between the Holocene and the Anthropocene. Currently, there are a series of candidates for the starting point of Anthropocene including the emergence of sedentary agriculture; the industrial revolution of the 18th and 19th centuries; and the so-called *great acceleration* of socio-economic development following the end of the Second World War (Castree, 2014). Other controversies are more *existential*, as scientists question the extent to which human intervention in the environment compares with geological processes of the past and future, and whether the Anthropocene should be designated at all (Scourse, 2016).

In light of these controversies, it is, perhaps, of little surprise to find out that the Anthropocene has had a mixed reception in the academic world. In relation to the physical sciences, there are those who support its adaptation and others who are vehemently opposed to the Anthropocene on both technical and empirical grounds (for an excellent review of these debates and divisions see Castree, 2014). In terms of the social sciences and humanities, the notion of the Anthropocene has been generally welcomed on the basis that it provides a context within which theories of human action can be brought into conversation with the natural sciences in order to address the environmental challenges of the 21st century (Whitehead, 2014). Notwithstanding this, there has also been resistance to the term within the social sciences and humanities. Some academics have questioned whether the concept really adds anything substantively new to how we understand human-environmental relations and problematise the often generic and simplistic way in which the figure of the 'the human' is understood within accounts of the Anthropocene (Johnson et al., 2014). In an attempt to move beyond some of the scientific uncertainties and controversies that surround the Anthropocene, some writers have argued that the value of the Anthropocene as a concept is partially independent from its formal acceptance within geology. As Castree (2014, p. 441) states:

> [I]t is possible for the Anthropocene to remain an informal concept in geology even as it becomes a normal part of the vocabulary of many environmental scientists . . . the Anthropocene concept can be seen as a new, more graphic way to frame an existing idea, namely that of 'global environmental change' caused by human activities and extending in causes and effects *beyond* climate change.

In this chapter, the Anthropocene is a presented as a multifaceted concept with, at least, four sets of meanings. First, it can be understood in ontological terms as a factual statement of the significant extent of the (human) altered Earth we now live in (including, but not limited to, climate change). Second, the Anthropocene offers an epistemological framework within which it is increasingly necessary to combine knowledge of changing physical sciences, with insights into the human condition and social processes. The idea of the Anthropocene thus suggests it is arbitrary and unhelpful to try and separate out the natural and social sciences. Third, the Anthropocene presents a series of scientific challenges (to geologists, geographers, and environmental sciences) to develop frameworks for the measurement, recording, and designation of human impacts on the Earth system. Fourth, and finally, the Anthropocene can be seen as a moral imperative for humanity to take responsibility for the destabilising impacts that humans are having on the Earth. In this chapter, we focus primarily on the ontological and moral forms of the Anthropocene, as these have most valence for environmental planning.

Before considering the ontological and moral implications of the Anthropocene for environmental planning, it is necessary to introduce a concept with which it has become synonymous. The idea of *planetary boundaries* is deeply interconnected with the scientific and more philosophical discussions of the Anthropocene. The notion of planetary boundaries was proposed in 2009 as a way of unpacking the human impacts on global environmental systems that are associated with the Anthropocene, as well as assessing the scientific basis for the Anthropocene itself. The planetary boundaries model emerged primarily out of environmental science and seeks to convey the upper limits of human intervention within the global environment in nine key areas (Rockström et al., 2009a, 2009b). These nine planetary boundary areas are: biosphere integrity, land use change, freshwater use, biochemical flows (particular nitrogen and phosphorous), ocean acidification, atmospheric aerosol loading, stratospheric ozone depletion, climate change and novel entities (Stockholm Resilience Centre, 2017). These planetary boundary areas are now used by environmental scientists (if not geologists) to determine the extent to which the human impact on the global environment constitutions is a shift out of the Holocene into a new geological epoch. Part of the planetary boundaries model's purpose is to represent, across the nine category areas, the extent to which we are still in the 'safe operating space' of the Holocene, or at risk of slipping into new, Anthropocenic operating spaces. According to Steffen et al. (2015), we are currently witness to high-risk levels of human-induced environmental change in relation to the biosphere integrity (with particular relation to genetic diversity) and biochemical flows and experiencing increasing risk in relation to climate change and land use change. The planetary boundaries framework is significant in relation to environmental planning for two main reasons. First, because it seeks to represent the complex and interconnecting forms of environmental change that are associated with the Anthropocene. Second, in defining the notion of safe and unsafe operating spaces, it provides environmental planners with targets that their policies should be geared towards. As Castree (2014) observes, the idea of planetary boundaries is where the Anthropocene becomes less about politically neutral scientific classification, and more about the designation of normative routes for future human planning and action.

Environmental planning in the Anthropocene

This section outlines the key potential implications of the Anthropocene for environmental planning and planners. In doing this, it is important to recognise that while the Anthropocene has possible consequences for how we might think about and execute environmental planning; it is also clear that environmental planning can play an important role in how we collectively live in, and respond to, the Anthropocene era. This section will consider both of these issues in turn.

The implications of the Anthropocene to environmental planning

At, perhaps, the most basic of levels, the very notion of the Anthropocene suggests an elevated role for environmental planning in local, regional and global environmental affairs. In many ways, the Anthropocene is defined by *unintentional* human action. As a species, humans did not come to a collective decision to produce a new geological epoch and to test the ecological boundaries that define the Holocene. The Anthropocene is the product of a complex set of socio-economic, political, technical and cultural forces that have come together to produce a form of unintended global experiment (Castree, 2014). The identification of the Anthropocene does, however, appear to have initiated a collective desire (at least with political and scientific elites) for more intentional forms of intervention, management and stewardship within the environmental affairs (Whitehead, 2014). We will discuss the different *ecocentric* and *technocentric* destinations that such forms of intentional interventions within the Earth's history could produce in the section that follows. But at this point, it is important to note that recognition of the global, and potentially geologically significant impacts of humans on the environment, has been associated with a realisation that humans have both a responsibility and capacity to more deliberately shape our collective environmental futures. In his influential 2002 account of the Anthropocene, the 'Geology of Mankind', Paul Crutzen (2002, p. 23) reflected:

> Unless there is a global catastrophe – a meteorite impact, a world war or a pandemic – mankind will remain a major environmental force for many millennia. A daunting task lies ahead for scientists and engineers to guide society towards environmentally sustainable management during the era of the Anthropocene.

This chapter understands environmental planning – at an admittedly simply level – as a collection of legal, political and technical processes through which decisions over land use are made to balance the varied needs of nature and society, in order to secure long-term sustainability. While Crutzen does not mention environmental planners specifically (instead focusing his prose on scientists and engineers), it is clear that its frameworks and processes are going to be central to any attempts to orchestrate collective intentional responses to humans' increasing profound power over the environment.

Beyond the broad call of environmental planning to facilitate collective action related to the Anthropocene, the planetary boundaries associated with this potentially new geological epoch have varied levels of significance to the planning community. It is, for example, clear that environmental planning will have a pivotal role in producing safe operating spaces for biosphere integrity and land use change. In relation to the climate change planetary boundary, it is also evident that environmental planning has a key role to play in stabilising average global temperatures (particularly in relation to preserving carbon sinks and managing energy intensive patterns of suburban development). Significantly, however, it is also apparent that the notion of the Anthropocene may force environmental planning to broaden its focus from that which has taken on climate change over the last decade (as the primary target of planning regimes, see While, 2008) to recognise the complex set of environmental challenges the planet is now facing. It is likely that environmental planning (broadly defined) will also have an important role to play in managing freshwater use and biochemical flows but more limited responsibility when it comes to questions of ocean acidification and stratospheric ozone depletion.

In addition to forcing environmental policy makers to consider the connections between planning regulations and directives and planetary boundaries, it is also clear that addressing the Anthropocene necessitates planners to consider the complex connections that exist between different emerging environmental challenges. It would, after all, be foolish to establish a safe operating space in relation to climate change only to undermine the Earth's biosphere integrity. It is clear that there are many positive feedback loops to be exploited in relation to managing different planetary boundary areas: preserving carbon sinks, for example, could easily support in-situ ecological and biological diversity and integrity, while managing freshwater use could support desired land use practices. The issue is that in order to create safe operating spaces in the shadow of the Anthropocene, environmental planning must embrace integrated thinking and forecasting and an appreciation of the complex ways in which action in one area of the Earth's system impacts on others. Of course, the extent to which environmental planners can be reasonably expected to monitor such complex systems and thresholds is an open question.

All of these raise the question of the extent to which environmental planning in/for the Anthropocene is distinct from the established practices of sustainable development planning. While both approaches to environmental planning are clearly related – particularly in the context of their emphases on integrated thinking and a concern for socio-environmental systems – they have important lines of distinction. While sustainable development planning tends to assume various win-win-win scenarios for social, environmental and economic needs (Whitehead, 2006), planning in/for the Anthropocene would appear to require preparing for much harder environmental boundaries, which will often involve uncomfortable trade-offs between the needs of the Earth system and those of the social economy.

In addition to a tighter focus on planetary boundaries, environmental planning in/for the Anthropocene should also involve planning practices becoming ever more future oriented. The future orientation of Anthropocene thinking is emphasised by Castree (2014, p. 444) when he observes that related actions should be 'powerfully forward-facing: they invite us to consider making significant present-day decisions in light of their (non-trivial) effects long into the future'. While environmental planning has always, to a large extent, been future oriented (particularly with its concern with the environmental needs of future generations), the Anthropocene suggests that the temporal orientation of the planner must transcend the near-future. Environmental planning in/for the Anthropocene is in part about aligning planning framework with geological timescales and thinking about the longer-term impacts of environmental decisions on the preservation, or otherwise, of safe planetary operating spaces.

A final implication of the Anthropocene for environmental planning concerns the extent to which the 'achieved' Anthropocene epoch is established within or outside safe operating limits. If the former is the case, environmental planning will have an important role in ensuring that Anthropocene is maintained as a sustainable context for life on Earth. If the latter is the case, however, environmental planning will be confronted with two main challenges. The first is how to facilitate the return of the Earth system to a safe operating space. This is likely to involve much greater emphasis being placed on regenerative techniques of environmental improvement, as opposed to merely sustaining degraded environmental support systems (Wahl, 2016). The second is how to plan with a changed and increasingly uncertain global environmental system. In this context, environmental planning is likely to have to employ a kind of super charged precautionary principle in order to deal with the heightened levels of environmental instability and risk that an extreme version of the Anthropocene could produce.

The role of environmental planning in the Anthropocene

While the idea and realities of the Anthropocene clearly has implications for how we might think about and practice environmental planning, it is also clear that planning has significance for the Anthropocene. There have been a series of social and political criticisms of the Anthropocene idea that are relevant to environmental planning (Johnson et al., 2014; Yusoff, 2013, 2017). Some critiques have focused on how accounts of the Anthropocene can be overly totalising in its accounts of our human future. For some, the scientific accounts of the Anthropocene tend to too quickly assert a sense of collective responsibility for the production of this new geological epoch and to assume fairly homogenous impacts of new planetary operating spaces (Whitehead, 2014). These generalising tendencies are clearly linked to the geological framing of the Anthropocene as something that must be present in stratigraphically consistent forms around the world. But what such accounts tend to overlook (if only inadvertently) are the *rough geographies* of the Anthropocene (Whitehead, 2014). These rough geographies suggest differential forms of responsibility for the Anthropocene (particularly among more economically developed nations and regions), and the likely uneven impacts of this epoch (with risk and vulnerability being concentrated in less economically developed areas). As different national, regional and local planning authorities respond to the Anthropocene, they have a crucial role in working through the differing levels of responsibility for and susceptibility to a less certain environmental future.

Other critiques of Anthropocene have noticed a tendency for scientific elites, who have assumed the role of isolating and designating a putative Anthropocene epoch, to adopt a privileged position in the development and recommendation of policy responses to it. This fails to recognise the political questions and inevitable trade-offs that responding to the Anthropocene will involve. Nordhaus et al. (2012, p. 37) observe that:

> These costs and benefits [that are associated with addressing various planetary boundary issues] are unevenly distributed temporally, spatially, and socially – there are both winners and losers. Balancing these trade-offs is an inherently political question, and attempts to depoliticize it with reference to scientific authority is dangerous, as it precludes democratic resolution of these debates, and limits, rather than expands, the range of available choices and opportunities.

Environmental planning is one democratically constituted forum for public action within which the necessary trade-offs associated with the Anthropocene can be raised and contested. Environmental planning systems at transnational, national and more local levels are thus one of several key arenas where different approaches to Anthropocene management can be debated and tested. It is perhaps easy in this context to overstate the democratic credentials of the planning systems (Minton, 2012). Notwithstanding this, in the absence of an unlikely global system of

Anthropocenic supra-governance, it is clear that the formation of Anthropocene-oriented planning regimes, at varied scales, will provide important opportunities for geographically diverse, indigenous preferences and priorities to be expressed and implemented. In this context, environmental planning regimes are likely to offer important practical contexts within which to re-democratise the Anthropocene and recognise various inexpert and more-than-scientific perspectives on how societies should respond to its demands.

Technocentric and *ecocentric* planning responses to the Anthropocene

Understanding the implications of the Anthropocene for environmental planning involves an appreciation of the different political, economic and technological responses which could be developed in response to emerging global environmental challenges. As we have already intimated in this chapter, there is as yet no agreed upon response to the Anthropocene beyond the realisation that humans have the power and responsibility to minimise environmental uncertainty and insecurity in the future. There are presently two broad schools of thought concerning how to respond to the threats of the Anthropocene era. These approaches are best captured in the phrases *technocentric* and *ecocentric*. As we will see technocentric and ecocentric responses propose radically different strategies for addressing the challenges of the Anthropocene, which in turn have very different implications for environmental planning systems.

The technocentric Anthropocene

Technocentrism is best conceived of as a value system which supports the use of technological innovation to deal with the challenges of environmental planning and management (Pepper, 1996). While technocentric solutions to environmental problems have shaped local environmentalism for centuries, in the Anthropocene they tend to be associated with much grander scales of technological intervention. In the context of the Anthropocene, technocentric solutions to environmental problems do not only involve the search for small scale technological fixes (such as air pollution filters, or smart energy meters) but the technological manipulation of global environmental systems. The atmospheric scientist Paul Crutzen (2002, p. 23) captured the essence of technocentric responses to the Anthropocene when he observed: 'This will require appropriate human behaviour at all scales, and may well involve internationally accepted, large-scale geo-engineering projects, for instance to 'optimize' climate. At this stage, however, we are still largely treading on *terra incognita*'.

The geoengineering projects endorsed by Crutzen involve 'large-scale interventions in the Earth's natural systems to counteract climate change' (Oxford Geoengineering Programme, 2017). Geoengineering programmes that would have implications for environmental planning include attempts to increase the reflectiveness of the Earth's surface and carbon dioxide removal schemes. If planning systems are to facilitate geoengineering programmes, it is clear that they are also going to be endorsing a belief that part of the solution to human induced environmental change is more human induced technological intervention in the Earth system. This is, of course, the point at which technocentric approaches to the Anthropocene stop being purely technological and take on a normative inflection. Not all large-scale geoengineering programmes need to be technocentric of course. Large-scale afforestation schemes could help to address a series of the challenges associated with the Anthropocene (including climate change, biosphere integrity, and unsustainable land use change) but are clearly more ecocentric in form (see next section).

In addition to mitigating the worst effects of the Anthropocene, technocentrism could also inform the ways in which communities seek to protect themselves in this new epoch. From facilitating flood protection technologies and zoning techniques to the building of more extensive and elaborate sea defences, again environmental planning could play a crucial role in a technocentric future. Even within a technologically oriented Anthropocene, it is unlikely that humanity will be able to afford technocentric solutions to all of our ecological crises. In this context, environmental planning will be a key arena within which decisions about where the mitigating technologies of Anthropocene will be developed and where they will be withheld will be made.

The ecocentric Anthropocene

In contrast to technocentrism, ecocentrism is based upon a nature-oriented worldview, which does not place the needs of humans above those of the nonhuman natural world and looks for nature-based solutions to ecological and social problems (Pepper, 1996). Unlike technocentric responses to the Anthropocene, ecocentric approaches promote the use of adaptive responses to environmental change, which promote transformations in the way in which social, economic and political systems operate (Whitehead, 2014). Ecocentric responses to the Anthropocene thus support strategies that would see reductions in the human impact on the global environment such that the forms of ecological stability associated with the Holocene could be generated again. In this sense, ecocentric planning for the Anthropocene is not about managing the changing environmental dynamics of the Anthropocene but facilitating a return to the Holocene. While the intentions of ecocentrism are clear, the precise form that related policies could take is an open question. In contrast to the large-scale ambitions of technocentrism, it is likely that ecocentric planning would support the rescaling of human life around more local, ecologically determined geographical areas such as eco- and bio-regions. Related planning initiatives would look for more nature-based solutions to environmental planning challenges: perhaps supporting the preservation and extension of wetlands to support flood prevention and coastal inundation rather than simply developing more elaborate draining and flood defence systems. Perhaps the most significant aspect of ecocentric environmental planning would be a commitment to regenerative development. Regenerative development involves not only trying to minimise the local and global environmental impacts of new developments but to ensure that such developments contribute measurable improvements within the environmental systems within which they operate (Wahl, 2016). Regenerative planning embodies most clearly the ecocentric desire for a return to the Holocene through a multitude of micro and macroscales ecological fixes to the global environment system.

Ecocentric planning for the Anthropocene can take different forms. What unites each of these initiatives is an increasing role for environmental planning with our day-to-day lives. Planning for a speedy return to the Holocene (if this is indeed possible) will clearly involve more proactive forms of environmental planning and generally more interventionist planning regimes.

Conclusion

As this chapter has demonstrated, there is much debate about the nature and standing of the Anthropocene, and how humans should be responding to the challenges it presents. What is clear within these discussions is that the Anthropocene has tangible implications for environmental planning. Put simply, it requires that environmental planning must be able to think and act over much greater temporal and spatial scales than it has ever done before. It also suggests that environmental planning must be more integrative in its appreciation of the complex ways in which different environmental systems connect. Whether human responses to the Anthropocene involve more technocentric or ecocentric strategies (or, most likely, a mix of the two), adapted environmental planning systems will play a crucial role in facilitating collective human action. Although the Anthropocene challenges the work of environmental planning, planning regimes will also challenge emerging debates and actions in the Anthropocene. The democratic constitution of environmental planning (at least within liberal states) means that it will play a crucial role in facilitating the incorporation of more local, indigenous and inexpert responses to environmental change in the Anthropocene. It is also likely that large scale environmental planning will play a significant role in developing just responses to the uneven systems of risk and responsibility that the Anthropocene is likely to produce.

There is clearly much work to be done before the Anthropocene is ever formally accepted as a stratigraphic marker of geological time. In the meantime, however, it appears that environmental planning will have to respond to the emerging real-time effects of the Anthropocene and endeavour to produce a safe, secure and just operating space for humanity.

References

Castree, N. (2014). 'The Anthropocene I: the back story'. *Geography Compass*, 8(7): 436–49.

Crutzen, P. (2002). 'Geology of mankind'. *Science*, 415: 23.

Crutzen, P. J. and Stoermer, E. F. (2000). 'The "Anthropocene"'. *Global Change Newsletter*, 41: 17–18.

Johnson, E., Morehouse, H., Dalby, S., Lehman, J., Nelson, S., Rowan, R., Wakefield, S. and Yussof, K. (2014). 'After the Anthropocene: politics and geographic inquiry for a new epoch'. *Progress in Human Geography*, 38(3): 439–56.

Minton, A. (2012). *Ground control: rear and happiness in the twenty first century city*. London: Penguin.

Nordhaus, T., Shellenberger, M. and Blomqvist, L. (2012). *The planetary boundaries hypothesis: a review of the evidence*. Oakland: Breakthrough Institute, pp 1–42.

Oxford Geoengineering Programme (2017). *What is geoengineering?* Available at: www.geoengineering.ox.ac.uk

Pepper, D. (1996). *Modern environmentalism: an introduction*. London: Routledge.

Rockström, J., Steffen, W., Noone, K., Persson, A., Chapin III, F. S., Lambin, E. F., Lenton, T. M., Scheffer, M., Folke, C., Schellnhuber, H. J., Nykvist, B., De Wit, C. A., Hughes, T., van der Leeuw, S., Rodhe, H., Sörlin, S., Snyder, P. K., Costanza, R., Svedin, U., Falkenmark, M., Karlberg, L., Corell, R. W., Fabry, V. J., Hansen, J., Walker, B., Liverman, D., Richardson, K., Crutzen, P. and Foley, J. A. (2009a). 'A safe operating space for humanity'. *Nature*, 461(7263): 472–5.

Rockström, J., Steffen, W., Noone, K., Persson, Å., Chapin, III, F. S., Lambin, E., Lenton, T. M., Scheffer, M., Folke, C., Schellnhuber, H., Nykvist, B., De Wit, C. A., Hughes, T., van der Leeuw, S., Rodhe, H., Sörlin, S., Snyder, P. K., Costanza, R., Svedin, U., Falkenmark, M., Karlberg, L., Corell, R. W., Fabry, V. J., Hansen, J., Walker, B., Liverman, D., Richardson, K., Crutzen, P. and Foley, J. A. (2009b). 'Planetary boundaries: exploring the safe operating space for humanity'. *Ecology and Society*, 14(2): 32 [Online]. Available at: www.ecologyandsociety.org/vol14/iss2/art32/

Scourse, J. (2016). 'Enough Anthropocene nonsense: we already know the planet is in crisis'. *The Conversation*. Available at: http://theconversation.com/enough-anthropocene-nonsense-we-already-know-the-world-is-in-crisis-43082

Steffen, W., Richardson, K., Rockström, J., Cornell, S. E., Fetzer, I., Bennett, E. M., Biggs, R., Carpenter, S. R., de Vries, W., de Wit, C. A., Folke, C., Gerten, D., Heinke, J., Mace, G. M., Persson, L. M., Ramanathan, V., Reyers, B. and Sörlin, S. (2015). 'Planetary boundaries: guiding human development on a changing planet'. *Science*, 347(6223): 1–10.

Stockholm Resilience Centre (2017). *The nine planetary boundaries*. Available at: www.stockholmresilience.org/research/planetary-boundaries/planetary-boundaries/about-the-research/the-nine-planetary-boundaries.html

Wahl, C. (2016). *Designing regenerative cultures*. Axminster, UK: Triarchy Press.

While, A. (2008). 'Climate change and planning: carbon control and spatial regulation'. *Town Planning Review*, 79(1): vii–xiii.

Whitehead, M. (2006). *Spaces of sustainability: geographical perspectives in the sustainable society*. Abingdon: Routledge.

Whitehead, M. (2014). *Environmental transformation: a geography of the Anthropocene*. Abingdon: Routledge.

Yusoff, K. (2013). 'Geologic life: prehistory, climate, futures in the Anthropocene'. *Environment and Planning D: Society and Space*, 31(5): 779–95.

Yusoff, K. (2017). 'Geosocial strata'. *Theory, Culture and Society*, 34(2–3): 105–27.

Zalasiewicz, J., Williams, M., Smith, A., Barry, T. L., Coe, A. L., Bown, P. R., Brenchley, P., Cantrill, D., Gale, A., Gibbard, P., Gregory, F. J., Hounslow, M. W., Kerr, A. C., Pearson, P., Knox, R., Powell, J., Waters, C., Marshall, J., Oates, M., Rawson, P. and Stone, P. (2008). 'Are we now living in the Anthropocene?' *GSA Today*, 18(2): 4–8.

7

Understanding risk in environmental planning

Eoin O'Neill

Introduction

Risk is a feature of everyday life. Whether it is facing the hazards of daily traffic, or a common workplace hazard, people face risks from a variety of sources every day. Arising from advances in science and communications, there is now growing risk awareness across society (Aven and Kristensen, 2005). With advanced economies having transitioned away from an 'industrial society' towards what is suggested to be a 'risk society' (Beck, 1992), people have shifted from just being concerned about governance of risk to increasingly being governed by risk itself (Rothstein et al., 2006).

According to Beck, 'modern society has become a risk society in the sense that it is increasingly occupied with debating, preventing and managing risks that it itself has produced' (Beck, 2006, p. 332). Yet, modernity's claim to control risks is frequently undermined by natural forces. Rather than problems being created by nature, nature is being socially constructed as a risk in and of itself, with a societal focus on policing it and safeguarding itself. This has led to heightened fear of risks and greater demands for security (Davoudi, 2014).

While people are more inclined to accept common or familiar risks day to day, other risks that may be less familiar or which people believe are under control generate more concern. This can include people's exposure to, perhaps less frequent, natural hazards like flooding, wildfire and earthquakes or to technological hazards resulting from human activities, such as industrial emissions, nuclear accidents or infrastructure failure.

To enable people to function and manage their daily lives in the face of common risks, people employ rules-of-thumb, known as heuristics. For more harmful and less familiar risks that have potential to cause significant damage on people or property, more complex approaches to the analysis of risks have been developed. Whilst risks cannot be entirely eliminated, they can be assessed to aid decision making and the implementation of a risk management approach, with resources prioritised accordingly. This includes the analysis of environmental risks many of which have spatial characteristics and hence of particular interest to planners.

Whereas conservation or preservation provided early environmental motivations for planners, increasingly risk is becoming central to their decision making. With risk-based approaches established in environmental regulation, the spatial dimension of many environmental risks mean that a risk-based regulatory approach is also now being incorporated into planning to evaluate and manage risks. This is not a simple task. Deciding upon the acceptability and tolerability of various types and scales of environmental risks requires peoples' input and understanding of how people perceive risks and respond to them. These are crucial for implementation of risk management policies and programmes.

This chapter explores some of the concepts that lie at the interface of risk and environmental planning. It will discuss the meaning of 'risk' and risk management and explore the role of planning in assisting the management of environmental risks before considering implementation challenges.

The meanings of risk

People refer to risk routinely in conversation but the concept conveys multiple meanings and how it is understood varies among people. What is clear is that there is no universal definition of risk and how it is used in daily speech by people can be quite different to that of an expert in a technical or professional setting, including planners. People often use qualitative judgements and may over- or under-estimate risk (O'Neill et al., 2016) when compared with expert assessments (Siegrist and Gutscher, 2006).

A comprehensive review of the historical origins of 'risk' by Aven (2012) found that it has frequently included reference to danger, hazard or harm, or the possibility of damage. In some respects, 'risk' can also be associated with positive results, but in the course of everyday language people usually speak about negative outcomes (Althaus, 2005). Probability calculations are also not explicitly integral to the 'colloquial use' of the term risk (Althaus, 2005) whereas in the scientific field, risk usually includes reference to considerations such as, likelihood, probability, chance, uncertainty or expected value (Aven, 2012). Alternative definitions frequently incorporate the concepts of vulnerability, exposure and hazard. These three concepts are mutually dependent and together represent another construct of risk (O'Neill et al., 2016).

White (2010) provides a detailed account of a number of principal risk constructs. For example, risk can be defined as the function of hazard and vulnerability. In this context, hazard represents the potential for damage to be caused, and vulnerability represents the susceptibility of people or the local environment to the threat of a hazard. The degree of loss or harm varies by a community's capacity to cope and that capacity is influenced by prevailing social and economic conditions (Birkmann et al., 2013). Alternatively, risk can be defined as the combination of hazard, vulnerability and exposure (Jóhannesdóttir and Gísladóttir, 2010). Exposure is the added component for this risk construct. Here vulnerable subjects are exposed to the threat of a hazard. Then again, risk can be defined as comprising the product of probability and consequences (O'Neill et al., 2016). This latter construct of risk has been frequently used in planning practices such as statutory planning guidance on development and flooding issued by the British and Irish governments.

Given the range of risk constructs employed, it is not surprising that individual perspectives on risk are also influenced by professional or disciplinary backgrounds (Renn, 1992). For example, those from an engineering or natural science background tend to view risk as a calculable and objective reality. On the other hand, social scientists are interested in evaluation of the subjective aspects (Bradbury, 1989). Economists tend to take a decision-oriented perspective (Althaus, 2005; Aven and Kristensen, 2005). Other social scientists view risk as comprising cultural, social or behavioural constructions. Social science research on 'risk perception' often focuses on qualitative aspects of how people assess and interpret risk.

Risk perception

The multiple perspectives on risk is in part due to the fact that people perceive risk in different ways. In addition to the influence of socio-economic factors (for example, age, education, gender, homeownership, etc), how people perceive and act upon risk tends to occur in at least two ways: first, through affective responses such as worry, dread or outrage (*risk as feeling*) and second, through rational and purposeful responses (*risk as analysis*) (Slovic and Peters, 2006). Whilst the most important singular factor influencing risk perception is usually prior experience, cognitive factors such as risk attitude or perceived control also matter. More generally, reviews of psychometric studies have highlighted a myriad of factors that frequently influence people's concern for risks (Slovic, 1987) with prominent factors identified by Gibson et al. (2012) and summarised in Table 7.1.

Beyond general cognitive and socio-economic factors, it has also been shown that geographical factors, across a range of real and perceived measures of proximity to individual hazard sources, influence people's perception of their risk. This has been shown in cases of, for example, elevation relative to floodwater levels (Botzen et al., 2009) and distance to the flood hazard source (Zhang et al., 2010); real distance to a hurricane wind hazard zone (Peacock et al., 2005); perceived distance to nuclear power plant (Giordano et al., 2010); and perceived distance to a flood hazard zone (O'Neill, 2016). It is evident that 'risk' has a spatial dimension and this can be manifested in a number of ways:

- A risk can be geographically located and distributed over space;
- People's perception of a risk's location or distribution can have a spatial dimension;
- How risk is managed at different scales has spatial dimensions.

This is of direct relevance to planning. Planners influence the design of, and our interaction with, the built environment and potential hazards resulting from that. Table 7.2 provides a

Table 7.1 Factors shown to influence risk perception

Factor	*Low risk perception factors*	*High risk perception factors*
Benefits	High benefits	Low benefits
Exposure	Voluntary	Involuntary
Type of risk	Chronic – kills one at a time; persistent	Catastrophic – kills large numbers at once
Familiarity	Old risk	New risk
Catastrophic potential	Common risk - learnt to live with	Dread - evokes emotional fear
Visibility	Visibility	Invisibility
Individual control	Possible	Not possible
Origin	Natural	Man-made
Knowledge	Known to those exposed	Not known to those exposed
Uncertainty	Known to science	Not known to science
Manifestation	Immediate or reversible damage	Delayed or irreversible damage
Damage	Not fatal	Fatal
Distribution of damage	Equitably distributed	Not equitably distributed
Damage visibility	Anonymous victims	Victims identifiable
Victims	Adult males	Children and women
Social or scientific status	Consensus possible	Controversial

Source: Gibson et al. (2012, p. 10)

Table 7.2 Frequent European hazards

Natural hazards	*Man-made hazards*
Flooding	Industrial accidents
Extreme weather	Radiological incidents
Wildfire/forest fire	Infrastructure disruption
Seismic and volcanic activity	Cyber security
Pandemics	Terrorism
Epizootics/animal diseases	

Source: EC (2017, p. 70)

useful synopsis of the 11 most frequently occurring natural and man-made hazards in Europe as reported by the European Commission (EC, 2017).

Given the significance of planners' role in determining where and how people inhabit the planet, planning is increasingly being identified in hazard and disaster management literature as playing a role in reducing risk (O'Neill and Scott, 2011). It is argued that urbanisation and increasing disconnectedness with nature is insulating people from environmental stimuli and increasing vulnerabilities, with a false sense of security being provided by infrastructure (O'Neill, 2018). Significantly, the World Bank (2016) has estimated that by 2030, natural disasters alone may cost cities across the globe US$415 billion annually, unless there is major worldwide investment in cities to make them more resilient.[1] Being resilient means improving people's capacity to cope and function in the face of adversity, including exposure to the impact of environmental hazards.

Through their influence of the design and location of different land uses planners help determine how close people are, or perceive to be, located to various hazards. In addition to the potential dangers arising from actual proximity to a risk source, perceived proximity to risk sources also have real impacts on, for example, property prices (Bin and Polasky, 2004) and adoption of protective behaviours (Miceli et al., 2008). Failure to consider environmental risk perceptions in planning may undermine integration of evacuation routes into plans and the integration of risk management responses in the built environment (Chmutina et al., 2014). It is important to understand how the built environment influences people's spatial perception of risk (Curtis et al., 2014; Lennon et al., 2014; O'Neill et al., 2015) and, hence, their preparation and contribution to developing resilience. Decision making in planning needs to be informed by risk and supports the implementation of risk management.

Risk management and the spatial dimension

The analysis of risk is an integral feature of the assessment, evaluation and management of harm to people and the environment (Klinke and Renn, 2002). Already well established in fields such as, health and safety, finance, and road safety, risk-based regulation has been adopted by governments internationally (Rothstein et al., 2006) and for over 20 years has informed environmental regulation in advanced economies. The motivation for its introduction has been, in part, about 'better regulation' whereby regulatory resources are directed on the highest risks or worst performing firms, in proportion to the risk (Gouldson et al., 2009).

Greiving and Fleischauer (2006) elaborate upon the integration of risk assessment and management processes into planning. Risk assessment involves the identification of hazards usually employing scientific methods to calculate risks by analysing the frequency of occurrence (hazard probability) and magnitude of consequences (hazard impact). They provide a detailed account of the process of risk management which they refer to as a process of implementing decisions to achieve tolerable levels of risk. This includes, reducing the likelihood of risk and its potential consequences through land use policy decisions. They identify four stages in the risk management cycle:

(1) Mitigation – reducing or eliminating risk from hazard;
(2) Preparedness – taking appropriate precautions including household measures and/or evacuation;
(3) Response – emergency assistance coinciding with an event;
(4) Recovery – restoration of structures and communities.

In the European Union (EU), the recent emphasis on risk-based environmental regulations in the area of flood hazard builds upon earlier risk-informed regulations aimed at avoiding technological disasters. It is recognised that the development of risk management with a spatial dimension is an emerging feature in the management of flood risks (Hartmann and Spit, 2016; O'Neill, 2018). Strategic environmental assessment already provides a basis for undertaking risk assessment in planning processes (Prenger-Berninghoff and Greiving, 2015).

The case of managing flood hazard provides a useful example of the shift towards risk management and away from a singular focus on defensive protection (Hall et al., 2003; Schanze, 2006). Risk-based decision making that involves adopting a 'whole systems' approach has been adopted requiring consideration of all potential interventions that may assist in managing flood risk (Sayers et al., 2002; O'Neill, 2018). Table 7.3 provides a summary of the types of interventions, both structural and non-structural, that are now envisaged as part of a modern flood risk management system.

The EU Floods Directive (CEC, 2007) integrates this thinking. Through a risk-based approach, investment in structural and non-structural measures can be assessed and compared and resources prioritised on a proportionate basis to maximise reductions in overall flood risk (Sayers et al., 2002).

The EU Floods Directive requires each member state to: first, undertake a preliminary flood risk assessment to identify areas of significant risk; second, produce flood hazard and flood risk maps for various scenarios and flood probabilities for areas identified as being at significant risk; and third, prepare flood risk management plans at the level of the river basin and set out a prioritised set of measures to deliver prevention, protection, preparedness, emergency response and resilience (O'Neill, 2018).

Planning as a non-structural element which determines how and where land is used or developed is now recognised as playing a central role in managing flood risks from various

Table 7.3 Flood risk management interventions

Structural	*Non-structural*
All flood defence infrastructures, e.g. flood walls, levees, bunds, embankments, culverts, relief channels, gates, dams and man-made floodwater storage, etc.	Flood forecasting and early warning systems Household flood protection measures Coordination of critical services providers Community action Awareness raising Environmental planning and land use

Source: adapted from Kundzewicz (2013, p. 10)

sources. Referring back to the principal definitions of risk, planning can play a major role in addressing consequences by reducing potential damages (Greiving, 2004) and vulnerability of land uses (Prenger-Berninghoff and Greiving, 2015) and hence reducing overall risk.

Challenges to risk-informed environmental planning

Whilst risk-based approaches align with contemporary approaches to manage environmental hazards, its implementation is not without its challenges (White, 2015). For example, places can be exposed to varying degrees of risk from numerous hazard sources simultaneously, as shown in the example of London in Box 7.1. Environmental risks arise from multiple sources: flooding, contamination of groundwater, emissions to air, hurricanes, noise and infrastructure failure. This implies that planning needs to consider the potential for trade-offs between different risks and identify how they should be treated across a given spatial area (Greiving, 2004).

Moreover, there is a recognition of a shift in governance approach to the management of risks and its application in planning (Renn and Klinke, 2013, 2015). Consideration of different issues like uncertainty and risk acceptance must be deliberated to assist the implementation of a risk-based approach. Whilst collaborative and inclusive models of decision making are already well established in planning, more adaptive models may be necessary to cope with uncertainty in decision making.

Box 7.1 Environmental risks and London housing capacity

London's Strategic Housing Land Availability Assessment (SHLAA) provides insights to how London planners pro-actively strive to assess specified environmental constraints (hazards) when determining the quantity and suitability of land potentially available for the development of future housing across Greater London. The SHLAA establishes probability-based housing capacity estimates for each site. This provides a housing capacity on a site basis calculated on the notional housing capacity (using density and public transport access etc). The likelihood of each prospective site coming forward for development is, in part, informed by the number and severity (e.g. low, medium or high assigned constraint category) of planning policy, environmental and delivery constraints that impact upon an individual site.

The environment-related constraints account for the location of sites in or near, for example, flood risk zones, gasholders and hazardous installations, high voltage electricity lines (pylons), aircraft noise contour levels and contaminated land. Air quality and local road noise was initially considered but excluded on the assumption that they can be addressed at a site level. Whilst some constraints may be mitigated, in general, the existence of an environmental-related constraint reduces the likelihood of development occurring and, taken on an aggregate basis across London, informs the assessment of available housing capacity for London.

Source: GLA (2017, pp. 15–16)

There is also a growing consideration about handling uncertainty in planning particularly in relation to the impacts of climate change. Risk assessments may give false impressions about the completeness of knowledge when there may be a degree of uncertainty. Whilst probabilistic information can help inform decision making, this risk knowledge should be distinguished from uncertainty (Knight, 1921). We know, for example, that it is not possible to predict with precision how nature will behave, that there are gaps in our knowledge, and calculating probabilities may not always be possible.

Failure to acknowledge these uncertainties can lead to poor decision making and maladaptation. Instead, planners need to act in proportion to uncertainty (Zandvoort et al., 2018). In situations of high uncertainty, the application of the precautionary principle advocating preventative decision-taking is often sought (O'Neill, 2016). Another approach to addressing high uncertainty, that has gained policy attention, is 'robust decision making' which is about considering 'what actions should we take now given that we cannot predict the future?' (Lempert and Schlesinger, 2000, p. 391). Lempert and Schlesinger (2000) argue that even when faced with uncertainties about the future and surprise events, robust strategies may provide a solid basis to building consensus across multiple stakeholders.

Inclusive stakeholder participation is viewed as essential to successful risk management (Fiorino, 1990; Boholm, 2008; Challies et al., 2016). Whilst resolving complex technical and scientific issues involves deliberation among experts, on issues with high levels of uncertainty, both public and expert involvement is important in: assessing trade-offs between risk and benefit; identifying appropriate precautionary actions; and assessing trade-offs between different competing risks or pressures (Klinke and Renn, 2002). It is also important to engage with communities to:

- Explore their viewpoints on risk tolerability – what is safe? (Storbjörk, 2007);
- Discuss competing values and beliefs (Klinke and Renn, 2012);
- Clarify roles and debate responsibilities for protection in the management of environmental risks (White, 2010).

This can be difficult as engaging and communicating about risk is complex and citizens and experts may have different preferences regarding how probabilistic information is presented (O'Sullivan et al., 2012). The media contributes to risk discussions by framing risk in certain ways and influencing public opinion (Devitt and O'Neill, 2017). Sometimes what experts regard as minor or low-level risks may receive disproportionate attention due to social amplification of risk, as various societal and cultural processes interact with the hazard (Kasperson et al., 1988). Therefore, ways of involving the public in risk management is continuously evolving, as outlined in Table 7.4.

Part of the complexity associated with risk communication is found in another risk construct which is elaborated by Sandmann (1989). He refers to risk as a function of hazard and outrage

Table 7.4 Development stages in risk management

All we have to do is get the numbers right
All we have to do is tell them the numbers
All we have to do is explain what we mean by the numbers
All we have to do is show them that they've accepted similar risks in the past
All we have to do is show them that it's a good deal for them
All we have to do is treat them nice
All we have to do is make them partners
All of the above

Source: Fischhoff (1995, p. 138)

and suggests that various factors, which are frequently identified in risk perception research, combine to collectively produce outrage. Such factors include: voluntariness, control, fairness, process, morality, familiarity, memorability, dread and diffusion in time and space (Sandmann, 1987). Given the difficulty of responding to extreme outrage, it is essential that any process has legitimacy and is designed to promote mutual trust between citizens and experts.

Conclusion

Planning systems play a central role in how current and future generations experience environmental risks. Environmental planning can play a transformative role in managing and reducing future risk. To fulfil this role, it is essential that the public is fully engaged and their perceptions of risk and safety are taken into account. Adaptive planning and collaborative processes are key in advancing risk deliberations and promoting the development of more resilient places in which people feel safe to live and work.

Note

1 Technological risks can also be very costly, for example, the total financial damage (on a 30-year basis) caused by the nuclear disaster at Chernobyl is estimated to be $235billion (BFM, 2009).

References

Althaus, C. E. (2005). 'A disciplinary perspective on the epistemological status of risk'. *Risk Analysis*, 25(3): 567–88.

Aven, T. (2012). 'The risk concept – historical and recent development trends'. *Reliability Engineering and System Safety*, 99: 33–44.

Aven, T. and Kristensen, V. (2005). 'Perspectives on risk: review and discussion of the basis for establishing a unified and holistic approach'. *Reliability Engineering and System Safety*, 90(1): 1–14.

Beck, U. (1992). *Risk society: towards a new modernity*. London: Sage.

Beck, U. (2006). 'Living in the world risk society: a Hobhouse memorial public lecture given on Wednesday 15 February 2006 at the London school of economics'. *Economy and society*, 35(3): 329–45.

BFM (Belarus Foreign Ministry) (2009). *Chernobyl disaster*. Available at: chernobyl.undp.org/russian/docs/belarus_23_anniversary.pdf

Bin, O. and Polasky, S. (2004). 'Effects of flood hazards on property values: evidence before and after Hurricane Floyd'. *Land Economics*, 80(4): 490–500.

Birkmann, J., Cardona, O. D., Carreño, M. L., Barbat, A. H., Pelling, M., Schneiderbauer, S., Kienberger, S., Keiler, M., Alexander, D., Zeil, P. and Welle, T. (2013). 'Framing vulnerability, risk and societal responses: the MOVE framework'. *Natural Hazards*, 67(2): 193–211.

Boholm, Å. (2008). 'The public meeting as a theatre of dissent: risk and hazard in land use and environmental planning'. *Journal of Risk Research*, 11(1–2): 119–40.

Botzen, W. J., Aerts, J. C. and van den Bergh, J. C. (2009). 'Dependence of flood risk perceptions on socioeconomic and objective risk factors'. *Water Resources Research*, 45 (W10440): 1–15.

Bradbury, J. A. (1989). 'The policy implications of differing concepts of risk'. *Science, Technology, and Human Values*, 14(4): 380–99.

CEC (Commission of European Communities) (2007). *Directive 2007/60/EC of the European Parliament and of the Council 23 October 2007 on the assessment and management of flood risks*. Brussels: Official Journal of the European Communities 6.11.2007 L 288.

Challies, E., Newig, J., Thaler, T., Kochskämper, E. and Levin-Keitel, M. (2016). 'Participatory and collaborative governance for sustainable flood risk management: an emerging research agenda'. *Environmental Science and Policy*, 55(2): 275–80.

Chmutina, K., Ganor, T. and Bosher, L. (2014). 'Role of urban design and planning in disaster risk reduction'. *Proceedings of the Institution of Civil Engineers-Urban Design and Planning*, 167(3): 125–35.

Curtis, J. W., Shiau, E., Lowery, B., Sloane, D., Hennigan, K. and Curtis, A. (2014). 'The prospects and problems of integrating sketch maps with geographic information systems to understand environmental perception: a case study of mapping youth fear in Los Angeles gang neighborhoods'. *Environment and Planning B: Planning and Design*, 41(2): 251–71.

Davoudi, S. (2014). 'Climate change, securitisation of nature, and resilient urbanism'. *Environment and Planning C: Government and Policy*, 32(2): 360–75.

Devitt, C. and O'Neill, E. (2017). 'The framing of two major flood episodes in the Irish print news media: implications for societal adaptation to living with flood risk'. *Public Understanding of Science*, 26(7): 872–88.

EC (European Commission) (2017). *Overview of natural and man-made disaster risks the European Union may face*. SWD (2017) 176 final. Brussels: European Commission.

Fiorino, D. J. (1990). 'Citizen participation and environmental risk: a survey of institutional mechanisms'. *Science, Technology, and Human Values*, 15(2): 226–43.

Fischhoff, B. (1995). 'Risk perception and communication unplugged: twenty years of process'. *Risk analysis*, 15(2): 137–45.

Gibson, R., Stacey, N., Drais, E., Wallin, H., Zatorski, W. and Virpi, V. (2012). *Risk perception and risk communication with regard to nanomaterials in the workplace*. Luxembourg: European Agency for Safety and Health at Work.

Giordano, A., Anderson, S. and He, X. (2010). 'How near is near? The distance perceptions of residents of a nuclear emergency planning zone'. *Environmental Hazards*, 9(2): 167–82.

GLA (Greater London Authority) (2017). *The London strategic housing land availability assessment 2017*. London: Greater London Authority. Available at: www.london.gov.uk/what-we-do/planning/london-plan/new-london-plan/strategic-housing-land-availability-assessment

Gouldson, A., Morton, A. and Pollard, S. J. (2009). 'Better environmental regulation – contributions from risk-based decision-making'. *Science of the Total Environment*, 407(19): 5283–8.

Greiving, S. (2004). 'Risk assessment and management as an important tool for the EU strategic environmental assessment'. *disP – The Planning Review*, 40(157): 11–17.

Greiving, S. and Fleischauer, M. (2006). 'Spatial planning response towards natural and technological hazards', in P. Schmidt-Thomé (ed.) *Natural and technological hazards and risks affecting the spatial development of European regions* (Special Paper 42). Finland: Geological Survey of Finland, pp 109–23.

Hall, J. W., Meadowcroft, I. C., Sayers, P. B. and Bramley, M. E. (2003). 'Integrated flood risk management in England and Wales'. *Natural Hazards Review*, 4(3): 126–35.

Hartmann, T. and Spit, T. (2016). 'Implementing the European flood risk management plan'. *Journal of Environmental Planning and Management*, 59(2): 360–77.

Jóhannesdóttir, G. and Gísladóttir, G. (2010). 'People living under threat of volcanic hazard in southern Iceland: vulnerability and risk perception'. *Natural Hazards and Earth System Sciences*, 10(2): 407–20.

Kasperson, R. E., Renn, O., Slovic, P., Brown, H. S., Emel, J., Goble, R., Kasperson, J. X. and Ratick, S. (1988). 'The social amplification of risk: a conceptual framework'. *Risk Analysis*, 8(2): 177–87.

Klinke, A. and Renn, O. (2002). 'A new approach to risk evaluation and management: risk-based, precaution-based, and discourse-based strategies'. *Risk Analysis*, 22(6): 1071–94.

Klinke, A. and Renn, O. (2012). 'Adaptive and integrative governance on risk and uncertainty'. *Journal of Risk Research*, 15(3): 273–92.

Knight, F. H. (1921). *Risk, uncertainty and profit*. Chicago: University of Chicago Press.

Kundzewicz, Z. W. (2013). '15 floods: lessons about early warning systems', in *Late lessons from early warnings: science, precaution, innovation*. EEA Report No. 1/2013. Copenhagen: European Environment Agency, pp 347–68.

Lempert, R. and Schlesinger, M. (2000). 'Robust strategies for abating climate change'. *Climatic Change*, 45: 387–401.

Lennon, M., Scott, M. and O'Neill, E. (2014). 'Urban design and adapting to flood risk: the role of green infrastructure'. *Journal of Urban Design*, 19(5): 745–58.

Miceli, R., Sotgiu, I. and Settanni, M. (2008). 'Disaster preparedness and perception of flood risk: a study in an alpine valley in Italy'. *Journal of Environmental Psychology*, 28(2): 164–73.

O'Neill, E. (2016). 'The precautionary principle: a preferred approach for the unknown'. *Ethics, Place and Environment*, 19(2): 153–6.

O'Neill, E. (2018). 'Expanding the horizons of integrated flood risk management: a critical analysis from an Irish perspective'. *International Journal of River Basin Management*, 16(1): 71–7.

O'Neill, E., Brennan, M., Brereton, F. and Shahumyan, H. (2015). 'Exploring a spatial statistical approach to quantify flood risk perception using cognitive maps'. *Natural Hazards*, 76(3): 1573–601.

O'Neill, E., Brereton, F., Shahumyan, H. and Clinch, J. P. (2016). 'The impact of perceived flood exposure on flood-risk perception: the role of distance'. *Risk Analysis*, 36(11): 2158–86.

O'Neill, E. and Scott, M. (2011). 'Policy and planning brief: adapting to climate change – a European Union policy agenda'. *Planning Theory and Practice*, 12(2): 311–16.

O'Sullivan, J. J., Bradford, R. A., Bonaiuto, M., De Dominicis, S., Rotko, P., Aaltonen, J., Waylen, K. and Langan, S. J. (2012). 'Enhancing flood resilience through improved risk communications'. *Natural Hazards and Earth System Sciences*, 12(7): 2271–82.

Peacock, W. G., Brody, S. D. and Highfield, W. (2005). 'Hurricane risk perceptions among Florida's single family homeowners'. *Landscape and Urban Planning*, 73(2–3): 120–35.

Prenger-Berninghoff, K. and Greiving, S. (2015). 'The use of risk information in spatial planning in Europe: examples from case study sites in Italy and Romania with a focus on flood and landslide hazards', in G. Lollino, A. Manconi, F. Guzetti, M. Culshaw, P. T. Bobrowsky and F. Luino (eds.) *Engineering geology for society and territory – volume 5*. Cham: Springer, pp 737–41.

Renn, O. (1992). 'Concepts of risk: a classification', in S. Krimsky and D. Golding (eds.) *Social theories of risk*. Westport, CT: Praeger, pp 53–82.

Renn, O. and Klinke, A. (2013). 'A framework of adaptive risk governance for urban planning'. *Sustainability*, 5(5): 2036–59.

Renn, O. and Klinke, A. (2015). 'Risk governance and resilience: new approaches to cope with uncertainty and ambiguity', in: U. Fra Paleo (ed.) *Risk governance*. Dordrecht: Springer, pp 19–42.

Rothstein, H., Irving, P., Walden, T. and Yearsley, R. (2006). 'The risks of risk-based regulation: insights from the environmental policy domain'. *Environment International*, 32(8): 1056–65.

Sandmann, P. M. (1987). 'Risk communication: facing public outrage'. *EPA Journal*, 2(2) (November): 235–8.

Sandmann, P. M. (1989). 'Hazard versus outrage: a conceptual frame for describing public perception of risk', in H. Jungermann, R. E. Kasperson and P. M. Wiedemann (eds.) *Risk communication*. Julich, Germany: Nuclear Research Centre, pp 163–8.

Sayers, P. B., Hall, J. W. and Meadowcroft, I. C. (2002). 'Towards risk-based flood hazard management in the UK'. *Proceedings of Institute of Civil Engineers: Civil Engineering Paper 12803*, 150: 36–42.

Schanze, J. (2006). 'Flood risk management: a basic framework', in J. Schanze, E. Zeman and J. Marsalek (eds.) *Flood risk management: hazards, vulnerability and mitigation measures*, Dordrecht: Springer, pp 1–20.

Siegrist, M. and Gutscher, H. (2006). 'Flooding risks: a comparison of lay people's perceptions and expert's assessments in Switzerland'. *Risk Analysis*, 26(4): 971–9.

Slovic, P. (1987). 'Perception of risk'. *Science*, 236: 280–5.

Slovic, P. and Peters, E. (2006). 'Risk perception and affect'. *Current Directions in Psychological Science*, 15(6): 322–5.

Storbjörk, S. (2007). 'Governing climate adaptation in the local arena: challenges of risk management and planning in Sweden'. *Local Environment*, 12(5): 457–69.

White, I. (2010). *Water and the city: risk, resilience and planning for a sustainable future*. Abingdon, Oxon: Routledge.

White, I. (2015). *Environmental planning in context*. London: Palgrave Macmillan.

World Bank Group (2016). *Investing in urban resilience: protecting and promoting development in a changing world*. Washington, DC: World Bank. Available at: https://openknowledge.worldbank.org/handle/10986/25219

Zandvoort, M., Van der Vlist, M. J., Klijn, F. and Van den Brink, A. (2018). 'Navigating amid uncertainty in spatial planning'. *Planning Theory*, 17(1): 96–116.

Zhang, Y., Hwang, S. N. and Lindell, M. K. (2010). 'Hazard proximity or risk perception? Evaluating effects of natural and technological hazards on housing values'. *Environment and Behavior*, 42(5): 597–624.

8

Complexity, uncertainty and resilience

Jon Coaffee

> We live in a complex world. Anyone with a stake in managing some aspect of that world will benefit from a richer understanding of resilience and its implications.
>
> *(Walker and Salt, 2006, p. xiv)*

> Resilience is the ability of a system to survive and thrive in the face of a complex, uncertain and ever-changing future. It is a way of thinking about both short-term cycles and long term trends: minimizing disruptions in the face of shocks and stresses, recovering rapidly when they do occur, and adapting steadily to become better able to thrive as conditions continue to change. A resilience approach offers a proactive and holistic response to risk management'.
>
> *(Siemens, 2013, p. 10)*

Introduction

Ideas of complexity that are now present in many disciplines, including environmental planning, illuminates that the future cannot be determined accurately. Complex systems thinking has emphasised the dynamic and unpredictable nature of dependent systems with many feedback loops that move systems away from equilibrium. Small changes can have big effects and thus systems are forced to evolve, adapt and self-organise.

We are currently living in an epoch which many are terming the Anthropocene (see chapter by Whitehead in this volume) which began in the mid-20th century through a rapid increase in technological change, population growth and consumption, and is increasingly characterised by complex and dynamic system interaction, future volatility and ultimately an imperative to rethink the relationship of humans with nature, environment and technology. Concomitantly, in early 21st century – catalysed by the devastating events of 9/11 and the release of the fourth Intergovernmental Panel on Climate Change (IPCC) report in 2007 highlighting unequivocal evidence of a warming climate – ideas and practices of *resilience* have become a central organising concept within policy-making processes and the expanding institutional framework of national security and emergency preparedness. For many, resilience offers an integrated approach for coping with all manner of disruptive events, as well as a new way to engage with future uncertainty (Coaffee and Lee, 2016; Chandler and Coaffee, 2016; Coaffee et al., 2008; Walker and Cooper, 2011; Zolli and Healy, 2013; Chandler, 2014). As argued in this chapter, resilience thinking has subsequently been utilised to 'extend' established risk management approaches based on bouncing back to normal, and to advance ways of surviving and thriving in the future through adaptation and long-term transformative action – a more proactive 'bounce forwards' approach.

Despite concern over the diverse range of applications and flexibility in its terminology, resilience has begun to offer a new lexicon to make sense of a range of disruptive challenges facing environmental planning (Buckle et al., 2000). Some of the confusion over the vagueness of resilience concepts and discourse comes from its association with sustainable development or sustainability, with many proclaiming that resilience is replacing sustainability as the central organising concept of the age within environmental planning discourses as a result of increased volatility that requires a different framing and response (Zolli and Healy, 2013; Vale, 2014). In essence, where sustainability often assumes a present and future of equilibrium, resilience is based upon a change paradigm, which makes it particularly helpful for managing a complex and uncertain future: 'where sustainability aims to put the world back into balance, resilience looks for ways to manage in an imbalanced world' (Zolli, 2012, no page).

In many ways, it the spectre of unanticipated catastrophe that has driven the interest in resilience as a universal remedy for a range of 'natural' and man-made risks (Aradau and van Munster, 2011). Recent decades have been remarkable for the volume of high impact environmental and anthropocentric disasters which have highlighted the vulnerability, complexity and interdependency of contemporary life and have foregrounded the political prioritisation of enhanced security – often badged as resilience – as a political imaginary of 'insecurity by design' (Evans and Reid, 2014) and living in a world of 'persistent uncertainty' (Biermann, 2014).

The remainder of the chapter proceeds in three main sections. First, the new world of resilience processes and practices are illuminated by relating them to traditional notions of risk and risk management. Second, different ways in which resilience has been modelled are presented and their utility for understanding complexity and uncertainty in environmental planning are explored. Third, the chapter illuminates how emerging notions of evolutionary resilience are being utilised to better reflect the environmental and social uncertainties we are now living through and will increasingly experience in the future.

From risk to resilience

There is a paradox at the heart of understandings of complexity and resilience; whilst human activity is increasingly shaping the environmental conditions, the ability to do this in a conscious and deliberate way is hampered by our inability to tackle the complex interactions and interdependencies involved and, thus, the true nature of risk to global society.

Social scientists for many years have studied natural hazards and the need to make contingency against their impact (White, 1942; Kates, 1962), whilst accounts regarding the impact of complex risk only became prevalent in the late 1980s and 1990s, suggesting that concerns about such risks had become the defining societal characteristics (Beck, 1992; Douglas, 1994; Adams, 1995). This new range of 'risk theory' emerged primarily around concerns about global environmental hazards, the transnational nature of such risk and the effect of such risk in challenging existing political governance configurations. Most notable amongst this canon of risk theory was Ulrich Beck's 'Risk Society – Towards a New Modernity' (1992). Published in the wake of the Chernobyl nuclear catastrophe in Ukraine, 'Risk Society' considered what society might look like when disputes and conflicts about new types of risk produced by industrial society are fully realised. 'Risk Society' starkly illuminated the magnitude and boundless nature of the global risks, and how this is transforming the way in which risk is imagined, assessed, managed and governed but not eradicated. Beck's work provided the impetus for further academic thought related to the impact of the emergence of a set of newly defined and ubiquitous 'mega-scale' risks on the workings of Western society that 'cannot be delimited spatially, temporally, or socially' (Beck, 1995, p. 1). As Giddens (2002, p. 34) reiterated:

> [W]hichever way you look at it, we are caught up in risk management. With the spread of manufactured risk, governments can't pretend such management isn't their business. And they need to collaborate, since very few new-style risks have anything to do with the borders of nations.

'Risk Society' is a story of survivability. As Blowers (1999, p. 256) commented, 'Risk Society is a pessimistic and conflictual diagnosis of modern societies . . . that is exposed to risks from high technology . . . that imperil our very survival'. New risk theory also exposed the absence of social and cultural factors involved in discussions about risk that had been hidden beneath a preference for objective and rational approaches to risk assessment – the 'possibility of calculation' (Giddens, 2002, p. 28).

These new understandings of risk are echoed in more recent discourses in resilience studies which seek new ways of coping with increasingly non-linear outcomes; in effect, the acceptance of an unknown future has significantly contributed to the rise of *resilience* as the policy metaphor of choice for coping with and managing future uncertainty.

In order to unpack the usefulness of resilience in environmental planning, this chapter first draws attention to so-called *equilibrium* models of socio-ecological systems (SES) resilience that have dominated resilience discourse since the early 1970s and have focused upon theorising a bounce-back to a pre-defined state model of response to stress or perturbation. It will subsequently highlight more *evolutionary* approaches that have emerged within the resilience literature and which, by contrast, focus upon a bounce-forward and 'new normal' model of resilience, seeking to construct an approach more applicable to increasingly complex and non-linear systems which environmental planning regimes seeks to tackle.

The dominance of the equilibrium socio-ecological model

The use and application of the term resilience is broadly acknowledged to have emerged from C. S. 'Buzz' Holling's 1970s studies of systems ecology and his subsequent work with the Resilience Alliance (a research network comprised of scientists and practitioners from many disciplines who collaborate to explore the dynamics of social-ecological systems). Holling's ideas, best portrayed in the classic 1973 paper 'Resilience and stability of ecological systems', represented a paradigm shift in thinking, demonstrating a more dynamic process which he termed the *adaptive cycle* that was differentiated from earlier understandings of ecological systems which assumed a stable basis. The adaptive cycle focused attention upon processes of destruction and reorganisation, which were often marginalised in classical ecosystem studies in favour of growth and conservation. Including these processes sought to give a more complete view of system dynamics.

The adaptive cycle was essentially an equilibrium model in which system resources go through periods of production, consumption and conservation and in which resilience is the ability of a system to either bounce back to a steady state or to absorb shock events and persist under stress. The adaptive cycle of a resilient system was represented by four phases of system development, involving exploitation, conservation, release and reorganisation of resources. The exploitation-conservation phase represents a slow accumulation of system resources ready for a period of release-reorganisation. The phases of the adaptive cycle are depicted against two axes which represented the system potential and connectedness (see Figure 8.1). The connectedness and potential of the system increases during the exploitation-conservation phase as connections embed and increase system potential for resource production and release. Walker and Cooper (2011, p. 147) provide a summary of this process linking it to classical ecological study:

> Where classical systems ecology focused only on the phases of rapid successional growth (r) followed by the conservation phase of stable equilibrium (K), the Resilience Alliance argues that these phases are inevitably followed by collapse (Ω), and then a spontaneous reorganization that leads to a new growth phase (α).

In Holling's (1996, p. 33) words, such a dynamic adaptive cycle measures ecological resilience – the magnitude of change a system can absorb before it 'flips' and changes structure to another 'stability domain'. Ecological resilience in this definition focusses upon change and unpredictability *but* with a view to return to equilibrium.

Increasingly, such equilibrist approaches associated with the social-ecological systems (SES) theory have been critiqued as their influence has spread, particularly into the area of climate change adaptation. In general, such SES approaches are viewed as inherently conservative and, as noted above, have a core aim of system stability. They also tend to focus upon endogenous (internal) stresses with less attention given to exogenous (external) factors that might disturb or shock the system. More broadly, the development of SES based on equilibrium ideas is perhaps not best suited to modelling systems involving complex social dynamics which are less easy to conceptualise with in theoretical models (Alexander, 2013; Davoudi, 2012). For many, SES resilience approaches fail to account for political and power relations within a complex social system which affects whose needs are being met (resilience for whom?) and how distributional resource issues are mediated through the political processes of environmental planning. As Brown (2014, p. 109) highlighted, SES approaches to resilience 'promotes a scientific and technical approach akin to "imposed rationality" that is alien to the practice of ordinary people . . . is depoliticized

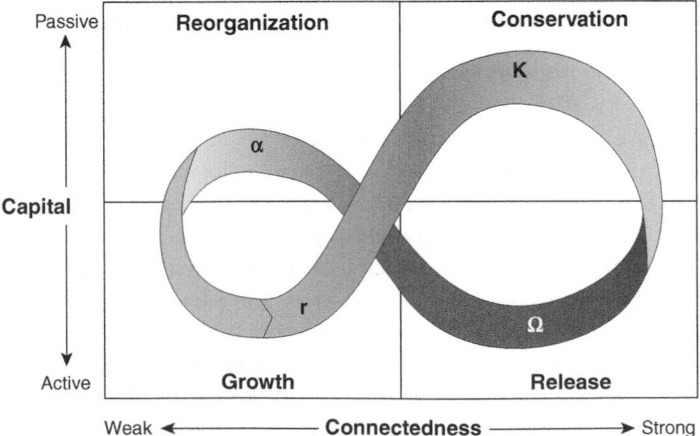

Figure 8.1 The adaptive cycle of a resilient system

Source: adapted from Holling (1973, p. 20)

and does not take account of the institutions within which practices and management are embedded' (see also Cannon and Muller-Mahn, 2010).

In an attempt to introduce greater social resonance and complexity theory into understandings of ecological resilience, the SES approach was 'updated', in particular through Gunderson and Holling's (2002) 'Panarchy' premised upon a hierarchy of adaptive cycles, and named after the Greek god Pan – 'the epitome of unpredictable change' (Holling, 2001, p. 396). This approach attempted to reconcile some of the limitations and contradictions of the earlier theories by providing a conceptual framework to account for the dual, and seemingly contradictory, characteristics of all complex systems; stability and change. The 'Panarchy' model outlined phases that are neither fixed nor sequential, rather operating as multiple, nested adaptive cycles that function and interact independently (Davoudi, 2012). 'Panarchy' also recognised that internal functions can introduce change, as in social systems, in effect working from the bottom up as well as top down (Figure 8.2).

The 'Panarchy' framework placed great emphasis on the interconnectedness of system levels between the smallest and the largest and the fastest and slowest. In this framework, the large, slow cycles set the conditions in which the smaller, faster cycles operate, although the smaller, faster cycles can impact upon the larger, slower cycles. Here there are many possible points of interconnectedness between these adjacent levels, with two of particular note. First, 'Revolt' which occurs when fast, small events overwhelm large, slow ones. Second, 'Remember' that facilitates a return to stability by drawing on the potential that has been accumulated and stored in a larger, slower cycle. Sometimes, though, this return is to a different path than the former state and is referred to as a hysteresis effect – referring to how a system responds to a loss of resilience or, more specifically, the return path taken following some disturbance or change due to cumulative effects which serve to increase resilience again (Ludwig et al., 1997).

Such a model, despite the acknowledgment of complexity and social systems, is still, for many, too divorced from reality of non-linear complex adaptive systems to be applied appropriately in a range of socio-economic and political policy spheres, failing to take into account the unevenness of space, inequality, power, or the agency of actors within social systems. Therefore, whilst the term resilience has its roots in physical (physics and engineering) and natural science (ecology, biology and biosciences), it is increasingly seen as a 'political, cultural, and social construction'

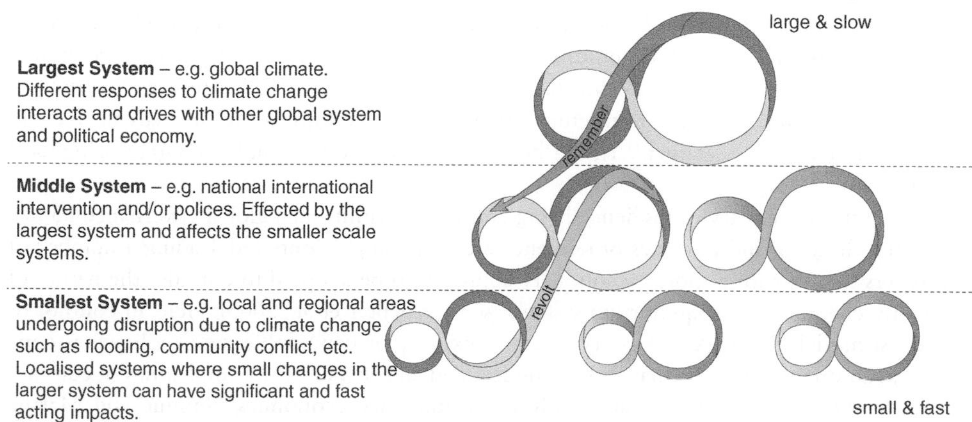

large & slow
Largest System – e.g. global climate. Different responses to climate change interacts and drives with other global system and political economy.
remember
Middle System – e.g. national international intervention and/or polices. Effected by the largest system and affects the smaller scale systems.
revolt
Smallest System – e.g. local and regional areas undergoing disruption due to climate change such as flooding, community conflict, etc. Localised systems where small changes in the larger system can have significant and fast acting impacts.
small & fast

Figure 8.2 A 'Panarchy' of nested systems

Source: adapted from Gunderson and Holling (2002, p. 10)

(White and O'Hare, 2014, p. 10) that is not amenable to wholesale transfer from the natural to the social sciences. As Cote and Nightingale (2012, p. 475) have further contended, resilience in SES has 'evolved through the application of ecological concepts to society assuming that social and ecological system dynamics are essentially similar', further arguing that resilience ideas have grown in 'remarkable isolation from critical social science literature' (cited in Brown, 2014). It is to this critical social science literature that this chapter now turns in exploring emerging 'evolutionary' approaches to resilience.

Towards an evolutionary approach to complexity and resilience

Resilience is everywhere today, rapidly becoming a principal framing device for political discourse to not only assess and understand the resistance to shock events of people, households and communities but also to describe the properties and ability of interconnected and complex ecological, technical, social and economic systems to adapt and change in the midst of failure.

In this sense, resilience represents a break from previous eras of environmental planning history where stability was sought through deductive and positivist methods that attempted to model the future from the past and fit planning requirements to those predictions. Resilience thus represents a new era for environmental planning where uncertainty and volatility predominate and where new and increasingly flexible repertoires of action are required (although it is often used interchangeable with sustainability or sustainable development). Notably, this has occurred in three core and integrated post-2015 dialogues – the Sendai Framework for Disaster Risk Reduction 2015–2030 which was adopted by UN Member States in March 2015 at the World Conference on Disaster Risk Reduction (WCDRR) held in Sendai, Japan; the UN Sustainable Development Goals (SDGs) released in September 2015; and the Conference of Parties (COP) that signed up to the UN Framework on Climate Change (UNFCCC) held the Paris Climate Conference in December 2015, also known as COP21, with the aim of achieving a legally binding and universal agreement on climate change adaptation.

These three agreements all highlighted, either explicitly or implicitly, the importance and utility of ideas and practices of resilience in tackling the integrated and complex environmental issues of reducing the risk of disasters, advancing sustainable development and mitigating and adapting to climate change, and their framing in resilience thinking that will ensure that resilience will be a vital area of study in environmental planning and sustainability for years to come. For example, the SDGs promise to promote climate change adaptation: 'strengthen resilience and adaptive capacity to climate-related hazards and natural disasters in all countries' (target 13.1); advance critical infrastructure resilience – 'develop quality, reliable, sustainable and resilient infrastructure . . .' (target 9.1); and 'build sustainable and resilient buildings utilizing local materials' (target 11c).

Despite the vocabulary of resilience being embedded within such global governance frameworks, the thinking and practices of resilience are increasingly contested, leading a number of researchers to argue for a more evolutionary approach to be adopted to consider the nature of constantly changing non-equilibrium systems (see Carpenter et al., 2005). Here, in contrast to equilibrist models that seek a recovery to a (pre-existing or new) stable state, resilience is considered as an ongoing process that seeks to understand and adapt to the complexities of constant change (Coaffee, 2013). Work particularly in evolutionary economics (Simmie and Martin, 2010) and environmental planning have adopted and modified important aspects of the adaptive cycle and 'Panarchy' models to broaden the description of resilience beyond bounce-back approaches (Folke et al., 2010) and to incorporate 'the dynamic interplay between persistence, adaptability and transformability across multiple scales and time frames' (Davoudi, 2012, p. 310).

In many ways, evolutionary resilience approaches have advanced as a reaction to the acknowledged limitations of equilibrium models. The malleability and flexibility in the use of the resilience concept has, however, made equilibrium approaches to resilience politically acceptable and durable. Such approaches seek to preserve system stability – to bounce back – after a perturbation and focus upon short-term and reactive measures predominantly concerning endogenous risk.

By contrast to equilibrium models, evolutionary approaches – often portrayed as the binary opposite for equilibrium approaches – focus upon adaptability and flexibility with the function of restoration to a new normality and an increasingly complex and volatile world: 'evolutionary resilience promotes the understanding of places not as units of analysis or neutral containers but as complex, interconnected socio-spatial systems with extensive and unpredictable feedback processes which operate at multiple scales and time frames' (Davoudi, 2012, p. 304). Such approaches tend to be proactive and focus predominantly on the medium to long-term exogenous risk. It can be contended that as a move towards resilience is always in a state of becoming, such constantly transformative approaches are vital in aiding our understanding of how society responds to volatility and unpredictability. However, in practice – and in its current guise – evolutionary resilience is rare and we have a noticeable implementation gap in environmental planning (Coaffee and Clarke, 2015). Such an implementation gap has been illuminated by the impact of Hurricane Sandy in New York in 2012 with commentators noting that 'the word "resilience" was everywhere – even on the sides of buses touting New Jersey as "A State of Resilience". *But evidence of actual planning for resilience was scant*' (De Souza and Parker, 2014, no page, emphasis added).

In seeking to address this implementation gap in environmental planning practice, we can see an ongoing paradigm shift occurring *from* equilibrium *to* evolutionary approaches. In this transition, resilience is as much about a set of transformative learning processes, as it is about outputs and outcomes. The transition and implementation of resilience is not without its challenges and critique – that it draws in an anticipatory and precautionary logic; that it is a depoliticising and reactive tool of government; and that its responsibilises professions, individuals and communities. Despite this many prefer to focus upon how its usage and the implementation of its 'principles' might be repoliticised so as to illuminate and change the uneven and problematic deployment of resilience policy and 'to ultimately recast resilience as a potential antidote, rather than complement, to perpetual neo-liberal vulnerability and insecurity' (Paganini, 2015, no page). The potentially positive force of resilience to transform the status quo and shine a light on how future vulnerabilities might be tackled is having a significant influence upon environmental planning and sustainability and can provide a proactive and optimistic set of frameworks and imaginaries for assessing and adapting to a range of contemporary and future risks.

Today, the principles of resilience carry tremendous influence in modifying international environmental planning agendas, whether this is dealing with the unique needs and characteristics of places, looking at the short, medium and long-term issues, advancing the knowledge, objectives and actions involved *and* recognising the wide range of stakeholders (who should be) involved in resilient planning or, ultimately, 'breaking planning out of its obsession with order, certainty and stasis' (Porter and Davoudi, 2012, p. 330).

But there remains a tension and distinction between a highly equilibrist and conservative approach to resilience, which focuses on reactive and short-term measures, and more evolutionary

Table 8.1 Aims and foci of equilibrist and evolutionary resilience for planning practice

	Equilibrist	*Evolutionary*
Aims	Equilibrist	Adaptive
	Existing normality	New normality
	Preserve	Transform
	Stability	Flexibility
Focus	Endogenous	Exogenous
	Short term	Medium to long term
	Reactive	Proactive
	Atomised	Abstract
Planning Approaches	Techno-rational	Sociocultural
	Vertical integration	Horizontal integration
	Building focus	Societal focus
	Homogeneity	Heterogeneity
	Deductive	Inductive
	Plan A (Optimisation)	Plan A and B (Redundancy)

Source: adapted from White and O'Hare (2014, p. 11)

socio-cultural and progressive approaches focused in the medium to long-term and with anticipation, adaptation and flexibility as core attributes. The binaries between these two approaches have been highlighted by White and O'Hare (2014, p. 11) and are shown in Table 8.1.

Current practice appears in a transition between these two meta-approaches that highlight both the regressive and progressive advance of ideas of resilience in environmental planning. Here transition is represented as a continuous process of change, from one relatively stable system state to another via a co-evolution of markets, networks, institutions, technologies, policies, individual behaviour and other trends. Whilst a transition can be accelerated by one-time events, such as large disruptions in continuity such as Hurricane Katrina, they are not caused by such events in isolation. Slower changes can also produce a persistent undercurrent for a fundamental change.

Summary and conclusion

Emerging since 2000 as an international policy rhetoric connected to countering crisis and disaster, resilience is now being fully embedded as a policy metaphor for envisioning future environmental planning activities and committing effort and resource towards strategic implementation of plans to reduce risk and vulnerability against shock events and slow burn stresses. The evolution of resilience and its embedding into environmental sustainability policy agendas is not only leading to fundamental questions about the reimagining of planning in a new era of risk, crisis and uncertainty but also illuminates government priorities and governmentalizing tendencies. On one hand, this highlights a top-down driven rhetoric where policy makers are being steered into taking resilience seriously. On the other hand, it also highlights the potential of resilience being acknowledged as the sustenance of a living civil society through enhanced citizen engagement, focused upon localised issues with the view to developing ways in which communities can better cope and thrive in complexity and adversity.

In his unpacking of resilience and complexity, David Chandler (2014) noted that there has been a rejection of modernist or liberal approaches based upon linear or deterministic knowledge assumptions of causality. Here, policy making was constructed in 'top-down' ways, determined by the 'known knowns' of established knowledge and generalisable assumptions of cause and effect. By contrast, contemporary resilience approaches are generally focused upon theories of general or emergent complexity in open systems linked to 'unknown unknowns', which are only revealed post-hoc through the appearance of the problem. In environmental planning practice, this has seen a shift in resilience thinking from techno-rational approaches towards socio-cultural understandings and from a focus upon standardised and top-down approaches towards diverse and locally integrated methods where there is more than just a 'Plan A'. Practically, this has meant that new and innovative approaches that are seeking to embed resilience into complex system are very different from traditional ways of working, with emerging approaches placing much greater emphasis on enhancing adaptive capacity, flexibility and agility and upon thinking through a variety of complex transitionary pathways from our current state to conditions of resilience.

References

Adams, J. (1995). *Risk*. London: UCL Press.

Alexander, D. E. (2013). 'Resilience and disaster risk reduction: an etymological journey'. *Natural Hazards and Earth System Science*, 13(11): 2707–16.

Aradau, C. and van Munster, R. (2011). *Politics of catastrophe: genealogies of the unknown*. Abingdon: Routledge.

Beck, U. (1992). *Risk society: towards a new modernity*. London: Sage.

Beck, U. (1995). *Ecological politics in an age of risk*. Cambridge: Polity Press.

Biermann, F. (2014). 'The Anthropocene: a governance perspective'. *The Anthropocene Review*, 1(1): 57–61.

Blowers, A. (1999). 'Nuclear waste and landscapes of risk'. *Landscape Research*, 24(3): 241–64.

Brown, K. (2014). 'Global environmental change I: a social turn for resilience?' *Progress in Human Geography*, 38(1): 107–17.

Buckle, P., Mars, G. and Smale, S. (2000). 'New approaches to assessing vulnerability and resilience'. *Australian Journal of Emergency Management*, 15(2): 8–14.

Cannon, T. and Muller-Mahn, D. (2010). 'Vulnerability, resilience and development discourses in context of climate change'. *Natural Hazards*, 55(3): 621–35.

Carpenter, S. R., Westley, F. and Turner, M. (2005). 'Surrogates for resilience of social-ecological systems'. *Ecosystems*, 8(8): 941–4.

Chandler, D. (2014). 'Beyond neoliberalism: resilience, the new art of governing complexity'. *Resilience*, 2(1): 47–63.

Chandler, D. and Coaffee, J (eds.) (2016). *Handbook of international resilience*. Abingdon/New York: Routledge.

Coaffee, J. (2013). 'From securitisation to integrated place making: towards next generation urban resilience in planning practice'. *Planning Practice and Research*, 28(3): 323–39.

Coaffee, J. and Clarke, J. (2015). 'On securing the generational challenge of urban resilience'. *Town Planning Review*, 86(3): 249–55.

Coaffee, J. and Lee, P. (2016). *Urban resilience: planning for risk crisis and uncertainty*. London: Palgrave Macmillan.

Coaffee, J., Murakami-Wood, D. and Rogers, P. (2008). *The everyday resilience of the city: how cities respond to terrorism and disaster*. London: Palgrave Macmillan.

Cote, M. and Nightingale, A. (2012). 'Resilience thinking meets social theory situating social change in socio-ecological systems (SES) research'. *Progress in Human Geography*, 36(4): 475–89.

Davoudi, S. (2012). 'Resilience: a bridging concept or a dead end?' *Planning Theory and Practice*, 13(2): 299–307.

De Souza, R. M. and Parker, M. (2014). 'The year that resilience gets real'. *New Security Beat*, 6 January [Online]. Available at: www.newsecuritybeat.org/2014/01/year-resilience-real/

Douglas, M. (1994). *Risk and blame – essays in cultural theory*. London: Routledge.

Evans, B. and Reid, J. (2014). *Resilient life: the art of living dangerously*. Cambridge: Polity Press.

Folke, C., Carpenter, S., Walker, B., Chapin, T. and Rockström, J. (2010). 'Resilience thinking: integrating resilience, adaptability and transformability'. *Ecology and Society*, 15(4): 20 [Online]. Available at: www.ecologyandsociety.org/vol15/iss4/art20/

Giddens, A. (2002). *Runaway world: how globalisation is reshaping our lives*. London: Profile.

Gunderson, L. and Holling, C. S. (eds.) (2002). *Panarchy: understanding transformations in human and natural systems*. Washington, DC: Island Press.

Holling, C. S. (1973). 'Resilience and stability of ecological systems'. *Annual Review of Ecology Evolution and Systematics*, 4: 1–23.

Holling, C. S. (1996). 'Engineering resilience versus ecological resilience', in P. C. Schulze (ed.) *Engineering within ecological constraints*. Washington, DC: National Academy Press, pp 31–43.

Holling, C. S. (2001). 'Understanding the complexity of economic, ecological, and social systems'. *Ecosystems*, 4(5): 390–405.

Kates, R. W. (1962). *Hazard and choice perception in flood plain management*. Paper 78. Department of Geography, University of Chicago.

Ludwig, D., Walker, B. and Holling, C. S. (1997). 'Sustainability, stability, and resilience'. *Conservation Ecology*, 1(1): 7 [Online]. Available at: www.consecol.org/vol1/iss1/art7/

Paganini, Z. (2015). *Underwater: the production of informal space though discourses of resilience in Canarsie, Brooklyn*. Paper presented to the session on Planning for Resilience in a Neoliberal Age at the 2015 Annual Meeting of the Association of American Geographers, 25 April.

Porter, L. and Davoudi, S. (2012). 'The politics of resilience for planning: a cautionary note'. *Planning Theory and Practice*, 13(2): 329–33.

Siemens (2013). *Toolkit for resilient cities*. Available at: http://w3.siemens.com/topics/global/en/sustainable-cities/resilience/Documents/pdf/Toolkit_for_Resilient_Cities_Summary.pdf

Simmie, J. and Martin, R. (2010). 'The economic resilience of regions: towards an evolutionary approach'. *Cambridge Journal of Regions, Economy and Society*, 3(1): 27–43.

Vale, L. J. (2014). 'The politics of resilient cities: whose resilience and whose city?' *Building Research and Information*, 42(2): 191–201.

Walker, B. and Salt, D. (2006). *Resilience thinking: sustaining ecosystems and people in a changing world*. Washington, DC: Island Press.

Walker, J. and Cooper, M. (2011). 'Genealogies of resilience: from systems ecology to the political economy of crisis adaptation'. *Security Dialogue*, 42(2): 143–60.

Welsh, M. (2014). 'Resilience and responsibility: governing uncertainty in a complex world'. *The Geographical Journal*, 180(1): 15–26.

White, G. F. (1942). *Human adjustment to floods: a geographical approach to flood problems in the United States*. Research Paper 29. Department of Geography, University of Chicago.

White, I. and O'Hare, P. (2014). 'From rhetoric to reality: which resilience, why resilience, and whose resilience in spatial planning?' *Environment and Planning C: Government and Policy*, 32(5): 934–50.

Zolli, A. (2012). 'Learning to bounce back'. *New York Times*, 2 November. Available at: www.nytimes.com/2012/11/03/opinion/forget-sustainability-its-about-resilience.html?_r=1

Zolli, A. and Healy, A. (2013). *Resilience: why things bounce back*. London: Headline.

Part 2
Environmental governance

Richard Cowell, Simin Davoudi, Iain White and Hilda Blanco

In this part of the Companion, we turn from concepts of 'environment' and 'nature' to consider how environmental problems are addressed, a subject we have labelled 'environmental governance'. Governance can be regarded as a descriptive label, useful for highlighting the changing nature of policy processes, and the way that governing is achieved (White, 2015; Richards and Smith, 2002). Probe beyond this general definition, however, and one encounters a complex and contested world in which questions abound: who is governing, over what spatial scale and with what effects? Analysts of governance are often especially concerned with shifts in the modes of governance – classically summarised as state, market and hierarchy – and the myriad connections between them, in the ways that influence is exerted. Environmental problems have themselves alerted to us to the challenges of effective governance, and created pressures to re-think the modes, scale and goals that governance arrangements should take. And as environmental governance has expanded, so we have seen increased interest in the social dimensions of such activities; asking not merely does environmental governance work for the environment, but are its outcomes and processes socially just?

The upsurge in interest in promoting sustainable development, visible around the world through the 1990s, coincided with wider debates around the changing nature of the state, and whether traditional state-centred approaches to the delivery of goals, often referred to as 'government', were shifting towards more networked, multi-actor modes of steering. Such shifts heralded a much messier institutional world for environmental planning but one that also offered greater sensitivity to the diverse, complex realities of steering societies towards greater sustainability. Non-governmental organisations (NGOs) being more engaged in policy making, efforts to align environmental and business interests and the cross-scale nature of much environmental action are just three examples to which one could point. States are still present in governance processes – often directly and coercively so – but also manifest in techniques by which they seek to coordinate the actions of others, rather than relying solely on traditional 'command and control'. Understanding the power of governance forms requires that we give greater attention to the actors – plus their tools and goals – that hold together these new governance systems and give them reach across time and space.

The chapters in this part intersect with issues of governance in diverse ways but have been organised into three themes: goals, actors and arenas. The first three chapters examine the goals and principles that society should pursue and considers how they can be better institutionalised in governance systems. Dimensions of justice feature prominently each.

Goals of intra-generational equity are central to **Chapter 9** by **Juliana Maantay**, which focuses on environmental justice. She charts the growing recognition that the poor and people of colour tended to be inequitably affected by environmental burdens but also the issues that arise in trying to cement such concerns into planning processes. Her review covers efforts to construct quantitative, spatial assessments of the social distribution of environmental goods and bads, and the challenges of linking spatial associations to actual health effects. Maantay examines environmental planning interventions that seek to raise the visibility of environmental justice problems (like 'Fair Share' guidelines and participatory geographic information systems [GIS]) but also those that exacerbate them, highlighting the injustices that can arise from zoning and 'green gentrification'. She observes how constructions of spatial scale attached to justice claims can distort our conceptions of fairness, with regional or national framings obscuring the displacement of harms to other countries. Climate change is the prime example, and dealing fairly with climate change is, says Maantay, the 21st-century's environmental justice challenge.

In **Chapter 10**, **John Barry** highlights the pathologies of unquestioning adherence to economic growth as a goal in modern society. He notes how obsession with increasing gross domestic product (GDP) has undermined planning in many countries, which is often reduced to a facilitator of market-based growth and competitiveness and trammelled by pressures for deregulation and accelerated development delivery. His chapter also traces how planning's growth dependence is intertwined with fossil fuel consumption, in assumptions of increased (car-based) mobility and carbon-intensive sprawl. For Barry, planning must play a central role in attaining any 'just transition' to post-growth and low carbon futures, and he makes the case for incorporating both ecological 'ceilings' and social justice 'floors' (after Raworth, 2017) as steering devices. Promoting such transitions entails a more pro-active and creative role for planning – and the state more widely – in coordinating and fostering creative responses; a vision of planning much broader than a mere corrective mechanism to be wheeled in when markets fail.

The underrepresentation of future generations is addressed by **Jonathan Boston**, in **Chapter 11**, in his review of the scope for anticipatory governance in the environmental field. From the recognition that anticipating future problems is essential to effective environmental planning, Boston explains why environmental problems are especially vulnerable to short-termist bias: arising from inter alia the effects of uncertainty, power asymmetries between present and future, and the relative intangibility of future environmental effects. To redress this, Boston explores sets of 'intervention logics' that might mainstream long-term interests and encourage timely preventative solutions: legal constraints on policy and decision makers; improving foresight capacity, and better accounting for environmental assets that underpin cross-generational welfare. In a reminder of the social and political embeddedness of environmental governance, he explains how hard it is to separate the efficacy of these interventions from the wider political culture, including qualities like trust and solidarity, which affect the capacity of societies to create and maintain durable agreements.

Boston's chapter makes clear that refining the goals for environmental planning needs to proceed in tandem with attention to how those goals get picked up and translated into practice. This brings us to the second theme in this part of the Companion – **actors**, which is addressed across six chapters.

The first of these, **Chapter 12** by **James Palmer**, helpfully brings together discussion of a key actor quality (expertise) with two factors that mediate the relationship between actors in environmental planning: knowledge and trust. Palmer articulates the importance of seeing knowledge and expertise not as pure, universal, science-based categories separate from the political worlds of environmental planning but as emergent, culturally specific phenomena, intimately bound up with social, political and economic concerns. He shows how if we take care to interpret how and when different forms of knowledge come to matter, and whose judgement it is that we come to trust and on what basis, we can see that what constitutes knowledge and expertise with environmental issues is constantly shifting, with qualities often reflecting the specific, situated controversies being addressed. Indeed, effective expertise can derive precisely from the capacity to fuse scientific knowledge with contextually specific concerns. Yet this also explains why experts face difficulties speaking authoritatively across the heterogeneity of environmental planning controversies.

In **Chapter 13**, **Nathalie Berny** and **Ellen Clarke** examine two sets of actors that have long been seen as central to the dynamics of environmental policy and planning: grassroots and environmental NGOs. They review the expansion of these categories of environmental mobilisations within and beyond the West, and stress the need to challenge simplistic dichotomies between 'local' or 'radical' grassroots organisations and 'reformist' NGOs. To do this, Berny and Clarke show the insights that can be gained by focusing on the transformative effects of environmental NGOs and grassroots organisations, and the actual strategies that each adopts – especially the relationships between them. They show how multi-scalar strategies have become a key feature of the shifting dynamics of environmental activism, as grassroots groups use strategies that (after Rootes, 2013) 'transcend the local', such as networking with actors in different locations and appealing to non-local environmental NGOs for support. An unfortunate indicator of the strength of this multi-scalar working, they note, is the growth of national legislation seeking to prevent transnational cooperation between NGOs.

As the limits of government capacities to resolve environmental problems have become clearer, so there has been an upsurge in interest in how the individual could be enrolled to co-produce environmental planning outcomes. This is the terrain for **Chapter 14** by **Erin Roberts**, which assesses the array of conceptual approaches to understanding sustainable behaviour. She begins with the 'cognitive paradigm', which locates agency firmly in the individual and explains behavioural outcomes from variations in dispositional factors. Roberts then outlines how 'contextual paradigm' approaches view the agency of individuals as shaped by diverse external structures, for example, rules, norms and material elements of social life. Seeking to transcend the agency-structure dualism are theories focusing on social practices ('practice theories'), which maintain an interest in material infrastructures and technologies but incorporate the agency of publics as embodied individuals, with bodies and minds that affect how practices are performed. As Roberts makes clear, the preferred conceptual approach matters greatly, as how individual agency is understood can bias for particular types of policy intervention, as well as according people very different scope for moral reflection.

Moral reflection on individual rights and responsibilities is central to debates about 'green citizenship', the subject of **Chapter 15** by **Bronwyn E. Wood** and **Kirsi Pauliina Kallio**. They explain why conceptualising citizenship appropriately for environmental challenges has created much contestation, with the dominant citizenship traditions – liberal and civic republican – both relying excessively on normative models and failing to account for the structures of injustice and power. These dominant traditions also overlook the multiple ways in which people actually live and act as citizens, such as actions that foment connections to environmental problems at spatial scales – local and transnational – outwith the confines of the nation state. Drawing on Isin (2008) and Latta (2007), Wood and Kallio note that citizenship is not something that needs to be in place *before* action on environmental issues but can emerge from existing struggles for greater sustainability, as people and communities become political. By acknowledging its practice-based dimensions, expressions of citizenship can be seen as diverse and spatially complex.

Growing interest in the networked nature of many spheres of environmental governance has prompted more attention to the elements that facilitate networks and hold them together. **Ross Beveridge** explores these 'in-between actors' – intermediaries – in **Chapter 16.** Intermediaries can be defined as having roles in facilitating, brokering, negotiating and disseminating knowledge, which are crucial to the functioning of networks, and they have expanded in significance where governments have retreated from direct policy delivery. But this relational definition of intermediaries – based on their role and position vis a vis other categories of actors – can also make them elusive, mutable and unstable. Thus, consultancies and regulators might be classic intermediaries but so too, depending on the situation, might be environmental NGOs or local government. Beveridge makes clear that intermediaries are not just neutral go-betweens but inevitably interpret and translate the tasks and knowledge that they handle, often profiting from their role, with implications for other actors and the outcomes achieved. This, he argues, makes them clearly political, raising questions of accountability and transparency.

This recognition of the interwoven nature of the actors involved in environmental planning is expanded to embrace nonhuman actors in **Chapter 17** by **Jonathan Metzger**. He questions the dominant, Western conception of environmental planning as exercises in correcting imbalances between human life and its surrounding but separate environment. Instead, the human predicament should be seen as fundamentally entangled with and dependent on the fate of innumerable other beings. Such 'more-than-human approaches' to environmental planning take us beyond commonplace notions of human dependence on ecosystems. They force us to recognise the sheer difficulty of ever fully knowing how the planet's life-sustaining webs unfold, the deep entanglement between human existence and the existences of other entities and the challenges of becoming morally sensitised to these attachments. In considering how more-than-human perspectives might reshape environmental planning, Metzger finds clues in experiments with expanding human needs analyses that include other beings, some drawing on indigenous peoples' ways of relating to the Earth, because of the way they highlight interdependencies and avoid encouraging fantasies of unitary, sovereign subjects.

The third theme, and a crucial dimension of environmental governance, concerns the **arenas** in which policy and planning debates play out, their spatial and institutional contours, and how they affect the exercise of agency.

A pivotal set of arenas is addressed in **Chapter 18** by **Patricia McCarney**. She summarises the expanding expectations being placed on cities to deliver more sustainability and resilience to their growing citizenry and three sets of challenges that they face. First, cities face challenges in pulling together governance processes that are rooted in different scales and sectors, as well as governing cohesively those processes that flow across jurisdictional boundaries – an endemic struggle as cities spread. Second are the challenges of bringing down to local level those sustainable development agendas configured by global institutions. McCarney pinpoints the high salience of urban governance to many of UN's 'Sustainable Development Goals' established in 2015 but also the relative invisibility of cities in this agenda. To this she adds a third challenge, the need to further develop globally standardised sets of sustainability indicators by which benchmarking, learning and progress can be achieved.

As well as the more tangible arenas of governments and business boardrooms, environmental planning also unfolds within the virtual, less clearly bordered arenas of media and communications, with wider implications for the cultural politics of the environment. This is the focus for **Chapter 19** by **Marisa B. McNatt**, **Michael K. Goodman** and **Maxwell T. Boykoff**. They explain the dynamic, multi-directional communication processes between science, policy, publics and the media: science and politics have clearly shaped media coverage of climate change, but mass media 'news' has also shaped ongoing scientific and political considerations, deliberations and decisions. McNatt et al. stress the need to consider shifting modes, with the rise in digital and social media, raising a host of important questions: how does this affect the quality and quantity of coverage or affect public awareness and engagement? How has this shift affected the distribution and power of 'claims makers', including the attribution of 'expertise'? Simply pushing for 'more media coverage' of the environment is no panacea for better environmental planning, as the form and style of coverage can affect beliefs about the efficacy of action.

One theme that connects many of the chapters in Part 2 is the recognition that our understanding of environmental governance – our theories, analyses and prescriptions – always need sensitivity to the heterogeneity of situated problems with which environmental planning is asked to wrestle (Marres, 2007). It is to these problems that we turn to next in Part 3, 'Critical environmental pressures and responses'.

References

Isin, E. (2008). 'Theorising acts of citizenship', in E. Isin and G. M. Nielsen (eds.) *Acts of citizenship*. London/New York: Palgrave Macmillan, pp 15–43.

Latta, A. (2007). 'Locating democratic politics in ecological citizenship'. *Environmental Politics*, 16(3): 377–93.

Marres, N. (2007). 'The issues deserve more credit: pragmatist contributions to the study of public involvement in controversy'. *Social Studies of Science*, 37(5): 759–80.

Raworth, K. (2017). *Doughnut economics*. White River Junction, VT: Chelsea Green Publishing.

Richards, D. and Smith, M. J. (2002). *Governance and public policy in the UK*. Oxford: Oxford University Press.

Rootes, C. (2013). 'From local conflict to national issue: when and how environmental campaigns succeed in transcending the local'. *Environmental Politics*, 22(1): 95–114.

White, I. (2015). *Environmental planning in context*. London: Palgrave Macmillan.

9
Environmental justice and fairness

Juliana Maantay

What is environmental justice? What is fairness?

Environmental Justice (EJ) is a concept now acknowledged internationally as a major factor in planning and policy, public health and human rights, and although affecting virtually every part of the globe, the initial recognition of EJ began in the United States. 'Three out of every five Black and Hispanic Americans lived in communities with uncontrolled toxic waste sites' (United Church of Christ, 1987, p. xiv). This was just one of the many shocking findings in the United Church of Christ's 1987 seminal report *Toxic wastes and race*, leading to charges of 'environmental injustice' and even 'environmental racism' in the United States. New cases of environmental injustice unfold every day. In 2007, a follow-up report at the 20th anniversary of the original 1987 publication painted an almost equally bleak picture of environmental justice in the United States (Bullard et al., 2007). Exactly what is environmental justice (or injustice), and how can we identify and address it?

A wide array of definitions is given for 'environmental justice', ranging from the official ones promulgated by governmental oversight agencies to ones developed by grassroots organisations and the communities most affected by environmental injustice to ones put forward by academics, researchers and expert advocates. Some definitions touch on just the basics, and others are more expansive, situating environmental justice in a context of associated human rights to housing, health care and economic equality (Bullard, 1994; Harvey, 1997; Hofrichter, 1993; US EPA, 1995). Although many definitions have focused on the disproportionate environmental burdens borne by the poor and communities of colour, expanded definitions of EJ include other vulnerable populations, such as the very young, the elderly, the infirm and immune-compromised, pregnant women, immigrants and future generations (Greenberg, 1993). At a minimum, EJ involves the right for all people to live in a healthy environment, one free from hazardous conditions and noxious substances, and adherence to the goal that no specific population should bear a disproportionate burden of the waste products of modern life and industry.

Although many definitions of EJ have similar aspirations at their core, some stress equal enforcement and implementation of environmental laws and regulations, and others take a more distributional tact, emphasising the uneven geographical locations of pollution as the root of the problem, concentrating pollution in some areas and not others, affecting some populations and not others. These two approaches are related but use different strategies in attempting to achieve environmental justice. The regulatory approach implies that if environmental laws were effective in preventing human and ecological harm and were applied with the same vigour everywhere, we would have less environmental injustice. The distributional approach implies that if pollution were not concentrated in some areas but rather distributed evenly, we would have less environmental injustice. Is either of these approaches capable of addressing the problem?

The concept of EJ is predicated on the ethics of 'fairness', but here is where things get complicated. We may think we understand the concept of 'fairness', but what is fairness in the context of environmental justice? For instance, would it be 'fair', and would environmental justice be served if everyone was polluted equally? The definitions of environmental justice often include the words 'equity' and/or 'equality' (Bryant, 1995). These words are frequently used interchangeably but have generally come to denote quite different aspects of justice. Equity and equality often are equated, respectively, with 'process' and 'outcome' forms of justice. Equity is said to pertain to 'fairness' in administrative and regulatory procedures, including equal opportunity to participate in decision-making processes, while equality connotes evenness of results (or the real potential for equality of results) (Renn et al., 1995). To avoid confusion, in this chapter the word equity is used in its broadest sense to encompass both procedural equity and outcome equality, meaning not only 'fairness' but also the potential for actually achieving equality of outcomes.

Nevertheless, the difference between process equity (or procedural fairness) and outcome equity is an important distinction to make since the way policy makers have responded to accusations of environmental injustice has primarily been to create additional public participatory processes, 'Fair Share' guidelines and other mechanisms to try to ensure 'fairness' in siting noxious facilities. However, recent history has shown us that process fairness has not necessarily resulted in equitable outcomes.

Unfortunately, many governmental agencies have framed the equity question in terms of fairness in distribution and siting, which assumes the existence of pollution as a natural and accepted part of industrial production and modern life, rather than focus on eliminating the pollution for everyone.

The beginnings of the EJ movement

The United States is an important starting place for understanding the emergence and evolution of the EJ movement. Developed nations in Europe and elsewhere were much slower than the United States to recognise the impacts of environmental injustice in their own countries. This was partially due to the way environmental injustice had been framed initially: primarily as a racial discrimination issue. However, since the populations of many developed countries contain a much smaller proportion of racial and ethnic minorities than does the United States, the fact of environmental injustice was not as readily apparent or supported by quantitative analysis due to the small numbers statistical problem. But once the definition of EJ was expanded to include other factors, such as class, poverty, immigrant status and overall deprivation, similar disproportionate impacts were found throughout the world (Adeola, 2009; Basu and Chakraborty, 2016; Carruthers, 2008; Chaix et al., 2006; Dunion and Scandrett, 2003; Grineski and Collins, 2008; Margai and Barry, 2011).

The EJ movement in the United States began in earnest in the 1980s. It was preceded by a strong wave of new-found environmentalism in the United States shortly after WWII, and the general public was becoming more aware of the dangers of pesticides; degradation of water and air quality; the problems of population growth, hazardous waste disposal, and environmental health impacts; and habitat destruction and the consequent loss of species. This heightened awareness was thanks in large part to publications that popularised new ways of thinking about environmental problems and the need to take immediate action, such as Rachel Carson's *Silent Spring* (1962), Paul Ehrlich's *The Population Bomb* (1968) and Buckminster Fuller's *Operating Manual for Spaceship Earth* (1968). These seminal books helped inspire the modern day mainstream environmental movement, culminating in the first Earth Day celebration and the creation of the federal Environmental Protection Agency, both in 1970. But it was some nationally notorious events (for example, the Cuyahoga River near Cleveland, Ohio, bursting into flames and the infamous Love Canal housing development in Niagara Falls, New York, built atop an illegal hazardous waste dump, with a concomitant plethora of rare diseases amongst the inhabitants) that galvanised public opinion.

This environmental movement was focused on conservation and preservation of the natural world and the desire to live in cleaner environments. It was assumed that bad environmental practices and pollution affected all of us more or less equally, and it was not until the 1980s that there was a growing recognition that poor people and communities of colour were much more adversely impacted by environmental burdens than was the rest of the country's population.

The 1987 *Toxic wastes and race* report by the United Church of Christ's Commission on Racial Justice quantified the disproportionate siting of hazardous waste facilities amongst communities of colour. This damning analysis was quickly followed by dozens of other studies by researchers and activists all over the United States, which looked at race and other socio-economic characteristics in relation to the location of a variety of environmentally burdensome facilities and land uses (White, 1998). Nearly all of the US-based research found a strong spatial correspondence between race and poor environmental conditions, including exposure to accidental chemical releases, Toxic Release Inventory facilities, Superfund, Petrofund and hazardous waste disposal sites (Goldman, 1993; Maantay, 2002a). There were also a number of critical commentaries on environmental justice, which helped make it a serious topic for academic study (for example, Pellow and Brulle, 2005; Pulido, 1996).

In addition to the surge in scholarly interest in EJ, dozens of grassroots and community-based organisations were formed and became active in the struggle for environmental justice (Perez et al., 2015; Taylor, 2011), some of which became important voices in the overall environmental movement.

The analysis of environmental justice – measuring EJ

The existence of environmental injustice was well known by many of those who were living it, and it was well documented anecdotally through verbal narratives and other qualitative means. But after the pattern was named and identified as characterising a certain condition, EJ needed to be measured using quantitative methods. It is difficult to solve a problem unless it is understood and quantified.

This coincided with the advent of more user-friendly geographic information systems (GIS), and it has been argued that without GIS, EJ research would not have exploded the way it did. GIS – computerised mapping and spatial analysis – made it possible to move EJ research beyond just anecdote and theoretical critical commentary. GIS allowed a systematic body of evidence to accumulate, using quantitative methods with duplicatable results, which meant EJ research could be considered 'scientific' and thus more 'useable' to policy makers and government officials (Maantay, 2002a).

For the next 15 years, most research surrounding EJ issues in the United States was undertaken essentially to 'prove' the existence of environmental injustices by looking at the locations of environmental 'bads' in relation to where minority populations and poor people lived. This was later expanded to examine the flip side of this – access/lack of access to environmental 'goods', such as healthy food options, parks and open spaces, and other health-promoting amenities (Maroko et al., 2009).

In nearly every study, race was a predictor for disproportionate environmental burdens (Maantay, 2002a). These results were echoed in many other studies in subsequent years, as well as studies finding curtailed access of minority and poor populations to environmental 'goods'. The cartographic and geostatistical evidence was overwhelming: poor people and minority people tended to live in much less salubrious places: places that were more polluted, with less access to things that make the environment good, and by extension, that could make their lives good.

Largely missing from these early EJ exposés, though, was an analysis of the actual impacts, health and otherwise, of living in areas with poor environmental conditions. Many of the EJ analyses conducted up to this point had looked at simple proximity to a hazard, and everyone captured within a certain distance radius was deemed to have been exposed. But what were the actual differences in people's health, life expectancy and wellbeing between those living near hazardous conditions and those who were not? One of the most important trends in EJ research has been the attempt to identify the populations affected and quantify the adverse health outcomes resulting from environmental injustice.

This type of health-based analysis was difficult to come to grips with, in terms of valid research design. Datasets on health outcomes are hard to obtain, especially at a patient record level, due to confidentiality restrictions, and aggregating health outcomes to larger geographic extents tends to obscure the correlations between environmental conditions and disease. Additionally, many diseases have long latency periods, and people are mobile, making it nearly impossible to ascribe living in a certain location to likelihood of exposure or acquiring a disease. Therefore, it's more complicated to prove the connection between adverse health outcomes and the disparate environmental factors in the home and work places (Maantay, 2007).

Actual exposure is, therefore, much less straightforward to measure and prove than simple proximity to a hazard, and there are many confounding variables and limitations in trying to definitively show correlation between adverse health outcomes and living in an environmentally hazardous area. However, we can now do more precise analyses to accurately delineate the geographical extent of the impacts and the numbers and characteristics of the potentially exposed populations. A wide variety of geostatistical methods have been applied to EJ studies, such as geographically weighted regression (GWR), spatial autoregressive (SAR) models, Bayesian modelling, cluster analysis, multi-level or hierarchical modelling, dasymetric disaggregation of population and modelling of actual and predicted environmental conditions (Chakraborty and Maantay, 2011).

Developments in life course dynamics and activity space studies have also become part of the standard toolbox to better assess the health-environment-EJ linkages. Once we can fully take into account people's intra-day movements, day-to-day travels, and their longer-term geographic mobility, we can more fully account for neighbourhood effects such as environmental and socio-demographic conditions (Kwan, 2013). Substantial work has also been done on creating indices for various purposes, such as ranking vulnerability to gentrification; differential risk from flooding and other natural and industrial hazards; the segregation/health relationship; and psycho-social stressors (Maantay, 2013; Maantay et al., 2010).

A relatively new development in EJ research is 'visual storytelling' or cartographic narrative techniques (Moore et al., 2017). These methods are less purely quantitative and more intuitive and involve mining data for exploratory purposes and communication. This kind of 'undisciplined' geo-visualisation can be a valuable tool in EJ research, and it can be effectively coupled with participatory GIScience processes, volunteered geographic information (VGI) and crowd-sourced data from social media. These are all excellent ways of portraying issues and bringing the story to the public's attention.

Through development of robust methodologies, we have become more adept at describing the problem, identifying the at-risk populations and demonstrating some of the health impacts. In a review of over 30 studies investigating the correlation between residential proximity to environmental hazards and adverse health outcomes (including pregnancy abnormalities, pre-term infants, childhood leukaemia, brain cancer, bladder cancer, lymphoma and various tumours, a majority of the studies showed a connection (Brender et al., 2011). But it should be cautioned that even this is no longer enough. Now the focus needs to be on using geospatial technologies to develop solutions, which brings us into the realm of policy and planning. In capitalist, market-based economies, how can at-risk vulnerable populations be protected? What is the best mechanism to address the issues of environmental injustice, health disparities and general inequalities?

Environmental justice policies and regulatory issues

Federal and state regulatory agencies

In the United States, EJ evolved from a social justice movement to real changes to the regulatory and legal environments. In 1982, grassroots protests and activism in reaction to the siting of a hazardous waste landfill in a majority Black community in North Carolina started a national conversation about 'environmental racism', which brought together two important and heretofore unconnected American movements: Civil Rights and Environmentalism. These protests helped to instigate the 1983 study on Environmental Justice by the US General Accounting Office and later the 1987 *Toxic wastes and race* study, two landmark documents in the history of EJ. Both of these reports confirmed the existence of disproportionate impacts of environmental hazards on communities of colour, specifically hazardous waste disposal sites. Virtually from the beginning, the EJ movement had the support of the faith-based community, similar to the impetus of the Civil Rights movement in the 1950s and 60s, thus giving the EJ movement a moral and ethical high ground that was difficult for politicians to argue against. In 1991, the First National People of Color Environmental Leadership Summit was held in Washington DC, which resulted in the delineation of the Seventeen Principles of Environmental Justice (Mohai and Bryant, 1992).

In 1994, President Bill Clinton signed Executive Order 12898 on Environmental Justice, which instituted the US federal government's official recognition of the need to assess the environmental justice implications of government programs, plans, policies and activities in accordance with Title VI of the Civil Rights Act of 1964. At the federal level, all governmental actions require an EJ assessment prior to new projects or policies being implemented. Federal agencies must consider environmental justice in their activities under the National Environmental Policy Act (NEPA).

The federal Environmental Protection Agency (EPA) also became active in EJ work in around 1992, creating an Office of Environmental Justice responsible for coordinating the Agency's efforts to integrate environmental justice into all policies, programmes, and activities, and has developed a new environmental justice mapping and screening tool called EJSCREEN (US EPA, 2016). This is based on nationally consistent data and is an approach that creates an index combining a number of environmental and socio-demographic indicators in maps and reports, in order to identify and prioritise what the EPA calls 'EJ communities'. Other offices within the EPA manage and disseminate data such as the Toxic Release Inventory (TRI) reporting, as well as information about locally used, stored and released hazardous substances under the Emergency Planning and Community Right to Know Act, and maintain various other publicly available datasets on environmental quality indicators that are mandated by law. These databases have played a crucial role in EJ research, allowing national level EJ analyses to be undertaken with consistent data, yielding more reliable results.

Environmental impact assessment

Most US states and many cities have their own local departments of environmental protection, within which environmental justice concerns may be addressed through the environmental assessment process, including a public participation component, although often the public has virtually no impact on the outcome despite involvement. Additionally, the entire environmental assessment process is advisory only, being used by the decision makers as one of many inputs in the approval process – its recommendations being non-binding.

Zoning and land use planning

Zoning and land use planning are potentially the most reliable controls municipalities have to combat environmental injustice but often have been used for the opposite effect. The ostensible purpose of zoning is to plan for the uses to which land will be put within a city or township, to protect the wellbeing of the residents, and allow commercial enterprises to flourish. Zoning is one of the most widespread and influential sets of regulations that exist in the United States and have the power to organise, shape and protect the character of their landscapes and welfare of their populations (Babcock, 1966). In fact, zoning acts as a gatekeeper in many places, dictating where the often necessary-but-noxious land uses can and cannot go (for example, waste water treatment plants, landfills, solid waste transfer stations, recycling centres) by specifying separation of certain uses, in effect ensuring environmental injustice by only allowing these uses in places that have been zoned for them. The places zoned for them are often in or near less affluent residential areas or communities of colour, thus having the effect of protecting the wealthier areas while sacrificing the less affluent. Zoning has been shown to be one of the root enabling causes of environmental injustice, and even revisions and updates to older zoning plans generally perpetuate environmental injustice by increasing the areal extent of industrial zones and the permissible level of environmental burden in less affluent and minority neighbourhoods, and reducing the areal extent of industrial zones and level of burden in 'better' neighbourhoods. This is done in such a normative way that it often escapes notice as a biased regulatory initiative or policy (Maantay, 2002b, 2002c, 2002d).

NIMBY-ism and 'Fair Share' guidelines

NIMBY-ism is a phenomenon related to zoning and land use planning. '*N*ot *I*n *M*y *B*ack *Y*ard' is usually a local protest effort to stop some undesirable project from locating in a specific area for which it has been earmarked. The NIMBY approach is often effective and in the past had tended to pertain to mostly middle- and upper-class neighbourhoods having the political clout and know-how to get things done within the established system. The contentious land use would then often be shifted to a poorer or more minority neighbourhood, where (at least in the past) the people were less politically powerful and, for a variety of reasons, were less able to combat the project (Heiman, 1990; Lake, 1993).

Some cities, notably New York, instituted 'Fair Share' guidelines to help address this inequitable state of affairs after it was pointed out that poor neighbourhoods were usually the recipients of unwanted land uses (NYC DEP, 1991). These guidelines recommended a procedure for examining the distribution of unwanted land uses within a community and trying to ensure that any one area was not overly saturated with them. Zoning still effectively limits the possible locations of many facilities, and physical requirements dictate their locations (waste water treatment plants, for instance, need to be on the waterfront for operational reasons). But it did put private developers on notice (as well as city planners) that EJ would need to be taken into consideration more comprehensively now and that it would be procedurally more difficult to ignore the wellbeing of people in poor neighbourhoods, just as it had always been (politically) impossible to ignore the wishes of the wealthier ones.

Community-led planning and public participatory GIS

Many cities and even some counties have instituted procedures supporting community-led plans and planning initiatives. There are provisions for communities to engage consultants and develop their own proactive planning (rather than the reactive planning often engendered by NIMBY-ism).

Community-led planning often includes Public Participatory GIS (PPGIS), which involves community members in the planning process by utilising GIS as 'a way of enhancing local peoples' abilities to share and analyze their knowledge of lifestyles and conditions' (Chambers, 1994, p. 3) and has proven to be an effective way to inject a degree of community input and control into the planning process. PPGIS has been effectively used all over the world, from urban to rural to tribal and indigenous environments (Harris et al., 1995), and has been particularly successful in counter-mapping – the production of maps and spatial analyses for advocacy and activist purposes to challenge the official viewpoint of the community and to offer a measure of empowerment to local areas trying to plan and take charge of their own futures. This idea of counter-mapping and PPGIS has been instrumental in the demonstration and rectification of environmental injustices (Aberley, 1993; Weiner et al., 1995).

Green gentrification – a new form of environmental injustice

Gentrification also influences environmental injustice. When private or municipal development schemes are proposed for less affluent or minority neighbourhoods, the result is often displacement or worsening economic conditions for the poor or minority residents already living in the community. 'Green' gentrification occurs when private or public improvements in green amenities (parks, open space networks, waterfront improvements, etc) result in rising property values and subsequent marginalisation and displacement of the original residents (Gould and Lewis, 2016). Even community-led improvements such as community gardens can have this effect by improving property values, enhancing neighbourhood aesthetics, reducing crime rates and thus inadvertently encouraging outside investment in their neighbourhood, whereby through their own hard work the community has sown the seeds of their own destruction.

'Greening' plans need to be contextualised to ensure that this new greener future will benefit all the residents of the city and not follow the all too common trajectory of exclusion, displacement and expulsion of the poor and otherwise marginalised people from the improved areas. This, unfortunately, only results in a different but equally pernicious type of environmental injustice.

International implications of environmental justice

Environmental justice has been studied in locations worldwide, with similar results as those in the United States. However, beyond the EJ ramifications for individual locations, the global impacts transcend any given city or country, and it is essential to take the issue of scale into account. Environmental injustice is tricky to measure with any confidence because analyses can be skewed by the geographic extent of the study area as well as the unit of data aggregation used. What might appear as environmental injustice at the city level – using the census enumeration unit or postal code as the unit of analysis – might yield very different results at the state level using counties as the unit or at the national level using states as the unit (Maantay, 2007). Scale is an important consideration in any examination of EJ. Although it is practical and necessary to use pre-defined units of analyses, there really isn't any logical reason why jurisdictional boundaries are thought to be realistic in depicting EJ since environmental injustice does not occur in a geographic vacuum or in isolation from its surrounding/bounding areas. Highly precise quantitative analyses can often mask the true EJ outcomes and be misleading. This has been one of the main challenges in EJ studies.

For instance, in NYC, all landfills have been closed in the past decade due to public outcry against them and acknowledgement of their detrimental impacts, and have now mainly been turned into parks, so there is no longer a localised EJ concern for the residents proximate to the old landfills. However, the city's solid waste and sewage sludge hasn't just disappeared; it is exported to other states and countries, where governments are more willing to take a risk for economic gain, rather than prioritise safeguarding their residents' wellbeing. This enlarges the sphere of influence of NYC's garbage and sewage waste and thereby creates an environmental justice spill-over effect to areas far removed from the origin of the pollution.

Similarly, in 2006 there was an event which has been categorised as being amongst the worst man-made environmental disasters on record, namely the illegal and improperly handled dumping of more than 500 tonnes of toxic waste in the west African nation of Cote d'Ivoire, after which over 100,000 people became ill (Margai and Barry, 2011). The waste had originated in refinery operations of a Mexican state-owned petroleum corporation and had been refused at several European and other African ports. When the ship finally reached the Port of Abidjan, the toxic stew was presented as relatively harmless 'ship's slops' since it was known that the city had no proper facilities for storing toxic waste. It was off-loaded and spread across the city and surrounding areas, dumped at night in waste grounds, public dumps, and along roads in populated areas, over the course of three weeks. The substance gave off toxic gas and resulted in burns to lungs and skin, as well as severe headaches and vomiting, and the toxins also contaminated the food chain. The areas selected for the dumping tended to be locations inhabited by ethnic minority populations, who were also marginalised based on religion and language. This is only one instance in what has become a pattern of resource exploitation, toxic contamination and transnational pollution visited upon the less developed countries by corporations and governments in the developed world.

In fact, in terms of a global assessment of EJ, the United States and other developed nations perpetrate environmental injustice on the less developed countries, regardless of and in addition to the environmental injustice borne by sub-populations within their own countries (Westra and Lawson, 2001). It's a vast transference and shifting around of the pollution burden from the creators of the problem (who at least can be said to benefit materially from its creation) to the parts of the world whose populations are least able to absorb the tragic health and ecological impacts of this 'trade' and who benefit least from the rampant consumerism and concomitant natural resource extraction that makes over consumption in developed countries possible (Castleman and Navarro, 1987). 'We won't be able to achieve sustainable development until we get justice in environmental protection . . . not only within the border of the United States but in the policies that are being exported abroad' (Bullard et al., 2007, p. ix).

The latest in the long list of transferred environmental justice issues is that of global climate change. The 5% of the world's population living in the United States is responsible for 20% of the world's greenhouse gas contributions, yet it will be primarily the people who live in coastal and other low lying areas in the less developed countries and in other vulnerable at-risk locations who will feel the brunt of climate change impacts and have the fewest resources to deal with them (Maantay and Becker, 2012). This is the environmental injustice story of the 21st century and one that perforce must be tackled on a global basis.

References

Aberley, D. (1993). *Boundaries of home: mapping for local empowerment*. Philadelphia, PA: New Society Publishers.

Adeola, F. O. (2009). 'From colonialism to internal colonialism and crude socioenvironmental injustice: anatomy of violent conflicts in the Niger delta of Nigeria', in F. C. Steady (ed.) *Environmental justice in the new millennium: global perspectives on race, ethnicity and human rights*. New York: Palgrave Macmillan, pp 135–63.

Babcock, R. (1966). *The zoning game*. Madison, WI: University of Wisconsin Press.

Basu, P. and Chakraborty, J. (2016). 'Environmental justice implications of industrial hazardous waste generation in India: a national scale analysis'. *Environmental Research Letters*, 11: 125001.

Brender, J., Maantay, J. and Chakraborty, J. (2011). 'Residential proximity to environmental hazards and adverse health outcomes'. *American Journal of Public Health*, 101(S1): S37–S52.

Bryant, B. (ed.) (1995). *Environmental justice: issues, policies, and solutions*. Washington, DC: Island Press.

Bullard, R. (1994). *Unequal protection: environmental justice and communities of color*. San Francisco, CA: Sierra Club Books.

Bullard, R., Mohai, P., Saha, R. and Wright, B. (2007). *Toxic waste and race at twenty, 1987–2007: a report prepared for the United Church of Christ, Justice & Witness Ministries*. Cleveland, OH: United Church of Christ. Available at: www.nrdc.org/sites/default/files/toxic-wastes-and-race-at-twenty-1987-2007.pdf

Carruthers, D. V. (2008). *Environmental justice in Latin America: problems, promises and practice*. Cambridge, MA: MIT Press.

Carson, R. (1962). *Silent spring*. Boston, MA: Houghton Mifflin Harcourt Co.

Castleman, B. I. and Navarro, V. (1987). 'International mobility of hazardous products, industries and wastes'. *Annual Review of Public Health*, 8: 1–19.

Chaix, B., Gustafsson, S., Jerrett, M., Kristersson, H., Lithman, T. and Boalt, A. (2006). 'Children's exposure to nitrogen dioxide in Sweden: investigating environmental injustice in an egalitarian country'. *Journal of Epidemiology and Community Health*, 60(3): 234–41.

Chakraborty, J. and Maantay, J. A. (2011). 'Proximity analysis for exposure assessment in environmental health justice research', in J. A. Maantay and S. McLafferty (eds.) *Geospatial analysis of environmental health*. Dordrecht: Springer-Verlag, pp 111–38.

Chambers, R. (1994). 'The origins and practice of participatory rural appraisal'. *World Development*, 22(7): 953–69.

Dunion, K. and Scandrett, E. (2003). 'Environmental justice in Scotland as a response to poverty in a Northern nation', in J. Agyeman, R. Bullard, R. and B. Evans (eds.) *Just sustainabilities: development in unequal word*. London, UK: Earthscan.

Ehrlich, P. (1968). *The population bomb*. New York: Sierra Club/Ballantine Books.

Fuller, B. (1968). *Operating manual for spaceship Earth*. Carbondale, IL: Southern Illinois University Press.

Goldman, B. A. (1993). *Not just prosperity: achieving sustainability with environmental justice*. Washington, DC: National Wildlife Foundation.

Gould, K. and Lewis, T. (2016). *Green gentrification: urban sustainability and the struggle for environmental justice*. New York: Routledge.

Greenberg, M. (1993). 'Proving environmental inequity in siting locally unwanted land uses'. *Journal of Risk: Issues in Health and Safety*, 4(3): 235–52.

Grineski, S. E. and Collins, T. (2008). 'Exploring patterns of environmental injustice in the global south: Maquiladoras in Ciudad Juárez, Mexico'. *Population and Environment*, 29(6): 247–70.

Harris, T., Weiner, D., Warner, T. and Levin, R. (1995). 'Pursuing social goals through participatory GIS: redressing South Africa's historical political ecology', in J. Pickles (ed.) *Ground truth: the social implications of GIS*. New York: The Guilford Press, pp 196–222.

Harvey, D. (1997). *Justice, nature, and the geography of difference*. Oxford: Wiley-Blackwell.

Heiman, M. (1990). 'From "not in my backyard!" to "not in anybody's backyard!": grassroots challenge to hazardous waste facility siting'. *American Planning Association Journal*, 56(3): 359–62.

Hofrichter, R. (ed.) (1993). *Toxic struggles: the theory and practice of environmental justice*. Philadelphia, PA: New Society Publishers.

Kwan, M.-P. (2013). 'Beyond space (as we knew it): toward temporally integrated geographies of segregation, health, and accessibility. Space-time integration in geography and GIScience'. *Annals of the Association of American Geographers*, 103(5): 1078–86.

Lake, R. (1993). 'Planners' alchemy transforming NIMBY to YIMBY: rethinking NIMBY'. *Journal of the American Planning Association*, 59(1): 87–93.

Maantay, J. A. (2002a). 'Mapping environmental injustices: pitfalls and potential of geographic information systems in assessing environmental health and equity'. *Environmental Health Perspectives*, 110(S2): 161–71.

Maantay, J. A. (2002b). 'Zoning, equity, and public health'. *American Journal of Public Health*, 91(7): 1033–41.

Maantay, J. A. (2002c). 'Zoning law, health, and environmental justice: what's the connection?' *Journal of Law, Medicine, and Ethics*, 30(4): 572–93.

Maantay, J. A. (2002d). 'Industrial zoning changes in New York City and environmental justice: a case study in "expulsive" zoning'. *Projections: The Planning Journal of Massachusetts Institute of Technology (MIT)*, 63–108. [Special issue: Planning for Environmental Justice]

Maantay, J. A. (2007). 'Asthma and air pollution in the Bronx: methodological and data considerations in using GIS for environmental justice and health research'. *Health and Place*, 13(1): 32–56.

Maantay, J. A. (2013). 'The collapse of place: derelict land, deprivation, and health inequality in Glasgow, Scotland'. *Cities and the Environment (CATE)*, 6(1): Article 10.

Maantay, J. A. and Becker, S. (eds.) (2012). 'The health impacts of global climate change: a geographic perspective'. *Journal of Applied Geography*, 33: 1–106.

Maantay, J. A., Maroko, A. R. and Culp, G. (2010). 'Using geographic information science to estimate vulnerable urban populations for flood hazard and risk assessment in New York City', in P. Showalter and Y. Lu (eds.) *Geotechnical contributions to urban hazard and disaster analysis*. Dordrecht: Springer-Verlag, pp 71–97.

Margai, F. M. and Barry, F. B. (2011). 'Global geographies of environmental injustice and health: a case study of illegal hazardous waste dumping in Cote d'Ivoire', in J. A. Maantay and S. McLafferty (eds.) *Geospatial analysis of environmental health*. Dordrecht: Springer, pp 257–81.

Maroko, A. R., Maantay, J. A., Sohler, N. L., Grady, K. and Arno, P. (2009). 'The complexities of measuring access to parks and physical activity sites in New York City: a quantitative and qualitative approach'. *International Journal of Health Geographics*, 8(34): 1–23.

Mohai, P. and Bryant, B. (1992). 'Environmental injustice: weighing race and class as factors in the distribution of environmental hazards'. *University of Colorado Law Review*, 63: 921–32.

Moore, S. A., Roth, R. E., Rosenfeld, H., Nost, E., Vincent, K., Arefin, M. R. and Buckingham, T. M. A. (2017). 'Undisciplining environmental justice research with visual storytelling'. *Geoforum*, 11 pp. Available at: http://dx.doi.org/10.1016/j.geoforum.2017.03.003

New York City Department of Environmental Protection (1991, Rev. 1998). *Fair share criteria: a guide for city agencies*. New York: NYC DEP.

Pellow, D. N. and Brulle, R. J. (2005). 'Power, justice, and the environment: toward critical environmental justice studies', in D. N. Pellow and R. J. Brulle (eds.) *Power, justice, and the environment: a critical appraisal of the environmental justice movement*. Cambridge, MA: MIT Press, pp 1–19.

Perez, A. C., Grafton, B., Mohai, P., Hardin, R., Hintzen, K. and Orvis, S. (2015). 'Evolution of the environmental justice movement: activism, formalization and differentiation'. *Environmental Research Letters*, 10: 105002.

Pulido, L. (1996). 'A critical review of the methodology of environmental racism research'. *Antipode*, 28(2): 142–59.

Renn, O., Webler, T. and Wiedmann, P. (eds.) (1995). *Fairness and competence in citizen participation*. Dordrecht: Kluwer Academic Publishers.

Taylor, D. E. (2011). 'Introduction: the evolution of environmental justice activism, research, and scholarship'. *Environmental Practice*, 13(4): 280–301.
United Church of Christ, Commission for Racial Justice (1987). *Toxic wastes and race in the United States: a national report of the racial and socio-economic characteristics of communities with hazardous waste sites*. New York: United Church of Christ.
United States Environmental Protection Agency (1995). *Draft environmental justice strategy for Executive Order 12898*. Washington, DC: USEPA.
United States Environmental Protection Agency (2016). *EJSCREEN user guide*. Washington, DC: USEPA. Available at: https://ejscreen.epa.gov/mapper/help/ejscreen_help.pdf
United States General Accounting Office (1983). *Siting of hazardous waste landfills and their correlation with racial and economic status of surrounding communities*. Gaithersburg, MD: USGAO. Available at: http://archive.gao.gov/d48t13/121648.pdf
Weiner, D., Warner, T. S., Harris, T. M. and Levin, M. R. (1995). 'Apartheid representation in a digital landscape: GIS, remote sensing, and local knowledge in Kiepersol, South Africa'. *Cartography and Geographic Information Systems*, 22(1): 30–44.
Westra, L. and Lawson, B. (2001). *Faces of environmental racism: confronting issues of global justice*. Oxford, UK: Rowman and Littlefield Publishers, Inc.
White, H. L. (1998). 'Race, class, and environmental hazards', in D. E. Camacho (ed.) *Environmental injustices, political struggles: race, class, and the environment*. Durham, NC: Duke University Press, pp 61–81.

10

Planning in and for a post-growth and post-carbon economy

John Barry

Introduction: the condition our condition is in

How does the old end and the new begin? How do we identify the point beyond which a practice, idea or objective is no longer beneficial, productive or necessary but has now become the opposite? This is the conundrum facing planning in the 21st century, how to decouple societies and economies from fossil fuel energy but also the role of planning in decoupling societies from an uncritical focus on achieving, facilitating or coordinating the achievement of orthodox 'economic growth'. Planning here is understood mainly as the purposeful, political and public (and ideally democratic) steering of infrastructural and mainly urban development, encompassing, inter alia, land use, spatial planning and energy planning and includes the achievement of economic, cultural and environmental goals. This chapter mainly discusses planning debates within the UK and Irish contexts.

The first question which serves as the starting point of this chapter is to ask if the objective of economic growth is now ecologically unsustainable, socially divisive and has in many countries passed the point when it is adding to human wellbeing? The second asks how growth and planning are both currently dependent upon a fossil fuel energy system which, like the growth economy it fuels, is now ecologically unsustainable, socially disruptive, produces multiple problems from ill health to extractive injustice and the creation of 'sacrifice zones' and ultimately constitutes a risky energy basis for a sustainable economy? Simply put our societies and conventional planning processes are dependent upon (some might say addicted to) gross domestic product (GDP) measured and endless economic growth and carbon energy, both of which have passed thresholds indicating we need to replace them. Which, of course, raises the addition question: if we accept or consider the exhaustion of endless growth and our continuing carbon energy dependence, what replaces them? And what is the role for planning in both that transition and the possible purpose of planning in a post-growth and post-carbon context? These are the main concerns of this chapter.

The structure of the chapter is as follows. The first section outlines what is meant by 'economic growth' and identifies some of the main 'pro-growth' bias (and carbon energy dependence) within dominant understandings of planning. These include, inter alia, planning's role in promoting policies and discourses of international competitiveness, the privatisation of public space, support for pro-market urban regeneration and a view of planning as facilitating market-based economic growth. It then proceeds to discuss some features of 'post-growth' and 'post-carbon' planning. These include the central role of planning in any 'just transition' to a low carbon economy; the place of a more proactive state in that planning process; and the integration of social justice 'floors' and ecological 'ceilings' into any post-carbon, post-growth planning. The chapter concludes that we need new imaginaries and objectives for planning in the 21st century, for conditions very different from those of post-war and late 20th century. We need to replace economic growth with notions of prosperity and ideas of planning creating prosperous and flourishing communities where high quality of life is decoupled not just from high carbon use, resource intensity and pollution but also simplistic and out of date objectives of 'economic growth'.

The pro-growth bias of conventional planning

A question that is rarely asked in public or political discourse is 'why do we need economic growth?', so self-evident is it that 'we' need 'it'. There is a small 'heterodox' academic political economy tradition critical of growth, including seminal works by Daly (1973) and Georgescu-Roegen (1971), for example (and more recent work by Jackson (2017) and Reardon et al. (2018) and an overview of this tradition see Barry [2012, pp. 151–4]), and landmark reports such as *Limits to Growth* (Meadows et al., 1972), and in the wake of the 2007–8 global economic crisis there was a short-lived but high-level political interest in 'moving beyond GDP' (Stiglitz et al., 2008) to focus on measuring wellbeing. However, this 'post-growth' body of knowledge is marginalised within the academy and receives no serious and sustained public discussion or political advocacy (apart from Green parties and sections of the environmental movement). Despite climate science indicating we need to question economic growth (Anderson and Bows, 2011), despite growing social psychological research highlighting the negative consequences of growth for human flourishing (Kasser, 2003) and epidemiological and sociological analysis pointing out how growth reproduces and deepens socio-economic inequality (Wilkinson and Pickett, 2010), economic growth remains a hegemonic 'core state imperative' and 'commonsense' social goal (Barry, 2018). It is economic growth we are told that ensures pensions for when we retire. It is economic growth that creates the wealth we can then use to spend on social or medical infrastructure (such as a national health care system) and welfare payments, or indeed remedial environmental policies. In this way, economic growth is a means to these various ends, and in growth-oriented societies, policies and practices from education to planning are both shaped to help achieve that end and increasingly judged by their contribution to realising economic growth. What is rarely aired is whether (i) growth really is a means to those ends and has not become an end in itself, (ii) those ends could be met without economic growth, and (iii) there is a threshold beyond which growth is no longer a net positive for societies?

First, we need to answer the question: what do we mean by 'economic growth'? By economic growth is meant undifferentiated, GDP measured growth as a permanent feature of the economy (Barry, 2018). Simply put, GDP is a measure of the 'busyness' of the economy, the monetised measurement of the market value of all final goods and services produced in a period of time (usually measured and commented upon etc. in the media, quarterly and annually). It is best thought of as measuring the financial value and volume of market exchanges. That only market exchanged goods and services are counted has important implications, not least in terms of 'what gets measured' not only 'gets done' but also 'gets recognised'. On the one hand, negative impacts of economic activity such as pollution is not included in GDP or a whole host of other negative 'externalities' such as ill health, anxiety, stress, etc. On the other, activities such as gendered housework or volunteering are excluded; hence, GDP only values and recognises 'employment' not 'work' per se.

GDP measured growth is *undifferentiated* in that it does not distinguish between economic activities and expenditure that many would view as socially negative. So being forced to have a long commute from one's home to place of work or study, due in part to spatial planning decisions and strategies, is a disutility for the individual (and also increases pollution if it's carbon-fuelled transportation, especially car-based), but this is good for GDP measured economic growth. These 'defensive expenditures' are people spending income (and resources and energy) to protect themselves and their welfare from the unwanted side effects (health, social, economic) of specific productive activities or due to the structure of the economy as a whole. In short, people are forced to spend income *that does not increase their welfare*. But GDP is indifferent and blind to those expenditures; whether its expenditures that add or subtract from welfare, they all add to economic growth. Growth is growth is growth. It is this that leads Mark Anielski to provocatively note that:

> The ideal economic or GDP hero is a chain-smoking terminal cancer patient going through an expensive divorce whose car is totalled in a 20-car pileup, while munching on fast-take-out-food and chatting on a cell phone. All add to GDP growth. The GDP villain is non-smoking, eats home-cooked wholesome meals and cycles to work.
>
> *(Anielski, 2007, p. 30)*

Related terms and concepts often found in the company of this growth discourse include 'attracting foreign direct investment', 'promoting free trade', 'competitiveness', 'increasing labour productivity', 'encouraging innovation', 'cultivating entrepreneurship', 'removing red tape and bureaucracy', 'reducing regulation' and so on, all of which are familiar to and woven within recent planning practices and theorising, of what Rydin terms 'growth-dependent planning' (Rydin, 2013). Discussing the British planning experience (though arguably relevant to most carbon-based, capitalist economies), she states that: 'Currently the institutional framework of the planning system embeds growth-dependent planning as the only alternative' (Rydin, 2013, p. 11), and that this dependence upon growth has increasingly shaped the planning profession (Rydin, 2013, p. 15).

The connection between growth and planning is a long one and can be traced historically, for example, to the 'public health' focus of urban planning in the late 19th and early 20th centuries, to the post-WWII context of economic and urban reconstruction and regeneration and Keynesian welfare state building (Barry, 2018). While there is evidence of planning being concerned with non-economic and non-growth objectives, such as public health or social inclusion (thus importantly reminding us that there is no *necessary* as opposed to *politically contingent* connection), in recent decades under neoliberalism the relationship between planning and growth has become even more accentuated. An illustration of this is that the Royal Town Planning Institute for example has a section on its website entitled 'Planning and Growth' where we are told:

> The RTPI's work programme on the value of planning seeks to promote and inform a more considered, balanced and evidenced debate on the relationship between planning and economic growth. Contrary to the views of some critics, *planning can and does play a positive and proactive role in sustainable economic growth and development.*
>
> *(RTPI website, emphasis added)*

Then there is also planning's implicit or explicit (and often uncritical) dependence upon conventional, mainstream economic analysis (Rydin, 2013; Healey, 1998; Adams and Tiesdell, 2010). For Rydin, 'Behind the paradigm of growth-dependent planning lies a particular view of urban change rooted in neo-classical economics' (Rydin, 2013, p. 35) in which market actors, most prominently developers, are the key actors. In recent times, even when planning is expressed as achieving the 'triple bottom lines' of the social, economic, environmental dimensions of 'sustainable development', it is the economic that usually dominates. And a narrow, monetised GDP measure of growth at that. A sense of how GDP is not only an inaccurate measure but perverse (and also indicates its concern with capital as opposed to labour) is that 'jobless growth' is viewed as 'good' and positive. Again, growth is growth is growth. As Ellis notes, even where planning policy acknowledges the goal of wellbeing, this 'is almost exclusively linked to arguments for economic growth, *without really appreciating that the very type of development that prioritises growth, also tends to erode the very basis of healthy urban development*' (Ellis, 2016, p. 3, emphasis added).

Boland (2014) perceptively deconstructs this uncritical pro-growth bias within dominant conceptualisations of planning. He notes in his analysis of 'neoliberal competitiveness' and planning's role in promoting and facilitating the international competitiveness of cities and city-regions to boost economic growth via, inter alia, shape and place making to attract mobile capital and a talented labour force: 'neoliberal competitiveness is a "postpolitical strategy" and represents a "dangerous obsession" for spatial planning . . . [characterised by the] "prioritisation of economic growth", "privileging of competitiveness", "marketisation of planning", and "speeding up of planning decisions"' (Boland, 2014, pp. 770, 773).

While it would be going too far to say that dominant accounts of contemporary planning sacrifice much at the altar of growth or that political (and democratic) pro-growth planning is not possible, we do have to recognise that growth occupies not only a central but one could say 'constitutive' position within planning. Here we have only to think of the determined and largely successful and ideologically motivated and sustained attack on, or framing as, planning as the 'dead hand of the regulatory state'. Here not only do we find familiar right-wing/neoliberal tropes of the 'anti-business' and therefore 'anti-growth' character of a certain type of planning but the internalisation by the planning process itself that its main (or one of its main) roles is to 'facilitate', 'predict and provide' for market-led development, regeneration, housing, spatial development, etc.

Part of the logic for prioritising growth above other social, cultural, health or ecological aims (apart from the 'group think' and institutionalisation of growth within planning) is what might be called an 'Achilles lance' justification. Just as the ancient Greek hero had a mythical lance that could heal the wounds it inflicts, likewise with growth there is a belief that it can also heal the various economic, social and ecological damage it causes. And on the face of it, there is sound reasoning behind this long-standing assumption, namely that the additional wealth, jobs and tax or rates revenue that growth brings can compensate, clean up, ameliorate and otherwise produce (eventually perhaps) 'benefits' that offset its 'costs'. But like growth itself, this tells us nothing about the distribution of the costs and benefits and whether what is lost or damaged can be 'made good' by the economic wealth created. But such is the persuasiveness and 'commonsense' of promoting and aiming for economic growth that whether expressed as a modern form of mythic thinking (Achilles Lance) or more prosaically as 'you cannot make an omelette without breaking eggs', planning has become a 'pro-growth' discipline and institutionalised process (Rydin, 2013). And the problem with this is not necessarily that planning is pro-growth. Growth can be – and we have the historical evidence that it was, is, and can be – under a certain set of specific circumstances and conditions, positive, lift people out of poverty, create public goods, enhance human individual and collective wellbeing and so on (Barry, 2012). Hence, the argument here in this chapter should not be read as 'anti-growth' but rather questioning economic growth, as defined above, as a permanent as opposed to a temporary feature of the economy, or growth for growth's sake. Economic growth is, and should always be, a means not an end in itself, but sadly this is something forgotten within planning.

Growth and fossil fuels

There is a strong causal link between carbon energy (fossil fuels such as coal, gas and especially oil) and economic growth. In short, carbon energy is needed for economic growth, and access to cheap, secure, reliable sources of carbon energy is an essential development objective. But this reliance on carbon energy as essential to the 'commonsense' pursuit of growth (viewed as the 'normal' or 'healthy' condition of the economy), and therefore planning in the pursuit or service of growth is largely un- or under-recognised. But the reality is that the role of energy for planning and the economy goes beyond what is necessary as a material input or precondition for urban development, economic activity and the achievement of GDP growth. Fossil fuels, particularly oil, also shape how we think about and conceptualise and understand planning and the economy. Mitchell, for example, makes a persuasive case that the modern 'economy', and the 'economic knowledge' that developed around it in the late 1930s should be viewed as a form of 'petro-knowledge' central to which was endless economic growth (Mitchell, 2011). Modern economics and, by extension, planning based either on a growth dependence (Rydin, 2013) and/or neoclassical and neoliberal economics (Boland, 2014) are 'petro-knowledges' in the sense of being oriented around achieving and managing an endlessly growing economy, an economy which depends on cheap and securely available sources of oil.

A key issue to consider is how and in what ways the planning system can reinforce as opposed to disrupt and help overcome what Unruh terms 'carbon lock-in' (Unruh, 2000). An example of the latter would be the privileging, naturalisation and assumption of increased motorised car mobility in transport and spatial planning. As Driscoll puts it:

> It is overwhelmingly evident at this juncture that radical interventions will be necessary in order to escape carbon lock-in in the transport system. *As long as planners treat road, rail and non-motorized transport modes as fungible goods, then it is likely that the existing path dependencies will reinforce and reproduce a high-carbon transport system.*
>
> *(Driscoll, 2014, p. 318, emphasis added)*

Thus, in this case, planning may have to face head on the unsustainable 'induced demand' that accompanies new road builds, both the increased traffic that results directly from the provision of new infrastructure and the land use patterns that are also induced by privileging the private car over other mobility modes. While actively 'demarketing the car' might be beyond the ability of planning alone to achieve (not to say controversial), at the very least overcoming carbon lock-in requires planning to continue its uneven evolution beyond an outdated and unsustainable 'predict and provide' approach. Other examples would include unsustainable and carbon-intensive forms of suburbanisation and sprawl, out of town developments, etc. and for planning to add another 'post-' to its 'to-do-list', namely 'post-suburbanisation', to add to post-carbon and post-growth (Phelps et al., 2006). But equally, if we accept both the causal link between growth and carbon energy (i.e. growth is dependent upon carbon) and arguments around the negative social, economic and environmental impacts of orthodox economic growth (Barry, 2012; Wilkinson and Pickett, 2010), then to that extent planning uncritically promotes an unsustainable 'pro-growth' agenda. And therefore, it also 'locks in' not simply an unsustainable but an unjust and inequality producing socio-economic system. After all, it is not just sustainability transitions per se that should concern planning but its contribution to a 'just transition' away from 'actually existing unsustainability'.

Planning for a just transition: social justice floors and ecological ceilings

The term 'just transition' has emerged in recent years to conjoin social justice – more specifically, the equitable distribution of the benefits and costs of the transition away from high carbon and unsustainable development trajectories – with the environmental, climate, resource and energy reasons for that transition. Whereas the latter are largely biophysical limits, ceilings or 'planetary boundaries' (Rockström et al., 2009) within which human development is sustainable, the former represent the 'social floors' below which individuals and communities are not allowed to fall in terms of livelihoods, flourishing, human rights and basic standards of living (Raworth, 2018). This is what Kate Raworth terms the 'Safe and Just Space for Humanity' (Raworth, 2012), and it captures the context of planning in the 21st century, i.e. our climate changed, carbon constrained world.

A key feature of the 'just transition' approach to decarbonisation, climate change and sustainability transitions is the recognition that it is possible to have 'unjust' sustainability transitions. On the one hand, this is a possibility if one focuses narrowly on the environmental/ecological aspects (to the neglect of social dimensions, such as justice and equity concerns or procedural issues of democratic decision making for example). On the other hand, unjust transitions are also possible if one narrowly reduces sustainability transitions to 'greening' the status quo.

In terms of the first possible 'unjust transition' type, we can point to authoritarian and inequitable approaches to achieving a transition beyond 'actually existing unsustainability' (Barry, 2012), such as approaches which violate people's basic human rights or are indifferent to the unfair distribution of costs of such transitions. Here, countries like China are routinely presented as a real-world example of this type of 'unjust transition' (Swilling and Annecke, 2012). However, one could also point to a similar problem in strands of what might be termed 'apocalyptic' environmental discourses around climate change, where both the scale and the urgency of the proposed transition mean that since 'survival' (of the species, civilisation, or culture) is the objective, both democracy and justice are superfluous. Since neither democracy nor justice is strictly speaking necessary for survival, and/or integrating them within a sustainability transition will delay or render that 'survival sustainability transition' impossible, we must abandon or 'temporarily' put democracy and claims of justice on hold. Within such a sustainability framing, planning for 'bare life' would be akin to a military process of non-market, non-democratic and non-justice sensitive resource allocation, place and space creation. Such planning sadly bears an uncanny resemblance to dominant features of existing planning practice in its technocratic, statist, expert, top-down 'command and control' character (Marshall and Cowell, 2016).

The second form of 'unjust transition' relates to the questionable 'justice' of 'greening business as usual' as dominant approaches to sustainability transitions, such as mainstream 'sustainable development' discourse, 'ecological modernisation' and 'green growth strategies'. Here, the problem is the greening/decarbonisation/dematerialisation and otherwise rendering less unsustainable a capitalist socio-economic order that is structurally unjust and inequitable. But the question here is: do we want a low carbon version of an unjust system? On this point, there is merit in Friends of the Earth Scotland's call for a 'Just Transition Commission' to oversee and develop a strategy for the implementation and review of any managed and orderly sustainability that would ensure, for example, that as far as possible costs are equitably distributed, and those displaced from jobs or settled and valued ways of life are compensated (Crighton, 2017). Such high-level state-based (if not statist or 'state-centred') strategies to develop policies aimed at decarbonisation could provide the guidance and democratically agreed objectives and principles enabling planning to contribute and play its part in addressing the possible injustices and negative impacts such a socio-energy transition could generate (Healy and Barry, 2017). And the institutionalisation of a 'just transition' strategy, especially if democratised and participative, opens up the possibilities of objectives beyond orthodox economic growth, such as health or wellbeing for example, emerging to supplement or even replace growth as the main aim of planning.

Bringing the state back in

A longstanding justification for state/public planning and wider state intervention is on the grounds of coping with market failures and 'externalities', where market activities create social, ecological or intergenerational costs and benefits which are not reflected in the market price of the good or service. For example, climate change has been described as the greatest market failure the world has seen. However, while a 'market failure' argument does offer a common and powerful reason for planning and state intervention in private market behaviour or the structure of the market itself, there are flaws with any normative framing of such 'remedial' justifications for planning. That is, such remedial arguments in support of planning are of course premised on the idea that the preferred, superior and default position is for the market to provide goods, services and the infrastructure for good lives and a good society, with the state via planning a secondary, inferior and 'second best' institutional provider. And this ranking and conventional thinking around planning is something that is fundamentally challenged by the socio-ecological crisis where democratic planning rather than free market provisioning is the superior and most effective way to solve problems, allocate resources and create the infrastructure for resilient and sustainable communities and societies. At the same time as there being grounds for a more proactive planning system to cope with (adapt) or deal with the root causes (mitigate) of the negative effects of climate change, there are also justifications for planning in relation to the lack of positive externalities and adequate provision of a variety of public goods. Thus, a planned approach should also ensure the creation and sustaining of public goods, such clean water, air and accessible open spaces and the planetary public good of a stable climate system.

The climate crisis and the need for greater 'carbon control and management' has, and can, provide opportunities and obligations for the state to become 'greener' by becoming more interventionist (and possibly more creative and more democratic) as it steers, coordinates or directly manages society-wide sustainability, energy and economic socio-technical transitions (Ellis et al., 2018). As While et al. note, the climate crisis has enabled a new state-centred and 'distinctive political economy associated with climate mitigation in which discourses of climate change both open up, and *necessitate an extension of, state intervention in the spheres of production and consumption*' (While et al., 2010, p. 82, emphasis added). Such 'policy or governance opportunity structures' that moments of crises present for state reconfiguration, obviously also provide opportunities for the rethinking and reconfiguration of urban planning. These range, inter alia, from the long overdue integration of spatial and energy planning (Stoeglehner et al., 2016), viewing 'war time mobilisation' as a model for rapid decarbonisation (Delina, 2016) – an explicit commitment by planning to move beyond technological innovation to provide policy support and space for social experiments in new low carbon and post-growth forms of living – to seeing how addressing the carbon and climate crisis can open up planning to more democratic and democratised institutionalised forms. And democratised planning is understood explicitly not as consensus creation and agreement but seeing planning decision making as recognising and welcoming and indeed encouraging agonistic contestation and debate (Barry and Ellis, 2010).

While obvious, it is worth highlighting that this direction of travel for 'just transition' planning requires the (re)-politicisation of planning as a necessary precondition for its democratisation, and a decisive movement away from a 'top-down' and technocratic, expert-driven conceptualisation of planning. And of course, recognising that the responsibility for envisioning and implementing any post-carbon, post-growth just transition is not the responsibility of the planning system alone, but it does have a vital role, not least as a key dimension of 'bringing the state back in' as a necessary (if insufficient) institutional component in navigating our pathways to sustainable socio-technical futures in the coming century.

Conclusion: planning for post-growth and post-carbon socio-technical futures

The aim of this short chapter has been to suggest a new purpose for planning in 21st century, one which focuses on human wellbeing and other objectives beyond GDP measured economic growth and associated ideas of competitiveness, etc. and is also explicitly premised on moving away from a carbon-based energy system. But while new in terms of the context of planning in the 21st century viz. climate change, decarbonisation, the 'Anthropocene' and 'post-growth economics', there is also something old in the argument here around planning's role in creating post-growth and post-carbon futures as a return to the 'public purpose' of planning (Bowie, 2016). It also proposes a creative and what might be termed 'realistically utopian' form of planning. As the ancient wisdom contained in books such as the Bible tell us that 'without vision the people perish', or more modern versions such as research on socio-technical sustainability transitions say much the same in terms of the importance of an agreed and co-created and shared 'transition vision', we need to ask what is the vision of and for planning in a carbon constrained, climate changed world? As Briel puts it 'Creating visions of a post-carbon urban future can help generate a positive image of the transformed city in which urban life quality, economic and social vibrancy improve for citizens, while carbon emissions decrease' (Briel, 2016, p. 1).

There is both a responsibility on the discipline/inter-discipline of planning, (including professional planning training and practice), and an exciting opportunity for planning to contribute to the transition to post-carbon and post-growth urban futures. This is especially in relation to the provisioning and governance of space and energy, the intersection of infrastructure, space, place and the materialities of low-carbon transformations across socio-technical networks at different scales. Here as well as place making, planning has a part to play in 'just transitions' towards and experiments in new socio-technical modernities, of what Ellis terms 'new visions of future urban living' (Ellis, 2016, p. 2). At the very least, there are grounds for planning to go beyond and supplement aiming for growth with other aims such as a health-led planning system, as well as for planning to help overcome 'carbon lock-in'. Or for planning to focus on prosperity with the aims of creating 'sustainable and prosperous communities . . . understood as places that support people to flourish and thrive in diverse ways that go far beyond orthodox prosperity as wealth creation and economic growth' (Woodcraft and Smith, 2018, p. 72).

Hence, planning in the 21st century should be oriented around the following: how to design urban forms that produce high levels of human wellbeing and flourishing while using less energy and resources and where growth is a potential by-product, not the goal of planning. And the reality is the sooner policies change to prepare communities, cities and our built 'commons' for a decarbonised and post-growth future, the more effective (and cheaper) that transition will be. After all, it is wise to fix the roof when it is sunny, not when it is raining.

References

Adams, D. and Tiesdell, S. (2010). 'Planners as market actors: rethinking state-market relations in land property'. *Planning Theory and Practice*, 11(2): 187–207.

Anderson, K. and Bows, A. (2011). 'Beyond "dangerous" climate change: emission scenarios for a new world'. *Philosophical Transactions of the Royal Society A*, 369(1934): 20–44.

Anielski, M. (2007). *The economics of happiness: building genuine wealth*. Gabriola Island: New Society Publishers.

Barry, J. (2012). *The politics of actually existing unsustainability: human flourishing in a climate changed, carbon-constrained world*. Oxford: Oxford University Press.

Barry, J. (2018). 'From pre-analytic axiom to "core state imperative" and dominant "commonsense": a genealogy of the ideology of economic growth'. *New Political Economy*. Available at: https://doi.org/10.1080/13563467.2018.1526268

Barry, J. and Ellis, G. (2010). 'Beyond consensus? Agonism, contestation, republicanism and a low carbon future', in P. Devine-Wright (ed.) *Renewable energy and the public: from NIMBY to participation*. London: Earthscan, pp 29–42.

Boland, P. (2014). 'The relationship between spatial planning and economic competitiveness: the "path to economic nirvana" or a "dangerous obsession"?' *Environment and Planning A*, 46(4): 770–87.

Bowie, D. (2016). *The radical and socialist tradition in British planning: from Puritan colonies to garden cities*. London: Routledge.

Briel, M. (2016). *Visions for post-carbon urban futures: why they are useful and how to create them*. POCACITO Policy Brief No. 2. Available at: https://pocacito.eu/sites/default/files/POCACITO_PolicyBrief_No-2_Visions_for_low_carbon_cities.pdf

Crighton, M. (2017). 'Planning for a just transition has to be part of Scotland's climate strategy'. *Friends of the Earth Scotland* [Blog]. Available at: https://foe.scot/just-transition-scotland-climate-strategy/

Daly, H. (ed.) (1973). *Toward a steady-state economy*. San Francisco: W. H. Freeman.

Delina, L. (2016). *Strategies for rapid climate mitigation: wartime mobilisation as a model for action?* London: Routledge.

Driscoll, P. (2014). 'Breaking carbon lock-in: path dependencies in large-scale transportation infrastructure projects'. *Planning Practice and Research*, 29(3): 317–30.

Ellis, G. (2016). *Using the planning system to secure health and well-being benefits*. Northern Ireland Assembly, Knowledge and Exchange Seminar Series Briefing. Available at: www.niassembly.gov.uk/globalassets/documents/raise/knowledge_exchange/briefing_papers/series5/geraint-ellis-kess-breifing-june-2016.pdf

Ellis, G., Hume, T., Barry, J. and Curry, R. (2018). *Catalysing and characterising transition synthesis report*. Belfast: Queen's University, Belfast.

Georgescu-Roegen, N. (1971). *The entropy law and the economic process*. Cambridge, MA: Harvard University Press.

Healy, N. and Barry, J. (2017). 'Politicizing energy justice and energy system transitions: fossil fuel divestment and a "just transition"'. *Energy Policy*, 108: 451–9.

Healey, P. (1998). 'Building institutional capacity through collaborative approaches to urban planning'. *Environment and Planning A*, 30(9): 1531–46.

Jackson, T. (2017). *Prosperity without growth: foundations for the economy of tomorrow*. London: Routledge.

Kasser, T. (2003). *The high price of materialism*. Cambridge, MA: MIT Press.

Marshall, T. and Cowell, R. (2016). 'Infrastructure, planning and the command of time'. *Environment and Planning C: Government and Policy*, 34(8): 1843–66.

Meadows, D., Meadows, D. and Randers, J. (1972). *The limits to growth: a report for the Club of Rome's project on the predicament of mankind*. New York: Universe Books.

Mitchell, T. (2011). *Carbon democracy: political power in the age of oil*. London: Verso.

Phelps, N., Parsons, N. and Ballas, D. (2006). *Post-suburban Europe: planning and politics at the margins of Europe's cities*. London: Palgrave Macmillan.

Raworth, K. (2012). *A safe and just space for humanity*. Oxford: Oxfam. Available at: www.oxfam.org/sites/www.oxfam.org/files/dp-a-safe-and-just-space-for-humanity-130212-en.pdf

Raworth, K. (2018). *Doughnut economics: seven ways to think like a 21st-century economist*. London: Random House.

Reardon, J., Madi, A. and Cato, M. (2018). *Introducing a new economics: pluralist, sustainable and progressive*. London: Pluto Press.

Rockström, J., Steffen, W., Noone, K., Persson, Å., Chapin, III, F. S., Lambin, E., Lenton, T. M., Scheffer, M., Folke, C., Schellnhuber, H., Nykvist, B., De Wit, C. A., Hughes, T., van der Leeuw, S., Rodhe, H.,

Sörlin, S., Snyder, P. K., Costanza, R., Svedin, U., Falkenmark, M., Karlberg, L., Corell, R. W., Fabry, V. J., Hansen, J., Walker, B., Liverman, D., Richardson, K., Crutzen, P. and Foley, J. (2009). 'Planetary boundaries: exploring the safe operating space for humanity'. *Ecology and Society*, 14(2): 32 [Online]. Available at: www.ecologyandsociety.org/vol14/iss2/art32/

RTPI (no date). *Value of planning*. RTPI website. Available at: www.rtpi.org.uk/valueofplanning

Rydin, Y. (2013). *The future of planning: beyond growth dependence*. Bristol: Policy Press.

Stiglitz, J., Sen, A. and Fitoussi, J.-P. (2008). *Report by the commission on the measurement of economic performance and social progress*. Available at: http://ec.europa.eu/eurostat/documents/118025/118123/Fitoussi+Commission+report

Stoeglehner, G., Neugebauer, G., Erker, S. and Narodoslawsky, M. (2016). *Integrated spatial and energy planning: supporting climate protection and the energy turn with means of spatial planning*. London: Springer.

Swilling, M. and Annecke, E. (2012). *Just transitions: explorations of sustainability in an unfair world*. Claremont, South Africa: UCT Press.

Unruh, G. (2000). 'Understanding carbon lock-in'. *Energy Policy*, 28(12): 817–30.

While, A., Jonas, A. E. and Gibbs, D. (2010). 'From sustainable development to carbon control: eco-state restructuring and the politics of urban and regional development'. *Transactions of the Institute of British Geographers*, 35(1): 76–93.

Wilkinson, R. and Pickett, K. (2010). *The spirit level: why more equal societies almost always do better*. London: Penguin.

Woodcraft, S. and Smith, C. (2018). 'From the "sustainable community" to prosperous people and places: inclusive change in the built environment', in T. Dixon, J. Connaughton and S. Green (eds.) *Sustainable futures in the built environment to 2050*. Oxford: Wiley-Blackwell, pp 72–93.

11

Enhancing anticipatory governance

Strategies for mitigating political myopia in environmental planning and policy making

Jonathan Boston

> The future whispers while the present shouts.
>
> *(Gore, 1992, p. 170)*

Introduction

Anticipation is vital for effective environmental planning and sound environmental governance. Good governance, in other words, depends on the capacity of planners, policy makers and their advisers to think ahead, exercise foresight, contemplate future problems, identify risks, strategize and plan. Hence, good governance must be anticipatory; it requires not only effective planning processes and policies to tackle the urgent problems of today but also prudent measures to mitigate the looming problems of tomorrow. Failing to do so is likely to generate substantial long-term costs, unwelcome surprises and lost opportunities.

Yet encouraging sound anticipatory governance – whether through wise environmental stewardship, effective environmental planning, responsible fiscal management or prudent infrastructure investment – is challenging. Governments face formidable hurdles: incomplete and conflicting information, deep uncertainty,[1] competing priorities, powerful vested interests, populist pressures, limited budgets, complex intertemporal trade-offs, short electoral cycles and global collective action problems. Such constraints and pressures often result in policy makers favouring near-term interests over long-term interests and current generations over future generations. Put differently, contemporary democratic (and non-democratic) policy making is frequently marred by what is variously called 'political myopia', 'policy short-termism' or a 'presentist bias' (Boston, 2017a, 2017b; Healy and Malhorta, 2009; Heller, 2003; Jacobs, 2011, 2016; MacKenzie, 2013; Thompson, 2005, 2010). For multiple reasons, as discussed later, myopic decision making is particularly evident when long-term environmental interests are at stake. Very commonly, for instance, governmental efforts to mitigate environmental problems – such as climate change, ocean acidification, deteriorating air or water quality, declining fish stocks and biodiversity loss – are hesitant, sluggish, half-hearted or ineffective. And this applies even when the problems are local or national in nature and thus do not require international cooperation to address effectively.

This chapter briefly explores the nature and causes of political myopia in advanced democracies and the reasons why environmental interests, and by extension, environmental planning, can be particularly vulnerable to short-termist decision making. It then briefly surveys possible strategies for mitigating political myopia and enhancing the quality of anticipatory governance. Clearly, such topics are large and complex. They cannot be adequately canvassed in a short chapter. Necessarily, therefore, the following analysis is partial and incomplete. For instance, it focusses primarily on central government policy making and planning and largely ignores global and sub-national governance.

Political myopia and environmental policy making

Political myopia refers to a widespread and persistent temporal asymmetry in governmental decision making, namely a tendency for policy makers to focus on the near-term at the expense of the more distant future and to prioritise short-term interests over long-term ones. Alan Jacobs (2011, p. 266) refers to this phenomenon as a 'substantial policy tilt towards the short run'. Examples include a propensity for policy makers to ignore emerging policy problems and future societal needs, downplay long-term risks, fail to protect renewable resources and underinvest in disaster risk mitigation, public infrastructure, preventative health care measures and cost effective early intervention programmes. Political myopia appears to be particularly common when planners and policy makers are faced with intertemporal trade-offs involving non-simultaneous exchanges. The most problematic are those where short- to medium-term costs must be borne in order to realise long-term benefits. In such situations, governments often choose policies which minimise short-term pain even if alternative options are likely to generate greater long-term gains or avoid larger long-term costs. Such choices are deliberate and calculated; they are not random or accidental.

Of course, planners and policy makers are not universally or irredeemably myopic. On the contrary, there are many examples of far-sighted policy decisions, sometimes involving significant short-term sacrifices. But, overall, governments are more likely to favour short-term considerations over long-term ones. Hence, a presentist bias is more common than a long-termist one.

Such outcomes are hardly surprising. After all, there are multiple reasons why democratically elected governments might be expected to favour short-term interests. Among these are the following:

- *The human condition*: people are influenced by deep-seated psychological processes, cognitive biases and heuristics which tilt their preferences and behaviours towards the present (Jones and Baumgartner, 2005; Kahnemann, 2011). For instance, they tend to value losses more than equivalent gains, underestimate risk, especially low probability but high impact risk, and discount the future, at least moderately. Put simply, they are often impatient, preferring things sooner rather than later (Jacobs, 2011);
- *Uncertainty*: other things being equal, our capacity to assess causal pathways and predict policy outcomes diminishes as timeframes lengthen. Hence, if there is a temporal disjunction between the flow of costs and benefits of a policy such that the costs are generally front loaded while the benefits are generally back loaded (as for most long-term investments), the benefits will be more uncertain than the costs. This reduces the attractiveness of policies with such features;

- *Politically salient asymmetries*: there are many asymmetries besides those involving costs and benefits which affect intertemporal decision making and reinforce a presentist bias (Boston, 2017a). One of these is the voting asymmetry; current generations have voting rights while future generations do not. Hence, future generations have no capacity to protect their interests; they must rely on current generations. Similarly, there are well known interest group asymmetries: in many policy contexts, powerful, well-resourced and concentrated interests (for example, large fossil fuel companies) are ranged against much weaker dispersed interests (for example, environmental lobby groups). In such situations, the concentrated interests often exercise a disproportionate influence on planning and policy making. Additionally, there are various accounting asymmetries which affect intertemporal policy choices. For instance, under current accounting rules in most democracies, governments report regularly on their financial performance but are not required to report on changes to the stocks of natural, social or human capital or changes in the flow or value of ecosystem services;
- *The informational environment*: planning and policy problems vary greatly along multiple dimensions. These include their visibility, complexity and the immediacy of their effects, together with the quality, availability and comprehensibility of information about their causes and consequences (Jacobs, 2011). Where problems lack vivid early warning signals, are relatively invisible or intangible, are difficult for voters to grasp or are distant spatially and/or temporally in their likely impacts, it is much harder for governments to mobilise public support for effective policy interventions. Yet if such problems – many of which are 'creeping' or 'slow-burner' in nature (Olson, 2016) – are not addressed expeditiously there may be severe, even irreversible, long-term consequences;
- *Ideological polarisation*: where political parties are deeply polarised on important policy issues it is harder to negotiate durable political bargains (Mansbridge and Martin, 2013). Ideological polarisation also increases the likelihood of basic scientific facts being contested and exacerbates the risk of opportunistic behaviour, thereby undermining political trust and encouraging dynamic inconsistency;
- *Incommensurability*: policy problems often entail conflicts between 'goods' that transcend different systems of value (Sen, 2009). For instance, an economic or financial goal may be ranged against the goal of protecting threatened species or unique landscapes. Where the 'goods' in question are incommensurable (i.e. they cannot be readily compared using the same yardstick or a common standard of measurement), there is a risk that short-term interests (for example, near-term financial returns) will prevail over long-term interests (for example, maintaining biodiversity).

The foregoing list highlights that political myopia afflicts some planning contexts and policy problems more than others. Hence, not all long-term interests are equally vulnerable. As noted earlier, long-term *environmental* interests are particularly at risk from short-termist decision making (Brown et al., 2015). This is because many of the values at stake are intangible; the evidence regarding the nature of the problem is frequently incomplete, uncertain or conflicting; and the risks are often distant, both spatially and temporally. For many voters and policy makers, they are 'out of sight and out of mind'. Equally, whereas fiscally unsustainable policies generate adverse market reactions (for example, higher bond rates) which prompt policy corrections, environmentally unsustainable policies lack corrective devices of similar political potency. Often, too, important environmental problems are transboundary, thus requiring the cooperation of multiple governments, agencies and organisations to mitigate. Such cooperation can readily be thwarted by competing national interests or a silo mentality by governmental agencies or planning authorities.

Compounding matters, many environmental problems are 'creeping': they emerge gradually and sometimes imperceptibly and often lack critical thresholds or abrupt tipping points, at least during their early stages (European Environment Agency, 2013; Olson, 2016; Scheffer et al., 2003). Typically, too, there is a long time-lag between cause and effect. Hence, potentially negative impacts may be on the radar, and their capacity to generate significant long-term harm may be evident (at least to the relevant experts), but they generally lack vivid, dramatic or unmistakeable early warning signals. Examples of environmental creeping problems include: the gradual depletion of aquifers; the slow accumulation of toxic chemicals and microplastics in the environment; the declining populations of pollinators; the loss of valuable agricultural land due to urbanisation and soil degradation; the increasing damage to ecosystem services from pollution, invasive species, soil erosion and habitat destruction; and the loss of biodiversity and wilderness areas. Regrettably, creeping problems often exhibit substantial path dependence and/or produce cumulative – and non-linear – impacts. They thus become harder and more costly to alleviate over time. Delays in responding may therefore reduce or even eliminate the possibility of low-cost solutions, shifting the burden of adjustment onto future citizens and taxpayers. Accordingly, any attempt to improve environmental planning – and anticipatory governance more generally – must take creeping problems seriously and develop strategies to enhance the attention they receive from policy makers.

Mitigating political myopia and enhancing anticipatory governance

How, then, might it be possible to mitigate the risk of political myopia and better safeguard the interests of future generations, not least their interests in a diverse, healthy and sustainable environment? There is a burgeoning literature on such questions (Ascher, 2009; Boston, 2016, 2017a, 2017b; Fuerth and Faber, 2012, 2013; González-Ricoy and Gosseries, 2017; Guston, 2014; Hovi et al., 2009; Quay, 2010). Significant contributions have been made by major international organisations and leading think tanks, as well as scholars across multiple disciplines. Numerous 'solutions' have been proposed. Unsurprisingly, many of these reflect the disciplinary backgrounds of their advocates. Constitutional lawyers, for instance, focus on constitutional mechanisms, such as giving greater recognition to the interests, needs and rights of future generations in constitutional documents, including the right to a healthy environment (Palmer and Butler, 2016); planners, strategists and futurists emphasise the need for better decision-making tools, improved forecasting and a greater reliance on foresight methodologies (Fuerth and Faber, 2012); social psychologists and behavioural economists concentrate on countering cognitive biases through adroit issue framing, policy 'nudges' and 'choice architecture' (Thaler and Sunstein, 2008); and accountants highlight the importance of accounting rules and coherent multidimensional reporting frameworks, including sustainability accounting and integrated reporting (Gleeson-White, 2014). Meanwhile, ecological economists underscore the importance of developing new policy paradigms and changing mindsets, not least through properly valuing ecosystem services and natural capital (Costanza, 2014); political scientists often advocate the reform of electoral and legislative institutions (MacKenzie, 2013), altering governance arrangements (Ostrom, 2009) or using deliberative decision-making processes to tackle complex intertemporal issues (Curato et al., 2017); and yet other scholars argue for better analytical frameworks (Karacaoglu, 2015) and improved procedural and/or substantive commitment devices (Elster, 2000; Reeves, 2015).

Drawing on the relevant literature, Table 11.1 outlines a selection of the suggested ways of reducing political myopia and fostering sound anticipatory governance. Particular attention is

Table 11.1 Proposals for enhancing anticipatory governance and environmental sustainability

Type of proposal	*Examples of specific policy proposals*
Overarching principles	
1 Constitutional protection for future generations	• Enact constitutional provisions imposing a duty on governments to safeguard the interests (welfare, needs or rights) of future generations and/or protect a healthy, ecologically balanced environment
2 Sustainable management of natural and physical resources	• Enact legislative requirements for managing the use, development and protection of natural and physical resources in order to meet the reasonably foreseeable needs of current and future generations
3 The precautionary principle	• Enact legislation requiring decision makers at all governmental levels to give effect to the precautionary principle – ideally at the stronger end of the potential spectrum of possible interpretations
4 The exercise of stewardship by public bodies	• Enact legislation requiring all public bodies to exercise good stewardship of their organisations, including their assets and liabilities, their long-term sustainability, their overall health and capability, and their impacts on the environment
Representing future-oriented interests and countering short-term electoral pressures	
5 Independent public institutions to represent and protect future-oriented interests	• Establish or strengthen independent public institutions with an explicit mandate to represent future-oriented interests in governmental policy making or exercise long-term guardianship roles (e.g. commissions or ombudsmen for future generations, children, the environment, cultural heritage, etc.)
6 Countering short-term electoral pressures	• Strengthen the decision rights of independent institutions and transfer specific decision rights from elected bodies to independent institutions • Require elected officials to take advice from independent experts before making decisions on particular policy issues and/or constrain the policy choices available to elected officials to those recommended by independent experts
Foresight capability	
7 Foresight processes and institutions	• Enact legislation requiring governments to produce periodic reports on the future which identify major risks, vulnerabilities and creeping problems and outline their plans to mitigate these risks and problems • Establish dedicated foresight capability within both the legislative and executive branches
8 Early warning systems	• Strengthen the capacity of governmental agencies to identify new risks, evolving risk profiles and possible tipping points, and report results
Planning processes and adaptive management	
9 Protecting the interests of future generations	• Enact legislation requiring all public decision-making bodies, including those charged with environmental planning, to identify and give proper weight to the interests of future generations as part of their normal processes

Type of proposal	Examples of specific policy proposals
10 Planning of public infrastructure and other investments	• Enact legislation requiring governments to prepare long-term plans and strategies (for example, 30 years+) for the management of public infrastructure and periodic national investment statements assessing the shape, health and evolving value of their portfolio of assets and liabilities, including natural capital
Analytical, policy and regulatory frameworks	
11 Analytical frameworks – shifting conceptual paradigms	• Develop further and apply more holistic analytical frameworks for policy analysis (e.g. the New Zealand Treasury's Living Standards Framework)
12 Specifying long-term goals	• Enact legislation requiring governments to set periodically long-term targets in specific policy domains – such as long-term greenhouse gas emissions reduction targets or biodiversity targets
13 Assessing, monitoring and reporting performance	• Enact legislation requiring governments to publish regularly comprehensive data on performance across all policy domains • Develop further and apply the concepts of national wealth accounting, natural capital accounting, and the valuing of ecosystem services • Develop comprehensive national balance sheets incorporating most, if not all, capital stocks (i.e. financial, manufactured, human, social and natural)
14 Research and development	• Increase funding of scientific research (for example, with respect to conservation and the environment) to enhance the quality of evidence for decision making
Problem solving and consensus building mechanisms	
15 Participatory and deliberative processes	• Extend the use deliberative policy-making processes, multi-stakeholder forums and collaborative governance arrangements, especially for issues with significant intertemporal dimensions and incommensurable goals, and where solutions require non-simultaneous exchanges
Nurturing a future-focussed political culture	
16 Improving public understanding of long-term issues	• Enhance the quality of civic education programmes in schools, including exposure to foresight methods, governmental policy making and global issues
17 Enhancing societal trust	• Undertake initiatives to build societal trust and solidarity (for example, by reducing income and wealth inequality), thereby increasing the capacity of political leaders to negotiate long-term policy commitments and protect future interests
18 Countering short-termism in business	• Implement regulatory changes to discourage short-term decision making in business (for example, by changing executive reward systems, requiring integrated reporting, developing new metrics of the long-term health of companies, developing new business models such as benefit corporations, etc.)

Source: adapted from Boston (2016, pp. 18–20)

given to proposals designed to enhance environmental planning and sustainability. The list of measures is far from exhaustive. Many additional proposals could be added, such as those of relevance to different tiers of government, specific institutions or discrete policy domains.

Implicitly or explicitly each proposal is assumed to achieve the desired outcome of better protecting long-term interests through a series of logically connected steps or 'intervention logic'. Of course, at each point in the relevant causal sequence, the expected behavioural changes might not occur. Alternatively, they may be hindered by countervailing forces. It is impossible here to consider all the proposed reforms and their respective intervention logics. Nevertheless, in brief, each reform is expected to protect long-term interests through one or more of the following distinct causal mechanisms:

- By changing the *mindsets* and *motives* of planners and policy makers (i.e. values, norms, preferences, and priorities) and activating future-oriented concerns;
- By *incentivising* planners and policy makers to give greater weight to future-oriented interests (for example, via changes to public opinion/preferences, political culture, the balance of political forces, accountability mechanisms, outcome-based performance measures, etc.);
- By enhancing the *capacity* of planners and policy makers to anticipate future needs and risks, exercise greater foresight and improve their planning capability (for example, via better information and analytical resources, improved horizon scanning and scenario analysis, more holistic policy frameworks, better policy tools, etc.);
- By *constraining* the formal decision rights and discretionary powers of planners and policy makers (for example, via constitutional rules, procedural rules and substantive policy rules);
- By *insulating* planners and policy makers from short-term political pressures;
- By establishing new *coordinating* mechanisms – whether at the sub-national, national or global levels – to enable decisions to protect long-term interests that would otherwise not be possible (Boston, 2017a, pp. 187–90).

In effect, each of these intervention logics is expected to operate by altering the planning context or the choice architecture in which decisions are made. Having said this, it is unclear from the available evidence to what extent, or in what particular circumstances, each intervention logic might be efficacious. Much the same conclusion applies to many of the specific proposals in Table 11.1. Although most have been partially or fully implemented in one or more advanced democracy over recent decades, the available evidence regarding their effectiveness in protecting long-term interests, not least environmental interests, is partial and incomplete. It is even more difficult to assess the likely impact of those proposals which have yet to be implemented (for example, comprehensive national balance sheets).

Nevertheless, several things are readily apparent. First, some democracies protect their long-term interests better than others. For instance, environmental performance (based on multiple indicators) varies significantly across the OECD (Hsu et al., 2016). Much the same applies to countries' fiscal performance.

Second, despite many constitutional, institutional, administrative and policy-oriented reforms over recent decades designed to improve environmental, social and fiscal sustainability, short-termist pressures remain strong. The rise of populist political movements in long established democracies, such as Britain and the United States, following the global financial crisis has, if anything, intensified a presentist bias in policy making. And in the case of climate change in the United States, despite a world class research community and a highly educated population, securing abroad cross-party agreement even on the most basic scientific issues has remained elusive. As a result, cost-effective policy measures at the federal level to reduce greenhouse gas emissions have been constantly frustrated.

Third, political myopia constitutes a wicked and complex problem. Plainly, there are no complete, all-inclusive and lasting solutions. Lessening its impact may be possible, but permanent defeat is not an option. After all, short-term electoral imperatives are bound to weigh heavily on those who rely on a democratic mandate for their authority, legitimacy and incumbency. And short-termism also afflicts authoritarian regimes and business decision making (Kay, 2012).

Fourth, given that political myopia has many causes – some deeply rooted in human psychology – any attempt to protect a society's long-term interests almost certainly requires multiple strategies, the deployment of many different policy levers, and continuing vigilance. Planning efforts, in other words, must be systematic, comprehensive and persistent. They must also be suitably tactical: that is, they must be tailored to reflect the particular context – constitutional, institutional, political, administrative, cultural, social and environmental – of each democracy. What may be modestly efficacious and durable in one country may be less so in another. Likewise, the politically feasibility of particular reforms will vary over time and between countries. Opportunities must therefore be seized as and when they arise. Moreover, some kinds of reforms are inherently problematic. Constitutional amendments, for instance, are invariably difficult to implement, although they are easier in some countries (for example, Britain and New Zealand) than in others (for example, the United States). Likewise, some of the proposed reforms, such as natural capital accounting, are demanding conceptually, methodologically and analytically.

Fifth, whatever their possible efficacy, not all the proposals in Table 11.1 are equally desirable. Some involve insulating environmental planning and specific areas of policy making from day-to-day political pressures and relying more heavily on technocratic expertise. While there are undoubtedly situations where it is appropriate for decision rights to reside with independent bodies (for example, those which are primarily technical in nature), it is vital to retain democratic control and accountability. Otherwise, legitimacy may be undermined. Similar tensions arise regarding proposals to constrain the future decision rights of legislatures through constitutional provisions to protect future generations. Moreover, there can be no guarantee that the courts will interpret any new provisions in the manner intended by their drafters, nor that the courts will, on average, be better protectors of environmental interests than elected officials.

Sixth, many of the proposals in Table 11.1 constitute commitment devices (Boston, 2017a; Elster, 2000; Kyland and Prescott, 1977; Reeves, 2015). These are mechanisms designed to change the structure of intertemporal pay-offs and/or limit future discretion by binding a person, organisation or government to a particular course of action. In so doing, they help counter short-term expediency, reinforce self-restraint and reduce dynamic inconsistency. Commitment devices relevant to prudent environmental planning differ greatly in their design, bindingness, durability, and accountability mechanisms. They also vary in terms of the stage during the policy process to which they apply (i.e. policy initiation/early warning systems, policy formulation, decision making, implementation and evaluation). Many of those in Table 11.1 are either procedural in nature (i.e. they require planners and policy makers to undertake certain tasks, such as foresight exercises) or substantive (i.e. they oblige actions in accordance with specific criteria). Some devices combine both procedural and substantive provisions. This is typically the case with respect to the rules and processes governing district/regional planning and resource management. While there is a good case for using commitment devices to mitigate political myopia, their effectiveness and durability depends on a supportive political culture. Where such a culture exists, the devices themselves may contribute only marginally to protecting long-term interests. Where it is lacking, it must be nurtured. This is no easy task.

Finally, long-term interests are more likely to be protected if they are deeply embedded across the political system. A strategy of embeddedness implies concerted efforts to mainstream and normalise the consideration of future interests at every stage of the policy process, every level of governance and all forms of planning. In addition to deploying various procedural and substantive commitment devices, another component of such a strategy must be to bring long-term issues into sharper political focus by making the intertemporal implications of the available options more transparent and comprehensible. This entails, amongst other things, better and more systematic measurement of all those things which affect 'collective intergenerational wellbeing' (Karacaoglu, 2015). In particular, it requires supplementing the measurement of flows, like GDP, to include the measurement of vital capital stocks, and the adoption of a broader and more inclusive definition of 'wealth' (Arrow et al., 2012; Hamilton and Hartwick, 2014; Helm, 2015; Helm and Hepburn, 2014). Potentially, the latter could include all types of capital, thereby enabling the measurement of a nation's 'comprehensive wealth' (i.e. its total net worth). Such a measure could provide a broad aggregate indication of a nation's long-term sustainability and inform environmental planning and governmental priority setting. The critical point here is that measurement matters: what is measured tends to count; what is not measured is often overlooked.

Conclusion

Better environmental planning and policy making requires improved anticipatory governance. Such governance has many attributes. Above all, it means taking care of tomorrow today – a task implicit in the need for planning. Plainly, this is difficult. Governments face a formidable array of risks, incessant demands, complex trade-offs, and much uncertainty. Moreover, there is a constant risk of urgent problems diverting attention from, and thwarting efforts to address, the problems of tomorrow. As a result, the voice of the future can easily be muted and future generations may face needless and unjustifiable burdens.

To mitigate such risks, active countervailing measures are required – ones designed to bring long-term considerations more sharply, vividly and repeatedly into short-term political focus. Four types of measures of relevance to the broad context of environmental planning deserve underscoring. First, there is a need for more and better commitment devices that oblige policy makers to look beyond the immediate horizon, take early warning signals seriously, and identify and pursue desirable long-term goals. Second, sound anticipatory governance requires efforts to embed or mainstream long-term interests more comprehensively and compellingly in day-to-day planning and policy making via, for instance, reformed systems of governmental accounting, better analytical frameworks, and improved regimes of performance measurement, monitoring and reporting. Third, advanced democracies need well-funded public institutions and independent bodies with explicit mandates to speak on behalf of future generations and long-term interests, not least environmental interests. Finally, it is critically important to nurture a political culture that cares about the future, accepts biophysical limits, understands the value of ecosystem services, treasures the natural world, and values resilience and sustainability. Without such a culture, sound anticipatory governance and wise environmental stewardship are impossible.

Note

1 According to Lempert et al. (2003, p. xii), 'deep uncertainty' refers to situations 'where analysts do not know, or the parties to a decision cannot agree on, (1) the appropriate conceptual models that describe the relationships among the key driving forces that will shape the long-term future, (2) the probability distributions used to represent uncertainty about key variables and parameters in the mathematical representations of these conceptual models, and/or (3) how to value the desirability of alternative outcomes'.

References

Arrow, K. J., Dasgupta, P., Goulder, L. H. and Mumford, K. J. (2012). 'Sustainability and the measurement of wealth'. *Environment and Development Economics*, 17(3): 317–53.

Ascher, W. (2009). *Bringing in the future: strategies for farsightedness and sustainability in developing countries*. Chicago: University of Chicago Press.

Boston, J. (2016). 'Anticipatory governance: how well is New Zealand safeguarding the future?' *Policy Quarterly*, 12(3): 11–24.

Boston, J. (2017a). *Governing for the future: designing democratic institutions for a better tomorrow*. Bingley: Emerald.

Boston, J. (2017b). *Safeguarding the future: governing in an uncertain world*. Wellington: Bridget Williams Books.

Brown, M., R. Stephens, R. T. T., Peart, R. and Fedder, B. (2015). *Vanishing nature: facing New Zealand's biodiversity crisis*. Auckland: Environmental Defence Society.

Costanza, R. (2014). 'A theory of socio-ecological system change'. *Journal of Bioeconomics*, 16(1): 39–44.

Curato, N., Dryzek, J., Ercan, S., Hendriks, C. and Niemeyer, S. (2017). 'Twelve key findings in deliberative democracy research'. *Daedalus*, 146(3): 28–38.

Elster, J. (2000). *Ulysses unbound: studies in rationality, precommitment, and constraint*. Cambridge: Cambridge University Press.

European Environment Agency (2013). *Late lessons from early warnings: science, precaution, innovation*. Copenhagen: European Environment Agency.

Fuerth, L. and Faber, E. (2012). *Anticipatory governance: practical upgrades – equipping the executive branch to cope with increasing speed and complexity of major challenges*. Washington, DC: Elliott School of International Affairs, George Washington University.

Fuerth, L. and Faber, E. (2013). 'Anticipatory governance: winning the future'. *The Futurist*, 47(4): 42–9.

Gleeson-White, J. (2014). *Six capitals: the revolution capitalism has to have – or can accountants save the planet?* Sydney: Allen & Unwin.

González-Ricoy, I. and Gosseries, A. (eds.) (2017). *Institutions for future generations*. Oxford: Oxford University Press.

Gore, A. (1992). *Earth in the balance: ecology and the human spirit*. Boston: Houghton Mifflin.

Guston, D. (2014). 'Understanding "anticipatory governance"'. *Social Studies of Science*, 44(2): 218–42.

Hamilton, K. and Hartwick, J. (2014). 'Wealth and sustainability'. *Oxford Review of Economic Policy*, 30(1): 170–87.

Healy, A. and Malhorta, N. (2009). 'Myopic voters and natural disaster policy'. *American Political Science Review*, 103(3): 387–406.

Heller, P. (2003). *Who will pay? Coping with aging societies, climate change and other long-term fiscal challenges*. Washington, DC: IMF.

Helm, D. (2015). *Natural capital: valuing the planet*. New Haven: Yale University Press.

Helm, D. and Hepburn, C. (eds.) (2014). *Nature in the balance: the economics of biodiversity*. Oxford: Oxford University Press.

Hovi, J., Sprinz, D. and Underdal, A. (2009). 'Implementing long-term climate policy: time inconsistency, domestic politics, international anarchy'. *Global Environmental Politics*, 9(3): 20–39.

Hsu, A., Esty, D. C., Levy, M. A. and de Sherbinin, A. (2016). *2016 Environmental performance index*. New Haven, CT: Yale University.

Jacobs, A. (2011). *Governing for the long term*. Cambridge: Cambridge University Press.

Jacobs, A. (2016). 'Policymaking for the long term in advanced democracies'. *Annual Review of Political Science*, 19(June): 433–54.

Jones, B. and Baumgartner, F. (2005). *The politics of attention: how government prioritizes problems*. Chicago: University of Chicago Press.

Kahnemann, D. (2011). *Thinking, fast and slow*. London: Penguin.

Karacaoglu, G. (2015). *The New Zealand treasury's living standards framework: a stylised model*. Wellington: The Treasury.

Kay, J. (2012). *The Kay review of UK equity markets and long-term decision making: final report*. London: HM Stationery Office.

Kyland, F. and Prescott, E. (1977). 'Rules rather than discretion: the inconsistency of optimal plans'. *Journal of Political Economy*, 85(3): 473–92.

Lempert, R., Popper, S. and Bankes, S. (2003). *Shaping the next one hundred years: new methods for quantitative, long-term policy analysis*. Santa Monica, CA: RAND. [MR-1626-RPC].

MacKenzie, M. (2013). *Future publics: democratic systems and long-term decisions*. PhD thesis, University of British Columbia.

Mansbridge, J. and Martin, C. (eds.) (2013). *Negotiating agreements in politics: taskforce report*. Washington, DC: American Political Science Association.

Olson, R. (2016). *Missing the slow train: how gradual change undermines public policy and collective action*. Washington, DC: Wilson Centre.

Ostrom, E. (2009). *Beyond markets and states: polycentric governance of complex economic systems*. Nobel Prize lecture, Workshop in Political Theory and Policy Analysis, Indiana University, Bloomington.

Palmer, G. and Butler, A. (2016). *A constitution for Aotearoa-New Zealand*. Wellington: Victoria University Press.

Quay, R. (2010). 'Anticipatory governance: a tool for climate change adaptation'. *Journal of the American Planning Association*, 76(4): 496–511.

Reeves, R. (2015). *Ulysses goes to Washington: political myopia and policy commitment devices*. Washington, DC: Centre for Effective Public Management, Brookings Institution.

Scheffer, M., Westley, F. and Brock, W. (2003). 'Slow response of societies to new problems: causes and costs'. *Ecosystems*, 6(5): 493–502.

Sen, A. (2009). *The idea of justice*. London: Allen Lane.

Thaler, R. and Sunstein, C. (2008). *Nudge: improving decisions about health, wealth and happiness*. New Haven: Yale University Press.

Thompson, D. (2005). 'Democracy in time: popular sovereignty and temporal representation'. *Constellations*, 12(2): 245–61.

Thompson, D. (2010). 'Representing future generations: political presentism and democratic trusteeship'. *Critical Review of International Social and Political Philosophy*, 13(1): 17–37.

12
Knowledge, expertise and trust

James Palmer

Introduction

The roles played by knowledge, expertise and trust in planning processes relating to sustainability and the environment are notoriously complicated. Seemingly all those involved in planning processes agree on the benefits of evidence-based policy (or, increasingly, 'data-based governance'), and yet in practice these processes are routinely characterised by protracted, acrimonious and sometimes irresolvable disagreement. Sometimes these disagreements concern the best means by which to achieve a universally agreed upon objective, but more often contestation hones in on the objectives of these processes themselves and may be traced to fundamental incompatibilities in the approaches that different actors take to understanding sustainability or, more generally, to thinking about what precisely should constitute a 'good', or 'natural', environment.

For critical social scientists, an essential requirement for advancing understandings of the relationships between knowledge, expertise and trust in planning for environmental sustainability has therefore been to revisit the foundational principles by which these variables are conceptualised in the first place. Beginning from the same first principles, this chapter seeks to offer an introductory overview of some of the key conceptual frameworks for thinking critically about knowledge, expertise and trust in controversies around the environment and sustainability, drawing in particular on work emanating from geography, political science, science and technology studies and related sub-fields.

The ideas outlined in the following sections draw on research that has examined the interactions of knowledge, expertise and trust in a range of recent and historical environmental planning controversies, as well as on scholarship exploring the role and influence of advisory institutions recognised for their expertise in these matters. Recognising environmental planning as complex, future oriented and deeply situated in place specific contexts, the chapter contends that planning theory and practice could both benefit from a greater attunement to the diverse geographies of the practices not only through which knowledge and expertise are constructed, but also through which trust and credibility emerge and evolve over time.

Knowledge in environmental planning: searching for 'the facts'?

All research seeking to shed light on the relationships between knowledge and policy-making processes must begin from the fundamental question of what, precisely, is to count as knowledge.

In an influential paper, Radaelli (1995) suggests that knowledge, in contrast to data or information, embodies a line of thinking which has been rigorously tested, and whose reliability is therefore deemed capable of transcending a range of different contexts. Yet, in the context of environmental planning processes specifically, much knowledge is future oriented, being based not upon historical observations of what *has* happened in the past but on *ex-ante* assessments – of various levels of formality – of what *might* be expected to happen once a particular course of action has been chosen (Turnpenny et al., 2009). From this perspective, knowledge might manifest in forms as diverse as economic cost benefit assessments, impassioned public consultation responses, independent advisory reports produced by scientific expert bodies or more tacit and experiential accounts of the local, situated environmental contexts within which specific planning controversies unfold. Given this diversity, the present chapter adopts a generous interpretation of knowledge, as including 'formal (scientific and specialist) as well as tacit and experiential knowledge, together with belief systems, analysis, and ideas' (Owens, 2015, p. 5).

The danger of adopting a generous definition of any concept, of course, is that it permits multiple, potentially conflicting perspectives to co-exist, impeding rather than facilitating the development of any singular, all-encompassing framework. Yet, the scholarly dilemmas raised by this decision in fact bear much resemblance to the practical dilemmas facing environmental planners themselves, as they strive to think through the conflicting knowledge claims, ideas and perspectives mobilised in response to specific planning proposals, consultations and other environmental initiatives (Palmer, 2016). The dilemma, in short, is to establish how and when different kinds of knowledge should come to matter in the environmental planning process (Owens et al., 2004).

Perhaps the most common rejoinder to this dilemma is to suggest that policy makers and planners might take refuge, before making any kind of decision, in a rational assessment of 'the facts'. This 'technical-rational' (Owens, 2005) approach, as illustrated in Figure 12.1, implies not only that the realm of science and the realm of politics are mutually exclusive but also suggests that scientific evidence will in and of itself resolve otherwise contentious planning controversies.

As enticing as it may seem, however, this view of science as a 'natural handmaiden to policy' (Collingridge and Reeve, 1986, p. 8) is based on several questionable assumptions, both about science itself and about the ways in which it comes into contact with policy and political processes. To begin with, it presumes – *contra* the fundamental tenets of the sociology of scientific knowledge (SSK) (Barnes et al., 1996) – that science is a singular, homogenous enterprise, that it always contributes to the identification of truthful knowledge and that the processes through which such knowledge is generated are not influenced by wider social or political contexts. Yet most scholars – and indeed actors in the real world of planning – ultimately recognise that

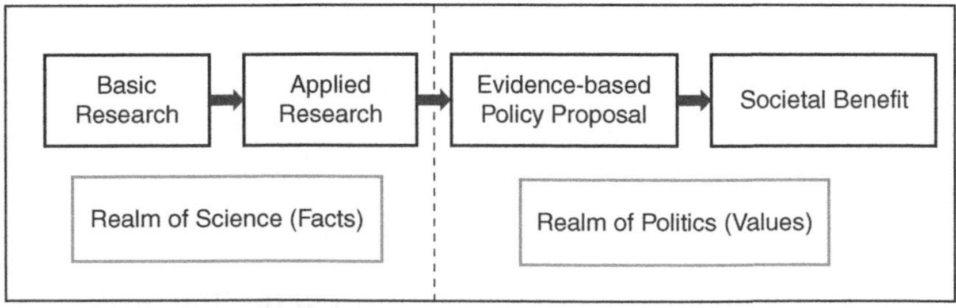

Figure 12.1 Schematic representation of the 'technical-rational' model of science-policy interactions
Source: Author's own

scientific evidence is almost always 'introduced at a specific point in [an] argument in order to persuade' (Majone, 1989, p. 10). Perhaps as problematically, however, this model also presumes that policy *can* be underpinned by science alone. A long tradition of work examining the politics of knowledge use in environmental planning processes suggests that the reality is considerably more complex.

Knowledge meets power: strategy, discourse and learning in environmental planning

One obvious problem is that knowledge can be – and frequently is – deployed selectively and strategically by actors within the planning process, either to help highlight and prioritise specific dimensions of a controversy ahead of others or to shape wider perceptions of the potential seriousness of the consequences likely to arise from a particular environmental decision (In't Veld and de Wit, 2000). For instance, in her seminal work exploring the proposed construction of a nuclear power plant on the shores of Cayuga Lake in New York State, Nelkin (1975, p. 48) highlights how research downplayed the potential environmental impacts of this plant by focussing 'almost entirely on the issue of thermal pollution' of the lake, rather than on the more uncertain question of impacts that might be exerted on the aquatic environment by radiation. Similarly, in his in-depth study of a long running transport planning controversy in the Danish town of Aalborg, Flyvbjerg (1997) documents how a daily newspaper editor's strategic interpretation of evidence enabled him to promote particular policy interests, in a quite different setting, as a member of the town's retail committee.

The most obvious upshot of such work, reinforced by more recent scholarship exploring diverse controversies, including those centred on sustainable biofuels (Palmer, 2014), fracking (Metze, 2017) and nuclear energy policy (Rough, 2011), is that the processes by which planners identify *and* interpret policy relevant knowledge and evidence are deeply politicised. Yet knowledge can also intersect with politics and power in ways less obviously attributable to the nefarious motives of individual actors – or sometimes individual institutions – within the planning community who are willing to act strategically. Indeed, for interpretive scholars, all knowledge claims – whether subject to conscious, strategic forms of deployment or not – actively shape perceptions of reality, thereby helping to 'determine what can and cannot be thought' in relation to specific environmental planning debates (Hajer and Versteeg, 2005, p. 178). Drawing inspiration from the ideas of French philosopher Michel Foucault (2002 [1969]), this approach recognises the performative quality of knowledge, evidence and ideas as cognitive variables which go far beyond the simple reflection, in neutral terms, of empirical reality. Instead, work in this vein makes use of the concept of discourse – defined by Hajer (1995, p. 44) as a set of 'ideas, concepts and categorisations that are produced, reproduced and transformed in a particular set of practices through which meaning is given to physical and social realities' – in order to draw analytical attention to processes through which competing schools of thinking around environmental controversies emerge and compete with one another over time (Owens and Cowell, 2011).

Work examining European Union (EU) debates about how best to respond to the indirect consequences of biofuel production, for example, has highlighted the predominance in this controversy of a discourse emphasising the need for more robust carbon accounting (Levidow, 2013; Palmer, 2014). Whilst this discourse served to sharpen the focus of planning efforts on developing analytical tools and techniques for more accurately measuring greenhouse gas emissions, it also detracted legitimacy from claims made by other actors and institutions who wished instead to draw attention to wider issues associated with biofuel driven land use change, such as biodiversity loss, food price impacts and land rights infringements. In a more wide-ranging study meanwhile, Jensen and Richardson (2004, p. x) have highlighted the profound influence, within EU mobility policy, of a discourse depicting 'Europe as monotopia', arguing that this discourse normalises a view of the EU as a single space over which it subsequently becomes both feasible and desirable to unfurl 'seamless networks enabling frictionless mobility'.

There are at least two analytical windfalls to be derived from these interpretive approaches to environmental planning. First, these perspectives help to pinpoint the effects exerted by ideas, knowledge and other cognitive variables in planning processes but without neglecting to pay attention to the wider institutional and place specific contexts within which those variables are situated and with which they form mutually constitutive relationships. Thus, although discourse is understood – in line with Foucault's original writings – as a 'complex entity that extends into the realms of ideology, strategy, language and practice' (Sharp and Richardson, 2001, p. 195), work in this vein nonetheless permits an examination of 'the fine grain level of policy making, where big ideas are reproduced through everyday planning activities' (Jensen and Richardson, 2004, p. 10). The second analytical gain, meanwhile, is an acknowledgement of the causal role played in planning processes by various kinds of policy *learning* (Heclo, 1974). Indeed, while many studies of planning processes drawing on discourse analysis are deeply critical of prevailing policy arrangements and ways of thinking about contemporary environmental issues, these approaches also offer a glimmer of hope that policy makers need not always inevitably function as 'players who unilaterally seek their own advantage in the political game' (Schön and Rein, 1994, p. 37). Instead, new knowledge and evidence can provoke learning on a conceptual, so-called 'double-loop' level (Argyris and Schön, 1978, p. 3), forcing productive reflection, on the part of individuals and wider groups of actors, not just on the approaches being taken to solving specific problems but also on the deeper assumptions underpinning existing problem framings themselves.

Confronting complexity and uncertainty: from scientific facts to expert judgements?

The imperative of engendering learning in the planning process has arguably been heightened, in recent years, by the increasing prevalence of complexity and uncertainty in the environmental sphere. Techno-scientific advances across a range of fields have begun to throw up new kinds of risks for planners to contend with, such as those associated with shale gas extraction, anthropogenic climate change and more recently geoengineering (Callon et al., 2009). In so doing, these advances have given rise to acrimonious debates among heterogeneous groups, all claiming to possess expert knowledge on the topics at hand. For Jasanoff (2005, p. 211), the profundity of these complexities and uncertainties is such that environmental planning controversies today can no longer be viewed as matters of deciding 'which scientific assessments are right, or even more technically defensible'. Instead, publics are faced with the thornier question of 'whose judgement should we trust, and on what basis?' (Jasanoff, 2005, p. 211).

Despite these developments, however, many scholars contend that it is necessary to preserve a special role for scientific and technical knowledge in planning processes. Collins and Evans (2007), for example, advocate postponing political judgements until *after* scientific and technical analyses of complex problems have been completed (seeing such separation as both feasible and desirable). In so doing, they not only reinforce some of the core assumptions of the technical-rational model encountered earlier in this chapter but also suggest that it is possible to identify all relevant holders of scientific and technical knowledge – whether formally certified and accredited by scientific institutions or not – in advance of deciding how to act on a given problem. Both sets of assumptions demand further scrutiny, paying specific attention to their consequences for our understandings of expertise itself and also of the processes through which expert status is attained in environmental planning controversies.

The first question to deal with concerns how citizens living in democratic societies recognise and define expertise. For Jasanoff (2005), as we have seen, it is judgements, rather than knowledge claims per se (whether scientific, technical or otherwise), that ultimately constitute expertise, and the role of the expert must therefore be viewed as distinct from that of the scientist. Research examining the practices of expert advisory institutions has made important contributions to this line of thinking by showing that judgements emanating from such bodies gain authority in part by situating ideas, evidence and knowledge within wider 'contexts of interpretation' (Jasanoff, 2011, p. 131). For instance, in their study of the *Gezondheidsraad* (the Health Council of the Netherlands), Bijker et al. (2009, p. 142) contend that the authority of this body is closely linked to its ability to produce recommendations that take the form of 'wisdom', defined as the 'well-argued reflection on the state of knowledge in relation to the state of the world'. Owens (2015, p. 166), not dissimilarly, traces much of the authority of the UK's Royal Commission on Environmental Pollution not to 'any technically oriented appraisal of "the facts"' but rather to its capacity to facilitate 'practical, public reasoning' (Weale, 2010, p. 266).

Geographers and political scientists, noting these features of expertise, have in recent years developed a range of typologies of expert advisory activity, and indeed influence, in policy and planning processes. Pielke Jr. (2007), for instance, proposes four ideal-typical kinds of expert advisor. The first, the 'pure scientist', refuses to engage with politics but might still inadvertently come to be viewed as an expert if the outputs of their research are made use of by planners. The second, the 'science arbiter', is a reactive expert who willingly contributes to politics by responding to policy makers' questions about the state of scientific knowledge, albeit only in an ad hoc fashion. The third and fourth types of expert, by contrast, are more proactive. Thus, the 'issue advocate' actively champions a specific approach to solving a policy problem, drawing on their scientific expertise to help substantiate their ideas. Where these types of expert claim to be doing nothing more than clarifying the state of existing scientific knowledge, they might be regarded as engaging in 'stealth issue advocacy'. Finally, the fourth ideal-type, the 'honest broker', seeks to expand the scope of choice available to policy makers by clarifying the *range* of policy options that would be compatible with existing scientific knowledge.

While Pielke's (2007) typology is best viewed as a heuristic tool, it nonetheless reinforces a view of expertise in which credibility can, under certain circumstances, accrue from an ability to actively *hybridise* scientific knowledge with wider social, political and economic concerns. Both Dunlop (2014), in her review of experts' engagement with ecosystem services policy, and Turnpenny et al. (2013), in their in-depth study of the workings of the UK's Parliamentary Environmental Audit Committee, similarly acknowledge the important role played by these kinds of hybridising activities, which blur – and bridge across – the ostensibly sharp boundaries between science and politics put forward by the technical-rational model. As the activities of experts come to be recognised as more diverse and complex than simply '[providing] "the facts" for policymakers to use' (Forsyth, 2003, p. 233), moreover, so too do the ways in which expert interventions might make a difference to planning processes. For Owens (2015), expert influence might take the form of 'direct hits' (where a specific recommendation leads directly to action in the planning community), or alternatively of 'dormant seeds', where ideas prove too challenging to be compatible with the political temper of their own time but nonetheless retain the potential to provoke 'enlightenment' (Weiss, 1977) at a later date. Alternatively, experts might also achieve a more 'atmospheric' form of influence (where their advice serves to contribute to a broader flow of ideas and ways of thinking about a specific issue) and can also 'do good by stealth', operating behind the scenes to provoke planners to rethink their approach to an issue before advice has even been formally delivered or published (Owens, 2015). Finally, of course, some efforts to influence policy can simply meet with outright failure, or 'non-influence'.

Placing expertise in context: geographies of knowledge, trust and credibility

By emphasising the status of expertise as a hybrid entity that involves interpreting knowledge claims within wider social, political, economic and ethical contexts, these ideas also pose a significant challenge to conventional thinking about how experts come to be identified, and to acquire credibility and trust, in contemporary democratic societies. Following seminal work in science studies, trust is invested in those figures fortunate enough to have directly *witnessed* the phenomenon on which knowledge claims are based. As Shapin (1998, p. 7) expresses it:

> [T]hat seventeenth-century English natural philosophers knew that there were such things as icebergs and polar bears was on no other basis than what they were told by those who had seen these things, for few, if any, of them had seen them for themselves.

Trustworthiness, in this view, therefore emerges from the societal perception of witnesses as honest, sincere individuals, who embody Robert Merton's classic principles of scientific enquiry, especially disinterestedness. While these ideas have been influential in highlighting the profoundly social nature of the processes by which scientific knowledge acquires credibility, however, their application to the environmental planning process is complicated by at least three further sets of issues.

The first, and perhaps most straightforward issue, concerns the Mertonian principle of disinterestedness. Whilst this may be a necessary quality of individuals charged with conveying knowledge claims under certain circumstances, such as for instance in the reporting of the results of a scientific laboratory experiment, its desirability is less self-evident in the context of environmental planning, where experts are charged with weighing up a range of knowledge claims within broader social, economic, political and ethical contexts. Indeed, in many circumstances experts sustain the trust of citizens precisely by virtue of conveying their intense interest in, and deep commitment to engaging with, the political dimensions of the issues on which they give advice.

The second issue arises from the contradiction between expert witnessing as a practice concerned with reporting what has happened in the past and the status of environmental planning as an endeavour concerned with anticipating what will happen in the future. Given this contradiction, adjudicating between the claims emerging from various forms of expert 'witnessing' becomes a deeply contested matter; indeed, it is not immediately clear what exactly should be witnessed in order for environmental planners to be adequately informed about the possible implications of their decisions. Recent geographical work on flood risk management has responded to this dilemma by seeking to enrol multiple forms of expert witnessing, each with distinct spatial and temporal characteristics, in a planning process relevant to this issue in the flood-prone North Yorkshire town of Pickering (Lane et al., 2011). These range from the witnessing of the inner workings of scientific models that manipulate various data inputs to 'map' flood risk spatially, through to the deeply embodied and emotive witnessing of the forceful nature and destructive consequences of flooding on the ground.

Finally, a third issue, closely related to the last, emerges from the fact that environmental planning controversies are always situated in local, place specific contexts. In order to gain the status of a trustworthy expert in these kinds of debates, therefore, one must go beyond the provision of purportedly transferable observation claims derived from *other* places and instead seek to build new observation claims within the specific context of the controversy at hand. In their work examining an ongoing controversy over the impacts of fracking in Pennsylvania on local air quality, for instance, Pritchard and Gabrys (2016) highlight how practices of 'citizen sensing', based on the use of personal air quality monitors by local residents, have begun to force critical reflection on prevailing, culturally specific ideas that shape the criteria against which the credibility of knowledge claims about air quality should ultimately be judged. An important corollary of these ideas is that trust and credibility should be seen not as 'fixed dispositions' but as 'negotiated outcomes borne of particular individual relationships, institutional settings, and social connections' (Withers, 2018, p. 13).

Conclusions: ecologies of environmental expertise in the planning process?

Rethinking trust and credibility as emergent, culturally specific phenomena reminds us that environmental planning controversies have complex epistemic geographies and that space and time play critical roles in shaping the processes through which all forms of knowledge and expertise are produced. Particularly when faced with environmental processes and phenomena which cannot always be directly observed – as is the case with air pollution, climate change or zoonotic diseases in agriculture, to name a few examples – this suggests that planners would be well advised to seek to better attune themselves to the multi-faceted knowledge production practices through which different groups involved in these controversies seek to render such processes visible and meaningful.

In pursuing this objective, planning theory and practice both stand to benefit from a continued and deepening engagement with recent scholarship at the interface between geography and science and technology studies. Here, planning processes underpinned by narrow, expert led forms of risk assessment are criticised not simply for their procedural exclusivity, wherein efforts to consult publics often comprise little more than 'constrained "top-down" exercises in legitimation' (Stirling, 2008, p. 268). More fundamentally, in their haste to resolve planning controversies, such processes are also deemed to inadvertently *accentuate* problems – including those created by a crisis of trust in experts and policy makers – by overlooking or downplaying important areas of uncertainty and ambiguity.

A less hubristic approach to planning, these thinkers argue, must begin from the recognition that controversies can never be reduced to discrete moments of conflict between two or more fixed sets of views or interests, capable of being resolved into a singular or definitive 'solution'. Instead, controversies should be viewed as dynamic, relationally constituted 'ecologies', within which a diversity of human and nonhuman actors, bodies, pressures and flows interact in inherently unpredictable and often transformative ways (Hinchliffe et al., 2017). Significant benefits to planning could therefore be derived by giving greater space (and time) to what Stengers (2005, p. 185) terms the 'virtual power' of these ecologies to 'make us think' in new ways. In short, this means that controversies should be embraced by participants on all sides as opportunities for 'slowing things down' (Bingham, 2008, p. 111) and thereby maximising the potential for deep, transformative kinds of learning and reflection.

In practical terms, this shift entails far more than simply making space for more wide-ranging, formal participation in environmental planning by diverse groups, for example, through deeper public consultation. Making greater use of such participatory tools would do nothing, after all, to counter a prevailing sense that they constitute little more than instrumental devices for legitimising a singular 'story of the world' whose broad thrust has already been predetermined.

Instead, the imperative should be to render planning processes more open ended and responsive to the 'eventful', indeterminate nature of controversies and to the power of all actors (human and nonhuman) involved in them to foster learning, whether that happens in formal participatory settings or in more informal, messy, and dynamic networks (Pallett and Chilvers, 2013). Relinquishing control over the nature and timing of participation in environmental planning may seem risky, of course. But it is only through embracing the full range of situated and emergent processes by which knowledge and expertise are produced in practice that planners can hope to build deep, lasting trust in environmental planning itself.

References

Argyris, C. and Schön, D. (1978). *Organizational learning: a theory of action perspective.* Reading, MA: Addison-Wesley.

Barnes, T., Bloor, D. and Henry, J. (1996). *Scientific knowledge: a sociological analysis.* London: Athlone Press.

Bijker, W. E., Bal, R. and Hendriks, R. (2009). *The paradox of scientific authority: the role of scientific advice in democracies.* Cambridge, MA: MIT Press.

Bingham, N. (2008). 'Slowing things down: lessons from the GM controversy'. *Geoforum*, 39(1): 111–22.

Callon, M., Lascoumes, P. and Barthe, Y. (2009). *Acting in an uncertain world: an essay on technical democracy.* Cambridge, MA: MIT Press.

Collingridge, D. and Reeve, C. (1986). *Science speaks to power.* London: Frances Pinter.

Collins, H. and Evans, R. (2007). *Rethinking expertise.* Chicago, IL: University of Chicago Press.

Dunlop, C. (2014). 'The possible experts: how epistemic communities negotiate barriers to knowledge use in ecosystem services policy'. *Environment and Planning C: Government and Policy*, 32(2): 208–28.

Flyvbjerg, B. (1997). *Rationality and power: democracy in practice.* Chicago, IL: University of Chicago Press.

Forsyth, T. (2003). *Critical political ecology: the politics of environmental science.* London: Routledge.

Foucault, M. (2002 [1969]). *The archaeology of knowledge.* London: Routledge.

Hajer, M. (1995). *The politics of environmental discourse: ecological modernisation and the policy process.* Oxford: Oxford University Press.

Hajer, M. and Versteeg, W. (2005). 'A decade of discourse analysis in environmental politics: achievements, challenges, perspectives'. *Journal of Environmental Policy and Planning*, 7(3): 175–84.

Heclo, H. (1974). *Modern social politics in Britain and Sweden.* New Haven, CT: Yale University Press.

Hinchliffe, S., Bingham, N., Allen, J. and Carter, S. (2017). *Pathological lives: disease, space and biopolitics.* Oxford: Wiley-Blackwell.

In't Veld, R. and de Wit, A. (2000). 'Clarifications', in R. In't Veld (ed.) *Willingly and knowingly: the roles of knowledge about nature and the environment in policy processes.* Utrecht: Lemmo, pp 147–57.

Jasanoff, S. (2005). 'Judgement under siege: the three-body problem of expert legitimacy', in S. Maasen and P. Weingart (eds.) *Democratization of expertise? Exploring novel forms of scientific advice in political decision-making.* Dordrecht: Springer, pp 209–24.

Jasanoff, S. (2011). 'Cosmopolitan knowledge: climate science and global civic epistemology', in J. S. Dryzek, R. B. Norgaard and D. Schlosberg (eds.) *The Oxford handbook of climate change and society.* Oxford: Oxford University Press, pp 129–43.

Jensen, O. B. and Richardson, T. (2004). *Making European space: mobility, power and territorial identity.* Abingdon: Routledge.

Lane, S. N., Odoni, N., Landström, C., Whatmore, S. J., Ward, N. and Bradley, S. (2011). 'Doing flood risk science differently: an experiment in radical scientific method'. *Transactions of the Institute of British Geographers*, 36(1): 15–36.

Levidow, L. (2013). 'EU criteria for sustainable biofuels: accounting for carbon, depoliticising plunder'. *Geoforum*, 44(1): 211–23.

Majone, G. (1989). *Evidence, argument and persuasion in the policy process.* New Haven, CT: Yale University Press.

Metze, T. (2017). 'Fracking the debate: frame shifts and boundary work in Dutch decision making on shale gas'. *Journal of European Environmental Policy and Planning*, 19(1): 35–52.

Nelkin, D. (1975). 'The political impact of technical expertise'. *Social Studies of Science*, 5(1): 35–54.

Owens, S. (2005). 'Making a difference? Some perspectives on environmental research and policy'. *Transactions of the Institute of British Geographers*, 30(3): 287–92.

Owens, S. (2015). *Knowledge, policy, and expertise: the UK Royal Commission on Environmental Pollution, 1970–2011*. Oxford: Oxford University Press.
Owens, S. and Cowell, R. (2011). *Land and limits: interpreting sustainability in the planning process* (2nd edn). Abingdon: Routledge.
Owens, S., Rayner, T. and Bina, O. (2004). 'New agendas for appraisal: reflections on theory, practice and research'. *Environment and Planning A*, 36(11): 1943–59.
Pallett, H. and Chilvers, J. (2013). 'A decade of learning about publics, participation, and climate change: institutionalising reflexivity?' *Environment and Planning A*, 45(5): 1162–83.
Palmer, J. (2014). 'Biofuels and the politics of land-use change: tracing the interactions of discourse and place in European policy making'. *Environment and Planning A*, 46(2): 337–52.
Palmer, J. (2016). 'Interpretive analysis and regulatory impact assessment', in C. A. Dunlop and C. M. Radaelli (eds.) *Handbook of regulatory impact assessment*. Cheltenham: Edward Elgar, pp 52–65.
Pielke Jr., R. A. (2007). *The honest broker: making sense of science in policy and politics*. Cambridge: Cambridge University Press.
Pritchard, H. and Gabrys, J. (2016). 'From citizen sensing to collective monitoring: working through the perceptive and affective problematics of environmental pollution'. *GeoHumanities*, 2(2): 354–71.
Radaelli, C. (1995). 'The role of knowledge in the policy process'. *Journal of European Public Policy*, 2(2): 159–83.
Rough, E. (2011). 'Policy learning through public inquiries? The case of UK nuclear energy policy 1955–61'. *Environment and Planning C: Government and Policy*, 29(1): 24–45.
Schön, D. and Rein, M. (1994). *Frame reflection: toward the resolution of intractable policy controversies*. New York: Basic Books.
Shapin, S. (1998). 'Placing the view from nowhere: historical and sociological problems in the location of science'. *Transactions of the Institute of British Geographers*, 23(1): 5–12.
Sharp, L. and Richardson, T. (2001). 'Reflections on Foucauldian discourse analysis in planning and environmental policy research'. *Journal of Environmental Policy and Planning*, 3(3): 193–209.
Stengers, I. (2005). 'Introductory notes on an ecology of practices'. *Cultural Studies Review*, 11(1): 183–96.
Stirling, A. (2008). '"Opening up" and "closing down": power, participation, and pluralism in the social appraisal of technology'. *Science, Technology and Human Values*, 33(2): 262–94.
Turnpenny, J., Radaelli, C. M., Jordan, A. and Jacob, K. (2009). 'The policy and politics of policy appraisal: emerging trends and new directions'. *Journal of European Public Policy*, 16(4): 640–53.
Turnpenny, J., Russel, D. and Rayner, T. (2013). 'The complexity of evidence for sustainable development policy: analysing the boundary work of the UK parliamentary environmental audit committee'. *Transactions of the Institute of British Geographers*, 38(4): 586–98.
Weale, A. (2010). 'Political theory and practical public reasoning'. *Political Studies*, 58(2): 266–81.
Weiss, C. (1977). 'Research for policy's sake: the enlightenment function of social research'. *Policy Analysis*, 3: 531–45.
Withers, C. (2018). 'Trust – in geography'. *Progress in Human Geography*, 42(4): 489–508.

13

Grassroots and environmental NGOs

Nathalie Berny and Ellen Clarke

The terms 'grassroots' and 'NGOs' usually refer to different patterns of collective action in terms of their level of operation (local versus general issues), form (informal versus permanently organised), activism (volunteers versus employees) and claims (radical versus reformist) (Horowitz and Watts, 2016). Behind this contrast lies a debate about the properties of social movements and their capacity to bring significant social and political change. The specific attributes of grassroots groups and non-governmental organisations (NGOs) still divide academics and actors. Local does not always equate with non-organised and radical, as revealed by critical work, for example, on the Transition Town Network (Smith, 2011). The purpose of this chapter is to show why both the grassroots and NGOs are significant and often concomitant elements in the struggles around the use of land and natural resources, although their boundaries sometimes blur. Together these mobilisations reflect the past and ongoing transformations of the environment.

This chapter's development is twofold. It first addresses the parallel expansion of environmental mobilisations and legislation, considering their extent beyond Western societies. It then focuses on the relationships between environmental NGOs (ENGOs) and the grassroots to question the difference in cause and strategy of contemporary mobilisations in the North and South.

Environmentalisms: causes, forms and outcomes

Grassroots groups and organisations have coexisted since the beginning of what is today labelled the 'environmental movement'. They contributed to a partial recognition of a right to participation in decision making for citizens and organisations advocating for the environment. While 'environmental movements in different countries are subject to the same global forces, they do not react in the same way' (Doherty and Doyle, 2006, p. 702). Some scholars thus prefer the term 'environmentalisms' (Doyle and MacGregor, 2014) to a 'global environmental movement' (McCormick, 1989; Dauvergne, 2016). These mobilisations can share similar concerns but not necessarily the same political ideologies, including within the same country (Rootes, 2007). Different dynamics have been identified to explain their emergence, shape and outcomes.

Environmentalism in Western countries and policy convergence

Social movement scientists identified two waves of environmentalisms in Western countries: the 1800s and the 1960–70s. They gave birth respectively to conservation and ecology organisations, differing in their claims, audience and modes of action (Jamison, 2001; McCormick, 1989). Planning and environmental regulation including provisions of public participation bear the mark of these social pressures.

The first environmental activists reacted to initial effects of the industrial revolution, including intensive extractive activities such as mining. Through their knowledge or leisure, enlightened amateurs, scientists, artists, hikers or hunters advocated for protecting a landscape, species or a natural space. Concerned by aesthetics, preservationists in the United States and open spaces movements in Britain challenged property rights (McCormick, 1989). Pollution in cities also became a concern for social elites. But only organisations committed to revealing impact on human health, such as the Manchester Association for the Prevention of Smoke created in 1842, challenged the faith in progress (Mosley, 2013).

Close connections with decision makers through social networks based on a hobby, notably hunting, proved useful for early policy initiatives. But for the same reason, their success varied, notably in Europe. The legal protection of the historic wood, Fontainebleau, which mobilised both artists and aristocrats in France, preceded the creation of the Yellowstone national park in 1872 in the United States. However, the European conservation movement encountered greater difficulties in establishing national parks so instead resorted to creating private nature reserves. Very early on, these organisations used international negotiations to secure protection status for wildlife. Regarding pollution in urban areas, the first legislation in France in 1814 and Britain in 1867 were mainly driven by the need to protect highly intensive capitalist industries and the well-being of the wealthiest. Similar patterns were observed in US urban regulations (Brosnan, 2014).

Environmental mobilisations emerging in the 1960s reflected a more widespread concern in a society facing swift development of industrial chemicals, transport infrastructure and nuclear energy. Groups that multiplied at national and local levels in North America and Europe (McCormick, 1989), made claims seeking mass support. Their modes of action drew upon contemporary movements of students, feminists and pacifists, with whom they shared defiance towards institutions, to solve the 'ecological crisis' being documented by scientists. Post-materialist values in wealthy societies explain why environmentalism was related to a 'new social movement' as well (Kriesi, 1996). Activists found inspiration in a wide array of ideologies. Anti-nuclear and local groups, such as the 'Groupes d'action municipale' in France or the 'Bürgerinitiativen' in Germany, presented their claims for local, participative modes of science and decision making as a cure for technocracy and modernity (Van Tatenhove and Leroy, 2003; Jamison, 2001). Urban planning had become a key dimension of the environmental struggle. Framed as 'quality of life' questions, environmental problems were a matter of individual rights rather than class interests.

During the 1970s, conservation organisations' membership expanded as well (McCormick, 1989). Ecology organisations later became more professionalised and bureaucratic, advocating for policy change. Used for both environmentalisms, the term ENGOs reflects their international expansion in terms of networks and public attention (Tamiotti and Finger, 2001). Fuelled by radical criticism against industrial societies, the ecological movement obtained mainly procedural gains without undermining the rationale behind economic growth (Kriesi, 1996). There is indeed a striking convergence in the way Western countries responded to social unrest. They provided claims to participation and transparency with right-to-know legislation and participatory procedures (Dryzek, 2013). The United States paved the way for environmental impact assessments with the 1970 National Environment Policy Act. But the burgeoning environmental legislation of the 1960s had resulted from a decade of environmental activism challenging decisions of federal agencies (Zelko, 2014). Across Europe, a number of land planning and environmental acts entitling ENGOs to be consulted and even rights of standing in the courts were adopted in the 1970s and thereafter.

The Aarhus Convention, adopted in 1998, reflects policy convergence at the international level. Domestic law principles have become enshrined in international law by this regional convention on access to information, justice and decision making mainly in Europe and Central Asia. Although this convention is usually presented as implementing Principle 10 of the Rio Declaration, it results from a joint initiative from governments of the European regional organisation of the United Nations. Environmental activists from Eastern European countries were especially active in advocating for a transparent and open process of decision making (Hochstetler, 2012). The negotiation involved ENGOs from the very beginning, and the extent of their participation remains a key feature of this convention.

Discursive analyses offer a stimulating framework for capturing the diffusion of green claims into society at the expense of their radicalism. Dryzek (2013) identifies the narrative behind this shift as 'democratic pragmatism'. Drawing from participative experiments in Northern Europe, Van Tatenhove and Leroy (2003) make a link between social mobilisation and the participative component of environmental policies. They also underline the pervasive influence of the 'ecological modernisation' discourse where the cooperation of actors, from civil society to firms, is supposed to reconcile economic growth and environmental protection. This assumption has since been expanded with the notion of sustainable development, driving the international agenda and raising controversies (Hollender, 2015; Dryzek, 2013).

Environmentalism of the poor and procedural environmental rights

Initially limited to scientific circles, the diffusion of the term 'environment' in the 1960s corresponded with a new understanding in Western societies of the complex relationship between human societies and nature. The term was also used to study social mobilisations in developing states (Guha and Martinez-Alier, 1997). These movements, related to environmental degradation and natural resources exploitation, formed in the 1970s in Latin America, Africa and Asia (Haynes, 1999). Opposed to modernisation led by social elites, which materialised in a development agenda in agriculture, transport and energy, they advocated for access to decision making and a recognition of their rights to access vital resources, paving the way for procedural environmental rights.

The two waves of environmentalism, already mentioned, have nevertheless coexisted: for instance, there are both Marxist ecologists and activists supporting ecological modernisation in India and Latin America. While in countries such as India local values praised nature itself, the influence of scientific conservation imported by the British remained pervasive in public policies as in other postcolonial states (Kashwan, 2017). Similar to Western societies, environmental struggles emerging since the 1970s took place inside local communities. Although they were mainly agrarian movements, they also promoted alternative models of development, based on local participation (Tamiotti and Finger, 2001). Several of such struggles – Chipko (India), Chico Dam (Philippines), Rubber Tappers (Brazil), Ogoni (Nigeria), Green Belt (Kenya) – became emblematic from their extent and international audience but remained paradoxically isolated from each other.

Because they primarily involved peasants and residents from the neighbourhood trying to protect their livelihood, Guha and Martinez-Alier (1997) oppose environmentalism of 'affluence' versus environmentalism of 'survival'. Diverging from the postmaterialist explanation applied to environmentalism in Western societies, they develop a political ecology approach focusing on distribution inequalities and access to key material resources (water, food, air) based on gender, ethnicity or class. Similar studies in Latin America and Asia showed 'the poor' suffer the most from environmental degradation, triggered by state development policy and global capitalism predating natural resources. Some came to use the concept of 'environmental justice' (EJ), first developed in the United States in the 1980s (Sikor and Newell, 2014). EJ has become a rallying point for both activists and academics who aim at demonstrating social inequalities in pollution and environmental degradation, a dimension also identified by European environmental history, as in Latin America (Mosley, 2013; Wakild, 2013).

While mobilisations triggered by the development agenda and subsequent intensive exploitation of natural resources have promoted participation at the local level, the principles behind 'democratic pragmatism' (Dryzek, 2013), such as right to information and public participation, have gained ground outside Europe (Mauerhofer, 2016). The precedence of social mobilisations on the adoption of environmental legislation and institutions varies between countries. International pressure is a significant force, especially in transition countries (Hochstetler, 2012). Comparative analyses give useful insights to explain the outcomes of such legislation. The existence of legal channels appears as a precondition but not sufficient, including in democratic states (Haynes, 1999).

In Latin America, where extractive industries have boomed, local communities and environmental activists trying to exert their rights for consultation or participation in decision making face violence and repression by the state or third parties (Hollender, 2015; Hochstetler, 2012). Most governments from both the left and the right maintain a largely anti-ecological stance in the 1990s and 2000s. Recently, 'neo-extractivism' promoted by governments concerned with defending their sovereignty in the management of natural resources has increased these trends (Laing, 2015). Despite looming democratic backlash, environmental activists still battle for legal grounds. In Ecuador, the ENGO 'Acción Ecológica' campaigned to get the right of the Earth included in the constitution (Martinez-Alier et al., 2016). The same organisation was threatened with dissolution by the Environment Minister in 2015 (The Economist, 2017). The Escazu agreement, adopted by 24 Latin American and Caribbean countries in March 2018, includes provisions very similar to the Aarhus convention regarding environmental procedural rights. It has become the first legally binding agreement which explicitly aims at preventing any 'threats, restriction and insecurity' against 'environmental human rights defenders' (article 9). The enforcement of this convention depends on a ratification process still ongoing in July 2019.

Since 2015, the NGO Global Witness has raised public awareness on violence against environmental activists, which primarily affects indigenous and local communities with most projects linked to extractive industries and agro-business (see Figure 13.1). The phenomenon remains underreported. Although there was a worldwide recognition of procedural environmental rights in domestic legislation and more than 100 constitutions, a growing wave of violence over the last decade, including murder, has prompted UN Environment to launch several initiatives around capacity building and information dissemination (UN Environment, 2018).

The focus on social movements and interest groups characterises the literature initially developed in the 1980s in Western societies, while studies from anthropologists, sociologists and geographers have flourished in political ecology and EJ to embrace issues crucial to the Global South. Both bodies of literature have addressed the transformative capacity of these movements. Definitional issues revolve around the same question: how far do the movements under study advocate for social and political change? Environmental organisations calling for mere policy change have been considered, analytically, as falling outside the social movement (Doherty and

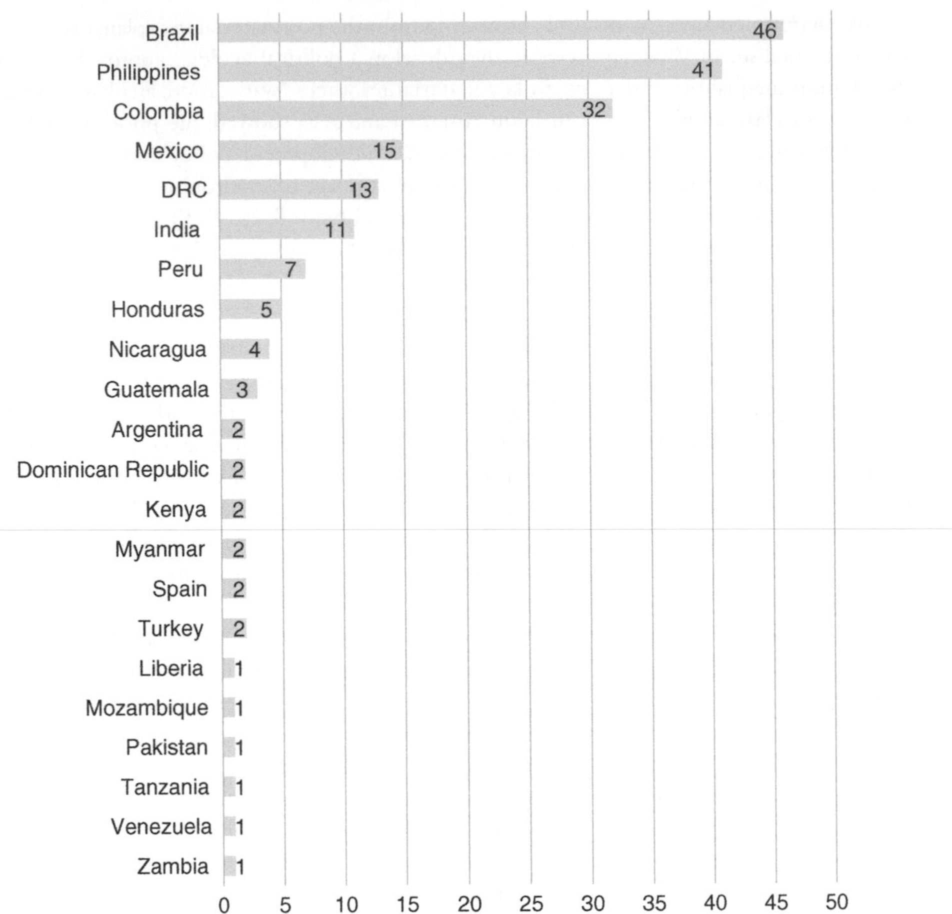

Figure 13.1 Environmental and land defenders killed by country in 2017

Source: Authors drawn from Global Witness data (methodology used available at: http://gu.com/p/6nae4)

Doyle, 2006), while an inclusive perspective captures the various forms of collective action within environmentalisms (Rootes, 2007).

Relationships between ENGOs and grassroots

Addressing the relationships between ENGOs and grassroots offers an opportunity to consider their respective transformative capacity on social norms and public policies. Local, 'grassroots organisations' (GROs) are more likely to involve the affected people impacted by environmental degradation and respect their ways and culture (Horowitz and Watts, 2016). GROs represent the DIY option, while ENGOs operate in the name of a cause, usually by employing staff and seeking the support of members. This section questions the common contrasts between GROs and ENGOs, their supposedly respective radical and reformist stance, and their origins in Western countries and the Global South.

Criticisms targeting them are ultimately not so different. Without the grassroots, ENGOs' agenda, and sometimes funding, become more dependent on their relations with decision makers and cooperation with firms than with their constituency. But without any organisation, grassroots mobilisations can hardly set up resistance and scale to bigger topics. The term 'not in my backyard' (NIMBY) has stigmatised the mobilisation of local residents that disappear once the threat has gone, oblivious of the impact of similar projects elsewhere. Likewise, environmentalism of the richest countries is accused of having displaced problems to less democratic or environmentally regulated countries (Dauvergne, 2016; Pellow, 2007)

GROs versus ENGOs

ENGOs multiplied in the international arena during the 1990s (Doherty and Doyle, 2006; Tamiotti and Finger, 2001) partially due to the initiatives launched in a new policy domain, including at the international level. Social movement scientists drew upon the concept of 'institutionalisation' to show that initial frontrunners tend to moderate their stance and modes of action as the general population and state authorities share a wider concern for their cause (Kriesi, 1996). They become policy players, trying to secure existing key legislation and promote new ones. Without necessarily giving up on unconventional modes of action, they seek to deploy lobbying efforts. Expertise is a key resource to frame claims and access decision makers. ENGOs thus played a crucial part in the implementation process, bringing essential information to domestic authorities and international institutions (Hochstetler, 2012; Tamiotti and Finger, 2001). They also monitored corporate practices within cooperative, soft law agreements inspired by ecological modernisation, or by using 'blame and shame' strategies targeting firms' reputations.

As institutionalisation shapes ENGOs' agenda and modes of action, their logics increasingly differ from the grassroots. Organisational growth, bringing more expertise and fundraising capacity, becomes a priority per se. In Britain, ENGOs, such as Greenpeace or Friends of the Earth, whose membership and capacity for action initially spectacularly increased, saw the number of supporters diminish in the 1990s because of the extension of 'chequebook participation', a term coining the marginal part left to members (Grant and Malloney, 1997). In Eastern Europe, most ENGOs survived thanks to international funding, but this isolated them from potential supporters, also creating distrust among the population (Hochstetler, 2012). GROs are exempt from neither defiance nor indifference from the people they speak for once they start lobbying the authorities and building extra-local coalitions, for instance, the Bolivian GRO studied by Perreault (2016).

By shifting the focus from conflict in situ or a policy problem to the organisations themselves, network analyses have illustrated the richness of environmental movements whose components are diverse, operating in more or less participatory and open rules, partially connected or, in contrast, ignorant of each other (for example, Saunders, 2013). Institutionalisation is far from having impoverished the variety of mobilisations involved in environmental struggles (Rootes, 2007). Activists can learn from overlapping membership in GROs/ENGOs, whilst defiance or inventiveness from the grassroots can spur ENGOs to internal reform.

This is also the case when the participation or contention of planning or environmental policy decisions come under scrutiny. The set of dichotomies between local/radical and national/reformist does not describe fully the strategies deployed by ENGOs and GROs.

Local versus national/global campaigns

In local issue-focused campaigns, activists have not necessarily desired to address systemic change or sustain action for others elsewhere. Such mobilisations play, however, a 'discovery role' vis-à-vis ENGOs on new or underreported problems (Carmin, 1999). For instance, grassroots action against airport expansion in the United Kingdom incorporated independent campaigns by large

ENGOs, such as the RSPB (Royal Society for the Protection of Birds) and its 'No Airport at Cliffe' campaign. This is only possible when the issue is consistent with the priorities of the ENGO. Moreover, ENGOs are reluctant to embrace local campaigns that are not supported by scientific evidence (Rootes, 2013). Alternatively, by relaying local causes, they increase their reputation vis-à-vis bystander audiences. Even ENGOs where local groups play a marginal part are eager to choose a symbolic local cause. From legal advice to publicity, ENGOs provide a platform for a community that displays them as connected to wider environmental debates.

The success or effectiveness of local issue-focused grassroots organisations is indeed broadly considered to be dependent on their ability to translate their concerns from a NIMBY to a 'not in anybody's backyard' (NIABY) narrative (Freudenberg and Steinsapir, 1991). Similarly, in non-democratic countries, protest movements which remain local seem doomed to failure in the face of repressive states (Haynes, 1999; Mohan and Stokke, 2000). Rootes (2013) suggests three ways to 'transcend the local': (i) representing the issue as an instance of national/transnational importance, (ii) networking with actors on the same issue but different locations, and (iii) appealing to non-local ENGOs. The mobilisations in situ of Larzac Plateau (France), Standing Rock (US) and Chiappas (Mexico) illustrate the first scenario. These places have become iconic, inspiring the discourse and symbols used by activists in defence of various causes, from global justice to indigenous people's rights.

Second, through networking local groups share knowledge, experience and publicity. 'Grassroots networks', without hierarchical management, can be extremely effective. For instance, the anti-road network in the United Kingdom in the 1990s helped coordinate the activities of local campaigners within a wider, more strategic context such as that of climate change or energy policy (Rootes, 2013). They refused the involvement of established ENGOs in order to set their own priorities. Another example of grassroots networks, the Global Anti-Incinerator Alliance (GAIA), brings together communities in the global North and South. GAIA developed from a successful campaign in the Philippines to establish a nationwide ban on incineration. It now incorporates both local groups and ENGOs to provide expertise and a common narrative for citizens otherwise isolated in their struggles. Although their modus operandi overtly departs from established ENGOs', such grassroots networks are not exempt from subsequent institutionalisation, such as occurred with GAIA.

The third option of appealing to non-local ENGOs has been widely documented in both Southern and Western countries. The campaign for 'the last wild river in western Europe' in France mixed legal and direct action. It brought together outside high profile supporters, trained activists from ENGOs, and determined local groups who succeeded in hampering the building of large dams on the French Loire (Hayes, 2002). Such examples of *transnational* activism across borders are also common for GROs in the Global South. The normative power of non-state actors has been documented by international relations academics. GROs built alliances with internationally active ENGOs in order to succeed in the face of unresponsive or repressive states (Keck and Sikkink, 1998). Local activists gained new skills in trying to achieve their goals in the domestic arena, developing 'bilateral activism' (Steinberg, 2001). They are all the more effective when they use the resources of the ENGO at the global level to serve their own strategy at the domestic level. The seminal study of toxic transnational campaigns is an example of such successful EJ collective action (Pellow, 2007).

North versus South

International ENGOs with national branches originate from Western countries, given the significant development of environmental movements in the 1970s. Due to their origin, they would often neglect the needs of local populations, reproducing Northern domination in post-colonial and post-development countries.

The conservationist agenda in post-colonial societies after WWII partially substantiates this claim. For instance, in Kenya and South Africa, conservation organisations supported projects implying the expulsion of local populations to protect wildlife (Haynes, 1999). Foreign hunters were, meanwhile, still able to enjoy their hobby as a source of income for the domestic social elites. Driven by environment protection priorities, conservation projects have not always considered local norms. In contrast, immersed social scientists, such as anthropologists in Amazonia, valued the local knowledge of indigenous populations and supported their legal recognition (Nicolle and Leroy, 2017). During the 2000s, the conservation agenda became more centred on the acknowledgment of these societies' culture and values regarding nature. This is not always mere rhetoric as illustrated by successful cases of cooperation (Martinez-Alier et al., 2016).

Beyond the conservation agenda, and across different disciplines, local participation and empowerment were presented as a way to achieve sustainability in post-development countries. For instance, in development studies, both neo-liberalism and post-Marxism perspectives converged on the positive links between the two (Mohan and Stokke, 2000). In the first approach, ENGOs are crucial actors to counterbalance state inefficiencies. They operate civil society capacity building programmes developed by international organisations such as the World Bank. They initiated or took part in certification procedures to ensure the protection of natural resources by means other than legislation, notably in non-democratic countries where implementation is jeopardised. In the second approach, as in political ecology, GROs are at the vanguard of the defence of local communities' livelihoods. By contrast, ENGOs are accused of providing an eco-narrative of 'sustainable development' or 'green economy' for unsustainable large corporations. Since 2000, several ENGOs also supported 'green grabbing' or the *commodification* of nature, which suggests that 'unsustainable practice "here" can be repaired by sustainable practice "there"' (Fairhead et al., 2012, p. 242). For instance, forestation projects result in carbon credits but are not necessarily favourable to biodiversity maintenance (Nicolle and Leroy, 2017). Environmentalism from a force for social change has thus become an agent of the neoliberal world order (Dauvergne, 2016).

However, approaching the variety of environmentalisms by focusing on ENGOs themselves gives another understanding of the conflicting views involved in planning issues. Doherty and Doyle (2013) track the exchange and conflicts that led Friends of the Earth International towards an EJ agenda driven by ENGOs from the South. As ascribed by Martinez-Alier et al. (2016), a number of these ENGOs that emerged here promoted alternative concepts. By making their way primarily in the international arena or academic circles, they reframed problems and their causes, such as 'ecological debt', 'land grabbing' or 'bio-piracy'. This confirms the importance of organised collective action in the long run and diffusion of claims as a sign of success. Finally, environmental activism takes shape in a changing world. Environmental injustice does not necessarily reproduce forms of North-South domination but also reflects South-South inequalities. In Africa, several local mobilisations resisted land grabbing projects promoted by Chinese or Indian firms, which reproduced forms of appropriation and repression in use at home (Temper, 2018).

Finally, the increase of legislation preventing transnational cooperation between NGOs reveals a contrario its political significance. 'Between 1993 and 2016, 48 countries enacted laws that restricted the activities of local NGOs receiving foreign funding, and 63 countries adopted laws restricting activities of foreign NGOs' (UN Environment, 2018). The phenomenon seems to have amplified since the mid-2000s (Matejova et al., 2018). Public authorities castigate NGOs as agents of 'colonialism', 'capitalism' or 'extremism' in respectively authoritarian, democratising and Western countries (Matejova et al., 2018).

The attention paid to multi-scale strategies thus remains a critical venue to explore the metamorphosis of activism related to planning and environmental issues. Following the reminder by Mohan and Stokke (2000) that the 'local is politics', shifting analysis to the actors involved and the dynamics of collective action is equally necessary to avoid any over-simplification in this regard. By focusing on macro-structural factors of conflicts, EJ, political ecology or post-Marxism studies tended to assume that the continuing degradation of the environment benefits the outside, non-local, hegemonic (corporate, elite or state) interests. Local communities thus become places for resistance against the global powers and an alternative explanatory factor of the outcome of environmental struggles.

Deconstructing the categories of NGO, GRO or community by analysing the dynamics of collective action and advocacy – who speaks for who – offers a necessary and complementary perspective in order to unveil the diverging interests and values behind discursive strategies. Local NGOs are not necessarily sympathetic to the needs of the local population when they are populated by social elites. Local communities are riven by conflicts of view and interest, and likewise the 'poor' are not necessarily in favour of the environment or participation as is often assumed (Martinez-Alier et al., 2016). Indigenous rights narratives sometimes serve the interests of the most powerful (Temper, 2018). Several studies selected here showed, on the contrary, a multiplicity of actors and coalitions involved, learning from each other, building alliances within the state and international arenas (Nicolle and Leroy, 2017; Hollender, 2015; Pellow, 2007; Keck and Sikkink, 1998).

Conclusion: resistance and alternatives

Questioning the similarities of land use conflicts across the globe offers stimulating perspectives on transformations shaping both domestic politics and global forces. Grassroots mobilisations often give birth to and cooperate with ENGOs. The latter appear as an agent of change in different contexts, impacting both institutional rules and specific decisions. However, their strength exposes them to limitations. The continuity of an organisation necessitates securing material resources (funding, offices, technologies), setting priorities, and taking credit for success. They have to build ties with other NGOs, public authorities, firms and/or individual members to raise resources and to get information relevant to their advocacy strategies. The assumption that an organisation has its own agenda is relevant (Temper, 2018), but that should not lead to the systematic conclusion that they develop at the expense of local interests. The breadth of studies embracing the subject shows a wide diversity of situations in this respect.

Different bodies of literature suggest that playing by the rules of the game undermines the capacity of social movements to bring radical transformation or systemic change. The procedural gains in right-to-know legislation, participative procedures and access to justice has been considered a Pyrrhic victory by many, academics and activists alike; they have not prevented expanding, unsustainable exploitation of the planet's finite natural resources. In the Global South, GROs usually promote democratic and wider justice principles, as to play their role they have to simultaneously fight against anti-democratic societies. Interestingly, local mobilisation displays similar patterns to the previous claims of Western environmentalists in the face of technocratic and central authorities.

In Latin America, activists called for local referendums to contest biased participative procedures. The focus on expertise to counter arguments and frame claims also built on both local and outside knowledge (Martinez-Alier et al., 2016). Several grassroots mobilisations and networks expanded on this method in order to prepare post-oil or post-growth alternatives (Hollender, 2015). Similar dynamics exist in Europe. For instance, the grassroots network 'Grands projets inutiles et imposées' ('Unnecessary imposed mega-projects') in the 2010s disputes the relevance of large-scale infrastructures as emblematic of an outdated economic model, while challenging the consultative procedures in use. As elsewhere, violent confrontation between activists and police forces reveal ongoing pressures on the environment.

References

Brosnan, K. A. (2014). 'Law and the environment', in A. C. Isenberg (ed.) *The Oxford handbook of environmental history*. Oxford: Oxford University Press, pp 514–41.

Carmin, J. (1999). 'Voluntary associations, professional organisations and the environmental movement in the United States'. *Environmental Politics*, 8(1): 101–21.

Dauvergne, P. (2016). *Environmentalism of the rich*. Cambridge, MA: MIT Press.

Doherty, B. and Doyle, T. (2006). 'Beyond borders: transnational politics, social movements and modern environmentalisms'. *Environmental Politics*, 15(5): 697–712.

Doherty, B. and Doyle, T. (eds.) (2013). *Environmentalism, resistance and solidarity: the politics of Friends of the Earth International*. Basingstoke: Palgrave Macmillan.

Doyle, T. and MacGregor, S. (eds.) (2014). *Environmental movements around the world: shades of green in politics and culture – volume 2: Europe, Asia and Oceania*. Santa Barbara, CA: Praeger.

Dryzek, J. S. (2013). *The politics of the earth: environmental discourses*. Oxford: Oxford University Press.

Fairhead, J., Leach, M. and Scoones, I. (2012). 'Green grabbing: a new appropriation of nature?' *Journal of Peasant Studies*, 32(2): 237–61.

Freudenberg, N. and Steinsapir, C. (1991). 'Not in our backyards: the grassroots environmental movement'. *Society and Natural Resources*, 4(3): 235–45.

Grant, J. and Malloney, W. (eds.) (1997). *The protest business? Mobilizing campaigns groups*. Manchester: Manchester University Press.

Guha, R. and Martinez-Alier, J. (eds.) (1997). *Varieties of environmentalism: essays north and south*. London: Earthscan.

Hayes, G. (2002). *Environmental protest and the state in France*. Basingstoke: Palgrave Macmillan.

Haynes, J. (1999). 'Power, politics and environmental movements in the third world'. *Environmental Politics*, 8(1): 222–42.

Hochstetler, K. (2012). 'Democracy and the environment in Latin America and Eastern Europe', in P. F. Steinberg and S. D. VanDeveer (eds.) *Comparative environmental politics: theory, practice, and prospects*. Cambridge, MA: MIT Press, pp 199–230.

Hollender, R. (2015). 'Post-growth in the Global South: the emergence of alternatives to development in Latin America'. *Socialism and Democracy*, 29(1): 73–101.

Horowitz, L. S. and Watts, M. J. (2016). 'Introduction: engaging with industry and governing the environment from the grassroots', in L. S. Horowitz and M. J. Watts (eds.) *Grassroots environmental governance*. London: Routledge, pp 13–42.

Jamison, A. (2001). *The making of green knowledge: environmental politics and cultural transformation*. Cambridge: Cambridge University Press.

Kashwan, P. (2017). *Democracy in the woods: environmental conservation and social justice in India, Tanzania and Mexico*. New York: Oxford University Press.

Keck, M. E. and Sikkink, K. (eds.) (1998). *Activists beyond borders: advocacy networks in international politics*. Ithaca: Cornell University Press.

Kriesi, H. (1996). 'The organizational structure of new social movements in a political context', in D. McAdam, J. McCarthy and M. N. Zald (eds.) *Comparative perspectives on social movements: political opportunities, mobilizing structures and cultural framing*. Cambridge: Cambridge University Press, pp 152–84.

Laing, A. F (2015). 'Resource sovereignties in Bolivia: re-conceptualising the relationship between indigenous identities and the environment during the TIPNIS conflict'. *Bulletin of Latin America Research*, 34(2): 149–66.

Martinez-Alier, J., Temper, L., Del Bene, D. and Scheidel, A. (2016). 'Is there a global environmental justice movement?' *The Journal of Peasant Studies*, 43(3): 731–55.

Matejova, M., Parker, S. and Dauvergne, P. (2018). 'The politics of repressing environmentalists as agents of foreign influence'. *Australian Journal of International Affairs*, 72(2): 145–62.

Mauerhofer, V. (2016). 'Public participation in environmental matters: compendium, challenges and chances globally'. *Land Use Policy*, 52: 481–91.

McCormick, J. (1989). *The global environmental movement*. London: Belhaven Press.
Mohan, G. and Stokke, K. (2000). 'Participatory development and empowerment: the dangers of localism'. *Third World Quarterly*, 21(2): 247–68.
Mosley, S. (2013). *The chimney of the world: a history of smoke pollution in Victorian and Edwardian Manchester*. London/New York: Routledge.
Nicolle, S. and Leroy, M. (2017). 'Advocacy coalitions and protected areas creation process: case study in the Amazon'. *Journal of Environmental Management*, 198: 99–109.
Pellow, D. N. (2007). *Resisting global toxics: transnational movements for environmental justice*. Cambridge, MA: MIT Press.
Perreault, T. (2016). 'Governing from the ground up? Translocal networks and the ambiguous politics of environmental justice in Bolivia', in L. S. Horowitz and M. J. Watts (eds.) *Grassroots environmental governance*. London: Routledge, pp 115–37.
Rootes, C. (2007). 'Environmental movements', in D. A. Snow, S. S. Soule and H. Kriesi (eds.) *The Blackwell companion to social movements*. Oxford: Blackwell, pp 608–40.
Rootes, C. (2013). 'From local conflict to national issue: when and how environmental campaigns succeed in transcending the local'. *Environmental Politics*, 22(1): 95–114.
Saunders, C. (2013). *Environmental networks and social movement theory*. New York: Bloomsbury.
Sikor, T. and Newell, P. (2014). 'Globalizing environmental justice'. *Geoforum*, 54: 151–7.
Smith, A. (2011). 'The transition town network: a review of current evolutions and renaissance'. *Social Movement Studies*, 10(1): 99–105.
Steinberg, P. F. (2001). *Environmental leadership in developing countries: transnational relations and biodiversity policy in Costa Rica and Bolivia*. Cambridge, MA: MIT Press.
Tamiotti, L. and Finger, M. (2001). 'Environmental organizations: changing roles and functions in global politics'. *Global Environmental Politics*, 1(1): 56–76.
Temper, L. (2018). 'From boomerangs to minefields and catapults: dynamics of trans-local resistance to land-grabs'. *The Journal of Peasant Studies*, pp 1–29.
The Economist (2017). 'Dying to defend the planet'. 11 February. Available at: www.economist.com/the-americas/2017/02/11/why-latin-america-is-the-deadliest-place-for-environmentalists
UN Environment (2018). *UN Environment calls on governments and business to promote, protect and respect environmental rights*, 6 March [Press release].
Van Tatenhove, J. P. M. and Leroy, P. (2003). 'Environment and participation in a context of political modernisation'. *Environmental Values*, 12(2): 155–74.
Wakild, E. (2013). 'Environmental justice, environmentalism, and environmental history in twentieth century Latin America'. *History Compass*, 11(2): 163–76.
Zelko, F. (2014). 'The politics of nature', in A. C. Isenberg (ed.) *The Oxford Handbook of Environmental History*. Oxford: Oxford University Press, pp 717–35.

14

Sustainable behaviour and environmental practices

Erin Roberts

There are a myriad of ways of conceptualising human behaviour and pro-environmental action within the social sciences, which map on to wider debates about the nature of social change. These can be categorised into two broad groups that are seemingly at odds with one another; at one end lies agency, also known as the cognitive paradigm, and at the other lies the contextual or structural paradigm (Burgess et al., 2003). While this chapter has been organised to reflect ongoing debates within the social sciences, it is important to consider that within each paradigm, interpretations of structure/agency vary widely. It is therefore more useful to think of the distinction not as a dichotomy but rather as a continuum along which various approaches can be plotted according to whether they are more or less structural/agentic. In what is to follow, this chapter will provide a brief overview of each paradigm, highlighting their relative strengths and weaknesses in relation to furthering our understanding of pro-environmental action. The chapter will then explore a relatively new, practice-based paradigm, which has garnered much interest over the last fifteen years for its promise of overcoming the structure/agency debate by inhabiting the 'middle-ground'.

The cognitive paradigm

Descended from neoclassical economic theory, cognitive perspectives are underpinned by the belief that people behave in a rational and consistent manner, thus agency resides firmly with the individual. Often dubbed 'rational choice' or 'expectancy-value' (EV) theories, such perspectives view individuals as 'utility maximisers' who calculate the costs and benefits of available choices, and act in ways that are optimal to them. Jackson (2011) illustrates this process in the example of a person deciding to travel to work by car, which largely rests upon the person's expectations that the journey will be cheaper and shorter by car than by public transport, and their positive evaluation of those outcomes. In order to make rational, utility-maximising decisions, people are assumed to be in receipt of complete or perfect information about the costs, benefits and impact of their actions (Jackson, 2005).

The notion of rational choice underpins numerous communicative campaigns devised by policy makers across Europe to fill a perceived 'information vacuum' among the population, which is expected to translate into pro-environmental action (Wilhite and Ling, 1995).

However, evidence from empirical studies has shown that greater awareness and concern about environmental degradation does not necessarily lead to pro-environmental action. This discrepancy has been called the 'value-action gap' (Blake, 1999), and numerous theoretical frameworks have since sought to explain it. In particular, a range of 'adjusted' social-psychological models have attempted to go beyond basic assumptions of rational choice by linking attitudes, values, morals or norms to individual behaviour.

One grouping of such 'adjusted' EV models are dubbed 'attitude-behaviour' or ABC models, which have long dominated social-psychological approaches of explaining pro-environmental action. The attitudinal component of these models is, however, heavily based on earlier *rational choice* calculations, in that an individual's beliefs about behavioural outcomes, and their evaluation of those outcomes are believed to determine their attitude towards a given behaviour. Bridging the gap between an individual's attitudes and behaviour is *behavioural intention*, which is the direct precursor of behaviour. Later models based on EV theory carried on this trend of including additional variables and, as they became more 'adjusted', so the relative influence of attitudes in predicting behavioural outcomes declined (Darnton, 2008). This pattern can be seen in Ajzen's (1991) 'Theory of Planned Behaviour' (TPB – see Figure 14.1), one of the most widely-used of these models, which incorporates an additional independent variable in the form of *perceived behavioural control*, that is the extent to which an individual perceives a given behaviour to be difficult to perform. To date, the TPB has been widely adopted to predict a variety of environmentally significant behaviours, including: car use (Bamberg and Schmidt, 2003), public transport use (Heath and Gifford, 2002), recycling and waste disposal (Mannetti et al., 2004) and domestic energy conservation (Harland et al., 1999). Many of these studies, however, tend not to measure behavioural outcomes, as their main focus lies in measuring the relationship between attitudes, intentions and perceived behavioural control (Jackson, 2005).

While attitude-behaviour models focus on the cognitive deliberation of information, another group of models focus more on the role of normative factors (i.e. morals and social norms) in shaping pro-environmental decision making. Norms-based approaches assert that it is through social comparison that people validate the correctness of their opinions and decisions, which in turn influences their behaviour (van der Linden, 2014). For example, in the context of energy consumption, people often alter their consumption patterns when provided with normative information about their neighbourhood average home energy use to conform to the in-group norm (Schultz et al., 2007).

Norm-based models usually distinguish between two types of norm: descriptive and injunctive (Cialdini et al., 1990). A descriptive norm is based on an individual's perception of what

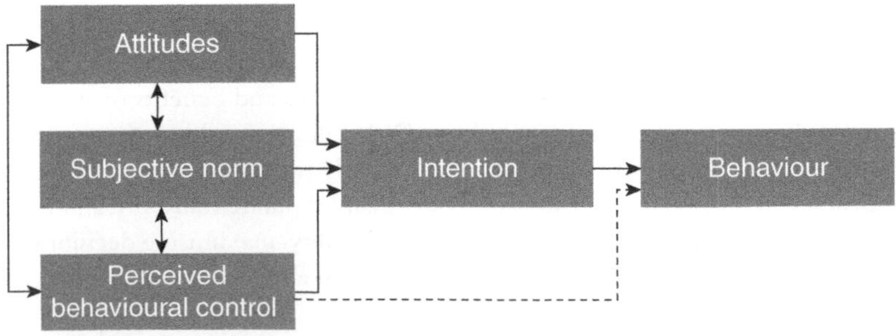

Figure 14.1 Theory of planned behaviour

Source: redrawn by author from Ajzen (1991, p. 182)

other people actually do, while an injunctive norm refers to an individual's perception of whether a given behaviour is socially approved within their culture. Descriptive and injunctive norms affect intentions and behaviour independently; therefore, understanding the relationship between these two concepts is vital for communicating social information. To illustrate, van der Linden (2014), gives the example of a communication campaign seeking to reduce frequent flying; if campaign materials merely state that CO_2 emissions are on the rise because people commonly choose to fly short distances rather than use alternatives (a descriptive norm), it is unlikely that they will have the desired effect given that it does not necessarily communicate that such behaviour is undesirable (an injunctive norm). Rather than discouraging this sort of behaviour, the message that is conveyed could easily be misread as: 'it's okay because everyone's doing it' (van der Linden 2014, p. 23). For communication campaigns to be effective then, descriptive and injunctive norms must align.

In addition to social norms, moral norms – which refer to the idea that some behaviours are inherently 'right' or 'wrong' regardless of their social consequences – are also believed to be important predictors of altruistic and pro-environmental behaviours. Social and moral norms are closely linked, as social and cultural learning play an important role in shaping moral beliefs (van der Linden, 2014). We learn what is right and wrong from those around us, and over time, these ideas are internalised and become personal or moral norms. Even though their origins lie in social norms, personal norms exercise influence over behaviour independently, and are believed to play an important role in the prediction of behaviours with a moral or ethical component, such as pro-environmental action (White et al., 2009). Steg and Vlek (2009) distinguish between three specific lines of research: those that are concerned with the value-basis of environmental beliefs and behaviour (for example, 'Value Theory'); those that have focused on the role of environmental concern (for example, 'New Environmental Paradigm'); and those that focus on moral obligations to act pro-environmentally (for example, 'Norm Activation Theory'). A particularly influential framework that combines the insights from all three lines of research is Stern and colleagues' (1999) 'Value-Belief-Norm' (VBN) theory.

While the VBN theory can explain relatively 'low cost' environmental behaviour and 'good intentions' such as willingness to change behaviour (Stern et al., 1999), its explanatory power weakens when in situations characterised by high behavioural costs or strong constraints on behaviour, such as reducing car use (Steg and Vlek, 2009). This suggests that behaviour does not result from internal processes alone, and that situational and contextual factors also have a role to play.

In light of the perceived shortcomings of earlier models, some researchers within the cognitive paradigm have sought to identify contextual and situational variables that disrupt the assumed linear transition from attitudes/values to behaviour. Within this literature, context has come to mean a variety of things; some take context to mean an individual's social network (for example, Olli et al., 2001), while others take it to mean facilitating conditions, such as the provision of recycling facilities, or the availability of public transport services for example (for example, Martin et al., 2006). In each case, contextual or situational variables have been shown to play a significant role in enabling or constraining individuals' pro-environmental behaviours (Steg, 2008).

As summarised in Table 14.1, cognitive models have added numerous determinants to explain pro-environmental behaviours and, in so doing, have gradually become more complex over time. However, due to their underlying assumptions regarding the nature of human action, the majority exhibit a degree of linearity.[1] Despite the evidence and growing recognition that contextual variables are important in shaping pro-environmental behaviour, the majority of social-psychological models continue to attribute variances in behaviour to predominantly dispositional factors. The narrow focus on attitudes or motivations in the theories above also blinds them to the role of non-deliberative influences, such as habit, in shaping behaviour. Human

Table 14.1 Major cognitive models

EV models	Expectancy-Value Theory/ Model(s)	This theory asserts that individuals behave in a predictable manner in which they prioritise their own self-interest through calculative action. Decision making involves *expectations* about a given behaviour and the *evaluation* of potential outcomes.
ABC	Attitude-Behaviour-Choice Model(s)	Similar to EV theory, this theory asserts that an individual's *attitudes* predict particular *behaviours* which they have *chosen* to adopt.
TPB	Theory of Planned Behaviour	This theory expands on earlier adjusted EV theories by asserting that behaviour can be predicted by behavioural intention and perceived behavioural control.
VBN	Value-Belief-Norm Theory	This theory proposes that personal norms are the direct antecedents of pro-environmental behaviour, which arise from beliefs regarding the individual's awareness of the consequences of their actions, and their feelings of personal responsibility for those consequences. In this theory, personal norms are dependent on more specific belief structures about human-environment relations, which are in turn dependent upon a relatively stable set of personal values.

behaviour is rarely the product of solely deliberative processes, which is why such models are often unable to fully explain pro-environmental behaviour.

The contextual paradigm

Within contextual approaches, individual agency is framed and constrained by external structures, which include: rules and standards that guide social behaviour, cultural norms and expectations, institutions, infrastructures and other material manifestations of social life (Jackson, 2005). In terms of illustrating this perspective's understanding of pro-environmental action, the most relevant lines of inquiry are studies of consumption and complementary strands of research on the coevolution of socio-technical systems and conventions of normality.

Since the mid-1980s, studies of consumption have been concerned with questions of identity and its related problematics, resulting in a focus on the communicative and symbolic aspects of consumption. However, more recent work suggests that the conspicuous and symbolic aspects of consumption have been grossly overemphasised, given that a great deal of everyday consumption is inconspicuous in nature, as part of the ordinary, mundane routines of millions of households (Gronow and Warde, 2001). These repetitive actions require little conscious thought and reflection, and are influenced not by cultural preferences and identity but by convenience, habit, practice, social norms and institutional contexts (Jackson, 2005). Much of our everyday consumption choices are thus rendered invisible – both to our peers and to ourselves – by these habitual routines.

Wilhite and Lutzenhiser (1999) identify four clearly relevant and interrelated social dynamics that they believe play a significant role in shaping pro-environmental action. The first of these relates to the *embeddedness of consumption patterns*, where they argue that people are compelled to consume in the conditions of late capitalism given that 'the entire social fabric is constructed in such a way as to encourage the association of consumption with the good life' (Wilhite and Lutzenhiser, 1999, p. 285). Lifestyles based on the logic of consumerism are thus, according to Evans and Jackson (2008), characterised by high levels of economic consumption, which translates into high levels of material consumption and accelerating environmental degradation.

A second social dynamic, and closely linked to the first, is that of *status marking and display*. Here, Wilhite and Lutzenhiser assert that an individual's standing in the community is displayed through culturally-appropriate arrangements of items that allow the person to differentiate themselves. This is often termed a process of 'distinction' in which class boundaries are defined and maintained through the display of tastes (i.e. cultural preferences) that are learned early in life, and which inform different life-styles (Bourdieu, 1984). The role of 'conspicuous' modes of consumption in the process of social stratification has long been recognised in the social sciences. While the use of material goods in the expression of culture and self-identity continues to be important, contemporary scholars now believe that in the conditions of late capitalism, pluralism has supplanted any hierarchical system of judgement (Shove and Warde, 2002). In this view, products and lifestyles are imbued with social meanings that vary both across and within each society (Wilhite and Lutzenhiser, 1999). While appropriating and using status-enhancing goods and services is a key component of modern consumerist lifestyles, a range of contemporary social movements concerned with ethical, 'green' or sustainable consumption go against the grain (Hards, 2013). By only consuming certain products (for example, local or organic food) and rejecting others (for example, mobile phones, computers and cars), individuals who ascribe to the values promoted by these social movements set themselves apart through their practices of 'green distinction' (Horton, 2003), 'conspicuous conservation' (Sexton and Sexton, 2011) or 'positional non-consumption' (Hards, 2013). Along with this plurality of taste and lifestyles, it has been argued by Shove and Warde (2002) that a rise in cultural 'omnivorousness' – described as an openness to appreciating everything – has significant implications in terms of the volume of goods consumed, thereby increasing the demand on finite resources and energy.

Wilhite and Lutzenhiser identify the dynamics of *sociality and conventionality* as a third driver of escalating resource consumption. They assert that in all cultures, people expect their social interactions to take place in certain ways; they understand what social conventions are and how they are expected to behave. Hards (2013), for example, asserts that conforming to expectations, or upholding the appearance of normality, plays a significant role in managing stigma-risk (in terms of mockery and embarrassment) – particularly in the presence of guests. Hards' contentions are not only supported by her own empirical work on energy consumption but also connect to insights from studies elsewhere; for example, in Scandinavia, for a guest to imply that they are cold constitutes a 'social crisis' for the host (Wilhite and Lutzenhiser, 1999), similarly those with cold homes in Britain are likely to be judged as stingy, poor or miserly (Hitchings and Day, 2011). Further cross-cultural empirical studies on space heating/cooling (Wilhite et al., 1996) and lighting (Linnet, 2011) have also demonstrated how resource demand is socially and culturally contingent. It has thus been suggested that living a worthwhile life in the conditions of late capitalism requires an ever-greater bundle of goods and services to meet the minimum standard (Jackson, 2011).

Security and convenience is the last of the social dynamics identified by Wilhite and Lutzenhiser. They argue that in contemporary lifestyles, convenience is one of the most important determinants of purchase and use patterns. For some, the turn to convenience is linked to the progressive erosion of collective spaciotemporal rhythms (Giddens, 1991), which has resulted in the commonly held perception that 'the pace of daily life is accelerating and that there is an increasing shortage of time' (Southerton, 2003, p. 6). Indeed, people have multiple competing demands on their time, which often leads to feelings of being rushed by a perceived time-squeeze (Southerton, 2003). Where people feel that they have insufficient time to accomplish things that are important to them, saving time through convenience becomes a matter of concern (Shove, 2003). A complementary but more defensive view of consumption focuses more on catering to just-in-case scenarios (Shove and Warde, 2002). According to Wilhite and Lutzenhiser, the over-dimensioning of objects, devices and appliances has to be understood in terms of risk management against an uncertain future. For example, having a large refrigerator that can hold a lot of food can accommodate unexpected visitors and reduce the frequency of shopping trips. They go on to argue that by consistently choosing oversized goods and devices, consumers redefine what is considered 'normal', and as standards gradually increase, so does resource/energy demand.

A linked body of literature, predominantly drawn from science and technology studies (STS), pays attention to how technologies and infrastructures shape our ability to act in certain ways. From this perspective, cultural conventions and systems of provision co-evolve in a process of socio-technical change that 'locks us in' to increasingly resource intensive consumption trajectories. Research conducted by Elizabeth Shove (2003) illustrates this by using the notion of 'comfort' in relation to the development of air conditioning systems; highlighting the mutually reinforcing developments in scientific classifications of what constitutes 'comfortable' indoor temperatures, building design and people's expectations of comfort. Such studies have, through a focus on socio-technical processes, explored 'how expectations and practices change, at what rate, and in what direction, and with what consequence for the consumption of environmentally critical resources such as energy and water' (Shove, 2003, p. 16).

Implicit within interpretation of the STS literature is the assumption that successful intervention within socio-technical systems results in change. However, introducing new rules or infrastructures does not necessarily create new practices, as socio-technical systems are not easily steered (Shove and Walker, 2007). Assuming that innovation leads to increased consumer demand, and consequently a system change, is therefore over simplistic (Shove, 2003). For example, studies tracing the historical emergence of showering as a popular activity demonstrate that despite having access to reliable water and electricity sources, showering remained unpopular for a long time (Southerton et al., 2004). Gradually, showering was accepted as it became associated with speed, convenience and conceptions of health and social respectability (Southerton et al., 2004, pp. 43–5).

Within this literature, agency is distributed throughout the socio-technical system, rather than residing purely with individuals. Consumption practices develop historically, informed by technological innovation, social contexts, and the temporal demands of everyday life, indicating how socio-technical systems mould the carbon intensity of social life in dynamic and complex ways.

Much like their cognitive counterparts however, contextual perspectives have been subject to much criticism, albeit for the opposite reasons. While cognitive approaches have been critiqued for their relative simplicity and linearity, contextual approaches have been accused of overstating the importance of structures in guiding everyday life. The ongoing structure-agency debate has since resulted in calls for alternative theoretical perspectives that acknowledge and account for the roles of both cognitive and contextual aspects in shaping pro-environmental action.

The practice paradigm

Among those that are keen to move beyond the agency-structure dualism, theories of social practice have offered an alternative, and increasingly popular, approach to understanding pro-environmental action. The plural is used to signify the diversity within this loosely connected group of theories, as there is no single, unified practice approach; instead, there are a variety of approaches that are united by some common ideas.

For practice theorists, the practical carrying out of social life takes centre stage. Giddens outlined the basic premise of the practice approach when he stated that 'the basic domain of study . . . is neither the experience of the individual actor, nor the existence of any form of societal totality, but social practices ordered across space and time' (Giddens, 1984, p. 2). In this view, our everyday actions are not seen as the result of people's attitudes, values and beliefs, nor are they believed to be shaped by structures and institutions; instead, they are understood to be embedded within and occurring as part of social practices (Hargreaves, 2011).

A practice can be seen as the coming together of interconnected elements to form a routinised pattern or 'block' of activity (Reckwitz, 2002). For Reckwitz, practices contain within them forms of bodily and mental knowledge that are embodied within practitioners, who, through performance, perpetuate and potentially transform the practices that they 'carry'. As such, there is a recursive relation between recognisable doings that are relatively stable (practice-as-entity) and the carrying out of a practice (practice-as-performance), in which people – through their embodied actions, and their knowledge and understanding about those actions – play a key role as carriers and performers of practice.

Along with the aforementioned embodied components, Reckwitz' (2002) description gives materials or 'things' a key role in the performance of practices. In developing his own 'ideal type' of practice theory, Reckwitz argued that earlier practice theorists had not adequately accounted for the material dimension of social practices. This inclusion of the material dimension is regarded as pertinent by later practice theorists, given the explosion of technical artefacts (for example, computers, mobile phones, and tablets) in contemporary society (Spaargaren, 2006). Within this conceptualisation, materials, things, technologies and infrastructures are conceptualised as active elements of practice in their own right. This has been of key importance to the development of practice theories as we know them today, and in particular, their application to studies of consumption. Practices are thus not purely social, given that much of social life is intertwined with material infrastructures, devices and artefacts that configure and co-constitute much of what we do (Shove, 2003).

Despite its usefulness, the above formulation of practice is difficult to apply empirically (Spaargaren, 2006). Shove and colleagues (2012), however, provide a somewhat more straightforward conceptualisation of practice, which is comprised of only three elements – meanings, materials and competencies – which can be seen in Figure 14.2.

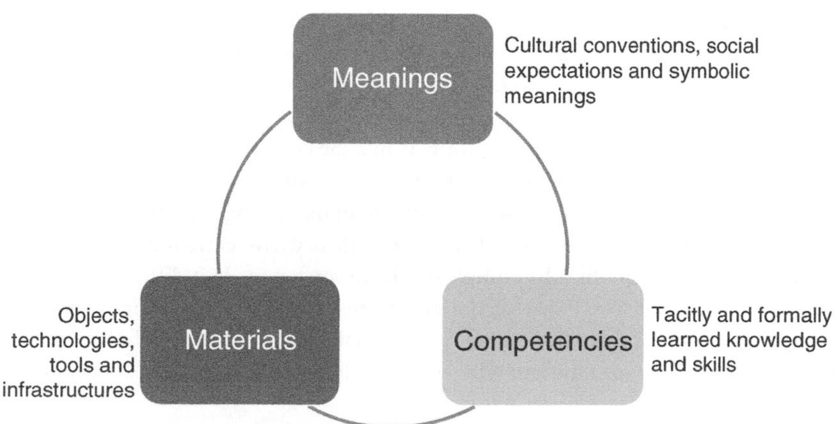

Figure 14.2 The three elements of social practice theory

Source: redrawn by author from Shove et al. (2012, p. 23)

To demonstrate this model at work, an illustrative example of an everyday practice will be used; cooking. The material element of cooking practices includes appliances and equipment (for example, oven, grill, hobs, microwave, food processor), consumables (for example, recipes, fruit and vegetables, meat and dairy) and domestic infrastructures (for example, plumbing and electrics). Knowledge of how to prepare food is also needed (competencies), and the necessary skills include basic knife skills and possessing knowledge about hygienic food preparation as well as how to operate appliances and equipment. Finally, these skills are intrinsically linked to cultural conventions of taking care of oneself and others (meanings), along with the related notions of convenience and health, which have gradually led to increases in resource consumption over time (see Shove, 2003). These elements are linked together by individuals when carrying out a practice. While cooking is a widely shared practice, however, not everyone cooks in the same way; practices are internally differentiated according to the particular configurations of different materials, meanings and competencies at hand (Warde, 2005).

The tendency to view individuals as merely the 'carriers' of practice has also garnered criticism. Sayer (2013), for example, argues that conceptualising the role of individuals in this way ignores their 'dynamic, normative or evaluative relation to practices', thus rendering them no more than 'passive' dupes (Sayer, 2013, p. 170). The use of shared practices as a unit of analysis has been another subject of criticism. Despite the focus on collective conventions, some forms of social interaction are under-studied and under-theorised in practice theories (Henwood et al., 2015; Warde, 2005). For example, little attention has been paid to how practices are transmitted between individuals, or how they are negotiated and performed in specific situations (Hargreaves, 2011). Finally, it seems that practice theory's complexity is both a strength and a weakness. Sahakian and Wilhite (2014), for example, argue that while practice theory offers a rich terrain of study, its complexity – even in its simplified three elements form – makes it decidedly harder to put the theory into practice in policy making.

Conclusion

This chapter has provided a brief overview of the various ways in which human action has been conceptualised within the social sciences. These were structured into three broad groups – cognitive, contextual and practice paradigms – highlighting the diversity of perspectives that researchers can call upon when examining sustainable behaviour and pro-environmental action. Despite this diversity however, policy makers largely focus on policy instruments aimed at influencing individual behaviour change to shape the transition to a more sustainable society. Such policy mechanisms are not always successful, however, given that they are underpinned by notions of rational choice that have been argued by proponents of the contextual (and indeed, some proponents of the cognitive) perspective to be overly simplistic. While the cognitive paradigm has been criticised for its relatively linear understanding of behaviour, the contextual paradigm has likewise been criticised for giving too much credence to social structure in the shaping of human action. In light of this ongoing debate, other perspectives such as practice theory have garnered much attention from academics and policy makers alike. Its complexity, however, has made it difficult to implement in policy making. This is not to say that such a perspective is not useful; practice theory's strength lies in its capacity to demonstrate how more or less sustainable practices take hold and persist over time (or not), and its ability to illustrate why certain policy instruments may not be successful.

Note

1 With the exception of those models such as the 'Theory of Interpersonal Behaviour' that recognise non-deliberative action.

References

Ajzen, I. (1991). 'The theory of planned behavior'. *Organizational Behavior and Human Decision Processes*, 50(2): 179–211.

Bamberg, S. and Schmidt, P. (2003). 'Incentives, morality or habit? Predicting students care use for university routes with models of Ajzen, Schwartz and Triandis'. *Environment and Behavior*, 35(2): 264–85.

Blake, J. (1999). 'Overcoming the "value-action gap" in environmental policy: tensions between national policy and local experience'. *Local Environment: The International Journal of Justice and Sustainability*, 4(3): 257–78.

Bourdieu, P. (1984). *Distinction: a social critique of the judgement of taste*. London: Routledge.

Burgess, J., Bedford, T., Hobson, K., Davies, G. and Harrison, C. (2003). '(Un)sustainable consumption', in F. Berkhout, M. Leach and I. Scoones (eds.) *Negotiating environmental change: new perspectives from social science*. Cheltenham: Edward Elgar, pp 261–92.

Cialdini, R. B., Reno, R. R. and Kallgren, C. A. (1990). 'A focus theory of normative conduct: recycling the concept of norms to reduce littering in public places'. *Journal of Personality and Social Psychology*, 58(6): 1015–26.

Darnton, A. (2008). *GSR behaviour change knowledge review. Reference report: an overview of behaviour change models and their uses*. London: HMT Publishing Unit.

Evans, D. and Jackson, T. (2008). *Sustainable consumption: perspectives from social and cultural theory*. RESOLVE Working Paper 05–08. Guildford: Research Group on Lifestyles, Values and the Environment, University of Surrey.

Giddens, A. (1984). *The constitution of society*. Cambridge: Polity Press.

Giddens, A. (1991). *Modernity and self-identity: self and society in the late modern age*. Cambridge: Polity Press.

Gronow, J. and Warde, A. (2001). 'Introduction', in J. Gronow and A. Warde (eds.) *Ordinary consumption*. London: Routledge, pp 1–8.

Hards, S. K. (2013). 'Status, stigma and energy practices in the home'. *Local Environment*, 18(4): 438–54.

Hargreaves, T. (2011). 'Practice-ing behaviour change: applying social practice theory to pro-environmental behaviour change'. *Journal of Consumer Culture*, 11(1): 79–99.

Harland, P., Staats, H. and Wilke, H. A. M. (1999). 'Explaining proenvironmental intention and behavior by personal norms and the theory of planned behavior'. *Journal of Applied Social Psychology*, 29(1): 2505–28.

Heath, Y. and Gifford, R. (2002). 'Extending the theory of planned behavior: predicting the use of public transportation'. *Journal of Applied Social Psychology*, 32(10): 2154–89.

Henwood, K., Pidgeon, N., Groves, C., Shirani, F., Butler, C. and Parkhill, K. (2015). 'Energy Biographies research report'. *Energy Biographies*. Available at: http://energybiographies.org/our-work/our-findings/reports/

Hitchings, R. and Day, R. (2011). 'How older people relate to the private winter warmth practices of their peers and why we should be interested'. *Environment and Planning A*, 43(10): 2452–67.

Horton, D. (2003). 'Green distinctions: the performance of identity among environmental activists'. *The Sociological Review*, 51(2): 63–77.

Jackson, T. (2005). *Motivating sustainable consumption: a review of evidence on consumer behaviour and behavioural change*. ESRC Sustainable Technologies Programme. Guildford: University of Surrey.

Jackson, T. (2011). 'Confronting consumption: challenges for economics and for policy', in S. Dietz, J. Michie and C. Oughton (eds.) *Political economy of the environment*. London: Routledge, pp 189–212.

Linnet, J. T. (2011). 'Money can't buy me hygge: Danish middle-class consumption, egalitarianism and the sanctity of inner space'. *Social Analysis*, 55(2): 21–44.

Mannetti, L., Pierro, A. and Stefano, L. (2004). 'Recycling: planned and self-expressive behaviour'. *Journal of Environmental Psychology*, 24(2): 227–36.

Martin, M., Williams, I. D. and Clark, M. (2006). 'Social, cultural and structural influences on household waste recycling: a case study'. *Resources, Conservation and Recycling*, 48(4): 357–95.

Olli, E., Grendstad, G. and Wollebaek, D. (2001). 'Correlates of environmental behaviors'. *Environment and Behavior*, 33(3): 191–208.

Reckwitz, A. (2002). 'Towards a theory of social practices: a development in culturalist theorizing'. *European Journal of Social Theory*, 5(2): 243–62.

Sahakian, M. and Wilhite, H. (2014). 'Making practice theory practicable: towards more sustainable forms of consumption'. *Journal of Consumer Culture*, 14(1): 25–44.

Sayer, A. (2013). 'Power, sustainability and well being: an outsider's view', in E. Shove and N. Spurling (eds.) *Sustainable practice: social theory and climate change*. London: Routledge, pp 183–96.

Schultz, P. W., Nolan, J. M., Cialdini, R. B., Goldstein, N. J. and Griskevicius, V. (2007). 'The constructive, destructive, and reconstructive power of social norms'. *Psychological Science*, 18(5): 429–34.

Sexton, S. E. and Sexton, A. (2011). *Conspicuous conservation: the Pirus effect and willingness to pay for environmental bona fides*. Unpublished Paper. Available at: http://works.bepress.com/sexton/11/

Shove, E. (2003). *Comfort, cleanliness and convenience: the social organization of normality*. Oxford: Berg.

Shove, E., Pantzar, M. and Watson, M. (2012). *The dynamics of social practice: everyday life and how it changes*. London: Sage.

Shove, E. and Walker, F. (2007). 'CAUTION! Transitions ahead: politics, practice and sustainable transition management'. *Environment and Planning A*, 39(4): 763–70.

Shove, E. and Warde, A. (2002). 'Inconspicuous consumption: the sociology of consumption, lifestyles and the environment', in R. Dunlap, F. Buttel, P. Dickens and A. Gijswijt (eds.) *Sociological theory and the environment: classical foundations, contemporary insights*. Lanham, MD: Rowman and Littlefield, pp 230–51.

Southerton, D. (2003). 'Squeezing time: allocating practices, coordinating networks and scheduling society'. *Time and Society*, 12(1): 5–25.

Southerton, D., Warde, A. and Hand, M. (2004). 'The limited autonomy of the consumer: implications for sustainable consumption', in D. Southerton, H. Chappells and B. Van Vliet (eds.) *Sustainable consumption: the implications of changing infrastructures of provision*. Cheltenham: Edward Elgar, pp 32–48.

Spaargaren, G. (2006). *The ecological modernization of social practices at the consumption junction*. Paper presented at the ISA-RC-24 conference Sustainable Consumption and Society, 2–3 June 2006, Madison, Wisconsin.

Steg, L. (2008). 'Promoting household energy conservation'. *Energy Policy*, 36(12): 4449–53.

Steg, L. and Vlek, C. (2009). 'Encouraging pro-environmental behaviour: an integrative review and research agenda'. *Journal of Environmental Psychology*, 29: 309–17.

Stern, P. C., Dietz, T., Abel, T., Guagnano, G. A. and Kalof, L. (1999). 'A value-belief-norm theory of support for social movements: the case of environmentalism'. *Human Ecology Review*, 6(2): 81–97.

van der Linden, S. (2014). 'Towards a new model for communicating climate change', in S. Cohen, J. Higham, P. Peeters and S. Gössling (eds.) *Understanding and governing sustainable tourism mobility: psychological and behavioural approaches*. London: Routledge, pp 243–75.

Warde, A. (2005). 'Consumption and theories of practice'. *Journal of Consumer Culture*, 5(2): 131–53.

White, K. M., Smith, J. R., Terry, D. J., Greenslade, J. H. and McKimmie, B. M. (2009). 'Social influence in the theory of planned behaviour: the role of descriptive, injunctive, and ingroup norms'. *British Journal of Social Psychology*, 48(1): 135–58.

Wilhite, H. and Ling, R. (1995). 'Measured energy savings from a more informative energy bill'. *Energy and Buildings*, 22(2): 145–55.

Wilhite, H. and Lutzenhiser, L. (1999). 'Social loading and sustainable consumption'. *Advances in Consumer Research*, 26(1): 281–7.

Wilhite, H., Nakagami, H., Masuda, T., Yamaga, Y. and Haneda, H. (1996). 'A cross-cultural analysis of household energy use behaviour in Japan and Norway'. *Energy Policy*, 24(9): 795–803.

15

Green citizenship

Towards spatial and lived perspectives

Bronwyn E. Wood and Kirsi Pauliina Kallio

Introduction

Few social issues have received greater attention in recent years than that of the environment. Most nations are aware of the challenges of environmental planning and sustainability in a context of heightened evidence, suggesting that the world is experiencing rapid rates of climate change, environmental degradation and declining resources. Almost two decades ago, the United Nations Development Programme (1998, p. 2) laid out the case for growing environmental problems facing our planet:

> The burning of fossil fuels has increased almost fivefold since 1950, the world's marine catch fourfold, and the consumption of freshwater twofold since 1960. The result is a severe stress on the capacity of the plant to absorb all of the pollution and waste produced, as well as a rapid deterioration of fresh water reserves, soil, forests, fish and biodiversity.

Since this time, the attention to environmental change has been fuelled by large international conventions such as Agenda 21, formed at the United Nations Conference on Environment and Development (UNCED) Earth Summit in Rio de Janeiro in 1992, the Copenhagen Accord resulting from the United Nations Climate Change Conference in Denmark in 2009, and the Paris Agreement in 2016. As countries have ratified and introduced policy reforms to address these conventions, it has become clear that a pool of active citizens is required to address these challenges and plan for sustainable environmental futures. This has, in turn, influenced the understandings and expectations societies hold toward citizens. Dean (2001, p. 491, emphasis added) refers to this as the 'greening of citizenship', outlining three ways of how it has occurred:

> First, environmental concerns have entered our understanding of the *rights* we enjoy as citizens. Second, the enhanced level of global awareness associated with ecological thinking has helped to broaden our understanding of the potential *scope* of citizenship. Third, emergent ecological concerns have added fuel to a complex debate about the *responsibilities* that attach to citizenship.

In parallel with these societal changes, the field of environmental citizenship has expanded over the past 25 years. As early as the 1970s, work within environmental education had articulated for a type of a 'green' citizen (Schild, 2016), whilst not specifically using those words, and indeed the Tbilisi Declaration in 1977 argued for the need to create an active and environmentally aware citizenry. One of the earliest essays which specifically attempted to integrate citizenship and environmental planning and studies was by van Steenbergen (1994, p. 142) who set out to bring together the two 'cultures' of citizenship problems and environmental concerns. Since this time, the field has diversified and deepened until such a time that some argue it has come of age (for example, Latta, 2007, p. 377), while others, such as Gabrielson (2008, p. 430), believe that the literature and theorising on green citizenship remains 'unnecessarily narrow' (see also Dobson, 2003; Dean, 2001).

This chapter maintains that the prevailing frameworks employed in green citizenship still involve limited, static and instrumental conceptions, somewhat failing to consider all pertinent spaces, scales and lived practices of citizenship. In response, drawing on feminist theorisation and citizenship conceptions introduced by Engin Isin (2008, 2012), we turn to more dynamic, transnational and inclusive notions of lived green citizenship. We begin by outlining the contested nature of green citizenship (and its various expressions through terms such as ecological, sustainable and environmental citizenship) to illustrate the multiple interpretations of scholarly debate. Recognising this, the term 'green citizenship' is adopted in the chapter, in keeping with Dean (2001), as a broad term which seeks to encompass and explore in its greatest sense the 'greening' of citizenship. After describing the shortcomings of traditional liberal and civic republican approaches, we argue for a widening and deepening of understandings, through a greater acknowledgement of space and the multiple scalar and practised dimensions of citizenship. The chapter concludes with a discussion of the future of green citizenship and environmental planning by examining the role of environmental education and the younger generation's uptake of these ideas.

The contested nature of green citizenship

Green citizenship is neither a neutral nor apolitical concept. Even the terms used to describe it are highly contested. While some use environmental, ecological, sustainable and green citizenship interchangeably, others underline and dispute the differences between them. For example, Dobson (2010) distinguishes between *environmental citizenship*, which he argues is driven by liberal citizenship traditions focusing on individual and personal rights and duties, whereas *ecological citizenship* captures a more global conception which he defines as 'the exercise of ecologically related responsibilities, nationally, internationally, and intergenerationally, rooted in justice, in both the public and private spheres' (Dobson, 2003, p. 206; see also Latta, 2007; Gabrielson, 2008; Scerri and Magee, 2012; Schild, 2016). Such debates about terms also mirror the conflicting and competing conceptions inherent in the idea of green citizenship. Not only is the concept of *citizenship* hotly contested, meaning multiple things to different groups of people (Faulks, 2000), but the concept of *environmentalism* is also contentious (Dean, 2001). We will examine two broad positions which green citizenship can fall into – that based upon a *liberal* tradition and that of a *civic republican* position. Much of the discourse surrounding green citizenship parallels these two frameworks.

The first of these traditions – the liberal framework – acknowledges the existence of citizens' environmental rights but focuses on the personal duties and obligations of citizens. The emphasis is often on personal lifestyle attitudes, choices, and the management of environmental problems through actions such as recycling and boycotting unethical products (Dobson, 2003;

Melo-Escrihuela, 2008). Latta (2007) argues that as a result of this prevailing approach, much of the focus has fallen on cultivating 'green' attitudes and practices of individuals, rather than more broadly on democracy or collective and societal action. As one concrete example, Dimick (2015) illustrates the prevalence of this (neo)liberal approach in education by examining how one teacher presented the idea of green citizenship to his class through a focus on their individual patterns of consumption. While this approach led his students to examine their own environmental consumer behaviour, Dimick cautions that this is a weakened form of green citizenship as it is 'disconnected from the contexts in which the decisions are made and from broader political activities' (Dimick, 2015, p. 396) and thus fails to challenge the root causes of global environmental injustice or challenge established social structures reproducing these injustices over time.

The second prevailing framework in green citizenship is that of civic republicanism, emphasising virtues, responsibilities and community concerns. The weight is on the common good; in this way, the approach attempts to restrain excesses of self-interest in the liberal tradition (Gabrielson, 2008). The virtues and character traits of green citizens are highlighted with appeal to a stewardship model to remind us of our interdependence on nature and its dependence on humans (Schild, 2016; Dobson, 2003). Education within this tradition assumes that individual actions alone are not adequate to address environmental concerns and that participatory political involvement by citizens in environmental planning and decision making is key. In terms of critique, Schild (2016) points out that such approaches can fail to explain why citizens would be motivated to take part in deliberative processes in the first place.

As a way to illustrate and advance upon how these competing conceptions map on to moral discourses which underpin or make possible competing conceptions of green citizenship, Dean (2001) suggests a possible heuristic model or taxonomy, shown in Figure 15.1.

Dean suggests that the axes in Figure 15.1 represent two normative conceptual continua. The horizontal axis relates to the liberal and civic republican traditions of citizenship, with more contractarian traditions at one end which highlight that to have freedom an individual must enter into a *contract* with society, and at the other end, more *solidaristic* traditions in which an individual develops close communal bonds to develop social cohesion. The vertical axis is a continuum between *equality* and *social traditions* in which there are more egalitarian notions about the relationships between individuals in society at one end and more hierarchical at the other. When intersected by environmental discourses, Dean suggests that four positions can be outlined:

- *Entrepreneurism*: compatible with economic liberalism and underpinned by economic rationale for environmental planning and decision making;
- *Survivalism*: compatible with moral authoritarianism, fundamentally inegalitarian as does not question unequal distribution of social power and resources;
- *Conformism*: aspires to social integration and belonging but accepts inequalities in social power and resources;
- *Reformism*: solidaristic, embraces the goal of greater equality in distributions of power and resources.

Dean acknowledges that this taxonomy is over simplistic as many positions combine elements of all four. However, his model helps to confirm the existence of multiple political interpretations of green citizenship and how these overlay deeper moral and political positions (see also Dobson, 2003).

While such frameworks are useful for positioning different perspectives and for considering the extent to which people are active or passive citizens, and for critiquing powerful social structures or ideologies, they do have limitations. Latta (2007, p. 378), among others, suggests that

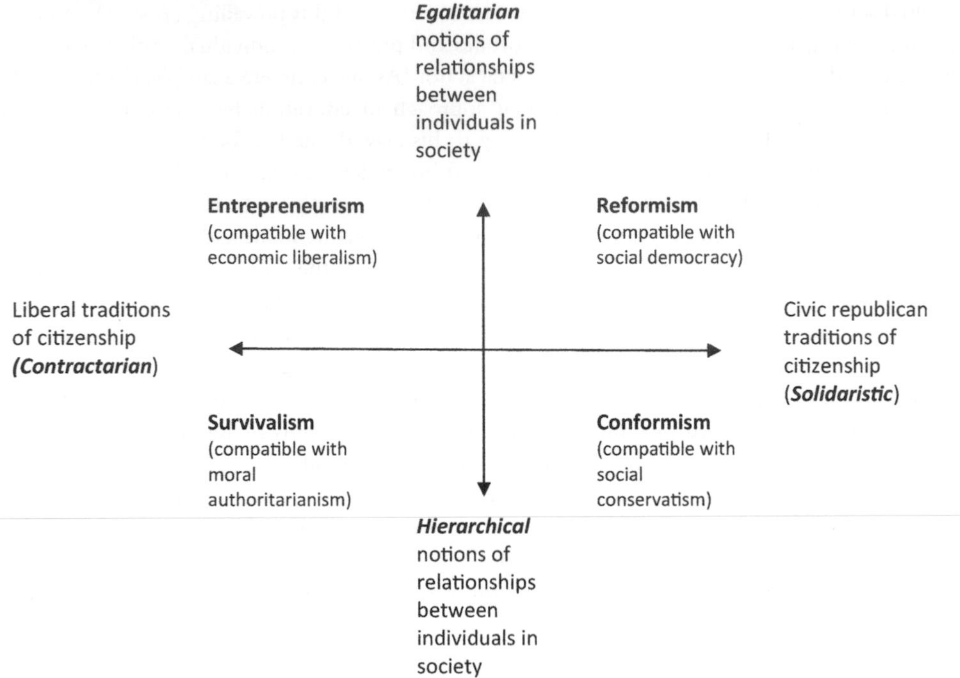

Figure 15.1 Taxonomy of conventional moral discourses
Source: adapted from Dean (2001, p. 494)

'while there is a strong democratic tendency in both citizenship studies and ecological political thought, existing theories of ecological citizenship seem to blunt the radical democratic edge of both traditions', and thus fail to account for existing structures of injustice and political agency. In addition, these traditional frameworks rely on narrow and normative conceptions of citizenship, overlooking the multiple ways in which people live and act as citizens and experience the environment. To address this, we turn our attention to an emerging body of work in citizenship and environmental studies which embraces more multi-scalar, dynamic and performative conceptions of citizenship with aims to expand its democratic and radical potential. Specifically, we will present two ways of deepening and expanding green citizenship by drawing on an emerging body of work applying critical citizenship theory. Both approaches stem from a dissatisfaction with the way citizenship has been understood in the traditional approaches outlined previously and both attempt to address the way in which environmental citizenship confronts traditional dichotomies between public and private and local and global.

Transnational and inclusive spaces of green citizenship

Critical citizenship literature aligns largely with feminist theoretical traditions – including queer, transnational and post-colonial perspectives – by drawing attention to the exclusionary and problematic nature of traditional understanding of participation and citizenship (for example, Lister, 2007; Yuval-Davis, 2007; Gabrielson, 2008; Staeheli et al., 2012). Feminist critiques have underscored the public-private dichotomy in society in which public participation has been profiled as a given status over the private domains of domestic and unpaid participation

(Mitchell et al., 2003). Despite claims to universalism, concepts of citizenship 'have been drawn to a quintessentially male template so that women's exclusion (and the chequered nature of their inclusion) was integral to both the theory and practice of citizenship' (Lister, 2007, p. 52). The critique involves a shift from a focus on the status of citizenship that many people fail to achieve – due to their age, sex, resources, political positioning and access to public space or institutions – to a focus on the experiences, acts and practices of being a citizen in both private and public spaces and with a range of scales.

These critiques relate to the spatial dimensions and scale of citizenship. Spatially limited understandings of citizenship are seen to align problematically with territorially defined nation states and the rights and duties of citizens that no longer match the global mobility of people and the border-crossing reach of environmental issues. Re-thinking the scales of citizenship has provided some insights to this problematic (for example, Hubbard, 2013; Staeheli, 2016). Among other critical citizenship scholars, we have developed in our own work transnational and relational understandings of citizenship, characterised by flexible and multiple notions of identity and connectedness beyond the nation state (Kallio et al., 2015; Kallio and Mitchell, 2016; Kallio, 2018a; Wood and Black, 2018).

Green studies, also, have played an important role in highlighting the global significance of citizenship issues, and the need for global actions. Evoking such ideas, Dobson (2003) argues for the necessity of *post-cosmopolitan* notions of green citizenship in which people see themselves as part of a wider, and indeed, global community, and are motivated by perceptions and actions which are based on virtue rather than self-interest (also Dean, 2001; Isin, 2012). In doing so, he extends the political space of citizenship to encompass not only other humans and societies known to an individual but also to strangers who have not yet been met. The networks of environmental connections and impacts which connect the human and nonhuman planet together are evoked through his post-cosmopolitan notion of green citizenship. Applying these ideas in research has significance for how we think about space, time, and citizenship, as it means we need to loosen our fixation on territories and their physical boundaries and widen our analysis of interconnections, networks and relationships.

Beyond status and practices: noticing acts of environmental citizenship

A further theoretical branch in critical citizenship literature draws from Engin Isin's (2008, 2012) influential work that distinguishes between three dimensions of citizenship. First, he identifies *citizenship as status* where the rights and responsibilities of people, as defined by the nation state or other established polities, hold the centre stage. Depending on their positions – be it birthright or gained – people hold different kinds of statuses and thus have more or less rights and responsibilities as members of the political communities where they live, including non-membership and very limited participation opportunities. Second, Isin defines *practices of citizenship* as formal or semi-formal activities that people can mobilise from their acquired positions in political communities, as collectives and individuals. These include various kinds of actions and customs, depending on the society and its political system, ranging from elections to demonstrations to public opinion statement. Third, a difference between practices and *acts of citizenship* is made by highlighting that not all politically influential activities are institutional, public, organised, broadly recognised or generally accepted. By associating citizenship closely with justice and liberal democratic ideals, Isin proposes that by negotiating, challenging and reworking the prevailing order – and thus calling into question the seeming naturalness of people's differing positions in a polity and participation opportunities as members of a political community – we can act as citizens beyond our given statuses and established practices. For example, as Isin (2008) argues, stateless people such as refugees often perform acts of citizenship whilst still failing to hold the status of citizenship (see also Häkli, 2017).

This theoretical approach has been picked up by some scholars in the context of environmental citizenship, yet not extensively. In his attempt to locate democratic politics in ecological citizenship, Latta (2007) has offered an Isinian perspective as a critique of the dominating approaches that, first, tend to focus on narrow concerns for the environment and, second, have limited relevance to progressive change in practical terms. He argues that the critical literature engenders 'appreciation for the way that ecological citizenship does not *precede* a politics of nature, as a kind of framework for progressive socio-environmental change but instead is an emergent property of *existing* struggles for sustainability and political-ecological rights' (Latta, 2007, p. 388, emphasis in original). Based on this idea, Latta suggests that, 'democratic tendencies in green politics should direct far greater attention to the actual spaces in which ecological citizens are daily being born in individuals' and communities' efforts to *become political vis-à-vis* nature' (Latta, 2007, p. 390, emphasis in original).

While in his recent work Latta (2007) has developed his ideas with reference to new materialist theorisation which does not fit unproblematically with Isinian thought that emphasises strongly human subjectivity (for a critique, see Häkli, 2017), others have taken them forward in more pragmatist and humanist manners. In her recent article on environmentally friendly food initiatives in Iran, Fadaee (2017) engages specifically with people's mundane acts of environmental citizenship. Drawing attention to how the everyday life of citizenship unfolds beyond the West and the North, she sets out to shed light on the pluralities of people's environmental engagements and subjectivities. Her analysis emphasises emotional and social experiences, alongside environmental awareness, as key elements of active environmentally oriented political agency. Writing from quite a different empirical context, Melbourne, Australia, Scerri and Magee (2012, p. 388) have also used Isinian thinking to formulate critique on what they call '"stakeholder" citizen-centred policies associated with what state theorists see as "weak" ecological modernisation', proposing instead a theoretical approach that is informed by pragmatism. Their argument builds on the nexus between political, ideological and cultural citizenship, a distinction they argue is key to understanding what is currently happening in our societies regarding sustainability.

A shift away from a focus on the status and formal/public practices of citizenship which many people fail to achieve or enact, for various reasons, acknowledges a broader spectrum of environmental agency and makes space for encouraging its different forms. These studies show that the Isinian three-partite notion of citizenship can be fruitful in broadening the conception of green citizenship, especially towards noticing mundane acts of citizenship that take place in people's everyday lives where the growing awareness of environmental issues is influencing their political agency (Kallio, 2018a). This would do justice to the plurality of citizenship, as Latta (2007, p. 328, emphasis in original) writes:

> Existing *in*justice is in part the product of asymmetries in citizens' abilities to exercise formal political rights, and also of the exclusive qualities of liberal universality embodied in 'politically just' democratic procedures relative to minority subject positions, dissenting visions of nature and divergent understandings of dialogue.

This is the direction which a large part of the critical environmental citizenship literature is also heading (for example, Gilbert and Phillips, 2003; Kurtz, 2005; Agyeman and Evans, 2006; MacGregor, 2006; Gabrielson, 2008; Harris, 2011; Repak, 2011; Fadaee, 2017).

One critical social group who should be better recognised as environmental citizens are young people. In traditional approaches, they typically appear as being socialised, informed, influenced and educated on environmental, ecological and sustainable politics *by* adults and institutions, not as co-learners and co-actors with and in such politics. In the last section, we will elaborate the case of youth citizenship in more detail as we consider it an important area in which green citizenship research and environmental planning ought to expand. Before that, we focus on how citizens can be 'greened' given the increasingly complex and widespread nature of environmental issues, the changing nature of communities, and the dynamic processes of socialisation that involve children and youth as active players. More specifically, we examine contemporary research with the younger generation to consider some of the specific challenges they face and the responses they may make.

Greening citizenship in a complex and transnational world

The focus of attention for the 'creation of green citizens' has inevitably come to lie strongly on the youngest citizens in societies. Paradoxically, they hold the greatest hope of solving complex environmental problems, while at the same time appear increasingly reluctant to do so through traditional political means (for example, Putnam, 2000; Circle, 2002; Kallio, 2018b; Bartos and Wood, 2017). There is some evidence that the current generation have less interest in environmental issues than earlier generations. For example, using data from a national survey of high school students in the United States since 1976, Wray-Lake and colleagues (2010) found that they showed declining levels of concern for the environment since the 1970s, although there were some increases in the 1990s. Individuals tended to place more responsibility on *governments* for pollution and environmental declines than on their *own* actions, and there were declining beliefs in the scarcity of *resources*. This led the research team to conclude that, 'clearly, the average high school senior across the past three decades has not viewed him or herself as the first line of defence in protecting the environment' (Wray-Lake et al., 2010, p. 82).

Amidst such fears, many Western nations have responded with a plethora of public policy initiatives to enhance environmental citizenship in youth. However, opportunities for young people to participate in environmental action through schooling and public programmes tend to be more cerebral and less experiential. This is despite the evidence suggesting that exposure to the natural environment itself is key to enhancing green citizenship (Chawla, 1998, 2007). In particular, prior lived experiences of environment have been found to be a crucial link in encouraging environmental awareness and action (for example, Bartos, 2013). Reflecting on this, Dobson (2003, p. 206, emphasis in original) surmises that:

> If this is right, then environmental and ecological citizenship will not be learned in the confines of the classroom – but given that these citizenships take us beyond environmental education, walks in the woods are not enough either. Ecological citizens are most likely to be created through what the French call *le vécu*, or 'lived experience'.

There are also concerns that students are more likely to receive narrow (neo)liberal experiences of citizenship through their schooling and less likely to receive civic republican or post-cosmopolitan approaches to environmental education (Schild, 2016). Westheimer and Kahne (2004) suggest that most schools do well in creating such 'personally responsible' or 'participatory' citizens, but these are rarely 'justice oriented'. Therefore, they create self-managing civic agents, neoliberal consumers and citizen-workers but rarely critical green citizens (Wong and Stimpson, 2003; Hayward, 2012; Dimick, 2015). Youth citizens therefore are likely to experience a 'thin environmentalism' in which they learn to address some of the symptoms of the current sustainability crisis but leave the drivers of sociological and social injustice unchallenged (Hayward, 2012). These narrow experiences of environmental citizenship are compounded by conceptions of political and citizenship participation that are conveyed within many citizenship curricula as a delayed act, thus reinforcing a view that the role of schools is to provide people for their *future* participation as citizens. Researchers also question whether children and young people in such forums can express 'dissident' perspectives from those of involved adults (Mathhews, 2001; Matthews and Limb, 2003; Kallio and Häkli, 2011), thus reinforcing a view of children and young people as citizens/subjects-in-waiting (Skelton, 2010).

However, this critique on schools is only partially fitting as it does not adequately acknowledge the active roles that children and youth play in the processes of socialisation and social learning, and how their relational living environments form a part of these learning process (for socialisation and learned citizenship, see Kallio, 2018a). In contrast to the findings above, several studies confirm that children and young people remain interested in and concerned with environmental issues. Studies in Australia (for example, Sargeant, 2008; Harris and Wyn, 2010), England (Holden et al., 2008), and across the OECD (Schulz et al., 2010) confirm that climate change and environmental degradation are perceived by young people as some of the most significant issues they face today. There is also some international evidence that young people are increasingly taking part in community-based action and in some internet campaigns concerning issues such as the environment and ethical consumption (Sherrod et al., 2010).

When examining young people's everyday and lived citizenship in New Zealand, Finland and England, we have also found that young people had a significant interest and concern for environmental issues. In an open-ended interview about 'important issues in our place', in Wood's (2011) study (n=122, 14–18 years), the young participants most frequently nominated *environmental sustainability* and *climate change* issues. Similarly, in Kallio and colleagues (Kallio, 2018c; Rinne and Kallio, 2017; see also Kolehmainen, 2017) studies in Finland and England (England n=134, Finland n=128, 10–12 and 14–17-year-olds), youth narratives about their lived realities brought up various connections with nature and environmental issues. Notably, both studies employed a specific focus on young people's lived and spatial experiences of being young citizens, as witnessed and experienced through their own lives and in this way developed a complex and inter-related understanding of green citizenship at the intersection of daily practices, relationships and global connections. Such evidence presents a deeper but arguably more complex, picture of green citizenship in the current and future generation and provides a way to break down the intractable positions accounted for in traditional moral discourses about the environment (Dean, 2001). It also speaks of the need for more in-depth research and flexible frameworks to account for the multiple expressions and spatial dimensions of green citizenship.

Conclusion

This chapter has established the inherently contested, political and debatable nature of green citizenship. It has argued for an approach to green citizenship which rests on deeper understandings of the spatial dimensions of environmental issues and environmental planning responses, and a greater recognition of the diversity of citizens represented in society, and their experiences and practices. Furthermore, the focus we have taken to spatial and lived expressions of green citizenship advocates for the importance of green studies to environmental planning, citizenship and political theory and the importance of trans-local and networked thinking when it comes to understanding the responsibilities, rights, activities, and lived experiences of citizenship. In turn, citizenship and political studies continue to shed light on the contested nature of green citizenship and have helped to highlight the importance of recognising the potential for wider political interpretations of this concept.

References

Agyeman, J. and Evans, B. (2006). 'Justice, governance, and sustainability: perspectives on environmental citizenship from North America and Europe', in A. Dobson and D. Bell (eds.) *Environmental citizenship*. Cambridge, MA: MIT Press.

Bartos, A. E. (2013). 'Friendship and environmental politics in childhood'. *Space and Polity*, 17(1): 17–32.

Bartos, A. E. and Wood, B. E. (2017). 'Ecological wellbeing, childhood, and climate change', in C. Ergler, R. Kearns and K. Witten (eds.) *Children's health and wellbeing in urban environments*. Oxon/New York: Routledge.

Chawla, L. (1998). 'Significant life experiences: a review of research on sources of environmental sensitivity'. *The Journal of Environmental Education*, 29: 11–21.

Chawla, L. (2007). 'Childhood experiences associated with care for the natural world: a theoretical framework for empirical results'. *Children, Youth and Environments*, 17(4): 144–70.

Circle (2002). *Youth civic engagement: facts and trends*. Medford, MD: Centre for Information and Research on Civic Learning and Engagement.

Dean, H. (2001). 'Green citizenship'. *Social Policy and Administration*, 35: 490–505.

Dimick, A. (2015). 'Supporting youth to develop environmental citizenship within/against a neoliberal context'. *Environmental Education Research*, 21(3): 390–402.

Dobson, A. (2003). *Citizenship and the environment*. Oxford: Oxford University Press.

Dobson, A. (2010). 'Democracy and nature: speaking and listening'. *Political Studies*, 58(4): 752–68.

Fadaee, S. (2017). 'Environmentally friendly food initiatives in Iran: between environmental citizenship and pluralizing the public sphere'. *Citizenship Studies*, 21(3): 344–58.

Faulks, K. (2000). *Citizenship*. London: Routledge.

Gabrielson, T. (2008). 'Green citizenship: a review and critique'. *Citizenship Studies*, 12(4): 429–46.

Gilbert, L. and Phillips, C. (2003). 'Practices of urban environmental citizenships: rights to the city and rights to nature in Toronto'. *Citizenship Studies*, 7(3): 313–30.

Häkli, J. (2017). 'The subject of citizenship – can there be a posthuman civil society?' *Political Geography*, 31 August [Online]. Available at: https://doi.org/10.1016/j.polgeo.2017.08.006

Harris, A. and Wyn, J. (2010). 'Special issue of *young* on emerging forms of youth participation: everyday and local perspectives'. *Young*, 10: 3–7.

Harris, L. M. (2011). 'Neo(liberal) citizens of Europe: politics, scales, and visibilities of environmental citizenship in contemporary Turkey'. *Citizenship Studies*, 15(6–7): 837–59.

Hayward, B. (2012). *Children, citizenship and environment: nurturing a democratic imagination in a changing world*. Oxon/NY: Routledge.

Holden, C., Joldoshalieva, R. and Shamatov, D. (2008). '"I would like to say that things must just get better": young citizens from England, Kyrgystan and South Africa speak out'. *Citizenship Teaching and Learning*, 4: 6–17.

Hubbard, P. (2013). 'Kissing is not a universal right: sexuality, law and the scales of citizenship'. *Geoforum*, 49: 224–32.

Isin, E. (2008). 'Theorising acts of citizenship', in E. Isin and G. M. Nielsen (eds.) *Acts of citizenship*. London/New York: Palgrave Macmillan, pp 15–43.

Isin, E. (2012). *Citizens without frontiers*. New York/London: Bloomsbury Academic.

Kallio, K. P. (2018a). 'Citizen-subject formation as geo-socialisation: a methodological approach on "learning to be citizens"'. *Geografiska Annaler, Series B: Human Geography*, 100: 81–96.

Kallio, K. P. (2018b). 'Not in the same world: topological youths, topographical policies'. *Geographical Review*, 108(4): 566–91.

Kallio, K. P. (2018c). 'Exploring space and politics with children: a geosocial methodological approach to studying experiential worlds', in A. Cutter-MacKenzie, K. Malone and E. Barratt (eds.) *Research handbook childhoodnature*. Singapore: Springer, pp 1–28.

Kallio, K. P. and Häkli, J. (2011). 'Tracing children's politics'. *Political Geography*, 30(2): 99–109.

Kallio, K. P., Häkli, J. and Bäcklund, P. (2015). 'Lived citizenship as the locus of political agency in participatory policy'. *Citizenship Studies*, 19(1): 101–19.

Kallio, K. P. and Mitchell, K. (2016). 'Introduction to the special issue on transnational lived citizenship'. *Global Networks*, 16(3): 259–67.

Kolehmainen, M. (2017). '"Isikin tykkää luonnosta, se on jotenkin rennompi siellä": lasten ja nuorten tärkeät luontopaikkakokemukset ja sosiaaliset suhteet'. University of Tampere: TamPub. Available at: http://urn.fi/URN:NBN:fi:uta-201711222764

Kurtz, H. E. (2005). 'Alternative visions for citizenship practice in an environmental justice dispute'. *Space and Polity*, 9: 77–91.

Latta, A. (2007). 'Locating democratic politics in ecological citizenship'. *Environmental Politics*, 16(3): 377–93.

Lister, R. (2007). 'Inclusive citizenship: realising the potential'. *Citizenship Studies*, 11(1): 49–61.

MacGregor, S. (2006). 'No sustainability without justice: a feminist critique of environmental citizenship', in A. Dobson and D. Bell (eds.) *Environmental citizenship*. Cambridge, MA: MIT Press, pp 101–26.

Mathhews, H. (2001). 'Citizenship, youth councils and young people's participation'. *Journal of Youth Studies*, 4(3): 299–318.

Matthews, H. and Limb, M. (2003). 'Another white elephant? Youth councils as democratic structures'. *Space and Polity*, 7(2): 173–92.

Melo-Escrihuela, C. (2008). 'Promoting ecological citizenship: rights, duties, and political agency'. *ACME: An International E-Journal for Critical Geographies*, 7(2): 113–34.

Mitchell, K., Marston, S. and Katz, C. (2003). 'Introduction: life's work: an introduction, review, and critique'. *Antipode*, 35(3): 415–42.

Putnam, R. (2000). *Bowling alone: the collapse and revival of American community*. New York: Simon and Schuster.

Repak, D. (2011). 'Mogu li se ekološki problemi izraziti kroz tradicionalne koncepte građanstva?' *Godišnjak Fakulteta političkih nauka*, 5(5): 215–28.

Rinne, E. and Kallio, K. P. (2017). 'Nuorten tilallisten mielikuvien lähteillä'. *Alue & Ympäristö*, 46(1): 17–31.

Sargeant, J. (2008). 'Australian children: locally secure, globally afraid?' in R. Gerber and M. Robertson (eds.) *Children's lifeworlds: locating indigenous voices*. New York: Nova Science Publishers, Inc., pp 119–33.

Scerri, A. and Magee, L. (2012). 'Green householders, stakeholder citizenship and sustainability'. *Environmental Politics*, 21(3): 387–411.

Schild, R. (2016). 'Environmental citizenship: what can political theory contribute to environmental education practice?' *The Journal of Environmental Education*, 47(1): 19–34.

Schulz, W., Ainley, J., Fraillon, J., Kerr, D. and Losito, B. (2010). *ICCS 2009: International report: civic knowledge, attitudes, and engagement among lower-secondary students in 38 countries*. Amsterdam: IEA.

Sherrod, L., Torney-Purta, J. and Flanagan, C. A. (2010). *Handbook on civic engagement in youth*. Hoboken, NJ: John Wiley and Sons.

Skelton, T. (2010). 'Taking young people as political actors seriously: opening the borders of political geography'. *Area*, 42(2): 145–51.

Staeheli, L. A. (2016). 'Globalization and the scales of citizenship'. *Geography Research Forum*, 19: 60–77.

Staeheli, L. A., Ehrkamp, P., Leitner, H. and Nagel, C. R. (2012). 'Dreaming the ordinary: daily life and the complex geographies of citizenship'. *Progress in Human Geography*, 36(5): 628–44.

UNDP (United Nations Development Programme) (1998). *Human development report 1998*. New York: Oxford University Press. Available at: http://hdr.undp.org/sites/default/files/reports/259/hdr_1998_en_complete_nostats.pdf

Van Steenbergen, B. (1994). 'Towards a global ecological citizen', in B. Van Steenbergen (ed.) *The condition of citizenship*. London: Sage.

Westheimer, J. and Kahne, J. (2004). 'What kind of citizen? The politics of educating for democracy'. *American Educational Research Journal*, 41(2): 237–69.

Wong, F. B. K. and Stimpson, P. (2003). 'Environmental education in Singapore: a curriculum for the environment or national interest?' *International Research in Geographic and Environmental Education*, 12(2): 123–38.

Wood, B. E. (2011). *Citizenship in our place: exploring New Zealand young people's everyday, place-based perspectives on participation in society*. Unpublished PhD, Victoria University of Wellington.

Wood, B. E. and Black, R. (2018). 'Globalisation, cosmopolitanism and diaspora: what are the implications for understanding citizenship?' *International Studies in Sociology of Education*, 27(2–3): 184–99.

Wray-Lake, L., Flanagan, C. and Osgood, W. (2010). 'Examining trends in adolescent environmental attitudes, beliefs and behaviors across three decades'. *Environment and Behavior*, 42(1): 61–85.

Yuval-Davis, N. (2007). 'Intersectionality, citizenship and contemporary politics of belonging'. *Critical Review of International Social and Political Philosophy*, 10(4): 561–74.

16
Intermediaries and networks

Ross Beveridge

Introduction

In very simple terms an intermediary operates in between other actors, making connections and reordering relationships between institutions, individuals and even 'things', like buildings and waste. Decisive to being an intermediary is the position within a wider network of actors: in the niche, at the meeting point, between the boundaries. As 'in-between' actors, intermediaries are seen to facilitate, broker, negotiate, and disseminate knowledge and other resources – all in the aid of other actors, or, more precisely, the relations between them. Looking across the literature any-*thing* can be an intermediary – it may be human or nonhuman, a business, a non-governmental organisation (NGO) or state actor. An intermediary can apparently operate in any field – energy efficiency, the chemical industry, participatory governance. And being an intermediary may be intentional, inadvertent, short or long term. Hence, intermediaries are defined by their relational and often difficult to discern work (Moss, 2009).

Why then the focus on intermediaries, given that they are rarely the most powerful or visible actors? They are seen as crucial to the functioning of wider networks of, for example, water management or eco-housing construction. They are involved in oiling the wheels on which networks depend. Their significance in environmental planning is bound-up with the rise of network governance as a political project and network approaches within social science to understand how complex social and political organisations work. This empirical and conceptual move to more displaced, decentralised notions of agency, power and effect has thrown light on the spaces between actors which need to be bridged for network objectives to be decided, pursued and achieved. To fill these gaps, a variety of – intermediary – functions may be required. As will be discussed, intermediaries do not only lie in between actors, they can be conceptualised as the points where structures and agency meet: where networks are animated and where intermediary agency has network effects (cf. Hermelin and Rämö, 2017). Hence, focusing on intermediaries throws light on those folds and crevices in networks, the points where agencies meet, where processes crucial to the entire functioning of the network take place.

This chapter will critically reflect on the notion of intermediaries. It provides an overview of the variety of intermediaries identified in the environmental planning and related fields, addressing: their roles, impact and influence on networks; their interests and motivations; and their importance. The chapter covers the following topics: the contexts of intermediation; an exploration of the definitions of intermediaries; an overview of the perceived impacts of intermediation in environmental planning; an examination of the conceptual limits of intermediation and intermediaries; and, finally, a brief consideration of future directions in intermediary research.

Contexts of intermediation

Studies of intermediaries and networks are apparent in a number of fields, for example, water infrastructures (Moss, 2009), urban and regional energy networks (Hodson and Marvin, 2009; Rohracher, 2009), strategic planning processes (Gunn and Hillier, 2012), environmental technology networks (Hermelin and Rämö, 2017) and in housing and the built environment (Fischer and Guy, 2009). However, studies of cultural intermediaries, market intermediaries, labour intermediaries, social intermediaries and welfare intermediaries tend to predate the work carried out in planning and related fields and indicate the greater concern for the facilitation of associations in these sectors (Medd and Marvin, 2008 in Moss, 2009, p. 1482). Intermediaries are perhaps most associated with the fields of finance and innovation. Howells (2006, p. 716) notes that the notion of an intermediary can be traced to 'middlemen', suppliers of knowledge and facilitators of relationships in agriculture, trade and textile industries in the 17th century. Thus, notions of financial intermediation are commonplace, reflecting the concern with relationships and knowledge in buying and selling and economic decision making more generally.

By contrast, intermediaries appear to be a relatively new concern in the planning research. An early reference to intermediaries in planning can be found in Simmie and French (1991), who draw on the political science term 'interest intermediation' to discuss corporatist relations between state, market and civil society actors in local planning authorities in London. However, a concerted concern for intermediaries first becomes apparent in attempts to understand the changes wrought by neoliberal reforms and moves from government to 'governance'. Hence, a substantial body of intermediary research emerges in the 2000s as the impacts of these shifts became increasingly apparent. Ultimately, the rise of intermediaries is bound-up with the growing influence of market logics and actors in the planning sector and the opportunities this has created for non-state (but not only market) actors.

Processes of privatisation, liberalisation, internationalisation and increased environmental regulation (Bakker, 2003; Guy et al., 2001; Moss et al., 2011) can be seen as generating network contexts in which intermediation became more of a 'need'. New forms of regulation (planning, financial, environmental), the unbounding of market forces, technological innovation and more intricate and individual patterns of production and consumption (Graham and Marvin, 2001) transformed socio-technical networks as well as the planning sector. Relations between actors became less stable, creating gaps in knowledge, service requirements and opportunities. Though a contested term, the generally agreed meaning of governance is that governing is increasingly defined by new structures in which the 'boundaries between and within public and private sectors have become blurred' (Stoker, 1998, p. 17). Political decision making is no longer seen as occurring only through the traditional institutional processes of political systems but through networks including actors from the private sector and civil society. Further, arrangements of governance do not rely on the formal authority of political institutions; instead, actions have to be coordinated and responsibilities between public and private actors negotiated (Stoker, 1998; Kooiman, 2003). For instance, climate change and pressing environmental challenges are seen to require a broader coalition of public and private actors and opportunities for intermediaries as governance networks encompass multiple scales, sectors and interconnected concerns (for example, attempts to create resilient systems or cities).

The re-imagined role of the state in governance networks has therefore created a context in which lines of authority and roles and functions of actors are less determined than before, more 'up-for-grabs'. Geographically, work on intermediaries has been quite focused on European contexts, perhaps reflecting the particular maelstrom of regulation and de-regulation which has characterised planning, particularly from the 1990s onwards. Exceptions include Kohl (2003) on the emergence of intermediaries in participatory planning processes in Bolivia from the 1990s onwards and, focusing on Vietnam, Pham et al. (2010) on the political implications of intermediary work in facilitating payment for environmental services. In terms of sectoral focus, much of the research on intermediaries has examined socio-technical networks, especially utility networks. This may reflect the extent of transformation that has occurred in Europe through liberalisation and environmental regulation. More generally, the preponderance of research conducted on the global north area – and the lack of international comparative research on intermediaries – makes it difficult to draw certain conclusions about the geographical diffusion of intermediaries.

Moss (2009) provided a very useful overview of the governance challenges emerging in the water sector from such transformations and how they were creating contexts of intermediation: '(a) between the state and utility companies, (b) between service providers and users, (c) between infrastructures and the localities they serve, and (d) between infrastructures and the natural environment' (Moss, 2009, pp. 1486–7). An emphasis, however nominal (see Davies, 2012), on 'network thinking' challenged hierarchies of authority. Hence, planning, policy making and implementation became characterised by more decentralised patterns of market, state and civil society relationships. While the emphasis has been on public – private partnerships, harmony should not be assumed – indeed, intermediaries may arise precisely where there is no harmony. More substantively, in terms of political power, the state's authority does not disappear, even as the market's authority increases. Indeed, scholars have argued that the apparatuses of the state are crucial to the dominance of the market within a complex and even formally inclusive model of governance (Swyngedouw, 2005).

The notion of intermediaries has, then, gained resonance as a way of capturing increasingly complex networks of governance, in which actors operate in interstitial spaces. Researchers have argued that new 'intermediary spaces' are emerging in between regulators, users and providers of services, such as water or energy, and that the intermediaries working within these spaces can be crucial to the configuration of relationships between economy, environment and society (Moss and Medd, 2005). Intermediaries are also seen to be active in production-consumption relationships, as well as in the 'governance gaps' of the water-energy-food nexus (cf. Weitz et al., 2017), but they also 'intermediate between different scales, between technologies and different social contexts, between different meanings and between different sets of interests' (Moss and Medd, 2005, p. 20). They are, therefore, crucial to understanding the disaggregation and general complexity of networks. A concern for intermediaries shows a range of actors and constellations of relationships often overlooked in more sweeping analyses. Hence, looking at intermediaries can throw light on, for example, contests over the meaning and functions of sustainability in planning in specific contexts, as well as often unpredictable planning and policy implementation processes.

Defining intermediaries

Who and what, then, is an intermediary? Somewhat bewilderingly, an intermediary can be a person, such as a consultant, an organisation, like a planning authority, a nonhuman entity, either 'natural' (for example, water) or manmade (for example, technology), a network providing information to other networks (Moss, 2009, p. 1489; compare to Beveridge and Guy, 2009) or a system combining potentially all of these elements, such as 'strategic planning' (Bryson et al., 2009, p. 175). This great mutability indicates that the first step to comprehending intermediaries is not to think in terms of particular types of actors but in terms of particular types of relational work (Grandclement et al., 2015, p. 215). Intermediaries and intermediation as concepts are always, either implicitly or explicitly, dependent on, even determined by, a sense of a wider network of actors.

Given the term's strong roots in economics, it is not surprising intermediaries are closely associated with innovation and entrepreneurship. Intermediaries are seen as being crucial to processes of Strategic Niche Management (SNM) (Geels and Deuten, 2006) and processes of innovation more generally (Howells, 2006). In most of the recent literature intermediaries are seen as more than just brokers but are facilitators of new arrangements in networks and even 'champions' of innovation, for example, facilitating change towards zero carbon housing (Martiskainen and Kivimaa, 2018, p. 17). Intermediaries' role in the network might also entail providing relationship guidance or orchestrating relationships between actors. In their study of clean tech regional networks in Sweden, Hermelin and Rämö (2017, p. 131) offer three intermediary functions: as *brokers* for business agreements; as *facilitators*, whereby they facilitate processes and coordinate arrangements; and as *legitimisers* involved in political and wider societal realms, such as the pursuit of sustainability (cf. Beveridge and Guy, 2009; Hodson and Marvin, 2009; Moss, 2009). However, we should be careful to avoid seeing intermediaries only in positive normative terms – they may just as well impede or slow down innovation (Moss et al., 2011).

Another literature concerned with innovation, especially in its initial iterations, Actor-network Theory (ANT) (for example, Latour, 2005), has been influential in thinking about intermediaries and networks in environmental planning. Studies have shown the ways in which intermediaries become crucial to how socio-technical practices emerge, are sustained over time and distance and become problematised through the (un)folding of 'actor-networks': the constellations of actors and entities through which human and nonhuman agency take form (Beveridge and Guy, 2009; Gunn and Hillier, 2012). Intermediaries are also often thought of in relation to other concepts, such as a 'Boundary Object' (Star and Griesemer, 1989), a thing that can have different meanings to different actors (Moss, 2009; Rohracher, 2009). This notion is useful for highlighting the importance of intermediaries working in relation to diverse actors and across multiple forms of social organisation. Intermediaries 'work in-between, make connections, enable a relationship between different persons or things' (Hodson et al., 2013, p. 1408). Some authors have sought to draw out a more encompassing concept of intermediation, in which intermediaries attend to multiple and not just bilateral relations. Van Lente et al. (2003) developed the concept of 'systemic intermediaries' to illustrate the work of intermediaries across energy policy networks, whereby they become central to long-term processes and coordinate the actions of industry, policy makers and researchers.

Intermediaries are best perceived in terms of place and position within wider networks. Taking inspiration from ANT, an intermediary can be defined as an entity 'that stands at a place in the network between two other actors and serves to translate between the actors in such a way that their interaction can be more effectively co-ordinated, controlled, or otherwise articulated' (Kaghan and Bowker, 2001, p. 258). Hence, intermediaries work in unstable and uncertain contexts, translating between actors. Intermediaries are, then, involved in the transfer or translation (Beveridge and Guy, 2009) of something, for example, specific technical knowledge, essential to the functioning of a wider network of sustainability in the chemical industry. Indeed, they may have to draw on different forms of expertise, both technical and social, to help engage a range of actors in processes of translation (Beveridge and Guy, 2009). Outcomes are thus uncertain and interests potentially mutable. These roles should not be seen as neutral but integral to how objectives are set and achieved, to who gains and loses (Moss et al., 2011).

Hence, intermediaries and intermediation are relational concepts, describing processes of interaction between actors. Janda and Parag's (2013) 'middle-out' approach, which draws on the intermediary concept, refers largely to scale, i.e. the middle as opposed to bottom or top. But in their discussion of how systemic or structural change occurs through actors' 'agency' and 'capacity' the merging or meeting point between structure and agency comes to the fore: agency and capacity are, of course, shaped by the structural conditions in which actors operate. Ultimately, we might interpret an intermediary approach to networks as denoting a concern for the agency of actors and the prospects for structural change related to that agency. In her fascinating account of nature-making experiments in Amsterdam, Kinder (2011) has shown that 'intermediary thinking' has entered planning and sustainability practice. Rather than directly controlling nature-making processes via fisheries or nurseries, she demonstrates that planners used urban infrastructure and building technologies to mediate the creation of ecosystems. Hence, by using nonhuman intermediaries in a strategic way, planners allowed for a degree of haphazard autonomy on behalf of nature, urban elements and the intermediaries themselves.

Thus, to summarise, the first and key step to understanding an intermediary is to think spatially and positionally. To find an intermediary, we have to look for those actors *in-between others*. They are always, then, defined relationally in terms of their relationships with other actors. Intermediaries are important not just in their actual practices within socio-technical networks but also analytically as they offer an entry point to examine networks, especially the often hidden work between larger more prominent actors. Moss (2009, p. 1481) has noted that the work of intermediaries like research organisations, non-profit agencies and information campaigns revealed much about the changed ways in which the 'main' actors – service providers, users and regulators – operated in water infrastructures.

Impacts of intermediation in environmental planning

Generally, the literature places the focus on intermediary practices rather than discovering the motivations and aims of the intermediaries themselves. Martiskainen and Kivimaa's (2018) examination of sustainable practices and new build housing argue that intermediaries are vital to the development of niches advancing zero carbon housing in a context in which both the state and market offer little general support. Fischer and Guy (2009) have shown that architects act as 'interpretative intermediaries'; in a context of a shifting regulatory system they work in between regulations and the design process, between regulators and urban designers, builders, etc.

In the broader field of energy politics, Hargreaves et al. (2013) identified the importance of intermediaries, such as national, regional and local government departments and organisations as well as NGOs (for example, Community Energy Scotland) and private sector actors (for example, consultants), to the development of grassroots innovations. An example is Hodson and Marvin's (2009) study of the London Hydrogen Partnership (LHP), established by the Greater London Authority, to promote a hydrogen economy in the city. They develop a sense of the spatial impacts of intermediary work as organisations seek to translate technological possibilities for energy transition into specific places in London. While these intermediaries are cross-sectoral 'transition managers', they perform spatial practices by developing place-based images of technological transitions and place-based networks of actors by working across scales and spaces, in the public as well as the private sector. Ornetzeder and Rohracher (2009) show that new types of intermediation in socio-technical systems emerge as new technologies, such as passive houses, prompt new networks to form and as a result needs arise for intermediary services to determine roles of actors to align technologies as well as planning practices (for example, setting new construction standards, etc.).

Certain shared characteristics are discernible, despite the diversity of intermediation practices and contexts in which intermediaries find themselves in planning and sustainability. Often intermediation work is 'hidden', concealed within the workings of networks (Moss, 2009) because they do not fit into, but work between, the standard categories of provider and user, regulator and market agent (Grandclement et al., 2015, p. 215). This hidden nature has important implications. In their case study of a small consultancy and the implementation of EU environmental regulation in the UK chemical industry, Beveridge and Guy (2009, p. 82) show that 'an intermediary is not a role that is defined necessarily by "size" or "strength", but by location in between actors and an ability to mediate across boundaries'. In the case of the consultants, reforming wastewater practices in the chemical sector entailed working in between the 'technology experts, their technology, the rogue chemicals, the inventor, his invention, the salesman, the on-site workers, their MD, the regional water company, the Directive itself' (Beveridge and Guy, 2009, p. 82). Hence, though individually and in terms of their roles intermediaries may be 'small', they are seen to be extremely important.

Hodson and Marvin (2010) provide one of the few really detailed reflections on what actors need to do to successfully perform as intermediaries. They are concerned with the translation of socio-technical change in urban energy transitions, but it is worth noting the more generic qualities they mention: ensuring a shared understanding amongst those working for them and those receiving their services of what successful intermediation is. This is crucial so they can develop self-purpose as well as convince others of the benefits of their work, which, as we have noticed, might be difficult to discern, given its displaced nature and subtle qualities.

From the intriguing insights offered by Hodson and Marvin (2010), a sense emerges that there is something entrepreneurial about intermediaries in the ways they operate in 'up-for-grabs' networks increasingly shaped by market logics. The increase in environmental consultants working for and within the state (Beveridge, 2012) and what we might call 'eco-preneurs' (Beveridge and Guy, 2005) is indicative of the extent to which new forms of regulation have opened up environmental planning and policy to private sector actors. Environmental consultants might be seen as exploiting intermediating opportunities for financial as well as environmental gain.

On such grounds, intermediaries should always be seen as political. They have agendas and operate in contexts of 'collective action', reordering relations, spaces where governance challenges are defined and dealt with (or not) (Moss, 2009, p. 1488; Moss et al., 2011). They may reshape and redefine the objectives of others, pursing their own. We should not view them as normatively 'good' or 'bad' in term of their impacts – both can be inherent to intermediation. In Grandclement et al.'s (2015) study of a household energy saving technology, politics emerged in relation to the conflicting goals of energy efficiency and household comfort and the intermediary's attempt to redefine objectives and relations. Hodson and Marvin (2010, p. 482) illuminate the political roles intermediaries have in shaping negotiations between actors in urban energy transitions. Their 'politicalness' lies in the influence they can have on the success of other actors – be they utilities or local authorities, NGOs or multi-nationals – in pursuing their objectives. They note (Hodson and Marvin, 2010, p. 483) intermediaries are crucial in representing and aligning interests. Ultimately, this can never be an entirely neutral process and, hence, intermediaries may (in)advertently develop political agendas. Sometimes intermediation has been articulated as actually bound-up in the political realm itself, with NGOs seen to be intermediaries between new legislation and citizens and potentially a force for democratic planning (Kohl, 2003). In contrast, Pham et al.'s (2010) examination of intermediaries (government agencies, international agencies, consultants, NGOs) and the payment for environmental services in Vietnam showed that intermediation practices were often seen as biased, partial and politically motivated.

The conceptual limits of an intermediary focus

The notion of intermediation is a concept with a clear but limited scope. This is a strength in the sense that it provides a focus and set of tools for understanding specific processes. It becomes more problematic when we attempt to consider the wider importance of intermediaries to networks. In epistemological terms, can a concern for intermediation alone really provide insights to the functioning of wider networks? While it can be argued that this might be the case empirically, and studies have shown how intermediaries are shaping aspects of wider networks, the notion of intermediaries alone is not able to conceptualise the broader impacts intermediation can have. It is a concept in need of other concepts. Intermediation must rely in the first instance on an understanding of how networks function in terms of, for instance, power and scale. Hence, in much of the literature previously discussed we see either a largely empirical assertion of how intermediaries shape governance (for example, Hodson and Marvin, 2010) or a largely conceptual assertion of their importance, drawing on another literature, such as Strategic Niche Management (Hargreaves et al., 2013) or ANT (for example, Beveridge and Guy, 2009). Of course, what happens across the rest of the networks in question cannot be shown or, in most cases, feasibly researched. A focus on intermediaries does, then, provide insights to specific processes in specific places at specific times. While this is an inherent and entirely normal limit to the analytical value of a concept, these limits should be reflected on when claims are made about intermediation.

Conceptually, intermediaries are intriguing because they lie at the meeting point of concerns for structure and agency. For this reason, the literature sometimes draws on complex social theory like ANT and, for instance, utilises notions of 'distributed agency' (Kinder, 2011) to explore intricate issues of intentionality and power in environmental planning in the city. In general, however, the planning literature has more often tended not to address these more theoretical questions, focusing instead on the impacts of practices of intermediation. Hence, the literature on intermediaries does not tell us much about some of the more conceptual questions intermediation raises. For instance, given their 'in-betweenness' and relational work, might we see intermediaries as 'binary-busters', operating at the points where agency equates to structure and vice versa, both product and productive of more complex and disaggregated socio-technical arrangements/networks?

While the literature focuses explicitly on the network effects intermediaries can have, there has been less concern for the effects of networks on intermediaries or, put differently, how intermediaries are shaped through the networks in which they become embedded. We learn little, for example, about the histories or biographies of intermediary actors, and the ways in which this changes over time and in relation to wider structural changes in a particular sector. Generally, this is seen as unimportant – what counts most, in the literature, is the effect an intermediary can have on those around it.

Future directions

Research on intermediaries in environmental planning has undoubtedly shown the importance of a diverse range of actors (things) that might, without the focus on the *in-between*, have gone under the radar. Intermediaries have been shown to have influence well beyond their own, often modest properties (of size, wealth, etc.). Indeed, one of the most important insights of the literature has been to show that positionality is crucial in networks; that at certain points in space/time gaps, needs and deficits emerge, or can, through entrepreneurial activity, be forged and addressed. Such insights are clearly crucial to environmental planning, showing the need for diverse actors to be enrolled in networks for objectives to be achieved. This raises the prospect that space and autonomy must be left for these intermediaries to do their work (Kinder, 2011). In other words, intermediaries might not just be necessary because of the inevitable gaps in knowledge and so on, which emerge in networks of actors. They might be crucial to the development of networks themselves. Future research on intermediaries might, then, consider the extent to which intermediaries can be 'planned' into networks. Or, in different terms, how planning processes might leave spaces for intermediaries both as necessary constituent parts of planning processes and as co-constructers of those processes.

One aspect of intermediaries not well accounted for in research is the 'why' of intermediaries' emergence. Although we should not associate the rise of intermediaries only with neoliberalism, it is surprising that the literature has not wondered what the presence of intermediaries in specific contexts means politically. The literature has not expressly questioned why networks need certain kinds of intermediation. Generally, the 'needs' for intermediation are taken as given. But by not really questioning the neoliberal drivers of fragmenting networks of governance, the literature might be seen to (in)advertently reify them. In short, through the very concern for intermediaries, and the resulting zooming into in-between places, the more general logics and consequences of networks might get lost. Hence, having provided a rich sense of the diversity of intermediaries and their practices, the literature might turn its focus to the politics, which shape their emergence, activity and impacts.

References

Bakker, K. (2003). *An uncooperative commodity: privatising water in England and Wales*. Oxford: Oxford University Press.

Beveridge, R. (2012). 'Consultants, depoliticization and arena-shifting in the policy process: privatizing water in Berlin'. *Policy Sciences*, 45(1): 47–68.

Beveridge, R. and Guy, S. (2005). 'The rise of the eco-preneur and the messy world of environmental innovation'. *Local Environment*, 10(6): 665–76.

Beveridge, R. and Guy, S. (2009). 'Governing through translations: intermediaries and the mediation of the EU's urban waste water directive'. *Journal of Environmental Policy and Planning*, 11(2): 69–85.

Bryson, J. M., Crosby, B. C. and Bryson, J. K. (2009). 'Understanding strategic planning and the formulation and implementation of strategic plans as a way of knowing: the contributions of actor-network theory'. *International Public Management Journal*, 12(2): 172–207.

Davies, J. (2012). 'Network governance theory: a Gramscian critique'. *Environment and Planning A*, 44(11): 2687–704.

Fischer, J. and Guy, S. (2009). 'Re-interpreting regulations: architects as intermediaries for low-carbon buildings'. *Urban Studies*, 46(12): 2577–94.

Geels, F. and Deuten, J. (2006). 'Local and global dynamics in technological development: a socio-cognitive perspective on knowledge flows and lessons from reinforced concrete'. *Science and Public Policy*, 33(4): 265–75.

Graham, S. and Marvin, S. (2001). *Splintering urbanism: networked infrastructures, technological mobilities, and the urban condition*. London: Routledge.

Grandclement, C., Karvonen, A. and Guy, S. (2015). 'Negotiating comfort in low energy housing: the politics of intermediation'. *Energy Policy*, 84: 213–22.

Gunn, S. and Hillier, J. (2012). 'Processes of innovation: reformation of the English strategic spatial planning system'. *Planning Theory and Practice*, 13(3): 359–81.

Guy, S., Marvin, S. and Moss, T. (2001). *Urban infrastructure in transition: networks, buildings, plans*. London: Earthscan.

Hargreaves, T., Hielscher, S., Seyfang, G. and Smith, A. (2013). 'Grassroots innovations in community energy: the role of intermediaries in niche development'. *Global Environmental Change*, 23(5): 868–80.

Hermelin, B. and Rämö, H. (2017). 'Intermediary activities and agendas of regional cleantech networks in Sweden'. *Environment and Planning C: Government and Policy*, 35(1): 130–46.

Hodson, M. and Marvin, S. (2009). 'Cities mediating technological transitions: understanding visions, intermediation and consequences'. *Technology Analysis and Strategic Management*, 21(4): 515–34.

Hodson, M. and Marvin, S. (2010). 'Can cities shape socio-technical transitions and how would we know if they were?' *Research Policy*, 39(4): 477–85.

Hodson, M., Marvin, S. and Bulkeley, H. (2013). 'The intermediary organisation of low carbon cities: a comparative analysis of transitions in Greater London and Greater Manchester'. *Urban Studies*, 50(7): 1403–22.

Howells, J. (2006). 'Intermediation and the role of intermediaries in innovation'. *Research Policy*, 35(5): 715–28.

Janda, K. B. and Parag, Y. (2013). 'A middle-out approach for improving energy performance in buildings'. *Building Research and Information*, 41(1): 39–50.

Kaghan, W. and Bowker, G. (2001). 'Out of machine age?' *Journal of Engineering and Technology Management*, 18(3–4): 253–69.

Kinder, K. (2011). 'Planning by intermediaries: making cities make nature in Amsterdam'. *Environment and Planning C*, 43(12): 2435–51.

Kohl, B. (2003). 'Non-governmental organizations and decentralization in Bolivia'. *Environment and Planning C*, 21(2): 317–31.

Kooiman, J. (2003). *Governing as governance*. London: Sage.

Latour, B. (2005). *Reassembling the social*. New York/Oxford: Oxford University Press.

Martiskainen, M. and Kivimaa, P. (2018). 'Creating innovative zero carbon homes in the United Kingdom – intermediaries and champions in building projects'. *Environmental Innovation and Societal Transitions*, 26: 15–31.

Moss, T. and Medd, W. (2005). 'Knowledge and policy frameworks for promoting sustainable water management through intermediation'. *EU project report: new intermediary services and the transformation of urban water supply and wastewater disposal systems in Europe*. Erkner: Institute for Regional Development and Structural Planning.

Moss, T. (2009). 'Intermediaries and the governance of sociotechnical networks'. *Environment and Planning A*, 41(6): 1480–95.

Moss, T., Guy, S., Marvin, S. and Medd, W. (2011). 'Intermediaries and the governance of urban infrastructures in transition', in S. Guy, S. Marvin, W. Medd and T. Moss (eds.) *Shaping urban infrastructures: intermediaries and the governance of socio-technical networks*. London: Earthscan, pp 17–35.

Ornetzeder, M. and Rohracher, H. (2009). 'Passive houses in Austria: the role of intermediary organisations for the successful transformation of sociotechnical system'. *Proceedings, ECEEE Summer Study*, Paper 7175: 1531–40.

Pham, T., Campbell, B., Garnett, S., Aslin, H. and Hoang, M. (2010). 'Importance and impacts of intermediary boundary organizations in facilitating payment for environmental services in Vietnam'. *Environmental Conservation*, 37(1): 64–72.

Rohracher, H. (2009). 'Intermediaries and the governance of choice: the case of green electricity labelling'. *Environment and Planning A*, 41(8): 2014–28.

Simmie, J. and French, S. (1991). 'Corporatism and planning in London in the 1980s'. *Governance*, 4(1): 19–41.

Star, S. and Griesemer, J. (1989). 'Institutional ecology, "translations" and boundary objects: amateurs and professionals in Berkeley's Museum of Vertebrate Zoology, 1907–39'. *Social Studies of Science*, 19(3): 387–420.

Stoker, G. (1998). 'Governance as theory: five propositions'. *International Social Science Journal*, 155: 17–28.

Swyngedouw, E. (2005). 'Governance innovation and the citizen: the Janus face of governance-beyond-the-state'. *Urban Studies*, 42(11): 1991–2006.

van Lente, H., Hekkert, M., Smits, R. and van Waveren, B. (2003). 'Roles of systemic intermediaries in transition processes'. *International Journal of Innovation Management*, 7(3): 247–79.

Weitz, N., Strambo, C., Kemp-Benedict, E. and Nilsson, M. (2017). 'Closing the governance gaps in the water-energy-food nexus: insights from integrative governance'. *Global Environmental Change*, 45: 165–73.

17

A more-than-human approach to environmental planning

Jonathan Metzger

Introduction

A common way of understanding humans' relation to other types of beings and existences in planning contexts is to imagine human life and its surrounding environments as two separate realms that sometimes affect, disturb or encroach upon each other. From such a perspective, the purpose of environmental planning becomes that of minimising or rectifying these perceived disturbances of what is supposed to be a natural balance between two fundamentally different domains. In contrast to such a traditional Western understanding of human-nature relations, this chapter introduces an approach to environmental planning that is founded upon a completely different intuition, namely that the destiny of humanity is – and always has been – fundamentally entangled with as well as dependent upon the fate and wellbeing of an innumerable myriad of other beings and existences, from the cellular to the planetary scale. If we follow such an intuition and accept – with Anna Tsing (2012, p. 144) – that 'human nature is an interspecies relationship', what would this imply for how we understand the role and function of environmental planning? And further, how can such a *more-than-human* sensibility be made into something more than just a vague idea to contemplate – how can it be translated into concrete planning practices that are made to matter and make a difference in the world?

This chapter will explore how one can approach environmental planning process from what can be called a more-than-human approach – which proceeds from a fundamentally relational understanding of human nature, as well as its conditions of existence and flourishing. The purpose of the proposed more-than-human approach is not to introduce some new 'off-the-shelf' methodology for best practice but rather to instil some humility regarding humanity's position in this world, and our fundamental existential dependence on a host of other beings and existences – many of which we often even have difficulties in grasping the nature of our relations to. Such an insight demands that planners constantly hesitate to ask themselves the daunting question 'what are we really busy doing here?' (see Stengers, 2005), which is neither pleasant nor 'efficient' in the crudely ignorant sense of that term but rather entails a cultivation of torment regarding the gravity of decisions made in environmental planning processes.

The chapter opens with a short introduction to the philosophical underpinnings of a more-than-human approach. After this, it moves on to discuss what the acceptance of these ideas might imply for how we understand the function and purpose of environmental planning. The subsequent question that is tackled is how such a new understanding of environmental planning can be translated into concrete methods and practice.

Finally, the chapter concludes by raising some of the daunting challenges of embarking upon a more-than-human approach to environmental planning. Even though this section comes at the end of the chapter, this is nonetheless perhaps the most important part.

More-than-human environmental relationality

In the 1990s, a new wave of ecological thinking developed within academic fields such as philosophy, geography and Science and Technology Studies (STS). This new, deeply relational understanding of humanity's conditions of existence has been labelled as a *more-than-human* approach by, for example, geographer Sarah Whatmore (2002) and as a 'new political ecology' by another prominent thinker in this vein, Bruno Latour (2004). One of the crucial aspects that differentiates this new political ecology from older variants according to Latour is that it does not for a second consider humans beings as separate from their living environments. Rather, it understands the human as a relational creature whose genesis, existence and development is irrevocably bound up with and dependent upon a limitless jumble of constitutive relations with 'significant others' of both recognisably human and nonhuman kinds (see also Haraway, 2008). This is thus a way of understanding ecology that distances itself from any notion about a pristine and harmonious nature set against a destructive humanity that supposedly needs to learn to adjust to some imagined natural order. Instead, it focuses on creative possibilities for co-becoming with a myriad of other creatures and beings by way of exploration and experimentation. Rather than contrasting mankind to nature and the rest of the world, this is an approach that perceives humans as relays entangled in a dynamic mélange of relations that can be more or less open, inclusive and stable over time – but without any preordained knowledge about how these relations may develop or change across time and space.

An important aspect of this new political ecology is that it does not only offer a new perspective on the relations between humans and others but also opens up the fundamental question of what it means to be human at all. Donna Haraway, one of the key thinkers of this vein, perhaps most clearly spells out the very thorough-going implications of this intuition when she notes that in fact the human genome is found in no more than ten per cent of the cells constituting a human body, while the remaining ninety per cent are made up of bacteria, fungi, protists, etc. Thus, Haraway concludes, even within the confines of their own bodies humans are 'vastly outnumbered' in relation to their 'tiny companions', which implies that even on the cellular level, 'to be one is always to *become with* many' (Haraway, 2008, p. 2).

What Haraway indirectly tells us is that our caring for the wellbeing of other beings and existences needs not to be born out of some form of noble altruism – but can just as well be grounded in the self-, kin- or species-preserving insight that the existence and flourishing of 'us' (however defined) is by necessity founded upon the continued wellbeing of a host of others – both near and far. Caring for 'one's own people', no matter how broadly or narrowly we define that grouping, thus inevitably also involves extending that care to a plethora of significant others. However, it can be very difficult – if not impossible – to grasp and fully comprehend exactly how all the life sustaining webs on this planet are woven. This means that humans be wise to tread delicately ahead, keeping in mind that we can do great – even existential – harm to others, which also risks coming back to 'ourselves' (however we define this fluid term) in ways that may have been difficult to predict before it was too late.

Thus, the entangled, relational nature of human existence focuses attention in two directions simultaneously: on the one hand, inwards – by opening up a questioning of the problematic bundle of tangled cultural and biological relations that we label as human – and on the other hand, outwards – towards a recognition of an innumerable array of co-affecting existences which includes humans of various kinds and allegiances as but one category of components. Perhaps the most crucial task of all becomes the question of fostering a capacity to think and act with both of these perspectives in mind at the same time. The take-home message here is thus that what any group of humans do at a particular time and place will both be dependent on, as well as have bearing upon, innumerable other beings and existences, both near and far – and it can be very difficult to ever attain a full grasp of how these existential threads of relations are spun. The questions 'who is affected by our actions?' and 'on whom and what do we depend for our wellbeing?' can, if diligently pursued, therefore always be predicted to be the opening of a process of tracing *relationally complex* entanglements, from the cellular to the planetary scale.

As Haraway (2016, p. 55) eloquently puts it, the art of 'getting along' in such a world comes to circle around:

> [O]ngoing multispecies stories and practices of becoming-with in times that remain at stake, in precarious times, in which the world is not finished and the sky has not fallen – yet. We are at stake to each other . . . human beings are not the only important actors . . . with all other beings able simply to react. The order is reknitted: human beings are with and of the earth, and the biotic and abiotic powers of this earth are the main story. However, the doings of situated, actual human beings matter. It matters with which ways of living and dying we cast our lot rather than others.

Some environmental planning scholars have begun to grapple with how different ways of planning effectively generate radically different, and more or less barbarous 'ways of living and dying' together for both humans and other beings and existences – which further, sometimes may come back to haunt 'us' in unexpected ways. Some of these attempts have approached this challenge through asking questions about how one can relate to 'nonhumans' such as animals in new ways in the planning process. This is indeed very encouraging since it highlights that there are more existences in this world than just humans. However, such a vocabulary nonetheless risks missing out on tackling the really constitutive and fundamental entanglements of that which often tends to be spontaneously separated into categories of 'human' and 'nonhuman'.[1] How can we find openings that enable us to learn to plan in ways that do not only recognise the existence of other beings and existences than humans but that actually enact a sense of the fundamentally entangled nature of human existence?

Towards a more-than-human approach to environmental planning

In direct relation to emerging insights about the fundamentally entangled fates of humans and nonhumans, Andrew Pickering (Pickering, 2007, p. 1) has argued that '[h]aving understood the world differently, something should follow for how we conduct our affairs in it'. Pickering thus urgently calls for '*other ways to go on*' than current mainstream business as usual practices of, for example, environmental planning (Pickering, 2007, p. 1, emphasis in original). The crucial remaining question is, of course: how? Should we understand the undoing of the hard boundary between the human and nonhuman as an invitation to even more thorough-going interventions into the life sustaining webs on Earth, perhaps on a planetary scale with the help of geoengineering megaprojects? Or should we simply modify our existing frameworks for decision making, to a large extent based on neoclassical economics, so as to also include the production of 'value' contributed by nonhumans, in the manner of, for example, the ecosystem services framework?

Alas, neither of these avenues of action can tackle the challenge of relational complexity in a satisfying manner, and are both quite troubling in that they are based on an assumption that humans can actually know the full effects of their actions. To the contrary, a more-than-human approach must be based on the recognition that we can probably never fathom the full effects of any action, in which ways this action will impact upon a range of entities both near and far, or reverberate back upon those who acted, sometimes in quite unpredictable ways (see further Metzger, 2018).

The thinker who has come closest to formulating a workable motto for a more-than-human environmental planning is perhaps the philosopher Michel Serres, who notes that the fate of what we call 'nature' is certainly in the reach of our 'mastery' in the present age, in the sense that in many aspects humans today have the capacity to dominate it. However, what is still missing is a 'mastery of mastery', in the form of the wisdom that is required to act responsibly in the face of this awesome power (Serres, 2006). For Bruno Latour, the practical path towards enacting such a new sensibility is for humans to find ways to make themselves constantly cognisant on 'where we reside and on what we depend for our atmospheric condition', thus generating a more responsible and responsive attitude towards 'the fragile envelopes we inhabit' (Latour, 2017, p. 83).

This appears to imply a turn towards what Helga Nowotny has termed the politics of 'the expansive present' (Nowotny quoted in Marres, 2012, p. 144) in which the basic premise is not a vague threat of impending doom but instead the aim of broadening the set of creatures and beings we consider ourselves obliged to show concern and care for in our decision-making processes, here and now. This is a type of policy that is both dependent on and a result of what Emilié Hache and Bruno Latour have described as a 'moral sensitisation' towards also other beings and existences than recognisable humans (Hache and Latour, 2010). It requires that we as humans highlight and scrutinise the ecological systems that we are a part of, that shape us, and that we, in our time, are dependent on – a way of living and thinking ecologically here and now, not solely a response rooted in the fear of some vague climatic holocaust in an uncertain future.

So what to do with such a radically expanded scope of beings and existences that potentially 'count' and which humans may therefore also become accountable before (Whatmore, 2002)? If one follows geographer Noel Castree (2003, p. 207), such an approach would, to begin with, demand an abandonment of the traditional idea that political rights, entitlements and deserts only apply to people; second, it would require facing up to the very real problem of defining political subjects in a world where the boundaries between humans and nonhumans are hard to discern; and third, that the domain of political reasoning is expanded to include nonhumans, without resorting to the idea that the latter exist 'in themselves'. What is nonetheless crucial to point out is how this perspective diverges from other superficially similar strands of thought that morally privilege the supposedly intrinsic value of nonhuman life. In contrast to such 'deep ecology' thinking, more-than-human perspectives are instead fundamentally relational in their ontological and ethical approach, with a particular interest in the entangled situatedness of human/nonhuman relations (see, for example, Latour, 2004; Haraway, 2008) and a refusal to a priori morally privilege any component in heterogeneous 'lively assemblages' – instead generally emphasising the importance of situated ethical judgement based on the principle of 'it depends' (Puig de la Bellacasa, 2010). The challenge posed is thus not only about appreciating that humans are not the only actors on stage in the unfolding drama of life on the planet known as 'the Earth' – but also relates to the need to enact a sense of 'wicked' relational complexity, dispelling all pipe dreams about existential autonomy or independence for humans and humanity through recognising that humanity as a species, often without even being conscious about it, in variegated ways is inextricably entangled with and co-dependent upon a myriad of 'significant others' – from the cellular to the planetary scale.

Even if we accept this type of reasoning as valid, the question remains how it can be translated into some form of guidelines for action and policy. Elsewhere (Metzger, 2012), it has been suggested that some possible points for a programme of action that can be drawn out of a more-than-human sensibility are:

- The need to work towards broadening the scope of 'beings that count', i.e. the circle of recognised ethical, political, and legal subjects. This would call for an effort to further expand and deepen the solidarity between different types of beings, human as well as nonhuman;
- To work towards highlighting mutual dependencies and to stop nurturing fantasies about unitary, sovereign, and independent subjects – regardless of whether these are portrayed as a rational individual, a 'culture', a nation, or a species – so as to instead focus on processes of mutual co-dependence and becoming-together;
- To work towards an increased recognition of both differences and asymmetries, acknowledging that different life forms have different needs, which is an insight that further poses demands for a new emphasis on collective caring in systems of mutual dependence in which heterogeneous members never can become entirely equal but are nonetheless dependent on one another's wellbeing, care and responsibility;
- To work towards a mobilisation for collective action and a constantly ongoing re-examination of the dividing line between what is considered to be individual and shared concerns – what is 'my problem', 'your problem' or 'our problem' – without ever taking for granted where the line is to be drawn between what is included or not within those pronouns. This also involves collectively exploring and establishing political priorities that cut across traditional societal categories and sectoral divides, mobilising the public and experts, and every imaginable, and not yet imaginable, tool a representative and participatory democratic system has to offer.

This altered way of thinking does not necessarily need to first be established on a societal macro level; it could just as well be anchored in micro-practices of, for example, planning, through which new forms of thought and concepts may emerge – thus enabling a re-evaluation of previously held beliefs and understandings of the world. Such new practices could help facilitate new understandings of the places that humans and other existences hold on Earth – and to what extent 'we' are not only dependent on 'them', but how the wellbeing and fate of any 'we' is fundamentally and inextricably tied to the fate and wellbeing of numerous 'thems' such as other species, both near and far (see also Houston et al., 2018). This approach to planning is thus a political perspective that to some extent still puts humankind in the centre of things, or at least in the central focus of care – but in a world inhabited by a myriad of other creatures and beings to which our destiny as a species is intimately linked. In turn, they rely on us for their future existence as much as we rely on them. However, the particularly wicked aspect of this insight in relation to environmental planning and governance is of course that 'we' do not always know from the outset, or ever fully, exactly how these relational webs of co-dependence are woven, and what actions will affect whom in which ways.

Practices of more-than-human environmental planning

So if one takes a more-than-human sensibility to heart, how can it then be translated into concrete methods and practices for environmental planning? Given that most mainstream planning approaches, even to this day, can be seen to be based on quite strict separations between the realms of 'the human' and 'the natural', this might seem like nothing less than a daunting challenge (Metzger, 2018). Yet, there exists some diverse but inspiring attempts at enacting a more-than-human sensibility in practice within environmental planning processes. One such example is the experimental planning methodologies proposed by feminist planner/geographers Julie-Katherine Gibson-Graham for feminist 'experiments with living differently in the Anthropocene' (Gibson-Graham, 2011, 2015). Coming from a different angle and background, and instead focussing on the framing and inclination of scientific input into policy analysis for environmental planning, Jolibert et al. (2011) expand human needs analysis to also include other beings, in this case otters, to study how their needs and prerequisites for wellbeing both diverge and converge so as to be able to devise policy that respects the needs and conditions of wellbeing for both human and nonhuman species.

Much inspiration for practical enactments of a more-than-human sensibility in environmental planning can also be gathered from the ways of relating to the Earth and living environments of indigenous or aboriginal people around the world (see, for example, the contributions in Walker et al., 2013). An interesting example of this is, for instance, the exciting work of the Bawaka Country Collective consisting of researchers, activists and residents of the Bawaka community in North East Arnhem Land in northern Australia (see, for example, Lloyd et al., 2012). However, if the full potential of utilising indigenous ways of relating to the Earth as inspirations for sensitising planners and other planning stakeholders to more-than-human interdependencies is ever to be realised, this would in turn demand that persistent, deeply seated Western philosophical preconceptions about for instance what is 'real' and not, as well as what is 'proper' must be actively challenged and overcome.

A method that specifically pertains to enacting a more-than-human sensibility in planning processes is the so-called 'reconnecting ceremony' or the *Council of All Beings* that was adopted by planning consultant Wendy Sarkissian in 2000 as part of an urban renewal project in Eagleby, a so-called deprived area outside Brisbane in Australia. The set-up of the exercise was the following: twenty participants – Eagleby residents, consultants and client representatives – arrived by boat to an ancient *corroboree*, an Aboriginal meeting and ceremonial site. A series of 'consciousness-raising exercises' were performed to generate a sense of interconnectedness with the history of humanity, the Earth and each other. The then ensuing central exercise involved people going off individually, asking to be chosen by some nonhuman entity so that they could embody it in a Council-type situation. Everyone left the circle for half an hour and waited for something to 'ask to be embodied'. They then proceeded to make masks for themselves that represented the nonhuman life forms they were invited to embody. After this, they sat in a circle and spoke about Eagleby's parks and open spaces while speaking on behalf of the entities they purported to represent.

Represented at the Council were – among other beings and entities – an eagle, the river, an egret and fire. The Council situation itself started out with each participant speaking on behalf of 'their' entity in a rather formal 'policy position' way. Then discussions became more spontaneous. When it was over, the masks were buried so as to symbolise that the participants neither could, nor desired to continue to represent the Beings in any formal sense. They then moved from the more meditative state into an analytic mode and Sarkissian facilitated a discussion about lessons and recommendations based on the previous experience, and asked participants to record their commitments on a large banner.

In a later analysis of the event, Sarkissian explained that the banner and the masks were tangible representations of abstract things that people rarely give voice to. She explained that as a maverick planner, her role in the process becomes that of a 'technician' of 'some other way of hearing', trying 'to find a way to hear the other part of the story', after which a community can decide on more solid grounds 'whether the rights and values of the river should be given more eminence than the rights of fishermen' (Sarkissian, 2005, p. 116). Even though the 'deep ecology' philosophy that underpinned this ceremony to quite some degree diverges from the more-than-human sensibility proposed in this chapter, the *Council of All Beings* from the perspective of a more-than-human sensibility nonetheless appears as a powerful ritual for articulating affect and attachments that otherwise can be difficult to render communicable. It is a device that can help connecting to and bringing in the complex entanglements of the variegated ecologies affecting and affected by human life into planning processes, in ways not afforded by existing approaches to deliberative planning.

The planning methodology experiment in Eagleby shines light on three important components to any putting-into-practice of a more-than-human approach to environmental planning: (i) the practical enactment of a relational sensibility focussing on complex and entangled situated co-becoming, beyond any idea of human exceptionalism; (ii) the opening up of the process for other voices, needs and desires than those readily identifiable as typically human; and (iii) relating an appreciation of more-than-human relations in the world to concrete interventions in the material fabric of places and discussing the future of these places in the light of a broad more-than-human scope of manifest needs and desires so as to engender a local environment that makes room for and is hospitable to a range of different existences.

Of course, in practice, all these aspects are inextricably intertwined with each other. An increased awareness of the relational ecological foundations of human flourishing will inevitably lead to discussions about the demands that such an insight will place on any concrete intervention into local environments. Conversely, any discussion about local development that also recognises other needs and desires than exclusively human will inevitably come to enact some form of more-than-human sensibility in practice.

Finally, again relating back to the Eagleby case, it isn't difficult to see the appeal and power of this type of 'playfully serious' experimental set-up. However, a more thorough going more-than-human planning process perhaps also needs to be complemented by a more sombre side. This would be constituted by a set of methods and practices that would focus on the daunting challenges of relational complexity, the lack of total knowledge and some degree of trembling before the risk of in hindsight quite unacceptable unintended consequences of action, no matter how thoughtfully measured. The enactment of such humility can perhaps be understood as a 'cultivation of torment', in the sense of going through the 'loop' of considering how the interventions that are proposed in any environmental planning process – no matter how well considered – will inevitably generate asymmetric trajectories of flourishing and withering away of different beings and existences (Metzger, 2016). An important grounding for facilitating the enactment of such a sensibility in really-existing planning processes would be to develop institutional designs that force the participants of environmental planning processes to openly articulate and directly face up to the wickedly tough choices that environmental planning interventions by necessity demand – and to further design planning procedures that force participants to directly face up to the difficult choices that must be made so as to engender an assumption of responsibility.

Challenges to a more-than-human approach to environmental planning

Even if one accepts the urgency of applying a more-than-human approach to environmental planning numerous formidable challenges remain before any such ambition can be realised on a broader scale. To begin with, most existing planning methodology today still constitutes a 'tightly woven modernist fabric' (Balducci et al., 2011, p. 489). In face of this, it is certainly an imposing risk that any ambition at putting a more-than-human sensibility into practices from such a vantage point will be dismissed as nothing but 'humbug' and 'hippie wish-wash', particularly in light of the pressure generated by constant 'reforms' to make planning practice more 'efficient' and 'scientific'. (However, to this an advocate for more-than-human planning could retort that we must keep in mind that the type of tools we generally see as guaranteeing rationality and underpinning rational action, were those that brought us into the series of ecological crises we are currently living through in the first place. So, we best try to do differently now.)

Another remaining thorny issue of a more philosophical nature concerns how to tackle the delicacy of exclusions. As Maan Barua's (2014) work on human-elephant coexistence in India so aptly illustrates, multispecies coexistence doesn't always go very smoothly, particularly not in situations of environmental and economic pressure. A more-than-human environmental planning must therefore be prepared to face up to wicked or even tragic choices of priorities. Such choices of action are often quite literally question of life and death, as Tom van Dooren (2011) shows in his discussion of the troubled coexistence of pygmy penguins and fur seals around Sydney's harbour.

Fully accepting the wickedness of relational ecological complexity across scales will inevitably also lead to the insight that a more-than-human approach to environmental planning will most certainly be anything but 'cosy' or 'cute'. To the contrary, it is a rabbit hole that will open up nexuses of ethical and political challenges that will demand a constant exercising of ethical judgement and responsibility in a very different manner to what planners are generally used to. This in turn demands the institutionalisation of waypoints in the planning process at which reflection is mandated on the choices made in the process and the asymmetrical preconditions of flourishing that they engender.

Finally, the development of practical methods for enacting a more-than-human sensibility in planning processes still remains an underdeveloped field. For instance, participatory methods need to be reinvented to also open up for the expression of other needs and desires than the narrowly defined human, while at the same time avoiding the naïve belief that any such expressions can be accessed without complex chains of mediation (see further, for example, Metzger, 2014).

To conclude, the philosophical insights that a more-than-human approach to environmental planning are founded upon are currently generating increasing amounts of curiosity in broader discussions about environmental issues. However, in their practical enactment in formalised environmental planning processes they still remain underdeveloped. Given this, there should still be ample room for informed and thoughtful experimentation to develop new ways of more-than-human environmental planning that hopefully can make some form of contribution to what Anna Tsing has aptly called 'the arts of living on a damaged planet' (see for example, Tsing et al., 2017).

Note

1 Please notice that, throughout the chapter, the relations between humans and vaguely defined 'beings and existences' will be discussed. The term of choice is thus purposely not 'nonhumans', specifically not 'animals'. The motivation behind this choice is that even though the more-than-human in geography to a large extent initially discussed human-animal relations (particularly relations between humans and

other mammals), the resultant sensibility actually challenges any such categorisation and also opens up for a relational understanding that goes way beyond this quite narrow focus of attention. Thus, in a thoroughly more-than-human perspective, the relational conditions of existence also of other entities that, for example, exist through and sometimes with the help of humans, but which also exceed human individuality and intentionality – such as cultures, identities, organisations and apprehensions of deities and spirits – are also taken into account. The more-than-human approach that is proposed in this chapter can thus be understood to propagate a planning for entangled bundles of more-than-human *assemblages*. There is no room to develop this insight further in the context of this short chapter but see, for example, De la Cadena (2015) and Müller and Schurr (2016).

References

Balducci, A., Boelens, L., Hillier, J., Nyseth, T. and Wilkinson, C. (2011). 'Introduction: strategic spatial planning in uncertainty: theory and exploratory practice'. *Town Planning Review*, 82(5): 481–501.

Barua, M. (2014). 'Volatile ecologies: towards a material politics of human-animal relations'. *Environment and Planning A*, 46(6): 1462–78.

De la Cadena, M. (2015). *Earth beings: ecologies of practice across Andean worlds*. Durham: Duke University Press.

Castree, N. (2003). 'Environmental issues: relational ontologies and hybrid politics'. *Progress in Human Geography*, 27(2): 203–11.

Gibson-Graham, J. K. (2011). 'A feminist project of belonging for the Anthropocene'. *Gender, Place and Culture – A Journal of Feminist Geography*, 18(1): 1–21.

Gibson-Graham, J. K. (2015). 'Ethical economic and ecological engagements in real(ity) time: experiments with living differently in the Anthropocene'. *Conjunctions. Transdisciplinary Journal of Cultural Participation*, 2(1): 44–71.

Hache, É. and Latour, B. (2010). 'Morality or moralism? An exercise in sensitization'. *Common Knowledge*, 16(2): 311–30.

Haraway, D. (2008). *When species meet*. Minneapolis: University of Minnesota Press.

Haraway, D. J. (2016). *Staying with the trouble: making kin in the Chthulucene*. Durham: Duke University Press.

Houston, D., Hillier, J., MacCallum, D., Steele, W. and Byrne, J. (2018). 'Make kin, not cities! Multispecies entanglements and "becoming-world" in planning theory'. *Planning Theory*, 17(2): 190–212.

Jolibert, C., Max-Neef, M., Rauschmayer, F. and Paavola, J. (2011). 'Should we care about the needs of non-humans? Needs assessment: a tool for environmental conflict resolution and sustainable organization of living beings'. *Environmental Policy and Governance*, 21(4): 259–69.

Latour, B. (2004). *Politics of nature: how to bring the sciences into democracy*. Cambridge, MA: Harvard University Press.

Latour, B. (2017). *Facing Gaia: eight lectures on the new climatic regime*. Cambridge: Polity Press.

Lloyd, K., Wright, S., Suchet-Pearson, S., Burarrwanga, L. and Country, B. (2012). 'Reframing development through collaboration: towards a relational ontology of connection in Bawaka, North East Arnhem Land'. *Third World Quarterly*, 33(6): 1075–94.

Marres, N. (2012). *Material participation: technology, the environment and everyday publics*. Basingstoke: Palgrave Macmillan.

Metzger, J. (2012). 'We are not alone in the universe'. *Eurozine*, 8 February. Available at: www.eurozine.com/articles/2012-02-08-metzger-en.html

Metzger, J. (2014). 'The moose are protesting: the more-than-human politics of transport infrastructure development', in J. Metzger, P. Allmendinger and S. Oosterlynck (eds.) *Planning against the political: democratic deficits in European territorial governance*. New York: Routledge, pp 191–214.

Metzger, J. (2016). 'Cultivating torment: the cosmopolitics of more-than-human urban planning'. *City*, 20(4): 581–601.

Metzger, J. (2018). 'Can the craft of planning be ecologized? (and why the answer to that question doesn't include "ecosystem services")', in M. Kurath, M. Marskamp, J. Paulos and J. Ruegg (eds.) *Relational planning: tracing artefacts, agency and practices*. London: Palgrave Macmillan, pp 99–120.

Müller, M. and Schurr, C. (2016). 'Assemblage thinking and actor-network theory: conjunctions, disjunctions, cross-fertilisations'. *Transactions of the Institute of British Geographers*, 41(3): 217–29.

Pickering, A. (2007). *Producing another world*. Paper presented at the "Assembling Culture" Workshop, 10–11 December 2007, University of Melbourne, Australia.

Puig de la Bellacasa, M. (2010). 'Ethical doings in naturecultures'. *Ethics, Place and Environment*, 13(2): 151–69.

Sarkissian, W. (2005). 'Stories in a park: giving voice to the voiceless in Eagleby, Australia'. *Planning Theory and Practice*, 6(1): 103–17.

Serres, M. (2006). *Revisiting the natural contract*. Talk given at the Institute for the Humanities at Simon Fraser University on 4 May 2006. Available at: www.sfu.ca/humanities-institute-old/pdf/Naturalcontract.pdf

Stengers, I. (2005). 'The cosmopolitical proposal', in B. Latour and P. Weibel (eds.) *Making things public: atmospheres of democracy*. Karlsruhe: Engelhardt and Bauer, pp 994–1003.

Tsing, A. (2012). 'Unruly edges: mushrooms as companion species for Donna Haraway'. *Environmental Humanities*, 1(1): 141–54.

Tsing, A. L., Gan, E. and Bubandt, N. (eds.) (2017). *Arts of living on a damaged planet*. Minneapolis: University of Minnesota Press.

Van Dooren, T. (2011). 'Invasive species in penguin worlds: an ethical taxonomy of killing for conservation'. *Conservation and Society*, 9(4): 286–98.

Walker, R., Natcher, D. and Jojola, T. (eds.) (2013). *Reclaiming indigenous planning*. Native and Northern Series No. 70. Montreal: McGill Queen University Press.

Whatmore, S. (2002). *Hybrid geographies: natures, cultures, spaces*. London: Sage.

18

Cities leading

The pivotal role of local governance and planning for sustainable development

Patricia McCarney

Global agendas on sustainable development and the ever-widening global environmental discourse, invariably present an expanding set of challenges for city leaders on the ground. This is because cities are increasingly pivotal in addressing global climate change challenges and must lead in building sustainable cities of the future where the majority of the world's populations reside.

As voiced a few decades ago and as part of the Local Agenda 21 platform, calls for local action on the environment are not new. However, the agenda has become even more challenging today, the local stakeholders are broader and spread across multiple sectors and the interconnectivity of local issues on the ground in cities are increasingly more complex. Cities and their households and businesses are increasingly vulnerable to climate related risk, environmental degradation, natural disasters and externally induced shocks and stresses. City leaders are therefore being challenged to design more resilient cities and improve emergency preparedness strategies to address risk. City leaders are also being tasked with a wider and deeper set of challenges – ranging from planning and zoning for more sustainable city building, planning for low carbon and more efficient mobility, combatting deteriorating air quality, creating and investing in healthier green environments, more sustainable and resilient infrastructure and creating more inclusive cities for currently escalating (and shifting) immigration trends.

There are three key themes presented in this chapter that point to the need for more robust policy frameworks if cities are to effectively lead in this context of global sustainable development. First is the challenge of governance and the complicated nature of sustainability planning at the local scale. Second is the challenge of localising (and making operational) what is often configured as a more remote global agenda for sustainable development. Third is the local challenge of effective management and planning where evidence-based policy development and investment decision making are increasingly operating in the absence of high-calibre city-level data.

The challenge of governance and the complicated nature of sustainability planning at the local scale

Considering scale for environmental governance in cities

Extensive research has been undertaken over the past decade on scaling and re-scaling environmental governance. This research has tended to focus on the centralisation and decentralisation of decision making (Larson and Soto, 2008; Cohen and McCarthy, 2015; Reed and Bruyneel, 2010; Young, 2002) and also on a multi-level governance framework (Hooghe and Marks, 2003; Marks and Hooghe, 2004) for environmental planning and action. The decentralisation of decision making which follows the core principle of subsidiarity often involves a transfer of powers to lower tiers of government (for example from national to provincial or state levels or more locally to regional and municipal governments). Shared responsibilities both for funding and control over policy are also seen as shared across all levels of government in a multi-level governance approach, and decision making is often fluid and can become duplicative and contested.

Responsibilities for environmental issues are spread vertically over multiple hierarchical levels of governance from global to national to state/provincial to local. Responsibilities are also fragmented horizontally at each of these levels: at the global level across UN agencies, development banks, special purpose bodies; at the national and provincial/state levels across ministries and departments; and at the municipal level across departments of city halls, commissions and local utilities and service providers. The horizontal relationships are also further characterised by public-private partnerships.

In considering the nature of sustainability planning in cities, therefore, the complex nature of scale and multi-level governance and the complicated vertical and horizontal relations and competencies must be taken into account. In this context, local governments are confronting new development challenges which necessitate the adoption of amended roles, immersion into new task environments and engagement with new sets of actors. Local government mayors and councillors, senior officials, bureaucrats and local planners are increasingly being required to mediate global-local dynamics; to address pent-up demand for service delivery made persistent by enduring urbanisation pressures; to balance local economic development policy aimed at both urban competitiveness and poverty alleviation; to take on poverty as a result of senior levels of government difficulties to operationalise strategic interventions; to engage in more open, transparent and mutually respectful state-society relations; and to forge new and reformed intergovernmental relationships. These significant roles are being added to the existing and extensive functions performed by local governments, and there are increasingly higher expectations being placed on local government officials to perform efficiently and creatively in this evolving task environment.

In poorer countries of the Global South, these challenges are made more acute by the triple pressures of rapid urbanisation, poverty, and weak local government capacity. Local government capacity to function in an effective and responsive manner in addressing these challenges is limited in many parts of the developing world since local government is often the 'neglected third tier' in governance considerations (McCarney, 1996a; McCarney, 1996b). Weak capacity in local governments results from their limited resources and power in an intergovernmental setting characteristic of highly centralised states. Even in countries like South Africa, where strong local governments have been constitutionally mandated to operate as an effective third tier of government, and in other countries where decentralisation processes have been undertaken (or are in process), and where serious attempts have been (and are being) made at strengthening local governments more generally, capacity issues are often still prevalent and local officials are at some disadvantage to respond to the pressures increasingly demanded of them. These weaknesses in local government pose serious limitations on the ability of local officials to respond to the challenges of sustainable development and environmental planning.

Considering jurisdiction and municipal boundaries in environmental planning

Cities across the world – whether Buenos Aires, Johannesburg, Amsterdam, Taipei, Toronto or Los Angeles – are facing challenges of governance resulting from growth and spatial expansion where urban populations spread out beyond their old city limits, rendering the traditional municipal boundaries, and by extension, the traditional governing structures and institutions, outdated. This expansion is not just in terms of population settlement and spatial sprawl but perhaps, more importantly, in terms of their social and economic spheres of influence. The functional area of cities has extended beyond their jurisdictional boundaries. Cities have extensive labour markets, real estate markets, financial and business markets and service markets that spread over the jurisdictional territories of several municipalities and, in some cases, over more than one state or provincial boundary. In a number of cases cities have spread across international boundaries. Increasingly, these functions demand more integrated planning, service delivery and policy decisions than these multiple but individually bounded cities can provide. Governing has therefore become much more complex since a decision taken in one municipality that is part of multi-jurisdictional territory or metropolitan area affects the whole city. This phenomenon introduces new challenges of governance and, in particular, metropolitan governance.

Empirical evidence (McGee and Robinson, 1995; Environment and Urbanization, 2000; Myers and Dietz, 2002; National Research Council, 2003; Rojas et al., 2005; Laquian, 2005; Public Administration and Development, 2005) shows that urban areas around the world continue relentlessly to expand both in terms of density and horizontal space (Angel et al., 2005). Cities, even those of intermediate size, continue to grow and many of them spread over different administrative units. There is a need to govern these large areas in a coherent fashion. The challenges associated with building effective metropolitan governance arrangements are increasingly complex and also integrated (McCarney and Stren, 2008).

Cooperation among cities, working together instead of in competition within the same metropolitan territory, supports a more integrated approach to environmental planning. Planners in a city addressing under-serviced transport systems, inadequate infrastructure, air quality, water conservation, carbon neutrality and density are of necessity looking beyond municipal boundaries and jurisdictions. The challenges of sustainable development are inherently without borders. For example, in Canada, the Provincial Government of Ontario has created an entity – Metrolinx – to improve the coordination and integration of all modes of transportation in the Greater Toronto and Hamilton Area, thereby spanning multiple independent municipal and regional jurisdictions. Metrolinx, for example, has created a regional transportation plan that bridges multiple jurisdictions for more coordinated planning.[1]

The challenge of localising what has been long configured as a global agenda for sustainable development

Global agendas on sustainable development have been driven by the United Nations and member states for several decades and reach as far back as 1987 with the release of 'Our Common Future' (UNWCED, 1987) detailing what was referred to then as 'A global agenda for change'. Formulated by the World Commission on Environment and Development (WCED), this report answered a call by the General Assembly of the United Nations to propose long-term environmental strategies for achieving sustainable development by the Year 2000 and beyond. This global agenda was to consider ways and means by which the international community could more effectively address environment concerns.

The next major global agenda that followed 'Our Common Future' was titled 'The Millennium Development Goals' (MDGs), established in the year 2000, and included eight goals that were clear and concise, with measurable, quantitative targets to the year 2015. The focus of the MDGs was improvement of the lives of the world's poorest people. Leaders of 189 countries signed this historic millennium declaration at the United Nations Millennium Summit in 2000.

Like any international agreement, there were both successes and shortcomings. The successes included widespread public awareness, dedication and mobilisation across the globe to combat poverty and its attendant challenges of hunger, health, education and inequality. The shortcomings of the MDGs – two in particular of note here for this report – include a core weakness of accurate, standardised measurement and time-sensitive data to support monitoring on progress towards the 2015 targets. Moreover, the MDGs were silent on the role of cities in mobilising support and action to realise the success of the eight goals, despite the fact that within the MDG time period (in 2007), the world was to pass the historic global population threshold, wherein half of the world's population was to become urban.

Following the MDGs (2000–15), the 'Sustainable Development Goals' (SDGs) were established in 2015 with a view to creating a more sustainable world by 2030. While agreeing that combatting poverty should continue beyond the 2015 target date and retirement of the MDGs, there was a growing consensus that the world also needed to move forward onto a more sustainable trajectory in recognition of our changing climate and other serious environmental challenges. The SDGs were established for the post-2015 'Development Agenda – 2015–30' and include 17 goals that encompass worldwide environmental objectives alongside, and very much connected to, the poverty reduction goals of the MDGs.

The SDGs have been developed to bridge and carry forward the poverty agenda established by the MDGs pre-2015 and build the sustainable development agenda post-2015. This new era reflects a commitment to overcome the two challenges arising with respect to the MDGs, namely, to recognise the critical role of cities in building a more sustainable planet and to confront the need for standardised data and methods of monitoring local progress towards each of the 17 SDGs, both deemed essential to drive successful achievement of the SDGs by 2030.

While cities were not represented specifically in the MDGs, one of the 17 goals for the new 2030 Agenda is dedicated to addressing cities – Goal 11: Make cities and human settlements inclusive, safe, resilient and sustainable. The representation of cities as a specific goal and focus in the post-2015 development agenda is recognition of the pivotal role of cities as a transformative influence in building sustainability.

However, the multi-faceted roles of city leaders and local administrations in achievement of the post-2015 agenda go well beyond Goal 11 and can be effectively considered across all of the 17 SDGs.

Localising the global sustainable development goals

City mayors, city managers, city planners and designers, city engineers and citizens in their daily work and routines are all directly or indirectly related to each of the 17 SDGs. City leaders are driving policy change for more sustainable futures and are recognising their crucial role in global SDG achievement.

Considerations on each of the 17 SDGs from a local perspective is essential since achieving the UN SDGs will be a battle largely waged in cities where, in most countries, a large and growing percentage of the population resides. What follows are a few examples, drawn from a selection of the 17 goals, that highlight the critical importance of localising each of the 17 SDGs, since the powers and responsibilities held by local governments are central to operationalising each goal.

For example, when considering Goal 3 – Good Health and Well-Being – 'To ensure healthy lives and promote well-being for all at all ages', cities are leading actors in this core aspect of sustainable development. Local governments are responsible for city-wide infrastructure, green space, public recreation facilities, slum improvement, affordable housing and ensuring access by the poor to basic services, all of which are key components of good health and wellbeing of urban residents. City planners and transport planners are responsible for ensuring its citizens have safe and reliable roads and transit that reduce road and traffic accidents, one of the targets of Goal 3. It is also local governments that plan and design for alternative energy and transport services that can help to reduce air pollution; by planning walkable cities and safe bike paths, they are also positioned to help foster healthy lifestyles, a core target of Goal 3. Increasingly local governments are also taking leadership on health issues such as substance abuse, epidemics and HIV/AIDs, providing education, outreach, emergency response, and information services across their cities.

Another example of the central role cities play in implementing the global SDGs is Goal 6 – Clean Water and Sanitation – 'Ensure availability and sustainable management of water and sanitation for all'. With many countries experiencing upwards of 80% of their population living in cities, the need to localise SDG 6 is essential to guide action on the ground. In many poorer cities, a large portion of residents are often not served with clean water, solid waste collection and sanitation services. It is local governments, the level of government closest to the people that are responsible for the provision of water and sanitation services. Public works officials in city governments are also responsible for reducing pollution in water, managing wastewater treatment, eliminating dumping and improving water use efficiency, all targets of Goal 6. Local leadership across and between cities and transboundary cooperation is also a rising policy agenda driving integrated water management across regions, a key target in Goal 6.

Goal 7 – Affordable and Clean Energy – 'Ensure access to affordable, reliable, sustainable and modern energy for all' is another global goal that, to be successful by 2030, requires cities to lead on. Local governments have control over land and buildings through planning and zoning codes that impact directly on investment in energy efficient buildings and compact design. Local governments also have the power and responsibility to inject sustainability criteria into their city procurement policies. In addition, local governments own and operate or simply regulate many public buildings in their jurisdictions, including: hospitals; schools; recreation facilities such as arenas, gymnasiums and sports facilities; government offices (city halls); and universities. These powers can be used to impact energy consumption; ensure adoption of renewable energy use in these buildings; and drive the formulation of energy efficiency guidelines across the city for buildings, street lighting, and all related city services. Cities also regulate automobile and transport services which can positively impact energy efficiency, ensure cleaner fossil fuel technologies adoption and lower carbon emissions. Local governments are therefore pivotal in 'moving the needle' on clean energy in cities globally.

Goal 11 – Sustainable Cities and Communities – 'Make cities and human settlements inclusive, safe, resilient and sustainable' is the so-called 'urban goal'. SDG 11 is already highly localised, being the stand-alone goal of the SDGs devoted to cities. The target, to by 2030 'ensure access for all to adequate, safe and affordable housing and basic services and upgrade slums', is a direct responsibility of local governments. Planners, urban policy experts and infrastructure officials work together in cities to implement this target on a daily basis. Access to affordable shelter is a challenge of all cities, regardless of income and cities are on the frontline to regulate land and housing and construction markets. Slums and homelessness pose two of the key challenges for city leaders throughout the world that are encompassed in Goal 11. Indeed, all of the targets in Goal 11 – from public transport, reduction of emissions, green space, public space, participatory planning, urban sprawl, solid waste management and recycling, resilience to risk and disasters, and urban rural cooperation – are core responsibilities of local governments. It is local governments that are driving a more dedicated inter-governmental arrangement for investment funding and legislative reform needed to support Goal 11 and all of the other SDG targets.

A final example of the need to localise the SDGs and recognise the essential role of cities in reaching this goal by 2030 is also demonstrated in Goal 13 – Climate Action – 'Take urgent action to combat climate change and its impacts'. Cities are at the centre of the need for climate action. It is estimated that cities are responsible for around 70% of global energy consumption and energy-related greenhouse gas emissions. The concentration of people, resources and infrastructure in cities is making societies more vulnerable to the adverse impacts of climate change, such as sea level rise and storm events. Cities are likely to bear a major share of the burden of the costs and risks associated with climate change adaptation, as well as the responsibility of establishing more resilient infrastructure. However, cities also represent a key opportunity for driving progress toward low carbon development and societal resilience, further enabled by the leadership being taken by mayors globally on the climate agenda. Cities can pursue a wide range of strategies to address climate resilience and low carbon development, including integrated measures across areas including planning, transportation, energy, and buildings.

In fact, every one of the 17 SDGs can be mapped to an informed understanding of the essential and pivotal role of cities. Localising the SDGs has become an important discussion globally and the success of the SDGs is dependent upon the role of city mayors, city managers, city planners and designers, city engineers and citizens in their daily work and routines. City leaders and environmental planners are driving policy change for more sustainable futures and are recognising their crucial role in global SDG achievement.

The challenge of effective city management and planning for sustainable development in the absence of high-calibre city-level data

In addressing global challenges and opportunities for sustainability, the need for globally comparable city data has never been greater. Cities need indicators to measure their performance in delivering services and improving sustainable development.

Globally standardised data: an essential tool for environmental planning and governance

The ability to compare data across cities globally, using a globally standardised set of indicators, is essential for comparative learning and progress in city development. City metrics guide more effective city governance.

City leaders worldwide want to know how their cities are doing relative to their peers. Globally standardised indicators allow city leaders to measure their performance and compare it with other cities. For instance, comparable city level data can help build collaboration and understanding by fostering information exchange and sharing of tested practices across cities. Comparative analysis and knowledge sharing is vital in the face of rapid urbanisation and the associated demand for larger scales of infrastructure investment and city services as well as the emergent global challenges including climate change and the associated demand for sustainability planning, resilience and emergency preparedness.

Over the past decade, policy responses to the most pressing challenges and opportunities for sustainable development have been hindered by a set of core weaknesses in current research and information at the city level. City indicators collected at varying levels of government within nations, by researchers and international agencies and by different cities globally, did not conform to a standardised set of definitions and methodologies that allowed for sound global comparison across cities, making globally comparative research and exchange impossible.

Table 18.1 Schematic themes for 100 key performance indicators in ISO 37120

Economy	Safety
Education	Shelter
Energy	Solid Waste
Environment	Telecommunication and Innovation
Finance	Transportation
Fire and Emergency Response	Urban Planning
Governance	Wastewater
Health	Water and Sanitation
Recreation	

The first international standard on city indicators was published 15 May 2014 by the International Organization for Standardization (ISO) – titled 'ISO 37120 Sustainable Development of Communities – Indicators for City Services and Quality of Life' (ISO, 2014). These indicators are being leveraged by city and business leaders, researchers, planners, designers and other professionals to build sustainable, inclusive and prosperous cities worldwide. This international standard on city metrics ensures scholars, decision makers and citizens will have access to more accurate and reliable data on cities. This new standard, ISO 37120, not only contributes to the field of statistics by standardising the way data is being reported and verified but also standardising the way data is being managed and analysed.

This new international standard published by the International Organization for Standardization (ISO) includes a comprehensive set of 100 indicators – of which 46 are required for conformity – that measures a city's social, economic and environmental performance. ISO 37120 is now part of a new series of International Standards being developed for a holistic and integrated approach to sustainable development and resilience. The 100 indicators with definitions and methodologies published in ISO 37120 are divided into 17 themes representing key performance management fields in city services and quality of life (Table 18.1).

The ISO 37120 Standard and a web-based reporting mechanism for cities globally to implement ISO 37120, together with a city certification protocol that is now built,[2] offers a tool for cities and a strategic foundation for measurement and benchmarking progress in the post-2015 development agenda. The World Council on City Data (WCCD) and the set of ISO 37120 city indicators, together with the city indicators on resilience about to be published (2019) as a new ISO Standard (ISO 37123), assists in the measurement of progress by cities in reaching the SDG targets.

Data for cities reported in conformity with ISO 37120 is now leading to responsible city building driven by evidence-based planning. Accurate reporting of key performance indicators improves accountability and fosters sound decision making by city leaders, especially in terms of urban planning and design for more sustainable futures.

Conclusion

A global transformation has occurred that positions cities at the core of the post-2015 development agenda. Urbanisation is one of the most significant trends of the past and present century, providing the foundation and momentum for global change. The shift towards an increasingly urbanised world constitutes a transformative force which can be harnessed for a more sustainable trajectory, with cities taking the lead to address many of the global challenges of the 21st century, including sustainable development, poverty, inequality, migration, unemployment, environmental degradation, and climate change. City leaders have become a positive and potent force for addressing sustainable development and economic growth and for driving innovation, consumption and investment in both developed and developing countries.

Economic growth and development have usually been closely associated with the performance of nations. In recent decades, cities have emerged on the global stage as economic powerhouses, engaging in world markets to create more jobs, to attract global talent and investment and to spur long-term, sustainable economic growth.

This chapter has traced three key thematic concentrations: the challenge of governance and the complicated nature of sustainability planning at the local scale; the challenge of localising what is often configured as a global agenda for sustainable development; and the local challenge of effective management and planning where evidence-based policy development and investment decision making are increasingly operating in the absence of high-calibre city-level data. Each of these three challenges points to the need for more robust policy frameworks if cities are to effectively lead in this context of global sustainable development.

Notes

1 Refer to the 2041 Regional Transportation Plan, available at: www.metrolinx.com/en/regionalplanning/rtp/.
2 See dataforcities.org.

References

Angel, S., Sheppard, S. C. and Civco, D. L. (2005). *The dynamics of global urban expansion.* Transport and Urban Development Department. Washington, DC: The World Bank.

Cohen, A. and McCarthy, J. (2015). 'Reviewing rescaling: strengthening the case for environmental considerations'. *Progress in Human Geography*, 39(1): 3–25.

Environment and Urbanization (2000). Special issue on 'Poverty reduction and urban governance'. *Environment and Urbanization*, 12(1): 3–250.

Hooghe, L. and Marks, G. (2003). 'Unraveling the central state, but how? Types of multi-level governance'. *American Political Science Review*, 97(2): 233–43.

ISO (2014). *ISO 37120:2014: Sustainable development of communities – indicators for city services and quality of life*. Geneva: ISO.

Laquian, A. A. (2005). *Beyond metropolis: the planning and governance of Asia's mega-urban regions.* Washington, DC: Woodrow Wilson Center Press/Baltimore, MD: The Johns Hopkins University Press.

Larson, A. M. and Soto, F. (2008). 'Decentralization of natural resource governance regimes'. *Annual Review of Environment and Resources*, 33(1): 213–39.

Marks, G. and Hooghe, L. (2004). 'Contrasting visions of multi-level governance', in I. Bache and M. Flinders (eds.) *Multi-level governance.* Oxford: Oxford University Press, pp 15–30.

McCarney, P. (ed.) (1996a). *Cities and governance: new directions in Latin America, Asia and Africa.* Toronto: Centre for Urban and Community Studies, University of Toronto.

McCarney, P. (ed.) (1996b). *The changing nature of local government in developing countries.* Toronto: Centre for Urban and Community Studies, University of Toronto.

McCarney, P. L. and Stren, R. E. (2008). 'Metropolitan governance: governing in a city of cities', in *State of the world's cities 2008/2009.* UN-HABITAT. London and Sterling, VA: Earthscan, pp 226–37.

McGee, T. G. and Robinson, I. M. (eds.) (1995). *The mega-urban regions of Southeast Asia.* Vancouver: University of British Columbia Press.

Myers, D. J. and Dietz, H. A. (eds.) (2002). *Capital city politics in Latin America: democratization and empowerment.* Boulder, CO: Lynne Rienner.

National Research Council (2003). *Cities transformed. Demographic change and its implications in the Developing World.* London: Earthscan for the National Academy of Sciences.

Public Administration and Development (2005). Special issue on 'Metropolitan governance reform'. *Public Administration and Development*, 25(4): 275–364.

Reed, M. G. and Bruyneel, S. (2010). 'Rescaling environmental governance, rethinking the state: a three-dimensional review'. *Progress in Human Geography*, 34(5): 646–53.

Rojas, E. J., Cuadrado-Roura, R. and Guell, J. M. F. (eds.) (2005). *Governar las metrópolis*. Washington, DC: IADB.

UNWCED (United Nations World Commission on Environment and Development) (1987). *Our common future*. New York: United Nations.

Young, O. (2002). *The institutional dimensions of environmental change: fit, interplay and scale*. Cambridge, MA: MIT Press.

19

Anthropocene communications

Cultural politics and media representations of climate change[1]

Marisa B. McNatt, Michael K. Goodman and Maxwell T. Boykoff

Introduction

Over the past years, the number of *Reuters* stories about climate change has declined. This trend has been consistent with trends across other media outlets globally (see Figure 19.1) due largely to political economic trends of shrinking newsrooms and fewer specialist reporters covering climate stories with the same frequency as before. In 2010, the *Wall Street Journal* and the *Christian Science Monitor* closed their environmental blogs. Three years later, in January 2013, the *New York Times* dismantled its environment desk, assigning the reporters and editors to other departments, and discontinued its 'Green blog' two months later. Yet, initially, *Reuters* had largely bucked those trends, continuing to employ top climate and environment reporters from around the globe, including Deborah Zabarenko (North America), Alister Doyle (Europe) and David Fogerty (Asia) who fed top media organisations with reporting comprised of a steady diet of climate and environment stories. So why this subsequent and precipitous drop in *Reuters* coverage of climate change? In July 2013, David Fogerty – who left *Reuters* in late 2012 – took to *The Baron* blog to explain why. He recounted that, after the appointment of editor Paul Ingrassia in 2011, editorial decisions were made to deprioritise climate stories and to shift these specialists to different beats. Fogerty, for example, was moved from the climate beat to instead cover issues around shipping in the Asian region. While climate stories had been already declining upon the appointment of Ingrassia, many argued that his revamping of the *Reuters* reporting priorities served to accelerate this drop.

Crucially, Fogerty and others asserted that Ingrassia's ideological and political leanings also played a detrimental part in continued coverage (Fogerty, 2013). As such, in the summer of 2013, *Reuters* climate reporting faced deep levels of criticism and dismay from journalism colleagues and media critics and consumers. For example, journalist Alex Sobel Fitts at *Columbia Journalism Review* attributed variations in content and quantity to this new editorial turn (2013). And Max Greenberg at *Media Matters* framed this drop in coverage as a consequence of a 'climate of fear' imposed by new contrarian editorial influences (2013).

Thus, while the specific situation with *Reuters* provides a worrisome glimpse into the contentious and high-stakes arena of global reporting on climate change in the 21st century, what

2004-2019 World Newspaper Coverage of Climate Change or Global Warming

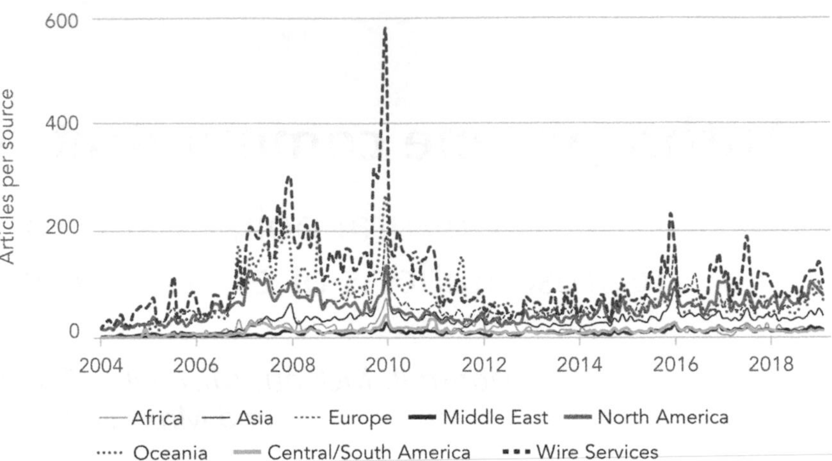

CIRES Center for Science and Technology Policy Research, University of Colorado Boulder, http://sciencepolicy.colorado.edu/media_coverage

Figure 19.1 World newspaper coverage of climate change 2004–19

Source: Authors' own

it shows more generally is the way that environmental communication in the context of climate politics is thoroughly enmeshed in a combination of large-scale social, political and economic factors connected up with smaller-scale power-laden editorial decision making, steeped in cultural economy and ideology.

Simultaneously, digital and social media are stepping into these spaces. This chapter explores key questions that arise regarding how the general decline in specialised reporting on climate change and rise in social media impacts overall quantity and quality of coverage as well as inputs into public awareness and engagement.

Most citizens around the world typically do not read peer reviewed literature. Instead, to learn about climate change, people in the public arena turn to media communications – television, newspapers, radio, new and social media – to link formal science and policy with their everyday lives. Over the past decades, the dynamics of science and politics have clearly shaped media coverage of climate change. Yet, it is also worth noting and considering how 'news' – generated by mass media – has, in turn, shaped ongoing scientific and political considerations, deliberations and decisions. In other words, it is instructive to account for how mass media have influenced who has a say, when and how in the public arena.

'The media' around the world are actually much more heterogeneous and varied than at first glance. In their multiple dimensions, media are constituted by many institutions, processes and practices that together serve as 'mediating' forces between communities such as science, policy and civil society. Media segments, articles, clips and pieces represent critical links between people's everyday realities and experiences and the ways in which these are discussed at a distance between science, policy and public actors. People throughout society rely upon media representations to help interpret and make sense of the many complexities relating to climate science and governance. Thus, media messages are critical inputs to what become public discourse on current climate challenges.

Yet, these media representations enter into an individual's pre-existing perceptions and perspectives and are taken up or resisted in varied ways. For example, Wouter Poortinga and colleagues (2011) have found that people's enduring values and existing ideologies strongly influence their understandings and behaviours as they relate to climate change. Indeed, as Lorraine Whitmarsh put it in summarising her research on climate contrarianism 'attitudes to climate change are relatively entrenched and . . . information about the issue will be evaluated and used in diverse ways according to individuals' values and worldviews'. She concluded:

> [S]imply providing climate change information is unlikely to be successful, as new information is often interpreted by people in line with their existing attitudes and worldviews . . . In other words, irrespective of how much information is provided, it is remarkably difficult to change attitudes that have become entrenched.
>
> *(Whitmarsh, 2011, p. 698)*

These dynamic science-policy-media-public interactions have been spaces where claims makers in the media have been changing (for example, Baum and Groeling, 2008; Fahy and Nisbet, 2011), and traditional media outlets have faced newfound challenges (Boykoff and Yulsman, 2013; Siles and Boczkowski, 2012) while shifts to new/digital/social media tools have recalibrated who has a say and how these claims circulate (Baek et al., 2012; Cacciatore et al., 2012). Traditional and legacy media organisations themselves have worked to adapt to these changing conditions and researchers have increasingly sought to make sense of the shifts (for example, Horan, 2013; Nielsen, 2012) and their implications (for example, Jacobson, 2012) in various cultural, political, social and environmental contexts (for example, Adams and Gynnild, 2013; Schuurman, 2013).

In recent decades, there has been significant expansion from traditional mass media into consumption of social and digital media. Essentially, in tandem with technological advances, this expansion in communications is seen to be a fundamental shift from broadcast, or 'one-to-many' (often one-way) communications to 'many-to-many' more interactive webs of communications (O'Neill and Boykoff, 2010; van Dijk, 2006). This movement has signalled substantive changes in how people access and interact with information and who has access.

Together, traditional/legacy and digital/social media spaces comprise a key part of what many now refer to as the 'cultural politics of climate change': dynamic and contested spaces where various 'actors' battle to shape public understanding and engagement (for example, Boykoff and Goodman, 2009). These are places where formal climate science, policy and politics operate at multiple scales through multiple media forms and are dynamic as well as contested processes that shape how meaning is constructed and negotiated. In these spaces of the 'everyday', cultural politics involve not only the discourses that gain traction in wider discourses but also those that are absent (Derrida, 1978). Contemplating climate considerations in this way helps to examine 'how social and political framings are woven into both the formulation of scientific explanations of environmental problems, and the solutions proposed to reduce them' (Forsyth, 2003, p. 1).

Media attention in the public sphere

Figure 19.1 appraises the trends in media coverage of climate change from 2004 into 2019 in 50 newspapers across 20 countries around the globe.[2] This visual representation has provided an opportunity to assess and analyse further questions of *how* and *why* there were apparent ebbs and flows in coverage. For instance, notably 2009 ended with soaring media coverage of climate change around the world and numerous studies have sought to better understand events and developments during this time period (see Boykoff, 2013 for example). At this time, climate news seemingly flooded the public arena and was dominated by the much-hyped and highly anticipated United Nations climate talks in Copenhagen, Denmark (COP15), along with news about the hacked emails of scientists from the University of East Anglia Climate Research Unit (referred to by some as 'Climategate'). These events also linked to ongoing stories of energy security, sustainability, carbon markets and green economies that were unfolding during this time.

Across this nearly 15-year look, increases in each of the regions have not been symmetrical. For example, there were a relatively low number of stories on climate change or global warming in the regions of South America and Africa throughout this period. This points to a critical regional 'information gap' in reporting on these issues and relates to capacity issues and support for reporters in these regions and countries (developing and poorer regions/countries).

Tracking media treatment of climate change and global warming through intersecting *political, scientific* and *ecological/meteorological* climate themes provides a useful framework for analyses of content and context. Such accounting helps to demonstrate how news pieces should not be treated in isolation from one another; rather, they should be considered connected parts of larger political, economic, social, environmental and cultural conditions.

Moreover, patterns revealed in the mobilisations of journalistic norms internal to the news-generation process cohere with externally influenced dominant market-based and utilitarian approaches that consider the spectrum of possible mitigation and adaptation action on climate change. Robert Brulle has argued that an excessive mass media focus merely on the debaters and their claims, 'works against the large-scale public engagement necessary to enact the far-reaching changes needed to meaningfully address global warming' (2010, p. 94). As such, examinations of the content of media treatment of climate change need to be considered within a context of larger political and social forces.

The cultural politics of climate change reside in many spaces and places – from workplaces to pubs and kitchen tables. 'Actors' on this stage range from fellow citizens to climate scientists as well as business industry interests and environmental non-governmental organisation (ENGO) activists. Over time, individuals, collectives, organisations, coalitions and interest groups have sought to access the power of mass media to influence architectures and processes of climate science, governance and public understanding through various media 'frames' and 'claims'. Questions regarding 'who speaks for the climate' involve considerations of how various perspectives – from climate scientists to business industry interest and ENGO activists – influence public discussions on climate change (Boykoff, 2011). 'Actors', 'agents', or 'operatives' in this theatre are ultimately all members of a collective public citizenry. However, differential access to media outlets across the globe are products of differences in power, and power saturates social, political, economic and institutional conditions undergirding mass media content production (Wynne, 2008).

In the highly contested arena of climate science and governance, different actors have sought to access and utilise mass media sources in order to shape perceptions on various climate issues contingent on their perspectives and interests. For example, 'contrarians', 'skeptics' or 'denialists' have had significant discursive traction in the US public sphere over time (Leiserowitz et al., 2013), particularly by way of media representations (Boykoff, 2013). Resistances to both diagnoses of the causes of climate change and prognoses for international climate policy implementation, in the United States more specifically, have often been associated with the political right: the Republican Party and more particularly a right wing faction called the 'Tea Party' (Dunlap, 2008). John Broder of the *New York Times* described this right-of-centre US political party stance as an 'article of faith', and polling data have shown that 'more than half of Tea Party supporters said that global warming would have no serious effect at any time in the future, while only 15% of other Americans share that view' (Broder, 2010, p. A1). Moreover, while carbon-based industry interests have exerted considerable influence over US climate policy, associated scientists and policy actors who have questioned the significance of human contributions – often dubbed 'climate contrarians' – have been primarily housed in North American universities, think tanks and lobbying organisations (Dunlap, 2013; McCright, 2007). In particular, US-based non-nation state organisations such as the 'Heartland Institute' have held numerous meetings to promote contrarian views on climate science and policy (Boykoff and Olson, 2013; Hoffman, 2011).

Contributions to climate storytelling through news

Climate change is a complex and multifaceted issue that cuts to the heart of humans' relationship with the environment. The cultural politics of climate change are situated, power-laden, media-led and recursive in an ongoing battlefield of knowledge and interpretation (Boykoff et al., 2009). Mass media link these varied spaces together, as powerful and important interpreters of climate science and policy, translating what can often be alienating, jargon-laden information for the broadly construed public citizenry. Media workers and institutions powerfully shape and negotiate meaning, influencing how citizens make sense of and value the world.

In various cultural, political, social, economic and environmental contexts, journalists, producers and editors as well as scientists, policy makers and non-nation state actors must scrupulously and intently negotiate how climate is considered as a 'problem' or a 'threat'. As part of this process, it has been demonstrated that media reports have often conflated the vast and varied terrain – from climate science to governance, from consensus to debate – as unified and universalised issues (Boykoff, 2011). As a consequence, these representations can confuse rather than clarify: they can contribute to ongoing illusory, misleading and counterproductive debates within the public and policy communities on critical dimensions of the climate issue.

To the extent that media fuse distinct facets into climate *gestalt* – by way of 'claims' as well as 'claims makers' – collective public discourses, as well as deliberations over alternatives for climate action, have been poorly served. For example, although scientific experts have reached the consensus that humans contribute to climate change, there remains some disagreement among climate science experts as to the severity of climate change impacts and when and where climate impacts will occur (Schmidt and Wolfe, 2008; see also Painter, 2013). Rosenberg et al. explain:

> Those that disagree that the problem [of anthropogenic climate change] is acute or in need of decisive action like to note points of disagreement among scientists to bolster their position that the science is unsure and not defined enough to use as a foundation for policy decisions.
>
> *(2010, p. 311)*

Media focusing on an area of climate change that contains scientific nuances and uncertainties, such as the degree to which an extreme weather event is the result of climate change, may result in a specious conclusion that more knowledge is needed before taking action on climate change. In another sense, a lack of media coverage on climate change solutions, or the idea that mere individual actions can make the requisite difference may also limit actions for climate change.

Regarding 'claims makers', efforts to make sense of complex climate science and governance through media representations involves decisions regarding what are 'experts' or 'authorities' who speak for climate. This is particularly challenging when covering climate change, where indicators of climate change – such as sea level rise, temperature shifts, changing rainfall patterns – may be difficult to detect and systematically analyse (Andreadis and Smith, 2007). Moreover, in the advent and increasingly widespread influence of new and social media (along with fewer 'gatekeepers' in content generation), the identification of 'expertise' can be more, rather than less, challenging. The abilities to quickly conduct a Google or Bing search for information is in one sense very liberating; yet, in another sense, this unfiltered access to complex information also intensifies possibilities of short-circuiting peer review processes (and determinations by 'experts') and can thereby do an 'end-run around established scientific norms' (McCright and Dunlap, 2003, p. 359). In other words, these developments have numerous and potentially paradoxical reverberations through ongoing public discourses on climate change.

There are many reasons why media accounts around the world routinely fail to provide greater nuance when covering various aspects of climate change. Central among them, the processes behind the building and the challenging of dominant discourses take place simultaneously at multiple scales. Large-scale social, political and economic factors influence everyday individual journalistic decisions, such as how to focus or contextualise a story with quick time to deadline. These issues intersect with processes such as journalistic norms and values (for example, Boykoff, 2011), citizen and digital journalism (for example, O'Neill and Boykoff, 2010), and letters to the editor (for example, Young, 2013) to further shape news narratives. Moreover, path dependence through histories of professionalised journalism, journalistic norms and values as well as power relations have shaped the production of news stories (Starr, 2004). These dynamic and multiscale influences are interrelated and difficult to disentangle: media portrayals of climate change are infused with cultural, social, environmental and political economic elements, as well as how media professionals must mindfully navigate through hazardous terrain in order to fairly and accurately represent various dimensions of climate science and governance (Ward, 2008).

Overall, media representations are derived through complex and non-linear relationships between scientists, policy actors and the public that is often mediated by journalists' news stories (Carvalho and Burgess, 2005). In this, multi-scalar processes of power shape how mass media depict climate change. Processes involve an inevitable series of editorial choices to cover and report on certain events within a larger current of dynamic activities and provide mechanisms for privileging certain interpretations and 'ways of knowing' over others. Resulting images, texts and stories compete for attention and thus permeate interactions between science, policy, media and the public in varied ways. Furthermore, these interactions spill back onto ongoing media representations. Through these selection and feedback processes, mass media have given voice to climate itself by articulating aspects of the phenomenon in particular ways via claims makers or authorised speakers. In other words, through the web of contextual and dynamic factors, the stream of events in our shared lives gets converted into finite news stories. Thus, constructions of meaning and discourse on climate change are derived through combined structural and agential components that are represented through mass media to the general public.

The rise of #climate news through digital and social media

Embedded in this dynamism is the burgeoning influence of digital and social media. With it comes numerous questions: does increased visibility of climate change in new/social media translate to improved communication or just more noise? Do these spaces provide opportunities for new forms of deliberative community regarding questions of climate mitigation and adaptation (for example, Harlow and Harp, 2013) and conduits to offline organising and social movements (for example, Jankowski, 2006; Tufekci, 2013)? Or has the content of this increased coverage shifted to polemics and arguments over measured analysis? In this democratised space of content production, do new/social media provide more space for contrarian views to circulate? And through its interactivity, does increased consumption through new/social media further fragment a public discourse on climate mitigation and adaptation, through information silos where members of the public can stick to sources that help support their already held views (for example, Hestres, 2013)?

Sharon Dunwoody has cautioned to not view various modes of media production equally. As she puts it:

> [B]ecause of their extensive reach and concomitant efficiencies of scale, mediated information channels such as television and newspapers have been the traditional channels of choice for information campaigns. But research on how individuals actually use mass media information suggests that these channels may be better for some persuasive purposes than for others.
>
> *(quoted in Boykoff, 2009, p. 2)*

Furthermore, Cass Sunstein (2007) offers a similarly complicating – and also less than rosy – perspective: he warned of the likelihood of the 'echo chamber' effect where this interactivity actually walls off users from one another by merely consuming news that meshes with their worldview and ideology.

Such considerations within these new media developments prompt us to reassess boundaries between who constitute 'authorized' speakers (and who do not) in mass media as well as who are legitimate 'claims-makers'. These are consistently being interrogated and challenged. Anthony Leiserowitz has written that these arenas of claims making and framing are 'exercises in power . . . Those with the power to define the terms of the debate strongly determine the outcomes' (2005, p. 149). These factors have produced mixed and varied impacts: journalist Alissa Quart (2010) has warned of dangers of mistaken (or convenient) reliance on '*fauxperts*' instead of 'experts' while Boykoff (2013) and Boykoff and Olson (2013) have examined these dynamics as they relate to amplified media attention to 'contrarian' views on various climate issues.

Conclusions

Connections between media information and policy decision making, perspectives and behavioural change are far from straightforward (Vainio and Paloniemi, 2013). Coverage certainly does not determine engagement; rather, it shapes engagement *possibility* in quantity, quality, depth and effect (Boykoff, 2008; Carvalho and Burgess, 2005). So, our explorations of media coverage of climate change around the world in this chapter seek to help readers better understand the dynamic web of influence that media play amidst many others that shape our attitudes, intentions, beliefs, perspectives and behaviours regarding climate change. As we have posited here, media representations – from news to entertainment, from broadcast to interactive and participatory – are critical links between people's perspectives and experiences, and the ways in which dimensions of climate change are discussed at a distance between science, policy and public actors (see also Doyle, 2011).

The road from information acquisition via mass media to various forms of engagement and action is far from straightforward, and is filled with turns, potholes and intersections. This is a complex arena: mass media portrayals do not *simply* translate truths or truth claims nor do they fill knowledge gaps for citizens and policy actors to make 'the right choices'. Moreover, media representations clearly do not dictate particular behavioural responses. For example, research has shown that fear-inducing and catastrophic tones in climate change stories can inspire feelings of paralysis through powerlessness and disbelief rather than motivation and engagement. In addition, O'Neill et al. (2013) found that imagery connected with climate change influences saliency (that climate change is important) and efficacy (that one can do something about climate change) in complex ways, in their study across the country contexts of Australia, the United States and United Kingdom. Among their results, they found that imagery of climate impacts promoted feelings of salience but undermined self-efficacy, while imagery of energy futures imagery promoted efficacy. Overall, media portrayals continue to influence – in non-linear and dynamic ways – individual to community and international level perceptions of climate science and governance (Wilby, 2008). In other words, mass media have constituted key interventions in shaping the variegated, politicised terrain within which people perceive, understand and engage with climate science and policy (Goodman and Boyd, 2011; Krosnick et al., 2006).

Over time, many researchers and practitioners have (vigorously) debated the extent to which media representations and portrayals are potentially conduits to attitudinal and behavioural change (for example, Dickinson et al., 2013). Nonetheless, as unparalleled forms of communication in the public arena, research into media representational practices remains vitally important in terms of how they influence a spectrum of possibilities for governance and decision making. As such, media messages – and language choices more broadly (Greenhill et al., 2013) – function as important interpreters of climate information in the public arena, and shape perceptions, attitudes, intentions, beliefs and behaviours related to climate change (Boykoff, 2011; Hmielowski et al., 2014). Studies across many decades have documented that citizen-consumers access understanding about science and policy (and more specifically climate change) largely through media messages (for example, Antilla, 2010; O'Sullivan et al., 2003).

Furthermore, mass media comprise a community where climate science, policy and politics can readily be addressed, analysed and discussed. The way that these issues are covered in media can have far reaching consequences in terms of ongoing climate scientific inquiry as well as policy maker and public perceptions, understanding and potential engagement. In this contemporary environment, numerous 'actors' compete in these media landscapes to influence decision making and policy prioritisation at many scales of governance. Multitudinous ways of knowing – both challenged and supported through media depictions – shape ongoing discourses and imaginaries, circulating in various cultural and political contexts and scales. Furthermore, varying media representational practices contribute – amid a complex web of factors – to divergent perceptions, priorities and behaviours.

More media coverage of climate change – even supremely fair and accurate portrayals – is not a panacea. In fact, increased media attention to the issue often unearths more questions to be answered and *greater* scientific understanding actually can contribute to a *greater* supply of knowledge from which to develop and argue varying interpretations of that science (Sarewitz, 2004). At best, media reporting helps address, analyse and discuss the issues *but not answer them*. And dynamic interactions of multiple scales and dimensions of power critically contribute to how climate change is portrayed in the media. As has been detailed previously, mass media representations arise through large-scale (or *macro*) relations, such as decision making in a capitalist or state controlled political economy and individual level (or *micro*) processes such as everyday journalistic and editorial practices.

The contemporary cultural politics of climate change thread through a multitude of rapidly expanding spaces. Within this, the media serve a vital role in communication processes between science, policy and the public. The influence of media representations as well as creative and participatory communications – nested in cultural politics more broadly – can be ignored or dismissed in shaping climate science and governance at our peril.

Notes

1 This chapter is adapted from Boykoff, M. T., McNatt, M. B. and Goodman, M. K. (2015). 'Communicating in the Anthropocene: the cultural politics of climate change news coverage around the world', in A. Hansen and R. Cox (eds.) *The Routledge Handbook of Environment and Communication*. London: Routledge, pp 221–31.

2 For monthly updates and the full list of sources go to: http://sciencepolicy.colorado.edu/media_coverage.

References

Adams, P. C. and Gynnild, A. (2013). 'Environmental messages in online media: the role of place'. *Environmental Communication: A Journal of Nature and Culture*, 7(1): 103, 113–30.

Andreadis, E. and Smith, J. (2007). 'Beyond the ozone layer'. *British Journalism Review*, 18(1): 50–6.

Antilla, L. (2010). 'Self-censorship and science: a geographical review of media coverage of climate tipping points'. *Public Understanding of Science*, 19(2): 240–56.

Baek, Y. M., Wojcieszak, M. and Delli Carpini, M. (2012). 'Online versus face-to-face deliberations: who? why? what? what effects?' *New Media and Society*, 14(3): 363–83.

Baum, M. A. and Groeling, T. (2008). 'New media and the polarization of American political discourse'. *Political Communication*, 25(1): 345–65.

Boykoff, M. (2008). 'The cultural politics of climate change discourse in UK tabloids'. *Political Geography*, 27(5): 549–69.

Boykoff, M. (2009). 'A discernible human influence on the COP15? Considering the role of media in shaping ongoing climate science'. *Copenhagen Climate Congress Theme 6, Session 53*.

Boykoff, M. (2011). *Who speaks for climate? Making sense of mass media reporting on climate change*. Cambridge: Cambridge University Press.

Boykoff, M. (2013). 'Public enemy no.1? Understanding media representations of outlier views on climate change'. *American Behavioral Scientist*, 57(6): 796–817.

Boykoff, M. and Goodman, M. K. (2009). 'Conspicuous redemption? Reflections on the promises and perils of the "celebritization" of climate change'. *Geoforum*, 40: 395–406.

Boykoff, M., Goodman, M. K. and Curtis, I. (2009). 'Cultural politics of climate change: interactions in everyday spaces', in M. Boykoff (ed.) *The politics of climate change: a survey*. London: Routledge/Europa, pp 136–54.

Boykoff, M. and Olson, S. (2013). 'Understanding contrarians as a species of "Charismatic Megafauna" in contemporary climate science-policy-public interactions'. *Celebrity Studies journal* (special issue editors M. K. Goodman and J. Littler).

Boykoff, M. and Yulsman, T. (2013). 'Political economy, media and climate change: the sinews of modern life'. *Wiley Interdisciplinary Reviews: Climate Change*, 4(5): 359–71.

Broder, J. M. (2010). 'Skepticism on climate change is article of faith for tea party'. *The New York Times*, 21 October: A1.

Brulle, R. J. (2010). 'From environmental campaigns to advancing a public dialogue: environmental communication for civic engagement'. *Environmental Communication: A Journal of Nature and Culture*, 4(1): 82–98.

Cacciatore, M. A., Anderson, A. A., Choi, D.-H., Brossard, D., Scheufele, D. A., Liang, X., Ladwig, P. J., Xenos, M. and Dudo, A. (2012). 'Coverage of emerging technologies: a comparison between print and online media'. *New Media and Society*, 14(6): 1039–59.

Carvalho, A. and Burgess, J. (2005). 'Cultural circuits of climate change in UK broadsheet newspapers, 1985–2003'. *Risk Analysis*, 25(6): 1457–69.

Derrida, J. (1978). 'Structure, sign, and play in the discourse of the human sciences', in J. Derrida (ed.) *Writing and difference*. Chicago, IL: University of Chicago Press, pp 278–93.

Dickinson, J. L., Crain, R., Yalowitz, S. and Cherry, T. M. (2013). 'How framing climate change influences citizen scientists' intentions to do something about it'. *The Journal of Environmental Education*, 44(3): 145–58.

Doyle, J. (2011). *Mediating climate change*. Surrey: Ashgate Publishing.

Dunlap, R. E. (2008). 'Climate-change views: republican-democrat gaps extend'. *Gallup*, 29 May.

Dunlap, R. E. (2013). 'Climate change skepticism and denial: an introduction'. *American Behavioral Scientist*, 57(6): 691–98.

Fahy, D. and Nisbet, M. C. (2011). 'The science journalist online: shifting roles and emerging practices'. *Journalism*, 12(7): 778–93.

Fitts, A. S. (2013). 'Reuters global warming about face'. *Columbia Journalism Review*, 26 July. Available at: www.cjr.org/the_observatory/reuterss_global_warming_about-.php?page=2

Fogerty, D. (2013). 'Climate change'. *The Baron*, 15 July.

Forsyth, T. (2003). *Critical political ecology: the politics of environmental science*. London: Routledge.

Goodman, M. and Boyd, E. (2011). 'A social life for carbon? Commodification, markets and care'. *The Geographical Journal*, 177(2): 102–9.

Greenberg, M. (2013). 'Reuters climate change coverage declined significantly after "skeptic" editor joined'. *Media Matters for America*, 23 July. Available at: http://mediamatters.org/blog/2013/07/23/reuters-climate- change-coverage-declined-signif/195015

Greenhill, M., Leviston, Z., Leonard, R. and Walker, I. (2013). 'Assessing climate change beliefs: response effects of question wording and response alternatives'. *Public Understanding of Science*, 22(3): 1–19.

Harlow, S. and Harp, D. (2013). 'Collective action on the web'. *Information, Communication and Society*, 15(2): 196–216.

Hestres, L. E. (2013). 'Preaching to the choir: internet-mediated advocacy, issue public mobilization, and climate change'. *New Media and Society*, 1(1): 1–17.

Hmielowski, J. D., Feldman, L., Myers, T. A., Leiserowitz, A. and Maibach, E. (2014). 'An attack on science? Media use, trust in scientists and perceptions of global warming'. *Public Understanding of Science*, 23(7): 866–83.

Hoffman, A. J. (2011). 'Talking past each other? Cultural framing of skeptical and convinced logics in the climate change debate'. *Organization and Environment*, 24(3): 3–33.

Horan, T. J. (2013). '"Soft" versus "hard" news on microblogging networks'. *Information, Communication and Society*, 16(1): 43–60.

Jacobson, S. (2012). 'Transcoding the news: an investigation into multimedia journalism published on nytimes.com 2000–2008'. *New Media and Society*, 14(5): 867–85.

Jankowski, N. W. (2006). 'Creating community with media: history, theories and scientific investigations', in L. A. Lievrouw and S. Livingstone (eds.) *The handbook of new media, updated student edition*. London/Thousand Oaks/New Delhi: Sage, pp 55–74.

Krosnick, J. A., Holbrook, A. L., Lowe, L. and Visser, P. S. (2006). 'The origins and consequences of democratic citizens' policy agendas: a study of popular concern about global warming'. *Climatic Change*, 77(1): 7–43.

Leiserowitz, A. A. (2005). 'American risk perceptions: is climate change dangerous?' *Risk Analysis*, 25: 1433–42.

Leiserowitz, A. A., Maibach, E., Roser-Renouf, C., Smith, N. and Dawson, E. (2013). 'Climategate, public opinion and loss of trust'. *American Behavioral Scientist*, 57(6): 818–37.

McCright, A. M. (2007). 'Dealing with climate contrarians', in S. C. Moser and L. Dilling (eds.) *Creating a climate for change: communicating climate change and facilitating social change*. Cambridge: Cambridge University Press, pp 200–12.

McCright, A. M. and Dunlap, R. E. (2003). 'Defeating Kyoto: the conservative movement's impact on US climate change policy'. *Social Problems*, 50(3): 348–73.

Nielsen, R. K. (2012). 'How newspapers began to blog'. *Information, Communication and Society*, 15(6): 959–68.

O'Neill, S. J. and Boykoff, M. T. (2010). 'The role of new media in engaging the public with climate change', in L. Whitmarsh, S. J. O'Neill and I. Lorenzoni (eds.) *Engaging the public with climate change: communication and behaviour change*. London: Earthscan, pp 233–51.

O'Neill, S. J., Boykoff, M. T., Day, S. A. and Niemeyer, S. (2013). 'On the use of imagery for climate change engagement'. *Global Environmental Change*, 23(2): 413–21.

O'Sullivan, T., Dutton, B. and Rayne, P. (2003). *Studying the media*. London: Hodder Arnold.

Painter, J. (2013). *Climate change in the media: reporting risk and uncertainty*. New York: I. B. Tauris & Co.

Poortinga, W., Spence, A., Whitmarsh, L., Capstick, S. and Pidgeon, N. F. (2011). 'Uncertain climate: an investigation into public scepticism about anthropogenic climate change'. *Global Environmental Change*, 21(3): 1015–24.

Quart, A. (2010). 'The trouble with experts'. *Columbia Journalism Review*, July/August: 17–18.

Rosenberg, S., Vedlitz, A., Cowman, D. F. and Zahran, S. (2010). 'Climate change: a profile of US climate scientists' perspectives'. *Climate Change*, 101(1): 311–29.

Sarewitz, D. (2004). 'How science makes environmental controversies worse'. *Environmental Science and Policy*, 7: 385–403.

Schmidt, G. and Wolfe, J. (2008). *Climate change: picturing the science*. New York: W. W. Norton and Company.
Schuurman, N. (2013). 'Tweet me your talk: geographical learning and knowledge production 2.0'. *Professional Geographer*, 65(3): 369–77.
Siles, I. and Boczkowski, P. J. (2012). 'Making sense of the newspaper crisis: a critical assessment of existing research and an agenda for future work'. *New Media and Society*, 14(8): 1375–94.
Starr, P. (2004). *The creation of the media: political origins of modern communications*. New York: Basic Books.
Sunstein, C. R. (2007). *Republic.com 2.0*. Princeton, NJ: Princeton University Press.
Tufekci, Z. (2013). 'Not this one: social movements, the attention economy, and microcelebrity networked activism'. *American Behavioral Scientist*, 57(7): 848–70.
Vainio, A. and Paloniemi, R. (2013). 'Does belief matter in climate change action?' *Public Understanding of Science*, 22(4): 382–95.
van Dijk, J. (2006). *The network society*. London: Sage.
Ward, B. (2008). *Communicating on climate change: an essential resource for journalists, scientists and editors*. Providence, RI: Metcalf Institute for Marine and Environmental Reporting, University of Rhode Island Graduate School of Oceanography.
Whitmarsh, L. (2011). 'Scepticism and uncertainty about climate change: dimensions, determinants and change over time'. *Global Environmental Change*, 21(2): 690–700.
Wilby, P. (2008). 'In dangerous denial'. *The Guardian*, 30 June: 9.
Wynne, B. (2008). 'Elephants in the rooms where publics encounter "science"'. *Public Understanding of Science*, 17: 21–33.
Young, N. (2013). 'Working the fringes: the role of letters to the editor in advancing non-standard media narratives about climate change'. *Public Understanding of Science*, 22(4): 443–59.

Part 3

Critical environmental pressures and responses

Iain White, Simin Davoudi, Richard Cowell and Hilda Blanco

> How did crisis, once a signifier for a critical, decisive moment, comes to be construed as a protracted historical and experiential condition?
>
> *(Roitman, 2014, p. 2)*

We live in a time of concurrent urban and environmental crises. The climate is changing, homes are unaffordable and ecosystems are under severe pressure. In many respects, the chapters here provide a partial snapshot of the diverse scope and substantial scale of the environmental planning challenges facing us and each one may be considered a 'crisis' in its own right. Significantly, as both the opening quote indicates and the chapters in this part will progressively reveal, crises have gradually transitioned from an urgent event or moment that demands immediate public and political attention to a *condition*. A part of the background hum of everyday life. A lived experience for many, with no real hope of early resolution. It also provides a challenging research agenda – one that is intrinsically linked to the material contained within the previous two parts. Beyond indications of criticality or action, the notion of crisis is a useful intellectual object as it serves to shed valuable light on the contrasting ideas of normalcy and business-as-usual that typically escape attention. What role has our understanding of the environment played in this? How effective are our governance processes at addressing these issues? And crucially for this part, how can an increased understanding of the pressures open up new political spaces or future imaginaries where effective responses can be developed and implemented?

The first two chapters direct our gaze towards how global problems produce uneven and inequitable pressures, particularly on parts of the world that may not typically receive much academic attention. In **Chapter 20**, **Darryn McEvoy** and **David Mitchell** begin by bringing together three issues that underpin much environmental planning focus: climate change, resilience and urbanisation. Considering these issues within the context of the Global South, they draw upon research from five different countries and reflect upon the complex interrelationships between security of land tenure and climate resilience, highlighting how marginalised communities face multiple challenges. In doing so, they reveal how these groups are not just the ones most exposed and sensitive to climate change but their restricted access to land, resources, and decision making means that they have less ability to adapt. Overall, by looking across and comparing different countries, they provide compelling evidence that the issue of land governance is a crucial mechanism to enable climate resilience in the Global South. There is a similarly strong undercurrent of social and environmental justice threading through **Chapter 21** by **John R. Campbell**. This chapter focuses on disaster risk reduction, where the author draws upon his extensive experience to provide a valuable overview of the key issues, various terminology and potential environmental planning responses. It highlights the different social and economic impacts between developed and developing nations, and emphasises throughout how disasters are not in any way 'natural'. This broad political-economy perspective provides a useful means to explore a number of important environmental planning issues influencing disaster risk reduction beyond the use of land, such as politics, investment and the need to adopt an integrative approach that reflects the multifaceted nature of risk.

Two far-reaching and widely relevant pressures then receive critical attention. In **Chapter 22**, **Matthew Cotton** examines the issue of planning, infrastructure, and low carbon energy. Given the core relationship to climate change, the need for a rapid decarbonisation of energy systems is a pivotal 21st-century environmental planning issue. The chapter provides a thorough overview of the state of the problem and the difficulties associated with the challenge of transition. The discussion is wide ranging touching upon the socio-technical nature of the area, where technology, spatial planning and social opposition to change can co-exist. The chapter stresses the value in adopting an encompassing view that goes beyond one-dimensional notions of technological or infrastructure 'fixes' to instead link the success of responses to governance, space, and place. This perspective provides a means to add analytical depth by considering how innovations may be subject to considerable public opposition and, as such, the research agenda needs to engage more strongly with issues of fairness and democracy. An equally complex issue is discussed in **Chapter 23** by **Nick Hacking**. Here we turn towards the problem of waste and the management of environmental resources. The discussion starts with a key conceptual aspect that underscores much debate concerning sustainable waste management: the idea of shifting from linearity to circularity, a dialogue that recognises how wastes 'flow' and move within a system, whether to landfill or to recycling and reuse. The chapter also highlights the importance of definitions, in particular how what we may simply posit as 'waste' is in fact socially constructed and dynamic. In reality, materials may undergo continual shifts in classification and in doing so provide diverse pressures and opportunities for environmental planning. To bring these issues to life, the author explores three common goals-based narratives in this field: ecological, eco-efficiency and short-term economic, demonstrating how each frame influences the nature of intervention differently. By linking theory to practice in this way, the critical aspects of power, politics and policy that underpin any transition of this nature are brought deftly to light.

The next two chapters touch upon the conceptual and practical difficulties in integrating the natural and built environments. In **Chapter 24**, **Kimmo Lapintie** and **Mina di Marino** discuss biodiversity and ecosystems services within environmental planning. They start with a brief historical perspective that helps situate the topic within both current debates and explains key points of difference to traditional notions of urban planning. It initially seeks to set the scene for ecosystems services by exposing and challenging the artificial dichotomy between the natural and human worlds before reflecting on its rather complex relationship to environmental planning. The perspective merges both thought and practice to provide a comprehensive view of the various reasons why biodiversity and ecosystems services offer substantial societal value but have struggled to realise this promise in the face of the existing ways of knowing and doing. They usefully sum up the discussion by providing insights into current and future challenges, and in doing so help reveal the knowledge gaps, conceptual difficulties, and practical constraints that together encompass the contemporary research agenda in this field. The following **Chapter 25** by **Mick Lennon** develops this theme by highlighting issues connected to the related area of environmental planning and green infrastructure, with a view to providing clarity to a research agenda that has also emerged rapidly within theory and practice. He begins by tracing its roots and development and uses this foundation to help explain the current positioning of the concept. To bring the topic to life, he then discusses examples that allow us to explore its application at the site, neighbourhood and city region scale. A key theme of note is the importance of integration. This is not just in the sense of developing a spatial network of green infrastructure but also regarding how the key principles of green infrastructure provide a common language that can cut across typical boundaries and siloes, such as those between administrative areas, disciplines or policy domains.

We end with two chapters that focus on thematic issues of central importance to environmental planning: water and air. In **Chapter 26**, **Sue Kidd** provides a thought-provoking discussion of the importance, pressures upon, and policy challenges presented by marine planning and coastal management. It introduces the major environmental, economic and social value of the habitat before outlining the various ways by which this is placed under enormous strain by human development and activities. Together this provides a persuasive argument for significant environmental planning attention. The chapter then develops this discussion by turning to the legal and policy realms, detailing and reviewing the main instruments and initiatives that hold relevance. The analysis reveals, however, that while there have been gains in some areas, the responses to the pressures are mixed and deterioration of the natural environment continues. It ends by charting a course ahead, recognising the critical role that governance can play in both elevating this as a core concern and enabling the integrated systemic approach that such a multi-scalar problem demands. In **Chapter 27**, **Anil Namdeo** and **Saad Almutairi** turn attention to transport and air pollution. They begin with some sobering insights into the growth of the transport sector globally and, in particular, the use of private passenger cars. This social and economic context then provides the route to understand the growing pressures and the need for an effective response. What follows is a breakdown of the links between transport and air pollution, in particular the effects on human health, that will have significance on an international scale. By explicitly connecting the various modes of transport to air pollution, we can also see the clear links to environmental planning. An undercurrent running throughout is that planning has a fundamental effect on the behaviour of residents and their personal exposure to pollution, so actions we take now will have a long resonance. Overall, it merges a detailed technical, scientific understanding of pressures and responses, such as via low emission technology, with a social science imperative regarding the ways we plan, such as by encouraging a modal shift to walkability or reducing the need to travel.

It is tempting in book parts relating to environmental planning pressures and responses to cover in a systematic way the various components or habitats that make up the natural world: the soils and forests, the rivers and lakes, or the flora and fauna. While there is value to such a perspective, the material here is compiled to emphasise the broader spatial challenges, interconnectivity of issues and cross-cutting questions that together help reveal and chart the research challenges ahead. For example, we can now apprehend more deeply how the problems of transition to low carbon energy, sustainable waste management or to a greater protection and enhancement of the marine environment are all centrally related to society, politics and governance. This strong interdisciplinary motif also serves to highlight the *limits* of environmental planning. It is important to note that the chapters all describe issues of significant complexity, which, while they are clearly within the remit of the discipline, are also beyond its power to resolve. This point allows us to return to our opening discussion concerning the changing nature of crises and better comprehend why problems can persist. It also helps explain why many authors in this section specifically highlighted the growing importance of issues relating to justice, fairness, advocacy or democracy within environmental planning. If resolution is not an option, emphasis inevitably turns toward the management of effects, which then logically directs attention to the perennial redistribution concerns of what, where and how and the importance of understanding who is privileged in decision making and why. These are issues developed further in the final part of this Companion.

Reference

Roitman, J. (2014). *Anti-crisis*. Durham, NC: Duke University Press.

20

Climate Resilient Land Governance in the Global South

Darryn McEvoy and David Mitchell

Introduction

It is now recognised that societies around the world will need to adapt to climatic changes that are unavoidable (IPCC, 2014). However, whilst there has been a recent accumulation of research and policy knowledge on planning for climate change (see for example: da Silva et al., 2012; Fünfgeld and McEvoy, 2014; UN-Habitat, 2014), as well as discussion of the opportunities and barriers to adaptation in a developed world context (Tompkins et al., 2010; Biesbroek et al., 2013), much less attention has been paid to the more complex challenges faced by low and middle income nations. Indeed, many developing nations in the Global South are experiencing a combination of rapid urbanisation and increasing climate impacts, compounded by development deficits and issues of restricted access to land, resources and shelter. In such cases, much less is known about the inter-relationships between security of land tenure and climate vulnerability and how these linkages potentially impact the viability of climate adaptation and environmental planning.

To address this gap in knowledge, this chapter highlights summary findings from a 2018 international review of 'land tenure and climate vulnerability' funded by the Global Land Tool Network (GLTN) (Mitchell and McEvoy, 2019). The methodology involved a comprehensive literature review as well as case study analyses of the inter-relationships between land tenure security and climate vulnerability in five developing countries in different regions of the world. These case studies were selected to account for a range of natural hazards and land tenure arrangements, as well as differing landscapes and land use. The case studies were the Solomon Islands (informal settlements in the capital city Honiara), Philippines (government responses post Typhoon Sendong), Syria (a combination of drought and armed conflict), Uganda (drought impacts on traditional pastoralism in the context of changing land rights) and the Caribbean (urban flash flooding in St Vincent and the Grenadines).

The international comparative analysis was framed according to five overarching research questions:

1 How does security of land tenure influence local exposure and sensitivity to climate-related impacts?

2 How does security of land tenure influence adaptive capacity and therefore the potential implementation of actions to increase climate resilience?
3 How do current and future climate impacts affect, or potentially affect, tenure security in different landscape contexts (for example, urban, agricultural, pastoral, etc.)?
4 How might climate adaptation actions impact on tenure security (in either a positive or negative way)?
5 What are the critical land administration issues that need to be addressed to enable more effective and equitable climate adaptation?

Findings from the study confirmed that insecure land tenure not only increases the exposure and sensitivity of marginalised communities to climate-related hazards but in many instances also adversely affects collective and individual capacities to adapt (though with some caveats which are highlighted later). As such, this chapter argues that land tenure arrangements and principles of good land governance need to be critical considerations when addressing climate-related challenges, particularly those being faced by marginalised communities in developing countries. Furthermore, the findings also highlight the need for pro-poor and gender-appropriate land policies, tools and approaches that better promote more equitable climate adaptation planning and implementation of resilience actions. A more integrated, and holistic, approach to regional, urban and environmental planning is therefore needed.

The importance of land tenure in planning for climate change

In simple terms, land tenure is the way in which interests in land are held or owned by people. In practice, it involves the legal or customary relationship between people with respect to land and natural resources, i.e. land tenure rules and systems define the ways by which land is held, how property rights to land are allocated, the security of those rights and ultimately how they are enforced. These rules vary in their legal recognition in different places and in some cases may involve very complex customary rights and associated dispute resolution institutions and processes (Feder and Feeny, 1991; FAO, 2012).

Property rights can also be complex, including both movable rights – such as livestock – and immovable rights such as buildings and trees and may be held by an individual or family, communal groups, or the State. They may also be part of open access communal regimes where specific rights are not assigned to individuals but rather to a group. Importantly, land tenure and property rights can also influence whether land and natural resources are misused, resulting in detrimental impacts on local – and even global – environments (Quizon et al., 2018).

In many developing countries, a range of tenure systems exist, each associated with different levels of security. These include registered and state-guaranteed freehold, usufruct (land use) rights, customary land and informal settlements established without formal government land records. It is therefore useful to frame land tenure rights as existing along a gradient of different levels of tenure security, as illustrated in Figure 20.1. The concept of a 'continuum of land rights' is particularly pertinent to climate or disaster-exposed areas, as it is those with least security of tenure who are often the most vulnerable to the impact of natural hazards (Dodman and Satterthwaite, 2008; Mitchell et al., 2015).

Customary land tenure systems predominate in many developing countries, especially in parts of Africa and the Pacific Island countries, often with complex arrangements for allocating land and accessing resources (Fitzpatrick, 2005). In such contexts, customary dispute resolution mechanisms are usually well understood by community members. However, these 'traditional' systems of land tenure are under increasing pressure because of contemporary population

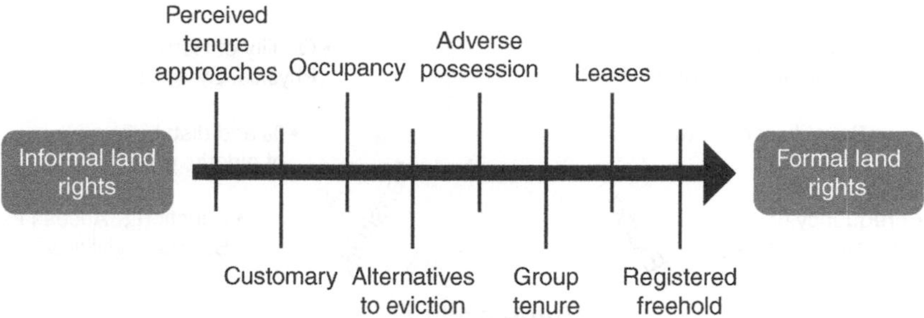

Figure 20.1 Continuum of land tenure rights

Source: UN-Habitat, IIRR and GLTN (2012, p. 12)

growth, environmental degradation, increasing competition for land, urbanisation encroaching onto peri-urban communal lands, conflicts and breakdowns in customary authority (Cotula et al., 2004; Fitzpatrick, 2005; Trundle and McEvoy, 2017).

An emerging evidence base of international literature argues that insecure land tenure exacerbates vulnerability to climate-related hazards, both directly and indirectly (see for example: Satterthwaite et al., 2018). Not only are those without secure land tenure often the most exposed to climate impacts, for example, being located in areas at high risk of flooding, storm surge, landslides, etc., they are also the most sensitive to climate impacts due to inadequate housing and restricted access to basic services (Dodman et al., 2013). Insecure land tenure can also have a negative impact on the adaptive capacity of households, a consequence of being disconnected from formal governance processes, lacking the knowledge and information to inform resilience decisions, as well as having restricted access to finance for implementing actions (McEvoy et al., 2019).

Responsible land governance is considered a key mechanism for improving tenure security for the approximately 70% of the global population who don't have any formal land rights (FAO, 2012). Bearing this in mind, the following section now provides a short synthesis of findings from each of the case study reviews, highlighting details of the inter-relationships between land tenure and climate vulnerability (framed according to exposure, sensitivity and adaptive capacity – see Figure 20.2) and how climate change adaptation and planning for disasters could impact peoples' actual and perceived security of tenure (Mitchell and McEvoy, 2019).

Synthesis of findings from five developing world case studies[1]

Context

Land tenure arrangements in all five cases are complex, with overlapping tenure systems (both formal and informal) subject to the remits of multiple agencies, laws and rules. As a consequence, instances of contested land ownership are a recurring theme across all cases. There are particular tensions between Western statutory legal systems and land tenure arrangements and customary systems of land tenure in three of the case studies, a legacy of British colonial times (Honiara, St Vincent and the Grenadines and Uganda). These tensions are well illustrated by the Honiara case study, with Western-style land tenure arrangements within the city boundary juxtaposed against customary systems that operate in the surrounding province of Guadalcanal. Causes of conflict evidenced in the other case studies included inequitable allocation of land to indigenous

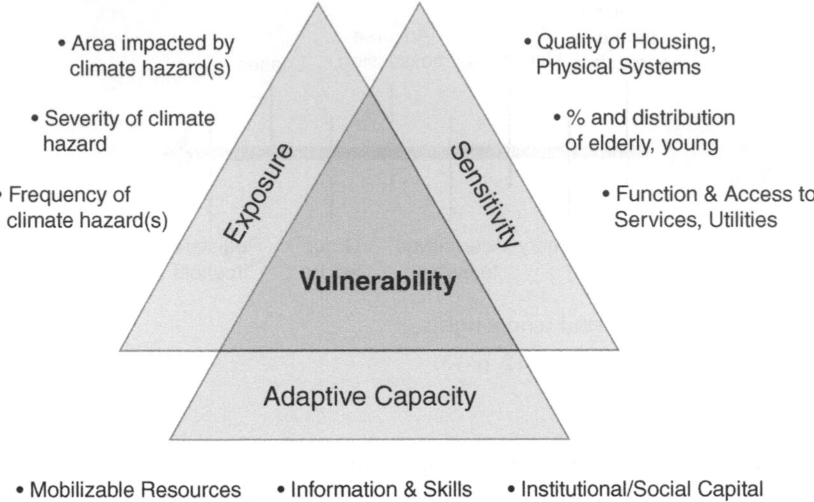

Figure 20.2 Vulnerability framework

Source: Authors' own, also published in Trundle and McEvoy (2017, p. 12)

communities under a new river basin management and development master plan (Philippines) and land privatization, which has also led to an informal, unequal and contested redistribution of land (Syria).

The inter-relationships between land tenure, climate vulnerability and adaptation

In the Honiara case study, increasing migration to the country's primary city from rural areas is resulting in rapid rates of urbanisation and an uncontrolled growth of informal settlements. People migrating into the city, and surrounding peri-urban areas, have commonly settled on the only land that is perceived as being available to them (taking into account the risk of eviction, access to urban services, etc.). However, this is often the most marginal land that has not been developed previously, with high levels of exposure to a range of climate-related hazards. Furthermore, housing that is built on land without a formal land title is often viewed as 'temporary' and as a consequence poor quality housing tends to be the norm (though acknowledging that there are exceptions often influenced by peoples' perceived security of tenure; see for example: Van Gelder, 2009; Kiddle, 2010). A lack of land title in Honiara's informal settlements also restricts access to essential services such as water, sanitation and electricity. In terms of adaptive capacity, informal settlers not only have restricted access to infrastructure and basic services, they also lack the information and financial resources that are necessary to better prepare for, and respond to, extreme events. Furthermore, being excluded from formal governance structures and processes, they also find themselves disconnected from important decision-making processes, for example, city-wide disaster risk reduction networks. However, it should be noted that there is a counter-argument that many urban communities in the city (sometimes described as urban villages – see: Asian Development Bank, 2016) retain strong socio-cultural networks and characteristics of endogenous resilience that are critically important for local adaptive capacity (Trundle et al., 2018).

The St Vincent and the Grenadines and the Philippines cases also examined the impacts of extreme weather events on urban poor communities. In the Caribbean example, flash flooding was identified as an issue for all three communities that were analysed in this case study. However, whilst all three were exposed to flood waters, their levels of sensitivity and adaptive capacity differed as a consequence of their land tenure arrangements. Those families with formal land titles (typically with higher income levels) were more likely to have protected their properties with river flood defences and/or had insurance cover in place to be able to recover after the flash flooding in 2013. Conversely, those with no formal land title had lower incomes, were living in poorer quality housing and had fewer resources/financial capacity to build back after the floods. They were also more likely to be reliant on the State, or other residents, for assistance with their recovery efforts.

The Philippines case focussed on the effects of Typhoon Sendong, which devastated Northern Mindanao, Philippines, in December 2011, resulting in 1,268 deaths. Cagayan de Oro City bore the brunt of the typhoon, both in terms of fatalities and the number of displaced persons. The impact was particularly severe on the informal settlements, many of which had been located in public spaces and ecologically fragile areas, with those living along the city's riverbanks suffering greatest impact. These informal communities were comprised predominantly of poor families with makeshift shelters, low incomes, little work security and no security of tenure. After the typhoon (and evacuation of residents), 'no-build' zones were declared along the riversides and coastlines that had previously been occupied by informal settlers. Very few of those displaced had proof of ownership or formal documents relating to the land; therefore; only a small percentage were able to receive compensation or reclaim their previous occupancy rights. The government's resettlement response has not met with universal approval, with complaints from resettled households that they are now located far away from sources of livelihood as well as having increased transport costs.

A severe drought in north-eastern Syria (2007–12), compounded by the ongoing civil war, significantly affected water supplies in the country with the agricultural sector being badly impacted. Land tenure was an influential contributing factor to the increased climate vulnerability of marginalised communities, with a decades-long process of privatising state agricultural lands (combined with informal and unequal redistribution of land) eroding customary laws over boundary rights, resulting in increasing land fragmentation and insecurity of tenure. This has led to rising tensions between farmers and pastoralists, with herders grazing farming lands which have become increasingly fragile as a consequence of drought and increasing desertification. Whilst some people have tried to resist the effects of the drought by adopting harmful coping strategies, for example, selling livestock below market prices, reducing their food intake and selling goods, others have been forced to migrate.

In the Ugandan case, pastoralists in the Karamoja region have traditionally grazed freely, subject to the rules of a hierarchy of cultural institutions involving elders, Kraal leaders and herders. Individuals have a communal right to use the land as long as they are members of a relevant clan or group. Migration during dry seasons was an important part of autonomous adaptation to local climatic conditions; however, changing rainfall patterns in recent times have led to enforced changes to their traditional migratory patterns for grazing (increasing variability of rainfall has altered the temporal availability of forage material and water and has resulted in the pastoralists migrating more frequently and for longer periods of time than previously). Changes to traditional migratory routes have also been affected by the increasing alienation of land and the individualisation of land rights, for example, migratory routes are blocked by fenced land, land protected for wildlife conservation or appropriation by companies that have legal concessions to extract minerals. The government response to date has been to emphasise strategies that attempt to advance more sedentary forms of agricultural practice in the region. However, critics of this policy have argued that pastoralism represents the most viable livelihood option for many of those affected and that there is a need for greater political recognition of the broader socio-cultural benefits of traditional grazing practices.

Towards an integrated approach: Climate Resilient Land Governance (CRLG)

Evidence from the case studies indicates that those with insecure tenure security are frequently the most exposed group to climate-related impacts, either being located in areas at high risk of flooding or being the most exposed to the impacts of drought. In all cases, they were also identified as being the most sensitive to climate impacts, for example, having poor quality housing and lacking access to services or being impacted by inequitable land reform, a reduced availability of land and access to resources and having limited alternative livelihood opportunities.[2] Insecure tenure security also constrained adaptive capacity, a consequence of being disconnected from formal governance processes and power structures, lacking the information for informed decision making and having restricted access to finance to implement resilience enhancing actions.

Whilst people-to-land relationships have long been recognised as an important factor for human wellbeing (i.e. secure access to land is needed for shelter, food security etc.), a combination of climate change and other contemporary human and natural drivers are now threatening to undermine this relationship. This chapter therefore argues that approaches to land governance, and environmental planning more generally, need to take more explicit account of the impacts of a changing climate and increased vulnerability of marginalised communities. Conversely, security of land tenure needs to be a central consideration in disaster risk reduction (DRR) and climate change adaptation planning.

From a land governance perspective, tenure security mapping based on the concept of the continuum of land rights would be a valuable starting point. This provides the necessary baseline data to inform tenure-responsive land use planning, using approaches to land administration that are fit-for-purpose, pro-poor and gender responsive. Land governance policy and practice should also explicitly factor in climate vulnerability at multiple scales in decision making (as well as aligning more closely with climate change adaptation and DRR agendas). Such an integrative approach establishes the necessary community baseline data for a comprehensive understanding of local vulnerabilities (climate and non-climate) that can then be used to inform appropriate multi-level and multi-actor resilience strategies.

The implications of climate change for future people-land relationships necessitate a reframing of land governance frameworks to make them fit-for-purpose in the context of contemporary challenges. First, knowledge of the complex interactions and feedbacks between land tenure security and climate vulnerability need to be better understood, and accounted for, by different communities of practice. This knowledge can then be used to frame a more holistic approach to policy and actions that aim to improve human wellbeing, particularly the needs of marginalised households. Such an approach will not only improve peoples' land tenure security but, at the same time, reduce their vulnerability to climate and natural hazards. Figure 20.3 highlights important links between land use and environmental planning and secure land tenure rights and the need for greater integration with climate change adaptation and disaster risk reduction agendas. A more integrated approach to addressing multiple inter-linked challenges is needed, one that involves new ways of working between different communities of practice, enhanced mutual understanding and closer alignment between respective institutions, policies, processes and tools.

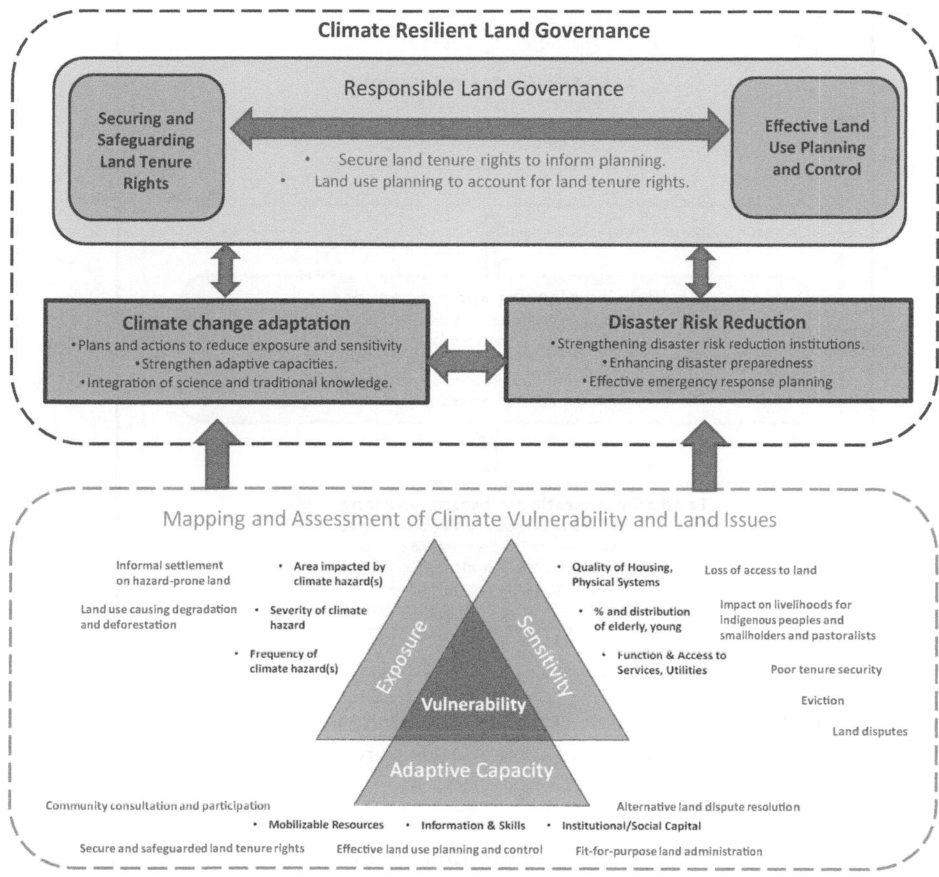

Figure 20.3 Inter-linked components – Climate Resilient Land Governance
Source: Authors' own

Three key entry points for closer integration and moving towards CRLG can be categorised as: (i) institutional arrangements; (ii) policies, planning and programmes; and (iii) land administration processes and tools (Figure 20.4). These components provide a conceptual and operational framework for addressing land issues in the context of increasing climate risks, as well as ensuring that land tenure is explicitly considered in climate change adaptation and disaster risk management planning.

Institutional arrangements

A number of global frameworks have been introduced in recent times to promote human wellbeing in a time of emerging shocks and stresses, including climate change, rapid urbanisation and inequitable access to land and resources. The most relevant international frameworks include the Universal Declaration of Human Rights, Sustainable Development Goals (SDGs), Paris Agreement, Sendai Framework, and the New Urban Agenda. Whilst sustainable development, climate change adaptation, disaster risk reduction, and land administration agendas share many common principles and goals, associated communities of practice have tended to act in

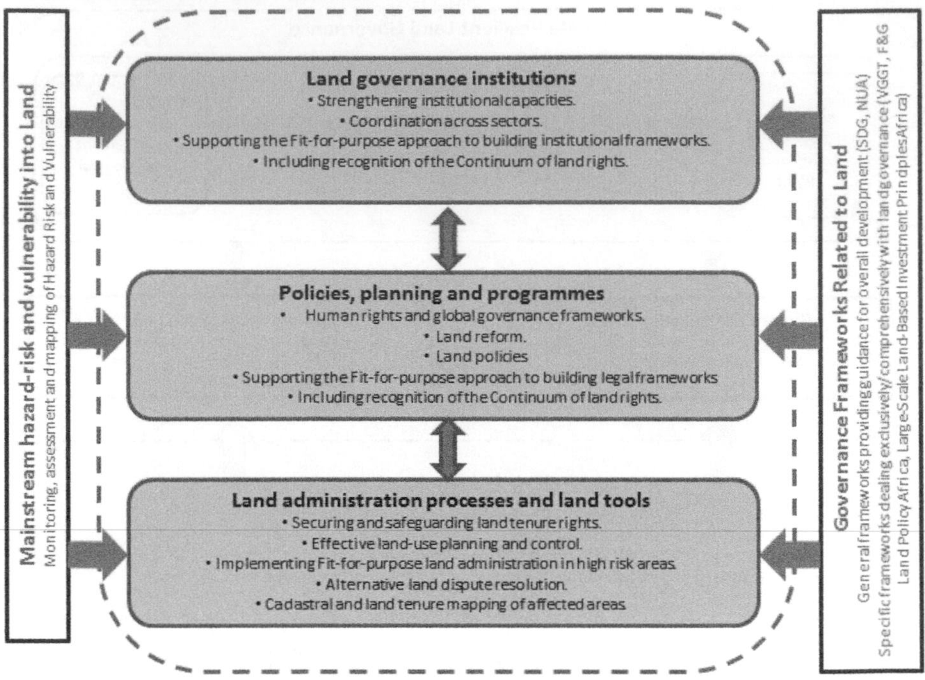

Figure 20.4 Entry points for Climate Resilient Land Governance

Source: Authors' own, also to be published as Mitchell and McEvoy (2019)

isolation. However, given the inter-connected nature of many of the Global South's resilience challenges, there is obvious value in increased collaboration and greater integration between different issues/agendas/communities of practice and new approaches to planning that build on respective institutional strengths. CRLG is an attempt to develop a framework that supports a more holistic approach. It advocates for fit-for-purpose land administration institutions – working in tandem with climate change adaptation, disaster management, and development agencies – to actively improve the tenure security of marginalised communities and at the same time reduce vulnerabilities to climate impacts.

At the local level, a lack of institutional capacity in many developing countries represents a key challenge for addressing land issues in the context of a changing climate. Capacity building and support for local actors is therefore critical to achieving more effective CRLG.

Policy, plans and programmes

Land policy, plans and programmes are the second point of entry for operationalising the principles of CRLG. As highlighted previously, mainstreaming climate considerations into processes of land governance is especially important. In urban and peri-urban areas, mainstreaming could be underpinned by the development of city-wide maps of hazard zones, overlaid with cadastral mapping and land tenure security assessments to inform planned relocation decisions, land readjustment and settlement upgrading. This also applies to municipal, utility and development plans. Environmental planning would benefit from the integration of hazard risk/vulnerability mapping and tenure-responsive land use planning, ensuring that all new development and settlement upgrading occurs in areas outside of high risk zones. Furthermore, adopting the continuum of land rights ensures that those with insecure land tenure rights are included in planned relocation and settlement upgrading strategies and actions. In rural areas, integration could involve empowering traditional approaches to land tenure and climate adaptation, and involve assessments that identify the impacts on vulnerable groups (including women, indigenous people, displaced persons and the poorest households).

Tenure-responsive land use planning will also help to address a range of problems faced by many developing countries, including unsustainable land use, weak or ineffective land use planning, non-existent or poorly enforced building codes and weak land administration practices that allows people to settle in high risk areas. These problems are difficult to address later on and may require future relocation of at-risk communities. If secure tenure is also provided at alternative locations, then those resettled are more likely to remain on their new land rather than returning to at-risk homes. Furthermore, if effectively enforced, tenure-responsive land use planning would not only prevent houses being constructed on hazard-prone land; it would support more effective use of building codes that improve the quality of housing constructed outside hazard-prone areas. Providing 'secure enough' land tenure rights for properties located outside hazard-prone zones can bring about positive changes, allowing those with informal tenures to participate in decisions about use of resources and plans for new settlements.[3]

Implementing climate resilient land policies at scale will require effective and 'fit-for-purpose' (FFP) land administration systems. CRLG therefore needs to be underpinned by pro-poor and fit-for-purpose systems that not only improve security of tenure rights though securing and safeguarding land tenure rights but also support land use planning and control. However, the implementation of FFP principles will often require amending spatial, legal or institutional frameworks to introduce low cost and more efficient processes (a major challenge is to improve tenure security in a way that is inclusive, cost effective and rapid).

In urban and peri-urban areas, a FFP approach to land administration can support:

- The recognition and recording of land tenure rights of properties exposed to hazards, address unsustainable use of upstream land and resources that impact downstream and support urban planning to better manage the growth of informal settlements;
- Informal settlement upgrading through identifying low cost ways of improving efficiency of land administration processes, as well as building local capacity;
- Management of planned relocation of high risk communities. FFP land administration and pro-poor land recordation can support the provision of 'secure enough' tenure at new destinations;
- More effective enforcement of no-build zones.

In rural areas, fit-for-purpose land administration can help address the over exploitation of natural resources through recording rights across the continuum at the scale needed. Secure land tenure rights make land disputes easier to resolve and can also help government control inappropriate land use.

Land administration processes and tools

Processes

Effective land administration supports the implementation of land policies, environmental planning and relevant sustainable development goals. Land administration processes worthy of note include: the continuum of land rights, land readjustment, mapping of tenure security, land recordation and the involvement of youth in land governance and climate change adaptation decision making.

The relationships between climate vulnerability and insecure land tenure reinforce the need to scale up the delivery of secure land tenure for all. This will require acceptance of the concept of a continuum of land rights, as well as fit-for-purpose and pro-poor approaches to land administration. This is important in addressing cross-cutting issues related to human mobility, food and water security, environmental impact, informal settlements, gender inequality, indigenous people and conflict. Policy and regulatory frameworks will often need to be amended to adopt the continuum of land rights.

One of the most important actions to reduce household vulnerability is improvements in housing conditions and access to infrastructure and services. Satterthwaite et al. (2018) argue that governments should recognise the many positive aspects of informal settlements and work with the inhabitants and community organisations to provide the necessary infrastructure, services and improved housing quality. However, as recognised in the Sendai Framework, in some cases planned relocation or land readjustment may be needed, provided this process also includes adequate compensation and provision of secure land tenure.

Information on tenure security – and any land disputes – is necessary where climate actions are planned. Country-level tenure assessments are important in identifying the types of land tenure that might need urgent attention and where conflict is occurring. This is useful for national priority setting. Assessment and mapping of tenure security is also critical for city level planning; providing evidence of existing rights to land, numbers of displaced and the landless, claims by indigenous peoples, impacts on women's rights and existing informal agreements. Where land agencies undertake an inventory of land tenure prior to disasters they also lay the foundation for DRR strategies. Inventories can identify land at risk of hazards, land that can be used for evacuation, potential sites for emergency shelter, post-disaster planned relocation, etc. (see for example, Caron et al., 2014).

While formal land administration systems will play a major role in providing tenure security, other community-based options such as pro-poor land recordation provide interim options for identifying and recording rights. The use of community-based systems for recording land tenure rights are emerging in response to the need to adopt low cost methods to record at scale, a desire for greater community participation and new mobile and internet-based technologies that are enabling different approaches for data collection and storage.

Supporting tools

There are many existing tools used by different communities of practice and reinventing the wheel needs to be avoided. What is needed for improved decision making is the adaptation of existing land tools to account for climate-related hazards or, alternatively, that a combination of complementary tools is used to better support decision making. This section highlights some of the most promising tools.

The UN-Habitat 'Planning for Climate Change' framework is a strategic, values-based, approach for urban and environmental planners (Ingram and Hamilton, 2014). It provides a suite of 42 tools, including a climate vulnerability assessment methodology based on a spatial analysis of natural hazard exposure data together with community and expert observations. This assessment provides the basis for a participatory development of climate adaptation action plans that focus on areas that are most in need. In the Honiara case, a climate vulnerability assessment was completed in 2014 followed by the development of the Honiara Urban Resilience and

Climate Action Plan (HURCAP) which was endorsed by national and local government in 2017. Land tenure issues were assessed and included in this action planning phase (Trundle and McEvoy, 2017).

Spatial information is fundamental to CRLG. Mapping at the parcel level informs the land administration functions of recording land tenure rights, enhancing tenure security and land use planning and control, as well as the protection of natural resources. The importance of spatial data and information is recognised by the Sendai Framework (United Nations, 2015) with specific mention of the role of geospatial data and technology, GIS, space data and technology, and risk mapping. Spatial information has become even more powerful due to advances in technology such as the increased use of drones, new opportunities from LiDAR and the use of innovative methods such as machine learning to identify slums and informal settlements and other features from imagery.

Other tools include the 'Social Tenure Domain Model' (STDM), which is one example of a pro-poor land administration tool (FIG, 2013). Whilst not developed for climate adaptation planning, it can be used to conduct participatory enumeration that subsequently informs climate vulnerability and disaster risk reduction planning. Importantly, community-driven enumeration documents land tenure arrangements that might not have been understood by outsiders, including local approaches that community members consider 'secure enough' tenure. In practical terms, STDM can be used on handheld devices and supported by recoding point location using handheld Global Navigation Satellite Systems and then related to cadastral mapping or imagery. 'Participatory and Inclusive Land Readjustment' (PILaR) involves a process where land units with different claimants are combined into a single area and redeveloped through a participatory and inclusive process that includes unified planning, re-parcelling and development. The strength of PILaR is that it allows local authorities and citizens to articulate their interests, exercise their land tenure rights, and mediate their differences. The GLTN 'Gender Evaluation Criteria' tool assesses the gender-responsiveness of policies, laws, and processes related to land and climate. One of the lessons after Typhoon Sendong in the Philippines was that land administration responses needed to be informed by the impact of the disaster on the poor and vulnerable. As women were disproportionally impacted, the criteria could be applied to land governance considerations in both pre- and post-disaster contexts. In the case of Syria, which is governed by Islamic laws, the GLTN publication 'Women and Land in the Muslim World' can help develop appropriate land responses (UN-Habitat, 2018).

Conclusions

This chapter has drawn from both a literature review and case study analysis to highlight the inter-linkages between people's security of land tenure and their vulnerability to climate and natural hazards. Importantly, insecure land tenure translates into increased exposure and sensitivity to climate hazards, as well as a reduced capacity to adapt. Given the documented evidence of linkages and feedbacks, it has been argued that improvements in land tenure security can be an important enabler of equitable climate change adaptation. It will not only lead to the increased resilience of poor and vulnerable communities that are currently marginalised but by explicitly considering land tenure issues in the development of adaptation strategies and actions, it will also increase the likelihood of acceptance (and ownership) of measures by affected communities, reducing adverse impacts on existing land tenure arrangements and potential conflict between adaptation 'winners' and 'losers'.

The chapter has also argued for a move towards a more integrated approach to addressing the myriad of challenges facing many developing countries, one that exploits the synergies and common goals of different communities of practice and leads to more effective land use and environmental planning. In addition to institutional arrangements that are fit-for-purpose, climate considerations need to be mainstreamed into land governance policies, programmes and plans, with appropriate processes and tools employed to support a more integrated approach to environmental planning. As countries in the Global South face multiple and increasing shocks and stresses, both climate and human driven, new ways of working are needed if their unique resilience and sustainability challenges are to be effectively managed.

Notes

1 This synthesis draws from case study briefs developed by regional experts for the GLTN study: Honiara (Trundle, Mitchell and McEvoy), St Vincent and the Grenadines (Browne), Philippines (Quizon, Don Marquez, Musni, Naungayan and the Xavier Science Foundation), Syria (du Parc Locmaria) and Uganda (Lwasa). See Mitchell and McEvoy (2019).

2 Although beyond the scope of this chapter, it is also important to note that there are many cross-cutting dimensions to the land tenure and climate vulnerability nexus. These include complex issues of human mobility, gender inequality, ecological integrity, food and water security, indigenous and tribal peoples land rights, increasing conflict over land and natural resources, to name but a few.

3 The GLTN toolkit 'Tenure Responsive Land Use Planning' (GLTN, 2016) provides valuable guidance on how land use planning can be made more tenure-responsive.

References

Asian Development Bank (2016). *The emergence of pacific urban villages: urbanization trends in the Pacific islands*. Philippines: Asian Development Bank.

Biesbroek, G. R., Klostermann, J. E. M., Termeer, C. J. A. M. and Kabat, P. (2013). 'On the nature of barriers to climate change adaptation'. *Regional Environmental Change*, 13(5): 1119–29.

Caron, C., Menon, G. and Kuritz, L. (2014). *Land tenure and disasters strengthening and clarifying land rights in disaster risk reduction and post-disaster programming*. Washington, DC: USAID Issue Brief.

Cotula, L., Camilla, T. and Hesse, C. (2004). *Land tenure and administration in Africa: lessons of experience and emerging issues*. London: International Institute for Environment and Development.

da Silva, J., Kernaghan, S. and Luque, A. (2012). 'A systems approach to meeting the challenges of urban climate change'. *International Journal of Urban Sustainable Development*, 4(2): 125–45.

Dodman, D., Brown, D., Francis, K., Hardoy, J., Johnson, C. and Satterthwaite, D. (2013). *Understanding the nature and scale of urban risk in low- and middle-income countries and its implications for humanitarian preparedness, planning and response*. Human Settlements Discussion Paper Series – Climate Change and Cities 4. London: International Institute for Environment and Development.

Dodman, D. and Satterthwaite, D. (2008). 'Institutional capacity, climate change adaptation and the urban poor'. *IDS Bulletin*, 39(4): 67–74.

FAO (2012). *Voluntary guidelines on the responsible governance of tenure of land, fisheries and forests in the context of national food security*. Rome: UN FAO. Available at: www.fao.org/docrep/016/i2801e/i2801e.pdf

Feder, G. and Feeny, D. (1991). 'Land tenure and property rights: theory and implications for development policy'. *The World Bank Economic Review*, 5(1): 135–53.

FIG (2013). *The social tenure domain model: a pro-poor land tool*. UN-Habitat, FIG Publication No. 52. Copenhagen: International Federation of Surveyors (FIG).

Fitzpatrick, D. (2005). '"Best practice" options for the legal recognition of customary tenure'. *Development and Change*, 36(3): 449–75.

Fünfgeld, H. and McEvoy, D. (2014). 'Frame divergence in climate change adaptation policy: insights from Australian local government planning'. *Environment and Planning C: Government and Policy*, 32(4): 603–22.

GLTN (2016). *Tenure responsive land use planning*. Nairobi: Global Land Tool Network. Available at: www.researchgate.net/publication/315483297_Tenure_Responsive_Land_Use_Planning_A_Guide_for_Country_Level_Implementation

Ingram, J. and Hamilton, C. (2014). *Planning for climate change: a strategic, values-based approach for urban planners*. Cities and Climate Change Initiative Tool Series. Nairobi: UN-Habitat and Ecoplan International.

IPCC (2014). *Climate change 2014: synthesis report*. Geneva: IPCC.

Kiddle, L. (2010). 'Perceived security of tenure and housing consolidation in informal settlements: case studies from urban Fiji'. *Pacific Economic Bulletin*, 25(3): 193–214.

McEvoy, D., Mitchell, D. and Trundle, A. (2019). 'Land tenure and urban climate resilience in the South Pacific'. *Climate and Development*. DOI: 10.1080/17565529.2019.1594666

Mitchell, D., Enemark, S. and van der Molen, P. (2015). 'Climate resilient urban development: why responsible land governance is important'. *Land Use Policy*, 48: 190–8.

Mitchell D. and McEvoy, D. (2019). *Land tenure and climate vulnerability*. Report for Global Land Tool Network, Nairobi, Kenya.

Quizon, A., Marquez, N., Naungayan, M. and Musni, D. (2018). *Climate change, natural disasters, and land tenure: case of Typhoon Sendong (Washi) in Cagayan de Oro City, Northern Mindanao*. Philippines: Asian NGO Coalition for Agrarian Reform and Rural Development (ANGOC).

Satterthwaite, D., Archer, D., Colenbrander, S., Dodman, D., Hardoy, J. and Patel, S. (2018). *Responding to climate change in cities and in their informal settlements and economies*. IIED and IIED-América Latina. Paper prepared for the IPCC for the International Scientific Conference on Cities and Climate Change in Edmonton, March 2018. Available at: https://citiesipcc.org/wp-content/uploads/2018/03/Informality-background-paper-for-IPCC-Cities.pdf

Tompkins, E. L., Adger, W. N., Boyd, E., Nicholson-Cole, S., Weatherhead, K. and Arnell, N. (2010). 'Observed adaptation to climate change: UK evidence of transition to a well-adapting society'. *Global Environmental Change*, 20(4): 627–35.

Trundle, A. and McEvoy, D. (2017). *Honiara urban resilience and climate action plan*. Fukuoka, Japan: UN-Habitat.

Trundle, A., Barth, B. and McEvoy, D. (2018). 'Leveraging endogenous climate resilience: urban adaptation in Pacific Small Island Developing States'. *Environment and Urbanization*, 28 December [Online].

UN-Habitat (2014). *Planning for climate change: a strategic, values-based approach for urban planners*. Asia Pacific Office, Japan: UN-Habitat. Available at: https://unhabitat.org/books/planning-for-climate-change-a-strategic-values-based-approach-for-urban-planners-cities-and-climate-change-initiative/

UN-Habitat (2018). *Women and land in the Muslim world: pathways to increase access to land for the realization of development, peace and human rights*. Nairobi: UN-Habitat. Available at: https://gltn.net/2018/02/22/women-and-land-in-the-muslim-world-2/

UN-Habitat, IIRR and GLTN (2012). *Handling land: innovative tools for land governance and secure tenure*. Nairobi: UN-Habitat. Available at: https://unhabitat.org/books/handling-land-innovative-tools-for-land-governance-and-secure-tenure/

United Nations (2015). *Sendai framework for disaster risk reduction 2015–2030*. Geneva, Switzerland: UNISDR.

Van Gelder, J.-L. (2009). 'Legal tenure security, perceived tenure security, and housing improvement in Buenos Aires: an attempt towards integration'. *International Journal of Urban and Regional Research*, 33(1): 126–46.

21

Planning and disaster risk reduction

John R. Campbell

Introduction

The costs of 'natural' disasters have grown steadily in recent decades in both developing and developed countries. While over the past century the numbers of fatalities have declined, there remains a massive contrast between the death tolls experienced in developing countries compared to developed ones (see Table 21.1). In terms of economic losses, the costs of disasters in developed countries often far exceed the losses in developing countries, but when the losses are expressed as a percentage of gross domestic product (GDP) the losses in developing countries are far more crippling (Table 21.1). Four decades ago, a group of geographers noticing these disparities asked why is it that similar events from a physical perspective had vastly different social and economic outcomes (O'Keefe et al., 1976). They challenged the notion that disasters were natural and pointed to the social, political and economic causes of disasters rather than the unpredictability of nature.

A few years later, another geographer, Kenneth Hewitt (1983), reflected on the characteristics of planning and research on natural hazards and concluded that three elements of what we now call disaster risk reduction (DRR) formed what he called the dominant view. These were the role of science and monitoring with a view to better understanding and predicting extreme geophysical events, the development of planning and managerial processes to control extreme events through measures such as engineering works or to keep people away from them through land use planning and finally emergency management including post disaster relief and rehabilitation. He observed that the social and economic causes of disasters were largely neglected. In the three and a half decades that have elapsed, and despite numerous efforts from the level of the United Nations to local governments to reduce disaster losses, little has changed. Massive investments have been made in measures, such as tsunami and flood warning systems, and there remains a tendency to favour measures to contain natural events through engineering works (for example, Kelman and Rauken, 2012), particularly in relation to river and coastal flooding, and post disaster responses continue to dominate the politics of disasters in both first and third world countries (Bankoff et al., 2004). Hewitt's seminal chapter was a call for disasters to be considered as reflecting everyday social, political and economic conditions and that there needed to be a much greater focus on the political economy that gives rise to disaster events. There have

Table 21.1 Selected disasters showing differences between developing and developed countries

	Country	*Year*	*Deaths*[a]	*Economic Losses*[a] (*$US million*)	*% of GDP*[b]
Earthquakes	Haiti	2010	229,549	8,000	120.80
	New Zealand	2011	182	18,000	10.68
Tropical Cyclones	Myanmar (Nargis)	2008	138,366	4,000	12.55
	USA (Katrina)	2005	1,973	159,060	1.21
Tsunami	Asia/Africa	2004	226,408	9,991	NA
	Indonesia	2004	166,604	4,583	1.78
	Japan	2011	19,976	212,520	3.45

a Data obtained from Centre for Research on the Epidemiology of Disasters (CRED EM-DAT) www.emdat.be/database
b Calculated using GDP (current US$) from World Bank country databases https://data.worldbank.org/country/

been notable advances in our understanding of the political economy of 'natural' disasters (for example, Wisner et al., 2004), but political uptake of these ideas has been limited. Consequently, we can see that planning approaches have the potential to address spatial inequalities that are likely to reduce disparities in disaster losses as well as keeping people away from hazardous areas.

This chapter reviews the approaches that are used (and some often neglected) by planners and others to reduce the social and economic losses caused by disasters. These are: (i) to fight the risks caused by natural extremes; (ii) to avoid the risks; (iii) to accept the risks; and (iv) to reduce the vulnerability of people exposed to risks. Before embarking on this task, the chapter will clarify the terminology used in the field of disaster risk reduction which has several terms with similar or overlapping meanings which can lead to confusion. The chapter will also briefly review global initiatives to tackle the losses that are incurred during disasters and conclude with a brief review of the implications of climate change for planning disaster risk reduction.

Terminology

A term that is often used in planning is hazard mitigation. It has generally been used to describe measures that seek to reduce people's exposure to the effects of extreme events either through measures to protect them by modifying the natural event or by keeping them away from areas that may be affected. The term hazard itself is particularly problematic. Its dictionary meaning is essentially associated with chance or risk. From this perspective, a hazard is the risk that a negative event may occur causing losses to those who are exposed to it. This has been at the basis of much planning to reduce the likelihood of disasters. Thus, for example, flood plains (with the exception of relic geomorphological features) are always at risk of being affected by flood events and planning either seeks to control the flood waters through engineering works such as levees (stop banks), channel straightening and flood gates, for example, or by managing land use so that development on the floodplain is limited. In comparison, the term disaster typically is used to refer to an actual event rather than the potential for the event to occur. The notion of disaster also has a connotation of magnitude. A flood that causes some minor losses which community members can repair with limited cost would be an inconvenience. In comparison, a flood that caused large numbers of homes to be destroyed or damaged, losses of life and injury and disrupted infrastructure may be considered a disaster, especially if the community affected has difficulty coping and recovering on its own account. Similar issues of magnitude apply to other extreme natural events such as earthquakes, tsunami, tropical cyclones, droughts, heatwaves and blizzards among others. High magnitude extreme events tend also to be characterised by low frequency of occurrence and are often described as rare or unscheduled events. However, all places can expect to experience high magnitude events of one kind or another at various frequencies and planning to reduce the risks associated with such events is important.

The terms hazard and disaster are often conflated in popular discourse, the media and indeed in some academic and planning publications. This has the potential to result in substantial confusion with serious implications for communication among different disciplines and between planners and politicians. One solution is to use the term hazard to refer to the natural extreme that is associated with disasters such as floods, tropical cyclones, earthquakes and droughts. In this approach, disasters only occur when a vulnerable entity (a place, community, city, country, etc) is exposed to the hazard event (Wisner et al., 2004). This moves a major component of the analysis of disasters away from an overemphasis on understanding the geology, meteorology or hydrology of extreme events (which is nonetheless still important) to examining the social causes, most of which, going back to Hewitt's work, lie in everyday social circumstances (Wisner et al., 2004).

There has been a significant effort at the international level to address the growing losses from natural disasters. The UN has had an agency responsible for addressing disasters since the establishment of UNDRO (the Office of the United Nations Coordinator for Disaster Relief) which later became UNOCHA, the UN Office for the Coordination of Humanitarian Affairs. Since the 1990s, the concept of disaster risk reduction has emerged, and the UN has established the Office for Disaster Risk Reduction (UNDRR) which has been part of a sustained international effort to reduce disaster losses, culminating most recently in the Sendai Framework for Disaster Risk Reduction which sets the agenda for countries across the work to improve their disaster management. Wisner (2016) points out that this concept is problematic as many disasters are created by so called development processes.

Fighting the risk, focussing on 'nature'

Attempts to control or contain nature have a very long history and go well back into antiquity. In the industrial era, perhaps associated with enlightenment thinking and a commitment to progress and mastery over nature, such measures became increasingly popular in the 19th and particularly 20th centuries. In terms of governmental measures, from national to local levels, to reduce disasters, activities such as flood control often had a much earlier genesis than land use planning. There remains a strong focus on understanding and monitoring nature to identify patterns of frequency and magnitude of extreme events to inform engineering approaches to infrastructure development and measures to contain their physical manifestations. Investments in disaster related science and monitoring far outweigh funding for research into the socio-economic causes of disasters.

Physicalist approaches (Pelling, 2001) are mostly used in both river and coastal flood management with the intention of controlling the natural hazards (see Table 21.2 for some examples of such measures). While there is a well-developed engineering paradigm for flood protection works, often contingent on cost benefit analysis and consideration of patterns of magnitude and frequency, these approaches are always prone to failure in the case of supradesign events. However, the perception that such works provide protection often leads to increased residential and commercial development in protected areas. When the supradesign event occurs, losses are much greater than would be the case if the works were not installed. Often flood control works also serve to contain sediments that would be deposited on flood plains during more frequent

Table 21.2 Controlling and containing floods

Large-scale measures	
Stop banks (levees)	
Channel improvements	Deepening
	Straightening
	Concretisation
Dams	Reservoirs
	Diversion
Watershed management	Reforestation
	Contour ploughing
	Terracing
Sea walls	
Small-scale measures	
Building elevation	Elevating land
	Elevating foundations
Flood proofing	Sealing all or part of structures

lower magnitude floods affecting soil fertility and building up of sediments in the river channel, raising the level of the river bed increasing the risk of the stop banks being topped. Known as the levee effect, this process, while reducing the incidence of smaller flood disasters, encourages development in areas that locks in communities to exposure to catastrophic events for decades.

A major problem with physicalist approaches such as these is the possibility (or perhaps the inevitability) of supradesign events that exceed the capacity of the engineering works to contain the extreme events such as, for example, a stop bank designed to contain a flood with a 1% probability of occurring in any given year failing to cope with a 0.5 % probability event of even greater magnitude. Because the flood control works are considered to protect areas previously prone to flooding, development in the protected areas intensifies. When the levees are breached, the loss is much greater than would have otherwise have been the case and major catastrophes can result.

Physicalist approaches tend to dominate disaster risk reduction practices almost throughout the world. They are not only used in relation to floods but include such techniques as cloud seeding, building walls to divert or halt lava flows and avalanches, and irrigation to reduce drought risk. Such approaches remain well supported in the neoliberal era with its strong commitment to modernist notions of progress, although there is often concern expressed that the costs of such measures often require substantial levels of government investment and place unfair burdens on taxpayers. Accordingly, there has been a move in some areas towards local funding of physicalist measures as central government subsidies have been reduced. Other approaches involving regulation of human activities (see section below) tend to conflict with prevailing free market philosophies. Political pressure to provide protection remains very strong, and following major disasters there are often increased demands for better protection. Where there is existing development because of the levee effect, demands for even greater protection are strong and difficult to deny. While there will always be a requirement for engineered and infrastructural approaches to disaster risk reduction, especially in areas that have already been developed, such measures need to be implemented in conjunction with planning approaches. In areas in which new development is proposed, planning to avoid exposure can be very effective both in economic and social terms. The next section outlines some of these measures.

Keeping people away from the risk

In contrast to measures to keep risk away from people are attempts that seek to keep people away from risk. There are three ways in which disaster losses may be reduced that involve keeping people away from the risk rather than keeping the risk away from people or that increase their resilience. First is emergency evacuation when a hazard event is pending. The second way of keeping people away from risk is to avoid investing, developing and living in areas that are exposed to hazard events. As Table 21.3, which lists some of the tools that are used to reduce losses from flood events, shows, these approaches usually involve land use planning. In many jurisdictions there is increasing use of such approaches although there is often resistance, for example, from land developers who do not want their activities restricted. Since the 1960s, there has been a stream of natural hazards research that examines people's perceptions of hazards to try and determine why they continued to expose themselves to risk. If 'misperceptions' of risk were the reasons for their exposure, then providing more accurate information would be a means of enabling people to make better informed choices about such things as residential location. This could be achieved through the production of hazard maps and placing hazard information on land title deeds for example. The extent to which such approaches work is not clear. Existing residential and commercial land owners often resist measures that demarcate their land as being at risk fearing that their properties will drop in value. In many local government settings, politicians place pressure on planners to ease restrictions on development, perhaps reflecting their own political dispositions, as a response to pressure from influential developers and in a desire to promote development within their areas. The reluctance to adopt measures to keep people away from hazards may also reflect their beliefs that protection measures are effective ways of avoiding disaster risk.

The third approach is to build resilience of communities and their assets to the effects of extreme events and to reduce damage and losses. Many jurisdictions, for example, have building codes that seek to strengthen buildings to withstand earthquakes and high winds and that include minimum floor levels in flood prone areas. Where communities are in areas that are disaster prone, retrofitting can also reduce damage and increase the resilience of structures.

An increasing number of countries have introduced national-level legislation aimed at bringing about sustainable development, that include elements to reduce disaster risk. There are problems with this approach as many aspects of development increase disaster vulnerability (Wisner, 2016). Nevertheless, these legal instruments often enable land use planning measures to be implemented at the level of local government. The degree of uptake varies considerably. In parts of the developing world, planning is limited for a variety of reasons and land use approaches to

Table 21.3 Keeping people away from floods

Information	Hazard maps
	Hazard information on title deeds
Encroachment lines	
Zoning	
Subdivision regulations	
Building codes	
Land acquisition	Demolition of at risk structures
	Relocation of occupants to new housing
	Relocation of at risk structures and occupants
Evacuation planning	

reduce exposure to disasters have even less uptake. International funding for disaster risk reduction is often predisposed towards 'concrete' measures that can be completed within a given time frame as opposed to measures that require ongoing commitment such as sustaining land use regimes and reducing the social elements of vulnerability. In many developing countries, urban areas are characterised by squatter settlements or large areas of informal housing that are often found in the most exposed locations such as unstable slopes or land that is frequently subjected to flooding. Innovative interventions notwithstanding, many of these areas fall outside conventional disaster risk reduction practices. For example, O'Hare and White (2018) develop the concept of flood disadvantage to describe the differential vulnerabilities to flood hazard among and within communities and interventions that may alleviate them. There are many social elements of planning that can address these problems including building inclusive communities, reducing spatial inequalities and enabling widespread access to services.

Integrated approaches to flood risk reduction

As will be apparent, most places are already committed to flood protection with many thousands of kilometres of river and coastline engineered to protect their communities. The costs of maintenance of protection schemes is very high, and indeed in some localities protection works have deteriorated under neglect. The costs of creating new areas of protection are even higher. This has resulted in moves towards reduced support for protection works and to let the costs be absorbed by those who are directly affected. An integrated approach has emerged such as multiscale efforts that combine watershed management and community risk reduction initiatives. Post disaster reconstruction activities increasingly consider the likelihood of the next extreme event and seek to incorporate measures to reduce losses. Coordination across the different levels of such approaches can be difficult. The source-pathway-receptor-consequence model in the United Kingdom is a management tool that attempts to provide an approach that enables the flood hazard to be assessed from the initial hazard event through to the eventual losses. Interventions to reduce risk may be made at different stages in this sequential model. Existing protected areas still need to be protected, but there is less support for new protection works. Accordingly, other approaches such as strong and integrated land use planning are needed. However, they also tend to be in opposition to prevailing neoliberal ideals that tend to oppose inflexible regulations and promote market-based approaches. This has led to information-based approaches (such as hazard mapping) becoming increasingly popular leaving individuals to choose where they may live. The risks and costs of disasters are then shifted towards individuals and away from governments. Such approaches neglect those for whom choice is impossible or limited because of financial restrictions. The outcome would be disaster ghettos where low value landowners or renters live in the most exposed locations.

Dealing with losses

Where measures to either contain nature or modify human behaviour, or a combination of both, fail, then losses will occur. These include losses of life (which can be massive in major disasters, most typically in the developing world – see Table 21.1), injury, loss of property, damage and destruction of infrastructure and disruption of economic and social activities. The most common way of dealing with losses for most people in the world, who have little choice, is to bear them and then face a difficult 'road to recovery'. Another approach is to share the losses. Insurance is one means by which the losses from disasters can be offset. Insured losses from disasters have increased steadily in recent decades but still fall well short of the total losses.

For example, in 2015 overall disaster losses in financial terms were estimated to be three times greater than insured losses (Munich Re, 2017). Most disaster or catastrophe insurance is in the developed world, and in developing countries, apart from micro-insurance schemes, insurance is the privilege of the wealthy. Moreover, there is little to suggest that insurance encourages behaviours that reduce risk, such as residential location choice. As well, recent disasters indicate that many insurance companies resist paying out and delay settlements following disasters losses. In Christchurch, New Zealand, some homeowners are still waiting for settlement (in 2018) following the damage or destruction of their houses caused by an earthquake in 2011. Many countries (including New Zealand) have government disaster insurance schemes that help spread the losses when disaster occur.

Political economy of disasters

This leads us to examine the political economy of disasters. It has long been clear that wealthy people on average tend to be less adversely affected by disaster events. The disparity between developed and developing country disasters has been known for a long time and has been the focus of many researchers examining the causes of disasters (see next section). However, as one observer noted in the 1970s, rich people never die from famine in Africa. Just as there are schisms between rich and poor in the developing world, so too is there a growing gap between rich and poor in the so-called first world. The term class-quake was coined to describe the differences between the middle and upper classes on the one hand and those living in informal settlements on marginal land in Guatemala City. It was the latter group that suffered inordinate losses of life and property during an event in 1976 (Wisner et al., 2004). For a long time, it was assumed that such inequality was confined to the developing world. The disparities between the losses of life in the Haiti and Christchurch earthquakes in 2010 and 2011, the Indian Ocean and Japan (Sendai) tsunami in 2004 and 2011 and Cyclone Nargis (Myanmar) and Hurricane Katrina (New Orleans) shown in Table 21.1 confirm the vast gulfs that exist between the rich and less wealthy nations. However, two of the events in the developed world hide some significant trends. Hurricane Katrina was marked by very significant patterns of mortality and hardship where people of colour and women (particularly those who were single parents) were subject to the greatest losses, not only directly from the flooding that was associated with the event but in the aftermath when their rehabilitation was seriously curtailed by governmental neglect. In Christchurch, the most severely impacted part of the city was the east, where the highest levels of social deprivation existed prior to the earthquakes that shattered much of the city. One of the main points of Hewitt's 1983 article was that disasters, rather than being outside the norm, are in fact reflections of the workings of societies. It follows that if we are to reduce disaster risk we need to examine those societal functions and seek to 'fix' those that give rise to these disaster inequalities. This led to the emergence of disaster vulnerability reduction as a key to disaster risk reduction.

Vulnerability versus resilience

The concept of vulnerability was broached by O'Keefe et al. (1976), who asserted that a disaster occurs when a vulnerable population is exposed to an extreme natural event. They didn't, however, explain in detail what exactly vulnerability was. Chambers' outline of the notion in 1989 has been widely accepted, and most use of the term reflects his early set of general principles.

Vulnerability here refers to exposure to contingencies and stress and difficulty in coping with them. Vulnerability has thus two sides: an external side of risks, shocks and stress to which an individual or household is subject and an internal side which is defencelessness, meaning a lack of means to cope without damaging loss (Chambers, 2006, p. 33).

From this perspective, the major differences in the impacts of disasters when comparisons are made through time and from place to place are not in the magnitude and/or frequency of extreme events (climate change notwithstanding) but differences in the capacity of those affected to cope and their potential to recover. This approach helps us to understand why such extreme differences in disaster losses (fatalities and economic losses) exist between developed and developing countries and between members of different class, racial, gender and ethnic groups within countries, including both developed and developing ones. There is also evidence that in the case of developing countries relatively low levels of vulnerability existed prior to colonisation but have increased since then, a process that has intensified in the context of post colonial neoliberal economics.

The most influential work using the vulnerability approach has been that of Wisner et al. (2004) who tackle the issue by seeking to identify the underlying causes of vulnerability which they locate in restricted access by people and communities to power, structures and resources, and predominant political and economic ideologies. They develop a pressure and release (PAR) model in which they identify a number of dynamic pressures that transform the underlying causes of vulnerability into the proximate causes of disasters: people living in 'unsafe conditions'. From this perspective, a major cause of vulnerability lies in the prevailing political economy. While they identify this as existing at an international or global level, with its manifestations being played out in actions by such agencies as the World Bank and by national governments, neoliberalism also influences local governments where there is a major push to reduce local taxes or land rates. Ameliorating local conditions that are unsafe and lead to disasters is possible but difficult to sustain in such a political economic environment where economic and fiscal policies both increase inequalities and restrict investment in programmes to address it. While planners are largely constrained by the political economic root causes within which they are located, planning can play an important role in addressing the dynamic processes and especially the unsafe conditions that exist within local communities. Planning that seeks to reduce local inequalities, give voice to community members in decision making (for example in post disaster recovery and reconstruction), improve local economic conditions and increase the safety of buildings can play a significant role in reducing disasters.

Perhaps, given that resilience is an antonym of vulnerability, there has been a growth in approaches to build the resilience of at risk communities. This has resulted in some tensions between what may be loosely termed the vulnerability camp which sees a key causative factor in vulnerability being political economic processes and resilience advocates who tend to focus more on what might be termed the proximate causes of disaster. Arguments against resilience also point to the concept's ecological roots that deflect attention from the political economic foundations of vulnerability. Communities in disaster settings often resist being described as resilient when they continue to suffer from disaster impacts. In comparison, the concept of vulnerability can lead to a discourse that suggests that victims of disaster lack agency and capacity in the face of disaster.

Climate change, planning and disaster risk reduction

Climate change is likely to influence patterns of disasters in two ways. First, it is anticipated that extreme climatic and hydrological events will occur more frequently and/or with greater magnitude. Second, some changes in mean conditions, particularly sea level, may have implications for disaster risk management. It is likely, for example, that in many places supradesign events may occur more often and areas that are occupied by residential or commercial premises that were relatively safe from hazards may become exposed. In such cases, decisions will need to be made as to whether increased or new investments in protection works will be necessary or whether relocation may be more efficacious. One approach is to identify areas that in the future may become exposed to increased river or coastal flooding to discourage further investment but where such an approach has been used there has been significant opposition from existing property owners. In addition, addressing the underlying causes of vulnerability will help build the capacity of communities to better cope with hazard events influenced by climate change and other environmental challenges created by a warming world.

Climate change introduces further uncertainty into planning for disaster risk reduction. While the effects of climate change are now beginning to emerge we still have little certainty about the likely patterns of magnitude and frequency of climate related extreme events. A precautionary approach would suggest that developments in areas likely to be at new levels of risk should be avoided and that planners should try to ensure that development does not take place. While adapting to climate change is likely to place perceived barriers to development in some areas, it is taking place in a political environment where there is increasing pressure for the reduction of 'red tape' and reduction of delays in implementing development activities. While there may be short term benefits for politicians in promoting development (and jobs) in some areas prone to climate change, increased risks are likely to be embedded for decades to come at the cost of future generations.

To the future: the roles of planners in disaster risk reduction

Given the complexities of disaster causation and the range of possible responses, what roles can planners play in reducing vulnerabilities and disaster risk? They will depend significantly on the mandates given to planners in their respective jurisdictions. If we use the Wisner et al. (2004) pressure and release model, it would appear that planners have little direct influence on the underlying or 'root' causes of vulnerability. On the other hand, they may play a significant part in reducing the numbers of people living in, or at the level of, unsafe conditions through numerous land use tools. However, the extent to which planners can play a role in improving the wellbeing of people that will enable them to reduce their vulnerability is probably limited. A key area for planners then is to address the dynamic processes identified by Wisner et al. (2004). This is likely to require innovative approaches to land use planning and increasing integration of planning with other elements of government and improved integration of national and local government policies and processes.

References

Bankoff, G., Frerks, G. and Hilhorst, D. (eds.) (2004). *Mapping vulnerability. Disasters, development and people*. Oxford: Earthscan.

Chambers, R. (2006). 'Vulnerability, coping and policy' [Editorial introduction]. *IDS Bulletin*, 37(4): 33–40. (First published in 1989 in the *IDS Bulletin*, 20(2): 1–7.)

Hewitt, K. (1983). 'The idea of calamity in a technocratic age', in K. Hewitt (ed.) *Interpretations of calamity*. Boston: Allen and Unwin, pp 3–32.

Kelman, I. and Rauken, T. (2012). 'The paradigm of structural engineering approaches for river flood risk reduction in Norway'. *Area*, 44(2): 144–51.

Munich Re (2017). 'Natural catastrophes 2016. Analyses, assessments, positions'. *Topics Geo*. Münchener Rückversicherungs-Gesellschaft, München.

O'Hare, P. and White, I. (2018). 'Beyond "just" flood risk management: the potential for – and limits to – alleviating flood disadvantage'. *Regional Environmental Change*, 18(2): 385–96.

O'Keefe, P., Westgate, K. and Wisner, B. (1976). 'Taking the naturalness out of natural disasters'. *Nature*, 260(5552): 566–7.

Pelling, M. (2001). 'Natural disasters?' in N. Castree and B. Braun (eds.) *Social nature: theory, practice and politics*. Oxford: Blackwell, pp 170–88.

Wisner, B., Blaikie, P., Cannon, T. and Davis, I. (2004). *At risk* (2nd edn). London: Routledge.

Wisner, B. (2016). 'Vulnerability as concept, model, metric and tool'. *Oxford Research Encyclopedia of Natural Hazard Science*. Available at: http://naturalhazardscience.oxfordre.com/view/10.1093/acrefore/9780199389407.001.0001/acrefore-9780199389407-e-25?print=pdf

22

Planning, infrastructure and low carbon energy

Matthew Cotton

Introduction: planning for low carbon energy systems

The Paris Agreement on Climate Change sets a global long-term temperature goal (LTTG) of 1.5°C of warming as a limit to dangerous anthropogenic interference with the climate system. Globally, the burning of coal, natural gas and oil for electricity and heat is the largest single source of global greenhouse gas emissions (collectively, the energy sector accounts for 25% of 2010 global greenhouse gas emissions) (Intergovernmental Panel on Climate Change, 2014). The rapid decarbonisation of energy systems for electricity, heating and cooling is therefore an urgent environmental policy priority for advanced industrial economies. The concept of 'sustainable transitions' is commonly employed to analyse the necessary changes to fundamentally 'socio-technical' energy systems. By this it is meant that the transitions literature examines not just how to implement 'techno-fixes' to the climate change problem but rather interrelated technical, social, institutional, governance and economic contexts that lead to complex and multi-dimensional adaptions towards sustainable modes of production and consumption. However, an emergent trend in transition analysis focuses specifically on *where* transitions take place, the spatial configurations and dynamics of the networks within which transitions evolve (Coenen et al., 2012), the social influence of new infrastructures on social practices (such as travelling on public transport or shopping for consumer electrical items) and, importantly, upon the governance frameworks that influence such socio-spatial and scalar dynamics.

We can understand EU-led sustainable transition drivers in the energy sector as a matter of *multi-scalar* governance. From the top-down are specific legislative and policy measures to reduce CO_2 emissions in the electricity sector. The Renewable Energy Directive (2009) sets legally binding targets for Member States to reduce emissions by 40% in the electricity sector by 2020. Though top-down in one sense, individual Member States have different available resources and unique energy markets and have consequently adopted different transition pathways to meet their obligations. These include plans for new renewable energy targets for the electricity, heating, cooling and transport sectors, research and development into new renewable energy technologies, alongside policy measures that link national, regional and local authority targets in compliance with the EU's sustainability criteria. The EU's supranational multi-scalar approach has shown success in transitioning Member States away from fossil fuel-generated electricity production (particularly coal) towards renewable and nuclear alternatives. However, commitment to low carbon energy (particularly renewable energy) generation is highly varied. Within the EU, Norway and Sweden perform strongly on renewable generation as a percentage of total electricity produced with 69.4% and 53.8% share of total gross generation from renewables respectively, based on 2016 figures (Eurostat, 2016). This compares favourably to Germany with 14.8% gross renewable generation and the United Kingdom on 9.3% (Eurostat, 2016).

The different levels of commitment to improving renewables capacity reflects the diversity of energy policy objectives and resources amongst Member States, which in turn produces very different energy systems. In Germany, for example, since 2011 the 'Energiewende' platform has reoriented energy policy from electricity demand to supply, with an emphasis upon renewable energy generation, phasing out of coal and nuclear, and a concurrent shift from centralised to distributed generation (using for example small-scale combined heat and power units, and electricity micro-grids). The 'Energiewende' aims to reduce electricity waste through overproduction and reduce energy consumption through domestic and industrial energy saving measures and efficiency savings. By contrast, the United Kingdom is performing poorly. Despite manifesto commitments to renewable energy generation, and legally binding CO_2 reduction targets, since 2015 the Conservative Government has made various cutbacks to renewable energy subsidies (for example, closing early the Renewables Obligation scheme to wind and solar, removing the exemption clause for renewable electricity from the Climate Change Levy (CCL) and cutting back the feed-in-tariff for small-scale wind and solar (Solorio and Fairbrass, 2017)). The government's alterative 'clean growth strategy' provides renewed backing for carbon capture and storage projects, support for a new generation of nuclear power stations (to follow the new Hinkley Point C project) and policies to promote the development of a market for electric vehicles. What we see in these two examples is a highly differentiated policy approach between comparable sized economies within Europe. Yet the common thread is the emphasis upon new technology development and marketisation for what are deemed emergent low carbon products (such as electric vehicles or carbon capture and storage (CCS) technologies) and/or the continued subsidisation (or conversely the removal of such subsidies) for established renewables technologies (such as solar photovoltaics and onshore wind turbines).

Developing new energy infrastructure

The common theme in sustainable energy transition policies is that of technology development – whether large-scale or micro-scale renewables, fourth generation nuclear power stations, smart electricity grids and/or increased infrastructural capacity to support electric vehicles or carbon capture and storage. Across Europe, energy policy frameworks have commonly focused upon investment strategies to bring low carbon technologies to market (what could be termed a 'technology-push' platform), or else upon policies that emphasise 'supply-push' to encourage research and development programmes in renewable generation, and 'demand-pull' regulatory mechanisms to improve the uptake of renewable energy technologies, such as feed-in-tariffs or subsidies (Strachan et al., 2006). Government are not only trying to bring new technologies to market, but they are also trying to *find sites* for energy technologies that must be integrated into electricity transmission and distribution networks (or 'grids'). As Wolsink (2007) argues, successful investments and the siting of low carbon energy infrastructures will eventually determine the success rate of national efforts in establishing low carbon energy capacity. Promoting different energy transitions is therefore inherently subject to 'spatial dynamics'. Whereas the fossil-fuelled economic generation of industrial societies required the development of electricity infrastructure close to sites of production (mainly near to coal reserves), new wind, biomass and nuclear generation capacity redeploys electricity technologies to areas of high amenity value (such as rural and coastal locations) or closer to cities to support an increasingly urbanised society. This means that one of the key areas of policy that requires development in producing successful low carbon transitions is that of 'spatial planning' to govern these rapid infrastructure changes.

The politics of spatial planning

Spatial planning is an arena in which the politics of low carbon infrastructure have been reimagined, reconfigured and deeply contested. The first factor of note is that processes of infrastructure planning have been subject to sustained policy innovation in the last 40 years. In the 1980s-90s, the market liberalisation and privatisation of energy systems in many European countries moved the socio-technical systems of energy production and consumption out of direct state institutional control (common after World War II). Their subsequent privatisation required new regulatory institutions and ways of managing costs, design, and siting procedures. The socio-technical systems of energy commonly exhibit strong path dependencies and high barriers for radical innovation. Market liberalisation, however, initiated a fundamental restructuring of such socio-technical systems (Markard and Truffer, 2006) involving new governance arrangements, including market integration across borders (Jamasb and Pollitt, 2005) and innovation in technology policy and regulation. In countries such as the United Kingdom and Germany, we see that low carbon energy infrastructure has been increasingly governed in the context of the 'regulatory state' (Eberlein and Grande, 2000), characterised by processes of privatisation and deregulation which replaced the 'dirigiste state' of the past (whereby the state exerts a strong directive influence over investment). Thus *regulation* rather than public ownership, planning or centralised administration has become the key context in how low carbon energy transitions have come to be governed (While et al., 2010).

The second key factor is a shift towards the strategic organisation of space at different scales (Albrechts, 2004) and a renewed emphasis upon spatial planning as a political process. With the development of the regulatory state, spatial planning ceases to be a passive, bureaucratic process of distributing physical resources to become something more overtly politically and ethically significant. Spatial planning has been described variously as a new space of governance, or a political resource (Allmendinger and Haughton, 2010), promoting new ways of thinking about space and place and the role of spatial strategies in contemporary governance contexts (Healey, 2004). Spatial planning has become an arena of political innovation. The so-called 'European Spatial Development Perspective', sometimes described as an 'inter-governmental' approach (Faludi, 2002), represents new forms of political practice across multiple scales of governance, whilst innovation in local and regional spatial strategies simultaneously provided new approaches that integrated land use policies with other forms of policies and programmes that influenced the nature of places (Baker et al., 2010). Across different scales and contexts, innovation in spatial planning and governance has considerable influence upon the types and mechanisms of infrastructure development. Yet despite innovation in new types of governance (including an increasing emphasis upon public participation), there still exist significant barriers to low carbon transitions. The most significant of these is the threat of public opposition to infrastructure projects.

Social opposition and place-protective action

The spatial planning of low carbon infrastructure is fraught with contestation at multiple scales and sites and governance. Barry et al. (2008) argue that public acceptability of renewable energy technologies cannot be taken for granted when the energy technology moves from abstract support to local implementation; and often the key challenges have little to do with the technology itself. We must consider the contextually, embedded qualities of particular social and physical landscapes (Cowell, 2010; Pasqualetti et al., 2002): how they become contested by multiple stakeholder actors and how they might be represented at different scales of governance in the same way as other energy policy considerations like economic and other resource availability, efficiency and technical feasibility. There is a mismatch between public support for renewable energy taken in the abstract and public support at a locally, geographically-situated level. This has been described as a 'social gap' between the high support for certain types of renewable energy (for example, wind) reported in social surveys/opinion polls and the relatively low success rate for planning applications – and an 'individual gap' whereby an individual supports renewable energy in general but opposes specific developments within particular locations (Bell et al., 2005). This 'individual gap' is commonly framed as a not-in-my-backyard (or NIMBY) problem. NIMBY commonly describes acceptance of the need for infrastructure but the rejection of specific proposals. Opponents of developments may deploy arguments that the facility is not needed locally, or should not be built anywhere (not-in-anyone's-backyard [NIABY]), that the siting procedures and decision making surrounding implementation of the facility construction are insufficient or else that the facility will bear unacceptable levels of risk in the form of harmful environmental and health effects. When opposing a specific technology on environmental, health and community cohesion grounds, specific impacts may be highlighted – including the decline in property values and decline in quality of life due to environmental impacts such as radioactive contamination (in the case of nuclear power), noise, traffic, avian collisions (in the case of transmission lines or wind turbines), odour or light pollution or more subtle issues such as the decline in community image and technological stigmatisation, the overburdening of community services and community budgets or the inability of the community to keep out other undesirable land uses once one has been sited.

When community actors engage in 'place-protective' action, this is commonly labelled as a *selfish* act – that opposing something that could potentially benefit the public good 'symbolises a perverse form of antisocial activism' (Hornblower, 1988, p. 44). For example, in 2009, in the United Kingdom when Ed Miliband MP was then Secretary for Climate Change, he stated that opposition to wind farms should be as socially unacceptable as not stopping at a zebra crossing or not wearing a seat belt (cited in Stratton, 2009). Research has shown that NIMBY is used almost exclusively by project managers, developers and policy makers as a blanket label to undermine the opposition tactics of a range of opponents (Burningham et al., 2006; Devine-Wright, 2009). A 'NIMBY' is characterised as a local home owner who is unwilling to think about the 'big picture' of infrastructure development for society; is unable or unwilling to grasp the complexities of the engineering practices, risk and policy dimensions; will not bear the costs of commercial production and yet are willing to reap the benefits; and is predominantly concerned with local house prices, poorly defined environmental degradation and loss of local amenity value rather than broader systemic environmental issues such as climate change or energy security (Devine-Wright, 2005). In short, opponents to low carbon energy technologies are *imagined* by proponents of the technology as selfish, and this in turn influences the types and forms of engagement that those planning and policy authorities employ (Barnett et al., 2012; Cotton and Devine-Wright, 2010).

The way in which opponents to low carbon technologies are imagined is important because it influences the social status and relationships within and between the communities affected by the actions of energy development organisations. Place-protective actions are deemed unethical by some commentators because opposition to energy technology projects is commonly dominated by white, middle class and politically empowered representatives of community organisations that may be successful in blocking planning applications. As Gerrar (1993) states, this effectively perpetuates environmental injustice – developers will tend, in the face of strong opposition, to move siting processes for unwanted facilities away from regions with higher levels of political, economic and social capital towards areas with less of these resources. This means that communities living under high levels of social deprivation or existing racial injustices become 'targets' for development and are less able to defend themselves. NIMBY positions are therefore potentially socially divisive by reinforcing class and racial divisions, economic marginalisation and unequal distribution of physical and social capital.

This position is not universally shared however. Devine-Wright (2009) and Burningham (2000) have shown empirically that 'NIMBY' is an inaccurate portrayal of why people react negatively to local environmental change. There are intangible socio-cultural factors that intensify localised opposition, and these cannot simply be ameliorated through community education campaigns or compensation schemes. Moreover, research has shown that place-protective action is frequently more scientifically grounded (and less locally-myopic) than project proponents imagine (van der Horst, 2007). Though it is true that opponents may perceive developments as risky, costly or visually unattractive, this often leads locally affected community members to question issues of community level fairness, energy strategy, utility and place identity. As Jay has argued, NIMBIES should be redefined as:

> [A]ny citizen, who tries to defend their home and their neighbourhood from plans which would destroy the view, pollute the environment, overload the transport network, upset the ecosystem and knock £50,000 off the value of their house. When it comes to our own back yard, we are all NIMBYs, every NIMBY deserves respect for standing up to corporate and government giants.
>
> *(Jay, 2005, p. 1)*

This is significant because a broadly *utilitarian* moral case is commonly made by energy developers to support their proposals (whereby the greatest good is maximised for the greatest number of people by implementing technologies that promote low carbon energy consumption). Affected site communities are therefore expected to think past their individual interests towards broader, national energy policy goals. However, this is problematic when individuals are asked to accept risks generated by a profit making industry without providing financial compensation or other benefits in kind, this is 'supererogatory' – it might be morally good for a community to accept such an imbalance of risks and benefits, but it is not morally required (Peterson and Hansson, 2004). What is needed for promoting social acceptance of low carbon infrastructure and the reduction of place-protective action is to ensure *fairness* in spatial planning processes in order to achieve socially just and publicly acceptable outcomes and thus facilitate rapid low carbon transitions.

Fairness and democracy in energy planning

The relative fairness of infrastructure planning is essential to the successful implementation in the context of the regulatory state. In the literatures on energy justice, the concept of *procedural* fairness (concerning how a technology and a site is chosen, what the alternatives might be, who regulates the industry, and who is involved in process) and *distributive* fairness (concerning how positive and negative outcomes are shared between those that profit and those that bear the impacts) are essential components. Justice is important because market liberalisation undermines the utilitarian argument for low carbon infrastructure as a *public good*. Low carbon infrastructures are privately owned with profits distributed to shareholders. If the state uses its planning powers to justify corporate interests over community place protection, then an inevitable conflict emerges. Ensuring that the process by which decisions over energy technology siting and choice is fair, and that where private industry profits communities are adequately compensated for bearing the risks and costs locally, is a measure of project success. This is because both procedural and distributive fairness aspects have been shown to be key drivers of public acceptability in energy project siting. Fairness is key because even if the outcomes of a siting process remain unwanted by opponents, they are more likely to be accepted if the process of deciding is perceived to be fair and transparent (Gross, 2007).

The push for more low carbon infrastructure tests the *democratic* nature of spatial planning processes. Governments have experimented with planning policy measures to balance the rights of the community and the national need for infrastructure. The most commonly cited solution to this problem is to increase the range and type of public participation processes available to local communities. This is sometimes referred to as a communicative (Healey, 1996) and participatory-deliberative turn in policy and planning (Saurugger, 2010). This is characterised by moving away from aggregative modes of decision making (such as voting), towards the direct involvement of citizens in plan-making through mechanisms such as citizens juries, consensus conferences or workshops. This move towards engagement has since become institutionalised across a range of different planning and governance contexts (Pløger, 2001).

Though participation has the capacity to improve the procedural fairness of energy project siting, it is by no means a panacea to all social opposition. When government action comes into conflict with community interests there are a number of ways in which opposition movements will try to stop such developments, irrespective of what types of public engagement process are offered by authorities (Fraune and Knodt, 2017). Direct action against developers at the sites of energy infrastructure siting remains a key tactic of social opposition movements, though under the governance of the regulatory state, it is common that formal processes of spatial planning (and the decisions made by planning authorities) become a focus of place-protective action. This means that spatial planning itself has been increasingly represented as 'a problem' in meeting infrastructure development (and specifically renewable energy) targets (Johnstone, 2014).

The response of many state authorities is a tendency towards 'streamlining' infrastructure planning processes. In practice, this means the closing down of opportunities for public participation in decision-making processes and outcomes, changing the emphasis from ensuring fairness for affected communities towards building as much as possible, as fast as possible. The justification is commonly one of urgency – that climate change targets need to be met by specific deadlines and so rapid decarbonisation is necessary. Though this has potential environmental benefits, it pits local interests against national interests (Cowell and Owens, 2006). This question of the scale at which decisions over energy technologies are made is fundamental to this question of fairness. What we find is that the concept of scale (either national or local) is something that is contested within the spatial planning domain. Under the conditions of the regulatory state, national authorities increasingly frame energy projects as major infrastructures of national significance, in order to try and improve their uptake in the market. In the United Kingdom, for example, the Planning Act 2008 and the Localism Act 2011 define major energy projects such as nuclear power stations or large onshore wind farms as nationally significant infrastructure projects (NSIPs). Infrastructure is defined in policy as important to the nation as a whole, and so decisions over 'siting' are taken first by private developers (who present infrastructure proposals to meet the needs set out in National Policy Statements), which are then approved at the level of central government. Johnstone argues that this type of policy creates and separates different categories of scale, which are then reified within planning practice, enacting a type of political separation between scalar boundaries. Whereas under the system of local planning control local authorities take the decisions over what gets built, a system of nationally significant infrastructure projects re-scales the significance of the technologies themselves, and thus the types of actors that are allowed to be involved in decision-making processes. Social actors such as opposition groups, that are then excluded from decision making, will try to challenge or reinterpret these scales in order to further their own strategic agendas (Johnstone, 2014). This is sometimes called 'jumping scales' (Swyngedouw, 2004), whereby opposition actors try to appeal to national level interests in order to be heard and public authorities try to characterise opponents as NIMBIES in order to shrink the problem down to a local scale. Moreover, although modern infrastructure planning policy maintains the *language* of public participation, there is increasing evidence that this is simply a form of 'deliberative speak' (Hindmarsh and Matthews, 2008) – a type of rhetorical device which emphasises concepts such as transparency, citizen involvement and community control in policy but without formal mechanisms of power sharing, partnership or voluntarism in order to make that participation happen. There is concern, therefore, that decisions over low carbon energy infrastructures are increasingly 'post-political' – whereby the methods of democratic decision making used in spatial planning do not lead to community empowerment but rather to the rubber-stamping of the interests of private developers (Johnstone, 2014; Cotton, 2014). This is, as Allmendinger and Haughton (2012) argue, because the rhetoric of participation, alongside vague objectives and nomenclature around sustainable growth, does nothing to remove conflict from planning processes, but rather it displaces or residualises conflict. In the post-political condition, decisions over infrastructure appear on the surface to be made through formal planning processes, but the opportunity to challenge the private interests of developers is curtailed and hidden behind a participatory façade.

Conclusion

In conclusion, changes in spatial governance have stimulated new forms of infrastructure planning, designed to improve the uptake of low carbon infrastructures and their rapid deployment and integration into electricity networks. The regulatory state creates conditions whereby major private infrastructure investors, such as large energy companies, have a powerful influence over energy policy. What gets built is ultimately a private investment decision, not something controlled centrally by the state. However, under concerns that local opponents will be too myopic to understand the broader issues at stake, developers and policy makers commonly imagine opponents of such infrastructure development processes as self-interested NIMBIES, incapable of having meaningful input in decisions. This has led to governments seeking to reform infrastructure planning processes and systems of governance to take power away from local people and re-scale the decisions upwards to the national level. However, the concern remains for many planning theorists and practitioners that these policy measures designed for rapid infrastructure deployment are, in fact, counter-intuitively fraught with more difficulties and planning delays. By removing the powers of local control over technology and siting choice, by drawing planning powers back to central authorities and by stripping away opportunities for public participation in decision making, local people have no choice but to block applications through other means. As planning becomes less fair, social acceptance of infrastructure falls. As it does so we see a rise in protest, direct action, trespass on sites, and political and social media campaigning to disrupt the activities of industry and the state. This ultimately increases developer and government costs, decreases local trust in authorities and slows the low carbon transition we desperately need in an age of rapid anthropogenic climate change.

References

Albrechts, L. (2004). 'Strategic (spatial) planning reexamined'. *Environment and Planning B: Planning and Design*, 31(5): 743–58.

Allmendinger, P. and Haughton, G. (2010). 'Spatial planning, devolution, and new planning spaces'. *Environment and Planning C: Government and Policy*, 28(5): 803–18.

Allmendinger, P. and Haughton, G. (2012). 'Post-political spatial planning in England: a crisis of consensus?' *Transactions of the Institute of British Geographers*, 37(1): 89–103.

Baker, M., Hincks, S. and Sherriff, G. (2010). 'Getting involved in plan making: participation and stakeholder involvement in local and regional spatial strategies in England'. *Environment and Planning C: Government and Policy*, 28(4): 574–94.

Barnett, J., Burningham, K., Walker, G. and Cass, N. (2012). 'Imagined publics and engagement around renewable energy technologies in the UK'. *Public Understanding of Science*, 21(1): 36–50.

Barry, J., Ellis, G. and Robinson, C. (2008). 'Cool rationalities and hot air: a rhetorical approach to understanding debates on renewable energy'. *Global Environmental Politics*, 8(2): 67–98.

Bell, D., Gray, T. and Haggett, C. (2005). 'The "social gap" in wind farm siting decisions: explanations and policy responses'. *Environmental Politics*, 14(4): 460–77.

Burningham, K. (2000). 'Using the language of NIMBY: a topic for research, not an activity for researchers'. *Local Environment*, 5(1): 55–67.

Burningham, K., Barnett, J. and Thrush, D. (2006). *The limitations of the NIMBY concept for understanding public engagement with renewable energy technologies: a literature review*. Manchester: Manchester University Press.

Coenen, L., Benneworth, P. and Truffer, B. (2012). 'Toward a spatial perspective on sustainability transitions'. *Research Policy*, 41(6): 968–79.

Cotton, M. (2014). 'Environmental justice challenges in United Kingdom infrastructure planning: lessons from a Welsh incinerator project'. *Environmental Justice*, 7(2): 39–44.

Cotton, M. and Devine-Wright, P. (2010). 'NIMBYism and community consultation in electricity transmission network planning', in P. Devine-Wright (ed.) *Renewable energy and the public: from NIMBY to participation*. London: Routledge, pp 115–30.

Cowell, R. (2010). 'Wind power, landscape and strategic, spatial planning – the construction of "acceptable locations" in Wales'. *Land Use Policy*, 27(2): 222–32.

Cowell, R. and Owens, S. (2006). 'Governing space: planning reform and the politics of sustainability'. *Environment and Planning C*, 24(3): 403–21.

Devine-Wright, P. (2005). 'Beyond NIMBYism: towards an integrated framework for understanding public perceptions of wind energy'. *Wind Energy*, 8(2): 125–39.

Devine-Wright, P. (2009). 'Rethinking NIMBYism: the role of place attachment and place identity in explaining place-protective action'. *Journal of Community and Applied Social Psychology*, 19(6): 426–41.

Eberlein, B. and Grande, E. (2000). 'Regulation and infrastructure management: German regulatory regimes and the EU framework'. *German Policy Studies*, 1(1): 39–66.

Eurostat (2016). Available at: https://ec.europa.eu/eurostat/tgm/table.do?tab=table&init=1&language=en&pcode=t2020_31&plugin=1.

Faludi, A. (2002). 'Positioning European spatial planning'. *European Planning Studies*, 10(7): 897–909.

Fraune, C. and Knodt, M. (2017). 'Challenges of citizen participation in infrastructure policy-making in multi-level systems – the case of onshore wind energy expansion in Germany'. *European Policy Analysis*, 3(2): 256–73.

Gerrar, M. B. (1993). 'The victims of NIMBY'. *Fordham Urban Law Journal*, 21(3): 495–522.

Gross, C. (2007). 'Community perspectives of wind energy in Australia: the application of a justice and community fairness framework to increase social acceptance'. *Energy Policy*, 35(5): 2727–36.

Healey, P. (1996). 'The communicative turn in planning theory and its implications for spatial strategy formation'. *Environment and Planning B: Planning and Design*, 23(2): 217–34.

Healey, P. (2004). 'The treatment of space and place in the new strategic spatial planning in Europe'. *International Journal of Urban and Regional Research*, 28(1): 45–67.

Hindmarsh, R. and Matthews, C. (2008). 'Deliberative speak at the turbine face: community engagement, wind Farms, and renewable energy transitions, in Australia'. *Journal of Environmental Policy & Planning*, 10(3): 217–32.

Hornblower, M. (1988). 'Ethics: not in my backyard, you don't'. *Time Magazine*, 27 June.

Intergovernmental Panel on Climate Change (2014). *Working group III: mitigation of climate change*. Geneva: World Meteorological Organization.

Jamasb, T. and Pollitt, M. (2005). 'Electricity market reform in the European Union: review of progress toward liberalization and integration'. *The Energy Journal*, 26(Special edn): 11–41.

Jay, A. (2005). *Not in our back yard: how to run a protest campaign and save the neighbourhood.* Devon: White Ladder Press.

Johnstone, P. (2014). 'Planning reform, rescaling, and the construction of the postpolitical: the case of the Planning Act 2008 and nuclear power consultation in the UK'. *Environment and Planning C: Government and Policy*, 32(4): 697–713.

Markard, J. and Truffer, B. (2006). 'Innovation processes in large technical systems: market liberalization as a driver for radical change?' *Research Policy*, 35(5): 609–25.

Pasqualetti, M. J., Gipe, P. and Righter, R. W. (2002). 'A landscape of power', in M. J. Pasqualetti, P. Gipe and R. W. Righter (eds.) *Wind power in view: energy landscapes in a crowded world.* San Diego: Academic Press, pp 3–17.

Peterson, M. and Hansson, S. O. (2004). 'On the application of rights-based moral theories to siting controversies'. *Journal of Risk Research*, 7(2): 269–75.

Pløger, J. (2001). 'Public participation and the art of governance'. *Environment and Planning B: Planning and Design*, 282: 219–41.

Renewable Energy Directive (2009). *Directive 2009/28/EC of the European Parliament and of the Council of 23 April 2009.* Official Journal of the European Union.

Saurugger, S. (2010). 'The social construction of the participatory turn: the emergence of a norm in the European Union'. *European Journal of Political Research*, 49(4): 471–95.

Solorio, I. and Fairbrass, J. (2017). 'The UK and EU renewable energy policy: the relentless British policy-shaper', in I. Solorio and J. Fairbrass (eds.) *A guide to EU renewable energy policy: comparing Europeanization and domestic policy change in EU Member States.* Berlin: Springer, pp 104–20.

Strachan, N., Lal, D. and von Malmborg, F. (2006). 'The evolving UK wind energy industry: critical policy and management aspects of the emerging research agenda'. *European Environment*, 16(1): 1–18.

Stratton, A. (2009). 'Opposing wind farms should be socially taboo, says Ed Miliband'. *The Guardian*, 24 April.

Swyngedouw, E. (2004). 'Scaled geographies: nature, place, and the politics of scale', in E. Sheppard and R. B. McMaster (eds.) *Scale and geographic inquiry: nature, society, and method.* Oxford: Blackwell, pp 129–53.

van der Horst, D. (2007). 'NIMBY or not? Exploring the relevance of location and the politics of voiced opinions in renewable energy siting controversies'. *Energy Policy*, 35(5): 2705–14.

While, A., Jonas, A. E. G. and Gibbs, D. (2010). 'From sustainable development to carbon control: eco-state restructuring and the politics of urban and regional development'. *Transactions of the Institute of British Geographers*, 35(1): 76–93.

Wolsink, M. (2007). 'Planning of renewables schemes: deliberative and fair decision-making on landscape issues instead of reproachful accusations of non-cooperation'. *Energy Policy*, 35(5): 2692–704.

23

Waste and management of environmental resources

Nick Hacking

Introduction: shifting from linearity to circularity

Wastes and resources governance is an evolving interdisciplinary domain long dominated by practice. Academic researchers from a variety of disciplines have contributed to emergent dialogue and debates about the key principles of sustainable waste management, in particular with the concepts of 'industrial ecology' (IE) and 'circular economy' (CE).

Wastes, like so many socio-economic materials, have become increasingly global in their flows over the last century (Schaffartzik et al., 2014; Krausmann et al., 2017a). Resource flows are currently reaching levels of production that are altering the planet's biogeochemical cycles and adding support to the concept that the planet is entering the Anthropocene era (Krausmann et al., 2017b). The global rise in waste production (see Figure 23.1) comes with a set of shifting normative narratives about reducing, recycling and reusing wastes and thinking of them more as *resources*. Actors in the wastes and resources sectors of both the developed and developing worlds are encouraged by institutions to abandon 'linearity' and make a 'sustainability transition' towards increased 'circularity' where less resources are buried or burned (and less energy is wasted) through improved product design, increased recycling and reuse. The idea is for improved waste governance to be driven by a mix of legislation and/or market incentives.

The latest conceptual thinking about such circular waste and resource principles suggests the CE is, however: 'still in [its] infancy and the literature is only emerging' (Korhonen et al., 2018, p. 545). At a time when the momentum for achieving sustainable development goals is perceived to have flagged amongst some public and private stakeholders, this chapter suggests that the ways that particular actors opt to 'buy-in' (or not) to particular knowledge frameworks will inevitably shape how moves towards CE unfold over time. In that context, some existing policy approaches to waste and resources are reviewed. These aims permit an assessment of the evolving nature of waste and resources governance which is closely linked to the delivery of waste and resources infrastructure via national planning systems.

The next section outlines definitions linked to various disciplinary literatures.

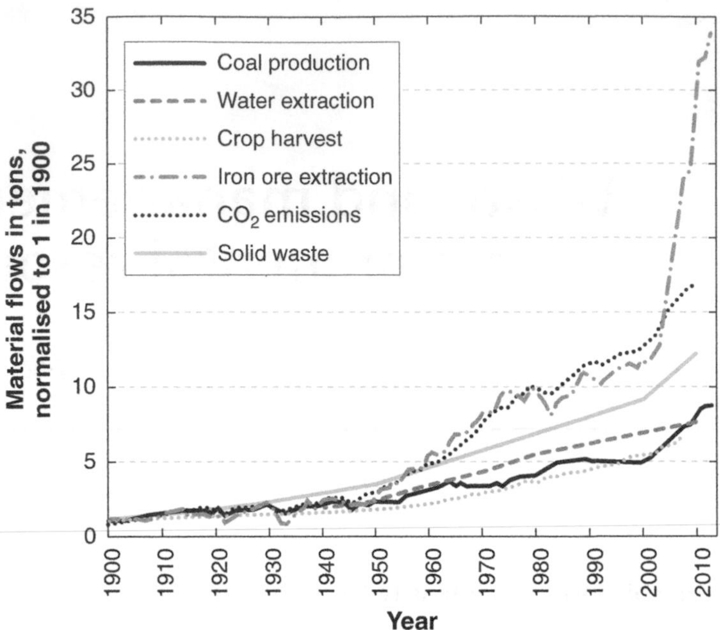

Figure 23.1 The rise of certain global material flows, 1900–2010
Source: adapted from Krausmann et al. (2017b, p. 648)

Definitions

Key to a normative shift from a linear to a circular system of waste and resources (at a range of levels) is the operationalisation of recasting wastes as reusable resources. Definitions are therefore critical, but 'waste' is hard to define (Wynne, 1987). Its materiality is contained through socially-constructed disposal practices from households through to recycling and disposal (Hird, 2012). This materiality is: 'a long ways from stuff that "just is" ... rather it *becomes*" (Gregson and Crang, 2010, p. 1028, emphasis added) and is ideologically, symbolically and culturally contained via institutions involved in legislation, surveillance, public education, health discourse and nation-building rhetoric (Hird, 2012). By contrast, 'resources' refer to inputs immediately before 'materials'. Where materials from a waste stream are equivalent in quality to virgin materials then that waste stream is a resource. Depending on the nature of the processes associated with them, some waste items are thus either materials or resources (Nakamura et al., 2007). Such definitions are also disputed given that they cover numerous waste and resources processes in a CE (Kirchherr et al., 2017). The most prominent CE definition comes from the Ellen MacArthur Foundation (EMF), a non-governmental organisation promoting CE activity amongst industry. The CE should be:

> [A]n industrial system that is restorative or regenerative by intention and design. It replaces the 'end-of-life' concept with restoration, shifts towards the use of renewable energy, eliminates the use of toxic chemicals, which impair reuse, and aims for the elimination of waste through the superior design of materials, products, systems, and, within this, business models.
>
> *(EMF, 2012, p. 7)*

In sum, materials and wastes are meant to undergo continual shifts in classification, or *becoming*, as they move through an idealised CE instead of reaching a point of 'end of use'. Yet, the materiality of wastes and resources produces a range of associated practices that further make defining and thinking about transition pathways towards CE contested and unclear (cf. Hird, 2012; Kirchherr et al., 2017). This means that waste and resources actors – including industry – are choosing to interpret CE activity in different ways as explored further in the 'Goals-based policy narratives' section.

Conceptual overview

In the 1960s and 1970s, 'industrial ecology' emerged as a response to reappraisals by natural scientists of the Earth as a *closed* chemical and biological system (apart from the input of solar irradiance) with a finite supply of resources (for example, Boulding, 1966; Meadows et al., 1972). Previously, the dominant, linear, end-of-pipe approaches to pollution control depended solely upon technical fixes. Natural resources, including water, energy, biomass, metals, and minerals, were treated separately along with waste disposal to the air, water, and soil (Weisz et al., 2015). Industrial ecology was radically different. Actors now needed to search for material value in relatively loosely governed, closed-loop biogeochemical systems and to operate at a range of scales from the local to the global.

With its biological metaphors – for example, 'positive feedback', 'homeostasis' and 'material flows' – the idea of an industrial ecology was structural, i.e. positivist in nature and advocated by practitioners and researchers: for example, economists, engineers, natural scientists and toxicologists chiefly in Japan, Germany and the United States (for example, Federal Waste Disposal Act, 1972; Stahel and Reday, 1976; Nishimura, 1977; Watanabe, 1994). Proponents built structural-functionalist systems models in which actors were assumed to have perfect knowledge and make rational decisions. Conceptions of agency and structure try to reveal the system's deeper structures and functions (cf. Nishimura, 1977). Other assumptions included the system's default orientation seeking equilibrium and sustainable system change only occurring incrementally and as a result of external, or *exogenous*, shocks to the system (cf. Geels, 2010).

Post-structural (i.e. post-positivist) shifts in the social sciences taking place from the 1960s onwards helped with critiques of the assumptions of structural-functionalism. In response, structural-functionalists updated their thinking. Actors, for example, acquired bounded rationality, i.e. they 'satisficed' rather than optimised (cf. Habermas, 1981). On this basis, a communicative approach, termed 'collaborative planning' (Healey, 1992), was developed and adopted by the UK's Labour government in 1999, for example. Collaborative planning has nevertheless been challenged for a lack of context dependency and unrealistic suppression of power and conflict (cf. Flyvbjerg, 1998).

By contrast to prior systems thinking, post-structural default model orientations are unstable (i.e. they remain in flux) and sustainable change can occur disruptively as a result of internal, i.e. *endogenous*, shocks to the system (cf. Geels, 2010). Emerging post-structural discourses then began stressing the utility of risk, trust, legitimacy, reflexivity, power and conflict, for example, via social constructions of knowledge and relational actor networks.

With this conceptual overview in mind, the following sub-sections briefly critically review three narratives promoting sustainable change in wastes and resources management in terms of their goals.

Goals-based policy narratives

The ways that waste and resources actors are hoped to move towards a CE, at a range of levels, will unfold over time depending upon the ways that particular actors advocate different frameworks of policy and practice. The academic and grey literature on wastes and resources governance can be broadly classified into three narratives, described below, which are linked to goals-based policies for sustainable change (cf. Cowell et al., 2017).

Ecologically-based goals

Ecologically-based goals are resource efficient. They progressively diminish the impact on the environment of development and growth via reductions in resource consumption (for example, energy, materials and water). Policies include reducing the use of land, water and air as waste sinks, designing out waste and handing future generations a robust stock of 'environmental capital' as well as reducing the disposal of waste to landfill. Ecologically-based goal-setting in the CE context typically focuses on physical fluxes of materials and energy, not money. Such goals are supported by an ecocentric worldview, for example, deep ecology (Naess, 1973) and a strong commitment to the precautionary principle (Stewart, 2002), i.e. a low tolerance for environmental risks.

Naess (1973) suggests that researchers using a deep ecology agenda should, when thinking about complex living systems, be pursuing a relational understanding of how actors can fight against pollution and resource depletion. Naess feels there is a need to adopt an 'anti-class posture' (Naess, 1973, p. 94), i.e. when assessing group conflicts, researchers should be able to recognise and unpick asymmetries of power between actors in networks. This approach to power relations is also necessary when analysing the relations between developing and developed nations which can be exploitative both in terms of waste dumping and materials extraction, recycling and re-use. In the context of power relations, a researcher using a deep ecology approach should show concern for any group's plans for the future which are not consistent with 'wide and widening classless diversity' (Naess, 1973, p. 97). Making systemic changes to fight pollution and reduce resource depletion, Naess (1973) argues involves increased decentralisation and local autonomy because energy consumption can be reduced at these levels. These points will be examined in more detail in the next section.

One influential risk heuristic, the social amplification of risk (SARF) model (Kasperson et al., 1988), suggests that certain elements of hazard events and the ways they are portrayed in the media (and elsewhere) relate to psychological, social, institutional and cultural processes in ways that perceptions of risk can increase or decrease and so shape behaviour. The utility of such an approach is to help provide improved policies better tailored to effective risk communication (Pidgeon and Henwood, 2010). SARF studies suggest that where individual actors fear for their health and wellbeing by the imposition of a new piece of infrastructure, such as nuclear power plants and waste incineration facilities, they are experiencing a form of 'dread risk' where little or no public engagement appears possible (Lima, 2004; Pidgeon and Henwood, 2010; Parkhill et al., 2010). Murdock et al. (2003, p. 157) critique the SARF model suggesting it: 'cannot offer a satisfactory account of risk communication and responses in contemporary democracies' due to a 'one-dimensional' view of how the media operates and a simplistic assessment of lay knowledge. More recently, Renn has updated SARF approaches and suggests that:

> [M]ost of the worries [that community members have about risky activities] are not related to blatant errors or poor judgement, but to divergent views about the tolerability of remaining uncertainty, short-term versus long-term impacts, the trustworthiness of risk regulating or risk managing agencies, and the experience of inequity or injustice with regard to the distribution of benefits and risks. They cannot be downplayed as irrational fears.
>
> *(Renn, 2017, p. 3)*

Analysing the nature of environmental risks and uncertainties of wastes and resource governance inevitably leads to thinking about the precautionary principle, a useful framework developed in parallel with the discourse on risk/uncertainty that can help to improve decision making regarding technology, science, ecological, as well as human health. The principle also offers the potential for improved regulation of risky activities. In the mid-1970s, precaution became a key plank of European environmental policy but was only a useful approach if a realistically irreversible environmental problem could genuinely be stopped. The early strength of the precautionary principle came from a number of innovative legal and regulatory approaches to environmental policy making.

The precautionary principle has been used by policy makers to justify discretionary decisions in situations where harm may arise from a particular course of action and when extensive scientific knowledge on the matter is absent. The precautionary principle also suggests that there is a social responsibility to protect the public from exposure to harm, when scientific investigation has found a plausible risk. These protections can be relaxed only if further scientific findings emerge that provide sound evidence that no harm will result.

Eco-efficiency-based goals

Another set of goals are eco-efficiency driven. The aim is to strike an 'optimal' balance between environmental and economic goals (cf. Huang et al., 1992). There is faith in markets to monetise costs and benefits and determine policy solutions. Adverse environmental impacts and their management are seen as costly. Proponents of mathematical modelling via this approach tend to specify the 'means' (for example, incineration, recycling and/or reuse) rather than the ends. These advocates become concerned if funds get allocated sub-optimally (see, for example, Hu et al., 2017). Some consider what is economically optimal as an appropriate way of identifying policy goals (i.e. the 'ends') (for example, Kinnaman, 2016) while others suggest that markets will help efficiently achieve ends based on other knowledge (for example, science).

This approach involves a plurality of ways of achieving eco-efficiency-based goals with wastes and resources management. These range from quantitative cost benefit analysis and cost effectiveness analysis studies, for example, to more qualitative and relational social constructivist analyses of the ways actors in networks help to create and continually recreate markets for wastes and resources (for example, Çalışkan and Callon, 2009; Kama, 2015).

In the context of analysis of markets and power relations, effective delivery of eco-efficient waste and resources policies demands that a centralised state achieves its policy objectives and governs successfully at a distance over the territory that it controls. Analysis via the governmentality perspective shows how and why a state must engage with the rationalities, agencies, institutional relations and technologies of governing (Foucault, 1991; Dean, 2010).

Short-term economic-driven goals

Lastly, there are short-term economic-driven goals for actors. These goals emphasise private profits and the reduction of costs to public and private sectors over a short time horizon. Advocates of such goals challenge interventions that interfere with either of these goals. Such perspectives may be opposed to government intervention in private affairs per se in a libertarian context. The justifications of such positions can entail actors downplaying environmental risks and/or playing up any uncertainty in knowledge about environmental risks. Short-term economic-driven goals are generally felt to hinder the development of CE practices. Markets are unlikely to support such normative sustainable change which requires a degree of investment. Instead, central and regional governments are relied upon to support and protect innovative practices in niches. A shift towards a CE, at a range of levels, is generally regarded as a long-term change.

The political power to implement policy narratives

Whichever goals are sought by waste and resources actors operating at a range of governance levels, centralised state power is typically deployed to offer (at a minimum) political support for a route map for sustainable change. Beyond this, certain neoliberal and authoritarian economies, such as in the EU and China, have shown a desire to more fully nurture/protect their own niche innovative CE activities (for example, McDowall et al., 2017).

Governmentality is one approach that suggests that normative moves towards greater sustainability (based upon whichever goals) require centralised calculations be made of statutory targets, performance indicators, audit processes, and funding mechanisms, for example, that come from distant territorial sites (Foucault, 1991; Dean, 2010; Davies, 2005; Bulkeley et al., 2005; Jeffreys and Sigley, 2014). Governmentality's top-down focus on the central state's imposition of waste and resources policies demands a commensurate bottom-up theory of localised dissent (cf. Hacking and Flynn, 2017). While centralised states try to impose their will on actors in the hope of governing economic relations, cross-cutting, bottom-up forums – so-called 'rhizomes' – create an acentred, nonhierarchical network of counter actors with the potential to communicate 'horizontally' (Deleuze and Guattari, 1987). Such opposing forces are typically held in tension as the optimal, or eco-efficient, balance between environmental and economic goals is weighed up by competing actors. This Deleuzoguattarian approach suggests that actors within certain communities with longstanding disaffection with the mechanisms of the state are reluctant to engage with less-than-meaningful engagement efforts (Hillier, 2017). Longitudinal studies suggest that, in certain places and over certain time frames, there are dynamic limits to the top-down governmentality-led approach to the delivery of waste and resources policies of central states (Hacking and Flynn, 2017, 2018).

It is in this context that the analysis of social, economic and technical barriers to and enablers of sustainable change, described in the next section, have taken hold in policy circles in the last two decades.

Barriers to and enablers of a CE transition

Moves towards greater CE activity, at various levels from the local to the global, are driven by a range of social, economic, technical, and institutional processes (cf. de Jesus and Mendonço, 2018; Kirchherr et al., 2018). The availability of financial funding is a universal enabler for actors pursuing the goals outlined previously. However, key transition pathways for actors with ecological-based goals favour greater upstream activity, for example, designing out waste, as well as downstream measures to close waste/resource loops, set standards and proactively enforce the waste hierarchy. Systemic eco-innovation is the preferred transition pathway for those with eco-efficiency-based goals (and those with short-term economic-driven goals to a lesser degree). CE policy analysis requires consideration of the processes identified in the following sub-sections which can act as both barriers to and enablers of greater CE activity.

Power relations

Case study evidence reveals structural asymmetries of power between actors at the local level in both developed and developing countries (for example, Walker, 1998; Davies, 2016; Johnson et al., 2018; Hacking and Flynn, 2018). Beck (1999, p. 5) suggests: 'the first law of environmental risks is: pollution follows the poor'. This immediately raises concerns about environmental justice in communities with existing relatively poor health (Petts, 2004). Structurally, the perpetuation of social inequalities in particular localities can reinforce landed interests and may be linked to dominant political parties (cf. Davoudi and Atkinson, 1999). Actors can be excluded from negotiating and bargaining by institutional barriers and/or the manoeuvres of other groups. Community actors quickly spot mismatches between what they know from first-hand experience and what is missing from an environmental statement (Petts, 2004).

Perceived trustworthiness of the regulator

A polluting actor may avoid paying enforcement fines or may avoid their share of responsibilities. The state regulator may not ensure proportionality with their regulatory action when setting gains against losses. In such cases, critical trust by communities in a regulator can quickly sour (cf. Pidgeon et al., 2003). Individuals living beside waste and resources infrastructure can be more distrustful of the regulator than a plant's operator (with whom they may well have an engaged and reflexive relationship) (Hacking and Flynn, 2017). Increasing the legitimacy of the decision-making process can theoretically contribute to increasing much needed trust. However, 'misconceptions' on all sides about the roles of developer-operators and the regulator are difficult, if not impossible, to overcome (Petts, 2008).

Institutional framing

There is a need to have a programme of CE infrastructure in place or in development. In this context, the public-private mix of CE developments varies depending upon the nature of particular economic regimes and their associated legal set-ups. A coherent, state-led strategic roadmap helps to avoid mismatches and contradictory incentives (European Commission, 2011; Gregson et al., 2015; European Parliament, 2017).

Technical solutions

Eco-efficiency-based goals place particular emphasis on research and development (R&D) spending to ensure the stimulation of the best available technology ('BAT') for increasing the efficiency of resource use. Preventing waste with ecological-based goals can be similarly technical but may come down more to process innovation/simplification.

Time frame for change

The financial case for making sustainable changes now rather than in the future has been made (cf. Stern, 2006; New Climate Economy, 2014). Similarly, there is a growing recognition in policy circles that sustainability transitions, such as with the CE, will realistically take place over very long time periods, for example, decades.

Publicly versus privately funded activity

Private actors in neoliberal, market-led economies will not fully support normative sustainable change if significant investment is required in CE niches. Instead, arrangements for sharing the burden of regulatory responsibilities between the state and industry is likely to be worked out via responsible intervention from accredited non-governmental organisations.

Contested views about the nature of sustainable change/uncertainty

Risk dialogue is now more fruitful when thought of in terms of ways that uncertainty needs to improve. In waste planning, a developer might negotiate over many iterations with the regulator what exactly is required to satisfy a political core's legal and procedural demands. The highly reflexive relationships that exist between developer/operators, local planning authorities and regulators, who share low-risk framings of waste activity, do not extend to community members where dread risk appears (cf. Hacking and Flynn, 2018). That process leads to consideration of the precautionary principle.

Pursuing the waste hierarchy

Focus as much attention on prevention of waste, at the top of the hierarchy, as on those tiers below.

Policy approaches

In sum, from reviewing the barriers/enablers to sustainable change, achieving the first two goals examined in the previous section (ecologically-based and eco-efficiency-based), could involve, for example, various degrees of the following policy mixes: (i) reuse, repair and remanufacturing; (ii) green public procurement and innovation procurement; and (iii) improving secondary materials markets (Milios, 2018).

Conclusion

Serious concerns for the governance of wastes, resources and energy appeared in the 1960s and 1970s via a new practice-led, systems approach: industrial ecology (IE). More recent discussions centre on so-called 'circular economy' (CE) principles. This practical guidance on reducing waste by closing material loops has been gaining traction with public and private actors in response to data indicating that the pressure on resources is increasing (see Figure 23.1) and the need to help get the normative drive towards sustainable development back on track (Kirchherr et al., 2018). Nevertheless, CE concepts – when considered beyond their practice origins – remain in their infancy (Korhonen et al., 2018).

Looked at via different policy goals, there are clearly different approaches to a CE transition at a range of scales each highlighting the need to overcome certain barriers in certain ways. Such policy analysis depends upon the theoretical assumptions made about the causal relationships involving agency and structure. The different theoretical approaches outlined here are not exhaustive.

Who chooses to buy-in to them or not as policy options will, in turn, impact upon how normative moves towards CE practice unfolds over time as it is scaled up, i.e. there are different transition pathways to be pursued by different networks of public and private waste and resource actors. In the analysis of the *becoming* of CE practice, academics are still playing 'catch up' in theorising this new governance territory but much of the most recent research, some of it cited here, points to the useful beginnings of different perspectives being used to contest what a CE can and should be and why.

References

Beck, U. (1999). *World risk society*. Wiley Online Library.

Boulding, K. (1966). 'The economics of the coming spaceship earth', in H. Jarrett (ed.) *Environmental quality in a growing economy*. Baltimore, MD: Resources for the Future/Johns Hopkins University Press, pp 3–14.

Bulkeley, H., Watson, M., Hudson, R. and Weaver, P. (2005). 'Governing municipal waste: towards a new analytical framework'. *Journal of Environmental Policy and Planning*, 7(1): 1–23.

Çalışkan, K. and Callon, M. (2009). 'Economization, part 1: shifting attention from the economy towards processes of economization'. *Economy and Society*, 38(3): 369–98.

Cowell, R., Flynn, A. and Hacking, N. (2017). *Assessing the impact of Brexit on the UK waste resource management sector*. Cardiff: Cardiff University. Available at: http://orca.cf.ac.uk/103020/

Davies, A. R. (2005). 'Incineration politics and the geographies of waste governance: a burning issue for Ireland?' *Environment and Planning C: Government and Policy*, 23(3): 375–97.

Davies, A. R. (2016). *The geographies of garbage governance: interventions, interactions and outcomes* (2nd edn). London: Routledge.

Davoudi, S. and Atkinson, R. (1999). 'Social exclusion and the British planning system'. *Planning Practice and Research*, 14(2): 225–36.

de Jesus, A. and Mendonça, S. (2018). 'Lost in transition? Drivers and barriers in the eco-innovation road to the circular economy'. *Ecological Economics*, 145(C): 75–89.

Dean, M. (2010). *Governmentality: power and rule in modern society* (2nd edn). London/Thousand Oaks/New Delhi: Sage Publications.

Deleuze, G. and Guattari, P. F. (1987). *Thousand plateaus: capitalism and schizophrenia*. Minneapolis: University of Minnesota Press.

EMF (2012). *Towards the circular economy, volume 1: an economic and business rationale for an accelerated transition*. Ellen MacArthur Foundation. Available at: www.ellenmacarthurfoundation.org/assets/downloads/publications/Ellen-MacArthur-Foundation-Towards-the-Circular-Economy-vol.1.pdf

European Commission (2011). *Communication from the Commission to the European Parliament, the Council, the European Economic and Social Committee and the Committee of the Regions, Roadmap to a resource efficient Europe. COM 2011 571 final*. Brussels: European Commission.

European Parliament (2017). *Circular economy package – four legislative proposals on waste*. Brussels: European Parliament. Available at: www.europarl.europa.eu/RegData/etudes/BRIE/2017/599288/EPRS_BRI(2017)599288_EN.pdf

Federal Waste Disposal Act (1972). *[West German] Federal Law Gazette I*, 11 June: 873.

Flyvbjerg, B. (1998). *Rationality and power: democracy in practice*. Chicago: University of Chicago Press.

Foucault, M. (1991). 'Governmentality', in G. Burchell, C. Gordon and P. Miller (eds.) *The Foucault effect: studies in governmentality*. Chicago: University of Chicago Press, pp 87–104.

Geels, F. (2010). 'Ontologies, socio-technical transitions (to sustainability), and the multi-level perspective'. *Research Policy*, 39(4): 495–510.

Gregson, N. and Crang, M. (2010). 'Materiality and waste: inorganic vitality in a networked world'. *Environment and Planning A*, 42(5): 1026–32.

Gregson, N., Crang, M., Fuller, S. and Holmes, H. (2015). 'Interrogating the circular economy: the moral economy of resource recovery in the EU'. *Economy and Society*, 44(2): 218–43.

Habermas, J. (1981). *Theorie des kommunikativen Handelns, volume 2*. Frankfurt: Suhrkamp.

Hacking, N. and Flynn, A. (2017). 'Networks, power and knowledge in the planning system: a case study of energy from waste'. *Progress in Planning*, 113: 1–37.

Hacking, N. and Flynn, A. (2018). 'Protesting against neoliberal and illiberal governmentalities: a comparative analysis of waste governance in the UK and China'. *Political Geography*, 63: 31–42.

Healey, P. (1992). 'Planning through debate: the communicative turn in planning theory'. *Town Planning Review*, 63(2): 143–62.

Hillier, J. (2017). *Stretching beyond the horizon: a multiplanar theory of spatial planning and governance*. Abingdon: Routledge.

Hird, M. J. (2012). 'Knowing waste: towards an inhuman epistemology'. *Social Epistemology*, 26(3–4): 453–69.

Hu, C., Liu, X. and Lu, J. (2017). 'A bi-objective two-stage robust location model for waste-to-energy facilities under uncertainty'. *Decision Support Systems*, 99(C): 37–50.

Huang, G., Baetz, B. W. and Patry, G. G. (1992). 'A grey linear programming approach for municipal solid waste management planning under uncertainty'. *Civil Engineering Systems*, 9(4): 319–35.

Jeffreys, E. and Sigley, G. (2014). 'Government, governance, governmentality and China: new approaches to the study of state, society and self'. *China Studies*, 18: 129–46.

Johnson, T., Lora-Wainwright, A. and Lu, J. (2018). 'The quest for environmental justice in China: citizen participation and the rural-urban network against Panguanying's waste incinerator'. *Sustainability Science*, 13(3): 733–46.

Kama, K. (2015). 'Circling the economy: resource-making and marketization in EU electronic waste policy'. *Area*, 47(1): 16–23.

Kasperson, R. E., Renn, O., Slovic, P., Brown, H. S., Emel, J., Goble, R., Kasperson, J. X. and Ratick, S. (1988). 'The social amplification of risk: a conceptual framework'. *Risk Analysis*, 8(2): 177–87.

Kinnaman, T. C. (2016). 'Understanding the economics of waste: drivers, policies, and external costs'. *International Review of Environmental and Resource Economics*, 8(3–4): 281–320.

Kirchherr, J., Reike, D. and Hekkert, M. (2017). 'Conceptualizing the circular economy: an analysis of 114 definitions'. *Resources, Conservation and Recycling*, 127: 221–32.

Kirchherr, J., Piscicelli, L., Bour, R., Kostense-Smit, E., Muller, J., Huibrechtse-Truijens, A. and Hekkert, M. (2018). 'Barriers to the circular economy: evidence from the European Union (EU)'. *Ecological Economics*, 150(C): 264–72.

Korhonen, J., Nuur, C., Feldmann, A. and Birkie, S. E. (2018). 'Circular economy as an essentially contested concept'. *Journal of Cleaner Production*, 175: 544–52.

Krausmann, F., Wiedenhofer, D., Lauk, C., Haas, W., Tanikawa, H., Fishman, T., Miatto, A., Schandl, H. and Haberl, H. (2017a). 'Global socioeconomic material stocks rise 23-fold over the 20th century and require half of annual resource use'. *Proceedings of the National Academy of Sciences of the United States of America*, 114(8): 1880–5.

Krausmann, F., Schandl, H., Eisenmenger, N., Giljum, S. and Jackson, T. (2017b). 'Material flow accounting: measuring global material use for sustainable development'. *Annual Review of Environment and Resources*, 42: 647–75.

Lima, M. L. (2004). 'On the influence of risk perception on mental health: living near an incinerator'. *Journal of Environmental Psychology*, 24(1): 71–84.

McDowall, W., Geng, Y., Huang, B., Barteková, E., Bleischwitz, R., Türkeli, S., Kemp, R. and Doménech, T. (2017). 'Circular economy policies in China and Europe'. *Journal of Industrial Ecology*, 21(3): 651–61.

Meadows, D. H., Meadows, D. L., Randers, J., and Behrens III, W. W. (1972). *The limits to growth: a report for the Club of Rome's project on the predicament of mankind*. New York: New American Library.

Milios, L. (2018). 'Advancing to a circular economy: three essential ingredients for a comprehensive policy mix'. *Sustainability Science*, 13(3): 861–78.

Murdock, G., Petts, J. and Horlick-Jones, T. (2003). 'After amplification: rethinking the role of the media in risk communication', in N. Pidgeon, R. E. Kasperson and P. Slovic (eds.) *The social amplification of risk*. Cambridge: Cambridge University Press, pp 156–78.

Naess, A. (1973). 'The shallow and the deep, long-range ecology movement. A summary'. *Inquiry*, 16(1–4): 95–100.

Nakamura, S., Nakajima, K., Kondo, Y. and Nagasaka, T. (2007). 'The waste input-output approach to materials flow analysis'. *Journal of Industrial Ecology*, 11(4): 50–63.

New Climate Economy (2014). *Better growth, better climate*. Washington, DC: New Climate Economy (World Resources Institute). Available at: https://newclimateeconomy.report/2014/misc/downloads/

Nishimura, H. (1977). 'Industrial ecology and its application to environmental assessment: a case of the Seto inland sea', in Y. Zhou (ed.) *Science for better environment: proceedings on the International Congress on the Human Environment (HESC), Kyoto, 1975*. Oxford: Pergamon, pp 362–74.

Parkhill, K. A., Pidgeon, N. F., Henwood, K. L., Simmons, P. and Venables, D. (2010). 'From the familiar to the extraordinary: local residents' perceptions of risk when living with nuclear power in the UK'. *Transactions of the Institute of British Geographers*, 35(1): 39–58.

Petts, J. (2004). 'Barriers to participation and deliberation in risk decisions: evidence from waste management'. *Journal of Risk Research*, 7(2): 115–33.

Petts, J. (2008). 'Public engagement to build trust: false hopes?' *Journal of Risk Research*, 11(6): 821–35.

Pidgeon, N. and Henwood, K. (2010). 'The social amplification of risk framework (SARF): theory, critiques, and policy implications', in P. Bennett, K. Calman, S. Curtis and D. Fischbacher-Smith (eds.) *Risk communication and public health* (2nd edn). Oxford: Oxford University Press, pp 53–67.

Pidgeon, N., Kasperson, R. E. and Slovic, P. (2003). *The social amplification of risk*. Cambridge: Cambridge University Press.

Renn, O. (2017). *Risk governance: coping with uncertainty in a complex world* (2nd edn). Abingdon: Routledge.

Schaffartzik, A., Mayer, A., Gingrich, S., Eisenmenger, N., Loy, C. and Krausmann, F. (2014). 'The global metabolic transition: regional patterns and trends of global material flows, 1950–2010'. *Global Environmental Change*, 26: 87–97.

Stahel, W. and Reday, G. (1976). *The potential for substituting manpower for energy, report to the Commission of the European Communities.* Geneva, Switzerland: Battelle, Geneva Research Centre.

Stern, N. (2006). *The economics of climate change: the Stern review.* Cambridge: Cambridge University Press.

Stewart, R. (2002). 'Environmental regulation under uncertainty'. *Research in Law and Economics*, 10: 71–126.

Walker, G. (1998). 'Environmental justice and the politics of risk'. *Town and Country Planning*, 67: 358–9.

Watanabe, C. (1994). *Industrial ecology and Japan's industrial policy in industrial ecology: US-Japan perspectives.* Paper presented at the Report on the US-Japan Workshop on Industrial Ecology, held 1–3 March 1993, Irvine, California.

Weisz, H., Suh, S. and Graedel, T. (2015). 'Industrial ecology: the role of manufactured capital in sustainability'. *Proceedings of the National Academy of Sciences*, 112(20): 6260–4.

Wynne, B. (1987). *Risk management and hazardous waste: implementation and the dialectics of credibility.* Berlin, Heidelberg: Springer-Verlag.

24

Biodiversity, ecosystem services and environmental planning[1]

Kimmo Lapintie and Mina di Marino

Introduction

During recent decades, discussions on urban nature have gone through several conceptual frameworks that highlight its different dimensions (Di Marino et al., 2018; Gunnarsson et al., 2017; Haaland and van den Bosch, 2015; Barbosa et al., 2007). The traditional dichotomy between the recreational values of blue-green infrastructure and the protection of selected species and landscapes with special natural and/or cultural value were reconceptualised in the sustainable development and urban ecology discourses, where nature was seen as part of an integrated urban and regional system with it ecological, cultural and economic dimensions. On the other hand, this early phase of the 1970s and 1980s also included more radical critique of speciesism, for instance, so-called deep ecology (Naess, 1989), according to which the human species does not have a legitimate status to use other species for its own purposes, thus optimising its own benefits from nature. This philosophical attitude was connected to animal rights movements and, more recently, to veganism as a consumer-driven rejection of the human exploitation of other sentient beings.

In urban and environmental planning, however, this radical standpoint never gained foothold, based mainly on the dominance of economic considerations. Traditionally, urban planning was – and still is – mainly based on a division of land into different functional areas for human beings, including the recreational green and blue elements of the urban fabric. Indeed, the environmental impacts of recreation in green areas depends on the level, type, frequency and duration of use for various human activities. In addition to this, urban parks, streetscapes and private gardens can also provide habitat and resources for fauna and plant species. However, within urban planning, the recreational areas are mainly acknowledged as places for social interactions and physical activities that support the psychological restoration of urban residents (Lee et al., 2015). This is one of the main purposes for which the green elements have proved to be effective planning (Kaplan, 1995; Wells, 2000; Berman et al., 2008; Lee et al., 2015; Grinde and Patil, 2009).

In this context, the protection of rare species and the later interest in biodiversity are exceptions to this human-centric understanding and management of green areas. Constructed and well-maintained parks, beaches and other blue-green elements are often not rich in biodiversity, although they are appreciated by residents for their recreational values. Rare species also do not reside in these artificial green areas compared to more natural ones. This has resulted in conflicts between compact urban development and natural values. For instance, the Siberian flying squirrel (Pteromys volans) is a strictly protected species in the European Union and US legislation, forbidding clearcutting and other major changes in the environment that could endanger its natural environment. A single protected species can thus prevent major development projects, something not easily understood by their proponents, politicians and even many citizens.

These conflicts reflect a major shift in environmental policies and planning, suggesting a more indirect assessment of the benefits of natural areas. However, the indirect benefits of green areas on climate change, water purification, biodiversity and formation of soil are often intangible and difficult to quantify. In contrast, environmental benefits directly provided by nature, such as food and forests, are more 'visible' and perceived by a larger audience of both experts and non-experts. However, the value of biodiversity cannot be derived directly from the human perception or even the restorative effects of forests and other natural areas; many of the rare species are seldom seen or appreciated by the urban populations. While it is taken for granted that biodiversity is not only beneficial but necessary for human beings in the long run, it remains a rather abstract concept that cannot so easily be adapted in the daily work of urban planning, particularly if it is confronted with major development of residential or other directly functional uses. The argument often is that local green areas can be sacrificed for development in order to preserve larger green areas outside the cities. But often no convincing cost-benefit analysis for such compensation measures can be given.

Thus, the aim of the chapter is to provide a critical understanding of the current conflicts between human/nature within the urban development. The chapter also focuses on the challenges to influence the current planning practices when developing the emerging concepts of green infrastructure and ecosystem services. Then, the chapter aims to contribute to the discourses on knowledge and socio-political legitimacy and the ways of communicating biodiversity and ecosystem services to large audience (of experts and non-experts). The discussion then shifts to visioning current and future challenges of the compact city by considering all these themes.

Dichotomy of human/nature: green infrastructure and ecosystem services

The most recent concepts in this area, green infrastructure (GI) and ecosystem services (ES), attack the traditional dichotomy of human/nature from two directions. GI can be defined as 'an interconnected green space network (including natural areas and features, public and private conservation lands and other protected open spaces) that is planned and managed for its natural resources and values and for the associated benefits to the population' (Benedict and McMahon, 2012, p. 3). The ES can be categorised as supporting services (for example, soil formation, biodiversity and habitat), providing services (for example, fish and wood), regulating services (for example, storing carbon and controlling flood) and cultural services (for example, recreation, wellbeing and inspiration from interaction with nature) (Millennium Ecosystem Services, 2005; TEEB, 2011). Infrastructure, meaning roads, pipes, ducts and wires, is the basic technical concept of urban development, and it is very easy to understand and appreciate. Services, on the other hand, are part of the functional understanding of the city, referring both to private services as part of the growing service economy and to public services as part of the welfare state. With respect to these, urban nature has traditionally been seen as a background, on top of which the 'hard' infrastructure is formed. The new concepts seek to change this by making the green-blue structure an active and connected part of the infrastructure, partly technical, as in the case of storm water management by retention pools, and partly ecological, as in the case of preserving areas rich in biodiversity. The 'soft' dimensions of cultural and aesthetic values can be included in the concept of ES, thus benefitting from this new framework.

However, the concepts used by scholars do not automatically find their way to mainstream planning practice. Even the most advanced cities have difficulties incorporating the new concepts into their daily work. In our study of Helsinki, Milan and Montreal (Di Marino and Lapintie, 2018) – all of which have adopted one or several of these concepts – we could clearly see that the planners and policy makers were unsure of how to address the ES and GI in different scales and types of plan. Furthermore, the traditional dichotomy of recreational values and protection of natural habitats still holds sway over the new conceptualisations.

For instance, ES are mentioned in the general requirements of the Helsinki City Plan (City of Helsinki, 2016), but they are not given any concrete interpretation in the plan. Instead of GI, Helsinki uses the concept 'network of green areas', thus losing the technical and functional connotation of the former. They are said to include 'many scenic and historic entities, large recreational areas, such as outdoor sports parks and neighbourhood parks, as well as protected areas, such as Natura areas'. Thus, the categorisation is still based on the traditional dichotomy, including recreation and the protection of nature. These two main categories are also depicted in two separate thematic maps on 'recreational network' and 'urban nature'. The latter includes 'nature conservation areas, Natura areas, the areas currently included in Helsinki's nature conservation programme, the woodland network, the meadowland network and the core areas of urban nature'. Only the first three categories, which already have a formal status, are included in the legally binding land use map of the City Plan. Thus, there is no analysis nor policies that would address the dynamic and systemic nature of the GI (Di Marino and Lapintie, 2018, p. 146).

To sum up this section, three key issues are identified as follows: the ability to inform planning practices, and therefore, select adequate planning tools; the dominance of existing discourses on recreational areas and protection of nature; and the difficulty in designing appropriate policies for the emerging concepts of GI and ES. In addition to the current practices in planning, the three issues are closely related to the knowledge gap between academics and practitioners. While planning practitioners are familiar with more established concepts of recreation and protection of nature, their understanding of ES, which are provided by, for instance, urban green areas (for example, improving physical and psychological health, filtering pollutants and dust from air, providing shade and lower temperatures) is still vague. It is evident that the mechanisms to transfer knowledge on GI and ES from research to practice are still complex and need further investigation.

Knowledge and socio-political legitimacy

The academic and planning discourses can be conflicting for several reasons. Urban and regional planning needs to deal with a variety of knowledge, some of which is produced outside its main domains (such as in ecology, sociology, economics, political science and philosophy), some of which is created through its own practice, where knowledge often has a tacit dimension (such as in design, communicative skills, creativity, visioning, meaning creation). There is no straightforward path from explicit knowledge to practice; existing knowledge does not automatically inform planning practice (even to the extent that it often does not 'exist' in the process). Therefore, it is necessary to understand what kind of inputs research could bring to planning practices in order to help planners and policy makers to deal with new challenges such as resilience, GI and ES within the land use changes.

Moreover, the inclusion of a multiplicity of epistemologies, often cherished in communicative and radical planning theories (Sandercock, 2000), fails to address the relevance of justification, which is still a necessary condition of knowledge (Gettier, 1963). The inevitable connectedness of knowledge and power (power/knowledge, i.e. through disciplinary silos and professionalism, political regimes, discourse coalitions) has to be taken into account (Foucault, 1980; Hajer, 1993). When thinking of disciplinary silos and professionalism, for instance, we need to reflect on the kinds of strategies, representations, arguments and theoretical inputs that have not yet found their way to the planning discourses. Theoretically, thus, it is necessary to understand this relationship between episteme and praxis, which often ends up in ignoring, disregarding, misunderstanding and tokenism of new knowledge that questions the existing conceptualisations of the professions responsible for planning.

Furthermore, as planning is an established social practice with a certain position in the political system, it can also be seen as a subsystem of wider social forces. It is a practice that needs to take an integrative approach, by using knowledge from different disciplines (for example, ecology, environmental sciences, nature conservation), and by developing specific skills (deliberation, tacit knowing, conflict resolution, strategic understanding, design skills). For instance, specific textual and visual representations of green can strongly influence the planning discourses. This is the case of the green fingers in Helsinki, which are broken in many places due to large roads cutting through them, but they are still represented as unity. The idealised map docs not allow policy makers and the larger public to understand that the movement of animals is threatened and people have difficulty to use them as recreational routes (Di Marino et al., 2018). In addition, people with different backgrounds and professional skills are differently emphasised and given different meanings, such as the city as a visual landscape, as an ecological habitat, as a social structure and as an economic entity. However, to maintain its legitimacy, planning cannot choose any of these perspectives but needs to present itself and its products as general and comprehensive. This is also one of the reasons why it is so difficult to develop strategic planning, which is exactly the opposite, concentrating on a few main targets and avoiding attempts at comprehensivity (Bafarasat, 2015; Albrechts and Balducci, 2013; Healey, 2007; Newman, 2007).

As this subsystem receives inputs from the outside, it needs to react to them but only if it considers them important for its social and political legitimacy. In fact, it seems to have established practices for non-responding to them: for instance, the common NIMBY (not in my back yard) argumentation aims at discrediting both local knowledge and local interests because they are considered to represent vested interests. In this context, planners think they represent both public interest and objectivity. They tend to inform the citizens and not to discuss or present arguments. Moreover, as the social and political pressure grows and the public discourses change, there comes a time when planning needs to change its own vocabulary and ways of argumentation. The ways of using, communicating and addressing the key environmental concepts are still generating several challenges and conflicts within planning strategies (resilience, urban growth and the compact city).

Many initiatives exist to improve the way to communicate biodiversity. Most of the time, however, a technocratic model of communication has been used, in which scientific facts are transmitted directly to policy advisers to solve problems (Young et al., 2014, p. 387). This way of communicating is, however, considered inadequate, and biodiversity is still not really used in decision making process and implementation. Discourses on biodiversity are often used to contain the urban growth, defend green areas and wildlife rather than ensure both new developments and opportunities to promote biodiversity and to maximise its contribution within the built-up areas and to people's life. Several discourses around biodiversity are also embedded in the environmental impact assessment of new urban developments and master plans. In this context, experts are often called to solve conflicts between green and grey infrastructures (Di Marino, 2016). For instance, wildlife eco-dots and green bridges are developed in response to new projects of infrastructures that are approved by policy makers. While there are several measures to solve environmental impacts, knowledge on biodiversity and related contents have not been effectively embedded within the early and middle stage of planning (for example when predicting new urban scenarios and developing master and detailed plans). Official practitioners and policy makers should further think about biodiversity and the ways to embed and communicate it. This process can also involve real estate developers. New residential constructions, for instance, emphasise the panoramic views to parks and lakes and the proximity to green areas, while the ecological benefits of these areas are rarely mentioned to the public.

This debate on the way we communicate biodiversity can be also extended to the concept of ES.

> Biodiversity and ecosystem services . . . include uncertainty, complexity, diverse values and the involvement of many sectors . . . The cross-sectoral nature of some conservation and environmental issues means that many policies are linked and contain multiple objectives, thereby adding to their complexity.
>
> (Young et al., 2014, p. 390)

On the other hand, the scholars argue that biodiversity is not visible to all publics or policy makers. People might consider the biodiversity an irrelevant issue to them. The biodiversity might be highlighted by focusing on ES, or the benefits of nature to people. This can result in an excessive 'commodification of nature'. The biodiversity loss should be communicated in a simple way, but it is caused by a complex set of issues working at different levels.

The ambiguity in the definitions of key terms such as ecosystem functions and services has been debated amongst scholars (Haase et al., 2014; Niemelä et al., 2010; Wallace, 2007). 'Which ecosystem services in a given scale are most relevant varies greatly depending on the environmental and socio-economic characteristics of each geographic location' (Gómez et al., 2013, p. 178). As the scholars stated, other factors, such as the economic evaluation of ES, are frequently requested by policy makers and practitioners as supporting information to guide decisions in urban planning and governance (Gómez et al., 2013). ES should be clearly communicated to policy makers, city planners and official practitioners, without any possibility of misinterpretation (Haase et al., 2014). The scholars stated that ES have to be easily understood by those actors, as well as non-scientific audiences and citizens, who should be able to understand the economic, physical, spiritual and wellbeing benefits. Nevertheless, a few studies have examined the ES generated by the land use mixtures and urban green areas, as well as the dynamics of urban ecosystems at a more detailed level (for example, functionality of existing green spaces within the urban fabric). In order to involve a wider number of stakeholders, Haase et al. (2014) argued that it is important: (i) to understand the relevance of the concept of ES in planning, as well as further investigate the ES; (ii) to develop a framework and selection of relevant ES and indicators; and (iii) to collect data and assess ES.

Recently, the concept of ES has been introduced in landscape and urban planning. However, the ways in which an ES framework could advance planning goals are still unclear (BenDor et al., 2017; Hansen et al., 2015). Scholars have identified several challenges in incorporating the concept in land use planning (Kaczorowska et al., 2016). Environmental planning goals are clear to larger audiences and related, for instance, to practical issues, such as the management of water and green quality as well as coastal hazards, while there has been a limited use of ES data and concepts. BenDor et al. (2017) gave an interesting picture of planning in which city and regional planning practitioners can contribute to the creation and implementation of policies. However, the connection of the planning profession and ES has been mainly unidirectional:

> Ecosystem service studies frequently reference planning efforts and the impacts of urban decisions on ecological functions . . ., however, it is rare for this information to be fed back into planning practice as a mechanism for development and land-use decisions.
>
> (BenDor et al., 2017, p. 260)

There are several well-known challenges in connecting landscape sciences with situated governance and place-based human values (Swaffield, 2013). The contextual application of knowledge and expertise and related questions of legitimacy, as well as the inevitable ambiguity of intents and uncertainty in decision making have characterised the democratic and deliberative processes. It is very difficult to apply the local contextual knowledge as well as transfer general principles from landscape ecology into real contexts of scientific credibility, political legitimacy, and social relevance (see the overview given by Swaffield, 2013).

To sum up, there are professional, cultural, planning and political contexts where the new knowledge on biodiversity, ES and GI is challenged by the status quo of expertise (Di Marino and Lapintie, 2018). The ways of representing our cities, communicating biodiversity and embedding ES and GI within planning practices are closely related to disciplinary silos and professional skills, as well as socio-political forces.

Current and future challenges

Since the 1990s, within the European policy framework, the compact city has been acknowledged as the most sustainable model for urban development (Naess, 2014). The concept of compact city has been dominated by the very dense model of populated core areas in Europe, based on the theory that we should provide an urban containment: a sustainable social mix of uses able to reduce the need to travel and the corresponding energy use and emissions (Jenks et al., 1996).

However, planners have expressed serious doubts about the positive effects of the compact city (Hofstad, 2012; Dempsey, 2010; Neuman, 2005; de Roo, 2000). As de Roo (2000) stated in the early 2000s, even if the compact city concept favours mixed use and a dense urban development, reducing travel needs might not be the primary solution to the environmental conflicts in urban areas. There are also conflicts between living and working functions. The urban growth that enables compact settlement is connected to a larger hinterland with urban sprawl and workplace locations outside the city as well as a corresponding need to travel long distances (de Roo, 2000; Elliott and Clements, 2014). In this context, by adopting the model of the compact city, politicians, policy makers and planners have often underestimated the conflicting priorities for the long-term protection of the environment (Dempsey, 2010).

Today, we are more aware that environmental conflicts can also be found in the complex spatial context of the compact urban core, and they can cause several consequences (for example, air pollution, soil contamination, climate change and loss of biodiversity). In the last two decades, the environmental conflicts have involved a large number of different interests and expertise. Today, some scholars have focused on a greener picture of compact cities, dealing with the climate change, for instance (Matthews et al., 2015; Jim, 2013). 'Cities could be compact as well as green, with meticulous attention to every aspect of the urban greening complex' (Jim, 2013, p. 755). As the scholars argued, however, this requires filling the existing knowledge gap on urban nature between researchers, practitioners and academics, while a successful synergy between the actors involved would help to go beyond the physical and institutional challenges.

In fact, planning GI is still facing several challenges in cities undergoing densification, such as provision of local green spaces, counteracting social inequalities, considering the resident perspectives, and avoiding the deterioration of recreational experience and compensation travels

(Haaland and van den Bosch, 2015). Compact cities have incurred inherent physical and institutional obstacles, restricting the quantity and quality of the urban green (Jim, 2004). Scholars have paid more attention to the fact that the densification of urban areas has exerted a high pressure on urban green spaces such as the loss of the green areas themselves (Tappert et al., 2018). In several cases, the increase of built-up areas has often limited the access to existing parks and other recreational areas (Naess, 2014).

Moreover, scholars have debated about the crucial role of blue and green areas in providing important ES in urban areas, the contact with nature and biodiversity, as well as the influence on physical and mental wellbeing (see for example, Gómez et al., 2013). Thus, this is not only a question of aesthetic, size and accessibility of the green areas (for example, closeness to home, offices and schools) and related provision of facilities and vegetation (Haaland and van den Bosch, 2015). The pressure on the green areas affects the everyday life and environment of urban dwellers (Gunnarsson et al., 2017).

Therefore, there is a need to further investigate the challenges amongst practitioners in using and addressing the key environmental concepts, such as biodiversity, ES, as well as GI and resilience within the planning policies and practices in contexts undergoing densifications (Di Marino and Lapintie, 2018). The current challenges are often related to the ways of communicating the key concepts of biodiversity, ES, GI and resilience (in terms of power and values) to a larger audience. Additionally, conflicts in participation and power/knowledge are also relevant to the cases where green areas are transformed to residential areas (for example, the influence of governance and institutions, and concerns for justice and fairness).

Note

1 The main research findings of this study were supported by the Academy of Finland, Strategic Research Council (SRC) on Urbanizing Society (Beyond MALPE-coordination: Integrative Envisioning – BeM-InE), grant number 13303549 STN.

References

Albrechts, L. and Balducci, A. (2013). 'Practicing strategic planning: in search of critical features to explain the strategic character of plans'. *disP – The Planning Review*, 49(3): 16–27.

Bafarasat, A. Z. (2015). 'Reflections on the three schools of thought on strategic spatial planning'. *Journal of Planning Literature*, 30(2): 132–48.

Barbosa, O., Tratalos, J. A., Armsworth, P. R., Davies, R. G., Fuller, R. A., Johnson, P. and Gaston, K. J. (2007). 'Who benefits from access to green space? A case study from Sheffield, UK'. *Landscape and Urban Planning*, 83: 187–95.

BenDor, T. K., Spurlock, D., Woodruff, S. C. and Olander, L. (2017). 'A research agenda for ecosystem services in American environmental and land use planning'. *Cities*, 60: 260–71.

Benedict, M. A. and McMahon, E. T. (2012). *Green infrastructure: linking landscapes and communities*. Washington, DC: Island Press.

Berman, M. G., Jonides, J. and Kaplan, S. (2008). 'The cognitive benefits of interacting with nature'. *Psychological Science*, 19(12): 1207–12.

City of Helsinki (2016). *Helsinki's new master plan (HELSINGIN YLEISKAAVA Selostus. Kaupunkikaava – Helsingin uusi yleiskaava)*. Helsingin kaupunkisuunnitteluvirasto. Available at: www.hel.fi/hel2/ksv/julkaisut/yos_2016-3.pdf

De Roo, G. (2000). 'Environmental conflicts in compact cities: complexity, decision making, and policy approaches'. *Environmental and Planning B: Planning and Design*, 27(1): 151–62.

Dempsey, N. (2010). 'Revisiting the compact city?' *Built Environment*, 36(1): 4–8.

Di Marino, M. (2016). 'Ecological networks and ecosystem services in urban regions. Implementation and planning practices', in S. Jombach, I. Valanszki, K. Filep-Kovacs, J. Gy. Fabos, R. L., Ryan, M. S. Lindhult and L. Kollanyi (eds.) *Greenways and landscapes in change – Proceedings of 5th Fabos Conference on Landscape and Greenway Planning* (Budapest, 30 June), pp 71–8.

Di Marino, M. and Lapintie, K. (2018). 'Exploring the concept of green infrastructure in urban landscape. Experiences from Italy, Canada and Finland'. *Landscape Research*, 43(1): 139–49.

Di Marino, M., Niemelä, J. and Lapintie, K. (2018). 'Urban nature for land use planning'. *Urbanistica*, 159: 94–102.

Elliott, J. R. and Clements, M. T. (2014). 'Urbanization and carbon emissions: a nationwide study of local countervailing effects in the United States'. *Social Science Quarterly*, 95(3): 795–816.

Foucault, Michel (1980). *Power/knowledge: Selected interviews and other writings 1972–1977*. Edited by C. Gordon. New York: Pantheon Books.

Gettier, E. L. (1963). 'Is justified true belief knowledge?' *Analysis*, 23(6): 121–3.

Gómez-Baggethun, E., Gren, Å., Barton, D. N., Langemeyer, J., McPhearson, T., O'Farrell, P., Andresson, E., Hamstead, Z. and Kremer, P. (2013). 'Urban ecosystem services', in Th. Elmqvist, M. Fragkias, J. Goodness, B. Güneralp, P. J. Marcotullio, R. I. McDonald, S. Parnell, M. Schewenius, M. Sendstad, K. C. Seto and C. Wilkinson (eds.) *Urbanization, biodiversity and ecosystem services: challenges and opportunities – a global assessment*. London: Springer, pp 175–251.

Grinde, B. and Patil, G. G. (2009). 'Biophilia: does visual contact with nature impact on health and well-being?' *International Journal of Environmental Research and Public Health*, 6(9): 2332–43.

Gunnarsson, B., Knez, I., Heldblom, M. and Sang, Å. O. (2017). 'Effects of biodiversity and environment-related attitude on perception of urban green space'. *Urban Ecosystems*, 20(1): 37–49.

Haaland, C. and van den Bosch, C. K. (2015). 'Challenges and strategies for urban green-space planning in cities undergoing densification: a review'. *Urban Forestry and Urban Greening*, 14(4): 760–71.

Haase, D., Larondelle, N., Andersson, E., Artmann, M., Borgström, S., Breuste, J., Gomez-Baggethun, E., Gren, A. X., Hamstead, Z., Hansen, R., Kabisch, N., Kremer, P., Langemeyer, J., Rall, E. L., McPhearson, T., Pauleit, S., Qureshi, S., Schwarz, N., Voigt, A., Wurster, D. and Elmqvist, T. (2014). 'A quantitative review of urban ecosystem service assessments: concepts, models, and implementation'. *AMBIO*, 43(4): 413–33.

Hajer, M. (1993). 'Discourse coalitions and the institutionalisation of practice: the case of acid rain in Great Britain', in F. Fischer and J. Forester (eds.) *The argumentative turn in policy analysis and planning*. Durham and London: Duke University Press, pp 43–76.

Hansen, R., Frantzeskaki, N., McPhearson, T., Rall, E., Kabisch, N., Kaczorowska, A., Kain, J. H., Artmann, M. and Pauleit, S. (2015). 'The uptake of the ecosystem services concept in planning discourses'. *Ecosystem Services*, 12: 228–46.

Healey, P. (2007). 'Relational complexity and the imaginative power of strategic spatial planning'. *European Planning Studies*, 14(4): 525–46.

Hofstad, H. (2012). 'Compact city development: high ideals and emerging practices'. *European Journal of Spatial Development*. Refereed Article No. 49, October 2012. Available at: www.nordregio.se/Global/EJSD/Refereed articles/refereed49.pdf

Jenks, M., Williams, K. and Burton, E. J. (1996). 'A sustainable future through the compact city? Urban intensification in the United Kingdom'. *Environments by Design*, 1(1): 5–20.

Jim, C. Y. (2004). 'Green-space preservation and allocation for sustainable greening of compact cities'. *Cities*, 21(4): 311–20.

Jim, C. Y. (2013). 'Sustainable urban greening strategies for compact cities in developing and developed economies'. *Ecosystem Services*, 16(4): 741–61.

Kaczorowska, A., Kain, J. H., Kronenberg, J. and Haase, D. (2016). 'Ecosystem services in urban land use planning: integration challenges in complex urban settings – case of Stockholm'. *Ecosystem Services*, 22(A): 204–12.

Kaplan, S. (1995). 'The restorative benefits of nature: toward an integrative framework'. *Journal of Environmental Psychology*, 15(3): 169–82.

Lee, A. C. K., Jordan, H. C. and Horsley, J. (2015). 'Value of urban green spaces in promoting healthy living and wellbeing: prospects for planning'. *Risk Management and Healthcare Policy*, 8: 131–7.

Matthews, T., Lob, A. Y. and Byrne, J. A. (2015). 'Re-conceptualizing green infrastructure for climate change adaptation: barriers to adoption and drivers for uptake by spatial planners'. *Landscape and Urban Planning*, 138: 155–63.

Millennium Ecosystem Services (2005). *Ecosystems and human well-being: synthesis*. Washington, DC: World Resources Institute. Available at: www.millenniumassessment.org/documents/document.356.aspx.pdf

Naess, A. (1989). *Ecology, community and lifestyle: outline of an ecosophy*. Cambridge: Cambridge University Press.

Naess, P. (2014). 'Urban form, sustainability and health: the case of Greater Oslo'. *European Planning Studies*, 22(7): 1524–43.

Neuman, M. (2005). 'The compact city fallacy'. *Journal of Planning Education and Research*, 25(1): 11–26.

Newman, P. (2007). 'Strategic spatial planning: collective action and moments of opportunity'. *European Planning Studies*, 16(10): 1371–83.

Niemelä, J., Saarela, S. R., Söderman, T., Kopperoinen, L., Yli-Pelkonen, V., Väre, S. and Kotze, D. J. (2010). 'Using the ecosystem services approach for better planning and conservation of urban green spaces: a Finland case study'. *Biodiversity and Conservation*, 19: 3225–43.

Sandercock, L. (2000). 'The death of radical planning: radical praxis for a postmodern age', in M. Malcolm, T. Hall and I. Borden (eds.) *The city cultures reader*. London/New York: Routledge, pp 423–39.

Swaffield, S. (2013). 'Empowering landscape ecology – connecting science to governance through design values'. *Landscape Ecology*, 28(6): 1193–201.

Tappert, S., Klöti, T. and Drilling, M. (2018). 'Contested urban green spaces in the compact city: the (re-) negotiation of urban gardening in Swiss cities'. *Landscape and Urban Planning*, 170: 69–78.

TEEB – The Economics of Ecosystems and Biodiversity (2011). *TEEB manual for cities: ecosystem services in urban management*. TEEB is hosted by the 'United Nations Environment Programme and supported by the European Commission and Various Governments'. Available at: http://doc.teebweb.org/wp-content/uploads/Study%20and%20Reports/Additional%20Reports/Manual%20for%20Cities/TEEB%20Manual%20for%20Cities_English.pdf

Wallace, K. J. (2007). 'Classification of ecosystem services: problems and solutions'. *Biological Conservation*, 139(3–4): 235–46.

Wells, N. M. (2000). 'At home with nature. Effects of "greenness" on children's cognitive functioning'. *Environment and Behavior*, 32(6): 775–95.

Young, J. C., Waylen, K. A., Sarkki, S., Abon, S., Bainbridge, I., Balian, E., Davidson, J., Edwards, D., Fairley, R., Margerison, C., McCracken, D., Owen, R., Quine, C. P., Roper, C. S. and Thompson, D. (2014). 'Improving the science-policy dialogue to meet the challenges of biodiversity conservation: having conversations rather than talking at one-another'. *Biodiversity Conservation*, 23(2): 387–404.

25

Grasping green infrastructure

An introduction to the theory and practice of a diverse environmental planning approach

Mick Lennon

Introduction

Green infrastructure (GI) can be a difficult to fathom concept. It is employed in different ways in different contexts: in some jurisdictions it is most commonly conceived as a tool for adjudicating on specific development proposals, while in others it is primarily employed in the formulation of policy that sets the broad framework for land use. Understanding such varying interpretations, or 'regional accents' is important as they influence the identification of problems, the forms of assessment used and the types of solutions advanced. Much of this variation results from how the roots of the concept straddle a number of disciplines, with the relative weight given to each of these roots reflected in the focus of GI initiatives in different contexts. Hence, in presenting an introduction and outline of the GI approach, this chapter first traces the diverse regional antecedents of the GI concept to understand how it shapes environmental planning activities in different contexts. This provides a platform upon which to review a selection of representative case studies that illustrate the most prominent regional accents characterising the approach and how these are related to the institutionalisation of the GI concept at different scales. The chapter closes by discussing commonalities between these differently accented approaches and outlines the shared achievement of the GI concept.

GI dialects

Whilst the roots of GI may tap deep into history, most academic commentators identify contemporary manifestations of the concept as grounded in a family of environmentally sensitive planning initiatives that emerged in the latter half of the 19th century. Prominent among these is the work of Fredrick Law Olmstead in Boston. Here, Olmstead responded to various health, aesthetic and flooding problems associated with a marshy area in the city by proposing a constructed wetland system. His proposal connected a string of metropolitan parks forming what became known as the 'Emerald Necklace'. The success of this intervention stimulated a series of similar planning initiatives in the United States ranging in environmental contexts from temperate Portland in the northwest to subtropical Memphis in the southeast. A key feature of each scheme was a desire to address pressing environmental problems while concurrently enhancing the local recreational offer by connecting public green spaces into a parkland chain (Fábos, 2004). Indeed, for several years the recreational aspects of such work enjoyed prominence and overshadowed Olmstead's more balanced attention to health, aesthetics and drainage issues. It is therefore not entirely surprising to note that, in North America, the first decade of the 20th century witnessed the drainage management dimensions that prompted Olmstead's work become increasingly decoupled from the parkland ideas that gave such work traction with the public. Hence, particularly in the USA, recent years have observed 'GI' being progressively used to reference an engineering approach specifically focused on greener drainage management techniques, rather than broader health and recreational issues. Correspondingly, the less tangible social dimensions of the approach's roots have increasingly become viewed more as ancillary benefits rather than core objectives of GI planning.

Across the Atlantic in England, a similar focusing of attention on the multifunctional potential of nature-based solutions emerged in the late 19th century. This was most prominently represented in thinking centred on the 'garden cities' concept. Initially advanced by Ebenezer Howard, the garden city was advocated as a new urban template that could address socio-economic, health and wellbeing issues associated with the rapid urbanisation of the industrial revolution. In his well referenced 'three-magnets' of town, country and town-country, Howard suggested that it was possible to combine the different benefits of urban and rural lifestyles into a garden city settlement that facilitated local food production, recreational space provision and ease of access to employment, as well as providing adequate drainage management, clean air and uncontaminated water (Lennon and Scott, 2016). Howard's ideas were physically applied in the planning of Letchworth and Welwyn Garden Cities in England and have had influence in North America and Australia. However, much of this influence has been confined to the creation of suburban extensions, many of which have morphed into dormitory settlements, thereby undermining the rationale of self-containment that underpinned the concept. Furthermore, although the garden city concept is a touchstone in many planning curricula, it largely remains more an aspiration of planners than an application of planning. It is against this backdrop that GI in the United Kingdom and Ireland has emerged and evolved as an approach heavily informed by ideas that seek to address many of the contemporary manifestations of those issues Howard was responding to over 100 years ago. Thus, the greening of the built environment advocated by proponents of GI in the United Kingdom and Ireland can be seen as set in arguments that advance the multiple health and wellbeing benefits of green spaces, ranging from drainage management, to the promotion of active transport and social cohesion.

A different inflection of these ideas took shape in continental Europe in the aftermath of World War II. Here, planning initiatives often emphasise green connections at the regional scale and are frequently themed together beneath the rubric of 'green structure' planning. Such structures generally comprise a series of green corridors extending from a city's central urban area into the rural hinterland. These corridors thereby serve to 'connect' the urban and the rural, rather than 'combine' both, as is the case with garden city morphologies. Each corridor may be characterised by a particular landscape type, reflecting different uses, vegetation and/or topographical attributes. Nevertheless, corridors are usually seen to serve many functions simultaneously, ranging from agriculture and conservation to recreation and commercial forestry. This regional focus and the attention allocated to uninterrupted physical connectivity have facilitated alignment with models operative in the field of landscape ecology. Specifically, the corridors of green structure planning have been viewed by many as offering opportunities to create 'ecological networks' that enable genetic exchange between species, as well as facilitating foraging and migration (Jongman et al., 2004). This perspective has enjoyed considerable purchase within the governing institutions of the European Union (EU), which seeks to encourage nature conservation within planning activities through the protection of a series of national and international ecological networks linked within and across national boundaries. In recent years, a complementary city-scale focus has been emphasised by the EU in which the protection and creation of smaller ecological networks within urban areas are promoted as a way to enrich local biodiversity. Hence, although always evident within GI thinking, a more explicit ecological focus has now been assimilated into GI discussions and activities across a wide range of jurisdictions both within the EU and beyond.

Regional accents

Given such varied roots and trajectories, it is unsurprising that GI planning is as varied as the contexts in which the concept is employed. Consequently, an appreciation of the range of intonations of GI evident in practice is best obtained by exploring representative interpretations of the concept as applied. However, it is important to note that no such review can comprehensively reflect the array of inflections currently given to the concept. Hence, the examples of GI planning provided below do not claim to be a complete survey of the concept's use. Instead, they are provided as illustrations of different articulations given GI in different contexts and at different scales. Nevertheless, these examples have been specifically chosen to supply the reader with a sense of the range of interpretations that characterise common use of the GI concept.

Site scale: North America – Seattle's Green Factor

Whilst recreation, mobility, urban heat island mitigation, air quality and ecological enhancement are all well represented in the North American inflection of GI, the common issue shared and generally anchoring GI initiatives here is drainage management. Much credit for this can be given to the US Environmental Protection Agency (EPA) which has invested significant resources in disseminating an interpretation of GI focused on using natural processes to manage water runoff (Mell, 2016). This has included the provision and extensive publicising of a suite of resources for designing, costing and managing drainage-focused GI initiatives. Such a model has been largely followed in Canada where GI discourses and activities are also moored by a concern for drainage management. Consequently, this water-centric GI approach has become increasingly institutionalised in North America as a progressive engineering solution to the problems and risks associated with managing drainage in the built environment. This has resulted in the formulation of methods and tools resonant with an engineering approach to environmental planning. Nonetheless, several planning authorities have sought to devise a way of marrying this engineering and water-centric focus with aesthetic, climatic and ecological issues as a means to facilitate a more holistic approach to enhancing the environmental quality of settlements. A pioneer in such efforts is Seattle in Washington State, United States.

Drawing upon the innovative but limited precedents of Berlin's Biotope Area Factor and Malmö's Green Space Factor, the planners in Seattle City sought to devise a development assessment tool that could combine the national and local prominence given to sustainable urban drainage management with aesthetic improvement and ecological enhancement. Thus, municipal staff working in collaboration with a private sector team of landscape architects and engineers progressively developed a series of 'score sheets' to facilitate flexibility for designers in addressing various environmental challenges. The finalised score sheet requires designers to meet a minimum threshold of multifunctional landscaping provision in new and retrofitted developments. This planning assessment tool thereby became the first of its kind in the United

States to oblige designers to meet a minimum score on a points-based system in order to secure development consent.

The assessment tool works by providing a downloadable score sheet that helps designers to determine the score for their proposal and to test alternative approaches (see Figure 25.1). Applicants are able to select from a conventional range of landscape features such as shrubs and small trees, as well as more advanced elements such as vegetated walls and bio retention facilities. The score sheet operates by computing an array of quantified landscape features proposed by a designer. The figure that emerges is subsequently divided by the parcel of land that is the subject of the planning application. Hence, a score of 0.3 equates to approximately 30% of the subject land parcel being landscaped. A key aspect of the tool is that the scores received for the various landscape elements are weighted relative to quantum (size and number), functional advantage, as well as ecological, community and aesthetic benefit. These relative weightings are specified in the score sheet. For example, different tree planting and preservation scenarios are weighted differently. Thus, whereas each newly planted medium/large tree with a canopy spread of 21–25 feet (calculated at 250 sq ft per tree) is multiplied by a factor of 0.4, each preserved large tree with trunks greater than 6 inches in diameter (calculated at 20 sq ft per inch diameter) is multiplied by a factor of 0.8. This is to reflect the comparable drainage, soil stability, biodiversity and aesthetic benefits received from each of these options. Furthermore, bonus scores are available for the use of drought-tolerant or native plant species, meeting at least 50% of irrigation needs by harvested water and providing landscaping that is visible from public spaces. In addition, designers can boost the scores for their proposals through layering vegetation, by for example, providing a tree with an understory of native shrubs rather than just a freestanding tree. This mechanism is included to encourage the formulation of more aesthetically and ecologically beneficial planting schemes. As such, the mechanism is used as a means to integrate the GI approach into planning activities by furnishing both an assessment tool for planners and a way of incentivising applicants to enhance the environmental credentials of their proposals. Furthermore, landscape improvements in right-of-way areas contiguous with the parcel of land subject to the calculation can also be included in the weighted assessment. This is aimed at providing an inducement to developers and their designers to invest in streetscape improvements.

The Seattle Green Factor (SGF) was adopted into the Seattle Municipal Code in December 2006 with a minimum score threshold of 0.30 for commercial zones. The SGF was subsequently expanded to include multifamily residential zones in 2009. This update also included measures to enhance design flexibility through a more nuanced system of weighted credits for permeable paving and green roofs, and introduced new credits for food cultivation spaces. The city planning department has invested considerable effort in disseminating and explaining how this guidance and assessment tool operates. This has included a series of workshops and technical design sessions on specific features included in the score sheet, as well as a regularly updated web portal providing information on the score sheet and case study examples of completed schemes (Seattle City, 2017). Whilst, the city authority claims that this mechanism provides clarity to applicants in how to design their schemes and transparency in the assessment process, some nevertheless 'lament that it burdens an already-complex approval process' (Rouse and Bunster-Ossa, 2013, p. 79).

The SGF illustrates a North American GI accent. Here, the ecosystem service emphasised in environmental planning activities is primarily drainage management. Yet this is done in a manner that provides for ecological and aesthetic enhancements in a way that seeks to work with natural hydrological processes rather than conventional infrastructural methods involving a heavy reliance on piping. The work of Seattle in this field has become a touchstone for operationalising a GI approach to planning in North America, particularly in the United States,

Revised 12/28/10

Green Factor Score Sheet

SEATTLE×*green factor*

Project title:

		enter sq ft of parcel			
	Parcel size *(enter this value first)*	5,000		SCORE	-
	Landscape Elements**	**Totals from GF worksheet**		**Factor**	**Total**
A	**Landscaped areas (select one of the following for each area)**				
1	Landscaped areas with a soil depth of less than 24"		*enter sq ft* 0	0.1	-
2	Landscaped areas with a soil depth of 24" or greater		*enter sq ft* 0	0.6	-
3	Bioretention facilities		*enter sq ft* 0	1.0	-
B	**Plantings (credit for plants in landscaped areas from Section A)**				
1	Mulch, ground covers, or other plants less than 2' tall at maturity		*enter sq ft* 0	0.1	-
2	Shrubs or perennials 2'+ at maturity - calculated at 12 sq ft per plant (typically planted no closer than 18" on center)	*enter number of plants* 0	0	0.3	-
3	Tree canopy for "small trees" or equivalent (canopy spread 8' to 15') - calculated at 75 sq ft per tree	*enter number of plants* 0	0	0.3	-
4	Tree canopy for "small/medium trees" or equivalent (canopy spread 16' to 20') - calculated at 150 sq ft per tree	*enter number of plants* 0	0	0.3	-
5	Tree canopy for "medium/large trees" or equivalent (canopy spread of 21' to 25') - calculated at 250 sq ft per tree	*enter number of plants* 0	0	0.4	-
6	Tree canopy for "large trees" or equivalent (canopy spread of 26' to 30') - calculated at 350 sq ft per tree	*enter number of plants* 0	0	0.4	-
7	Tree canopy for preservation of large existing trees with trunks 6"+ in diameter - calculated at 20 sq ft per inch diameter	*enter inches DBH* 0	0	0.8	-
C	**Green roofs**				
1	Over at least 2" and less than 4" of growth medium		*enter sq ft* 0	0.4	-
2	Over at least 4" of growth medium		*enter sq ft* 0	0.7	-
D	**Vegetated walls**		*enter sq ft* 0	0.7	-
E	**Approved water features**		*enter sq ft* 0	0.7	-
F	**Permeable paving**				
1	Permeable paving over at least 6" and less than 24" of soil or gravel		*enter sq ft* 0	0.2	-
2	Permeable paving over at least 24" of soil or gravel		*enter sq ft* 0	0.5	-
G	**Structural soil systems**		*enter sq ft* 0	0.2	-
		sub-total of sq ft =	*0*		
H	**Bonuses**				
1	Drought-tolerant or native plant species		*enter sq ft* 0	0.1	-
2	Landscaped areas where at least 50% of annual irrigation needs are met through the use of harvested rainwater		*enter sq ft* 0	0.2	-
3	Landscaping visible to passersby from adjacent public right of way or public open spaces		*enter sq ft* 0	0.1	-
4	Landscaping in food cultivation		*enter sq ft* 0	0.1	-
				Green Factor numerator =	-

** Do not count public rights-of-way in parcel size calculation.*

*** You may count landscape improvements in rights-of-way contiguous with the parcel. All landscaping on private and public property must comply with the Landscape Standards Director's Rule (DR 6-2009)*

Figure 25.1 Seattle Green Factor score sheet

Source: Seattle Municipal Authority (copyright permission obtained)

where several cities such as Chicago, Portland and Washington, DC are now developing similar assessment tools. What these initiatives share is the development of a quantitative assessment matrix that seeks to marry simplicity with clarity in an engineering-like building code that operationalises a decision support mechanism for valued ecosystems services in an environmental planning context.

Neighbourhood scale: United Kingdom & Ireland – Portmarknock's sustainable urban extension

Heavily influenced by the socially conscious 'garden city' movement and the desire to facilitate positive social-ecological interactions in the context of an ever-increasing urban population, the inflection given to the GI concept in the United Kingdom and Ireland often displays a more ostensible social orientation than that evident in North America. Thus, exemplars of the concept in this context, such as the Cambridgeshire and South Hampshire green infrastructure strategies in England, give equal weight to issues of recreation and economic development as to more conventional environmental planning themes, such as nature conservation, drainage management and climate change adaptation. Representative of this holistic interpretation of GI is the planning activities of Fingal County Council (FCC) in Dublin, Ireland.

Similar to many planning authorities operating on the periphery of major cities, the area administered by FCC encompasses a transition of land uses from an urban-suburban continuum extending from Dublin City to a rural coastal and agricultural landscape containing numerous nature conservation sites. In an effort to mitigate frictions between environmental protection and growth management in this transitionary land use context, FCC has been a pioneer in the deployment of the GI approach in Ireland. This has entailed a holistic perspective on planning that endeavours to increase the ecosystems services realised from the urban and rural landscape in a way that furnishes quality of life enhancements while simultaneously advancing ecological conservation (Lennon et al., 2016). How FCC has sought to realise this is illustrated in the innovative and interlinked local area plans for the contiguously located Baldoyle-Stapolin (FCC, 2013a) and Portmarnock South (FCC, 2013b) areas. These plans evidence a GI planning approach by seeking to integrate policy initiatives concerning sustainable drainage management, archaeology and built heritage, as well as recreation, biodiversity and landscape aesthetics. Partly drawing inspiration from garden city typologies, the plans include new residential areas integrated with parkland, sustainable urban drainage schemes, non-motorised transport routes and urban farming, that endeavour to promote local sustainability within a broader environmental planning strategy that seeks to consolidate and enhance a range of ecosystems services. Thus, in comparison with conventionally produced local area plans in Ireland, these plans are atypically detailed in the provision of design guidance. Accordingly, the plans detail mowing regimes, direction on how drainage management should be incorporated into the design of the public realm, and guidance on public lighting so as not cause undue interference to nocturnal animals. This multifunctional planning perspective extends into policy governing the construction phases of the local area plans. Here, FCC seeks to promote the use of development sites through the temporary use of undeveloped areas for ecological and social purposes, such as using them as ecologically rich temporary wildflower meadows and grassed recreation spaces.

Paralleling the attention allocated multifunctionality is the increased focus given to spatial and functional connectivity between land uses in local policy formulation and implementation. Prior to the advocacy of a GI planning approach, FCC had advanced habitat connectivity via ecological networks. Such networks render otherwise fragmented ecosystems biologically coherent by facilitating species movement and genetic exchange. Although promoting spatial and scalar integration, these networks focused primarily on 'ecological' connectivity. However, following greater acquaintance with GI theory and the consequent advocacy of a holistic approach to planning, FCC has sought to advance a more functionally integrated network of key sites that meet several social objectives while concurrently maintaining ecosystems integrity. This GI network is given graphic representation in a series of planning maps for the FCC administrative area that identify key sites of conservation and amenity value linked via a series of multipurpose corridors. Central to the planning of this GI network has been the use of spatial data analysis in identifying opportunities for enhanced connectivity. Using such evidence, efforts are made to produce comprehensive maps of GI assets from which to formulate site-specific initiatives that consolidate the broader GI network. This approach is reflected within the Baldoyle-Stapolin and Portmarnock South local area plans. Here connectivity is promoted both within the plan lands and with contiguous land uses through specific design guidance for particular locations. For example, direction on the integration of aesthetically appealing sustainable urban drainage schemes into new residential developments is provided at particular locations. Such guidance concurrently seeks to encourage the use of these schemes as routes for active transport (cycling and walking) and enhancements to existing ecological networks (see Figure 25.2). Importantly, the potential impacts of such guidance has undergone a rigorous process of environmental assessment that support their iterative development and help ensure their fitness for purpose in assisting planners evaluate the appropriateness of development proposals within the context of a GI approach applied at the neighbourhood scale.

City region scale: Europe – Copenhagen's finger plan

The deployment of the term GI has becoming increasingly common in a continental European context. It is now being used to describe a diversity of activities and spatially dissimilar initiatives from rooftop urban bee keeping to transcontinental ecological networks. Nevertheless, these initiatives often bear association through their alignment with an encompassing 'green structure' framework for green space provision, protection and use within a city region. Such frameworks often have a long history and enjoy currency with the public imagination that lends political support to both their maintenance and enhancement. Hence, green structure frameworks such as the 'green heart' of the Randstad Region of The Netherlands usually enjoy broad-based support for their preservation. Indeed, originally conceived as a means to counter the urban sprawl of several Dutch cities, the 'green heart' has since been reframed as a space where nature conservation and outdoors recreation activities can co-exist. Likewise, Helsinki and Stockholm have developed green structure frameworks focused on a series of green corridors radiating from the city centre that are designed to preserve green space connectivity between inner city districts and surrounding rural areas for nature conservation, recreation and drainage management. One of the best developed of such city region planning initiatives is that of Copenhagen in Denmark.

In the decades following World War II, Danish society experienced unprecedented growth. Much of this was centred on Copenhagen and the urban settlements within its hinterland. This placed considerable pressure on the landscape surrounding the city as an increasing urban population and a growing middle class resulted in urban generated residential encroachment on the rural environment within the expanding city region. Although a 'green network plan' had been developed prior to the World War II, its implementation had been effectively suspended during the conflict. However, renewed interest was given to this plan in the boom years immediately after the war. What resulted was the 1947 Finger Plan. This and all subsequent iterations of the plan, concentrates urban development along rail lines and roads radiating from the city centre area. The metaphor reflecting and informing planning in the city region became that

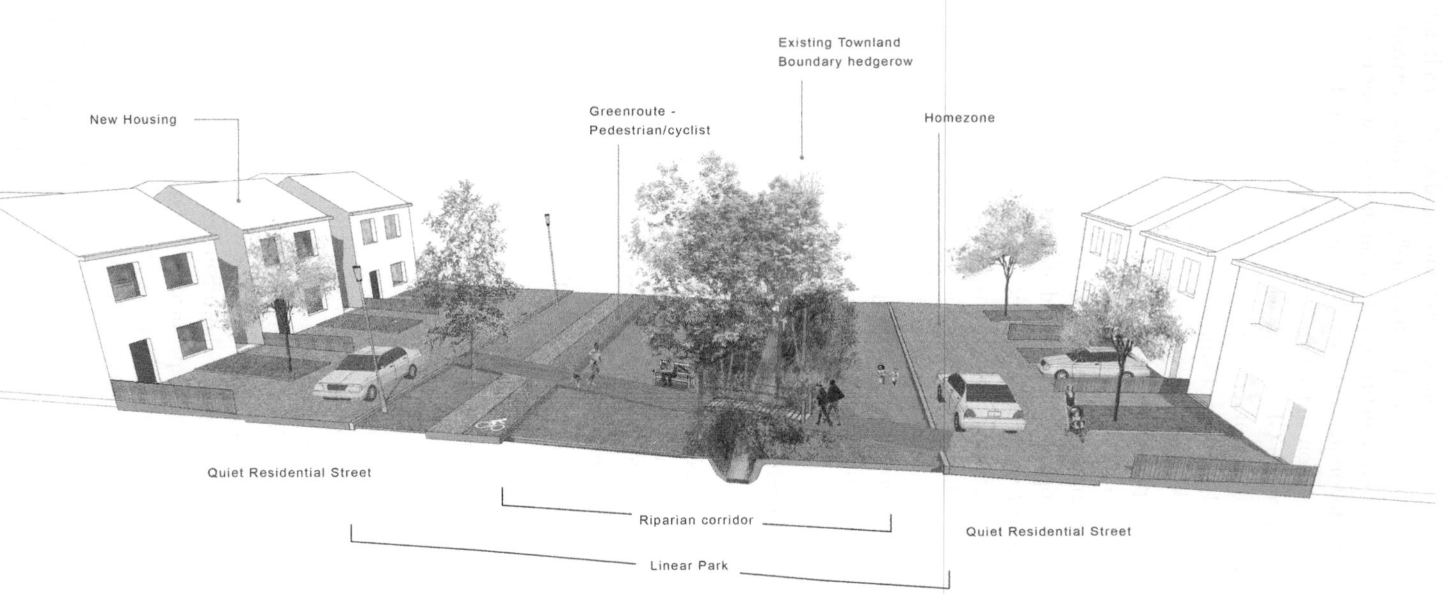

Figure 25.2 Multifunctional neighbourhood linear park concept from the Portmarnock South Local Area Plan

Source: Fingal County Council and Ait-Place Ltd (copyright permission obtained)

of a 'hand': the central city area being the 'palm' of the hand, while each of the five radial areas where development is concentrated comprising the 'fingers' (see Figure 25.3). Between each of these fingers are a series of 'green wedges'. As the fingers have extended, so too have the green wedges between them. The original intention of these wedges was that they would delineate the urban and rural environments, facilitate a compact city transit-orientated form of development, and supply urban residents with easy access to a rural landscape (Brüel, 2012). Although the initial versions of the Finger Plan focused primarily on coordinating urban growth, increasing attention has been given to protecting the green wedges since the 1980s, when permissive and uncoordinated encroachment by local municipalities located within the Western fingers threatened to erode the extent and environmental quality of some such wedges. Today, the inner green wedges close to the central urban area (the 'palm' of the 'hand'), primarily comprise sports grounds, allotments, parkland and other green space typologies commensurate with that of most cities. However, contiguous to these, and radiating out from the central city area in parallel with the 'fingers', are forests and agricultural lands. These provide ecological, hydrological and

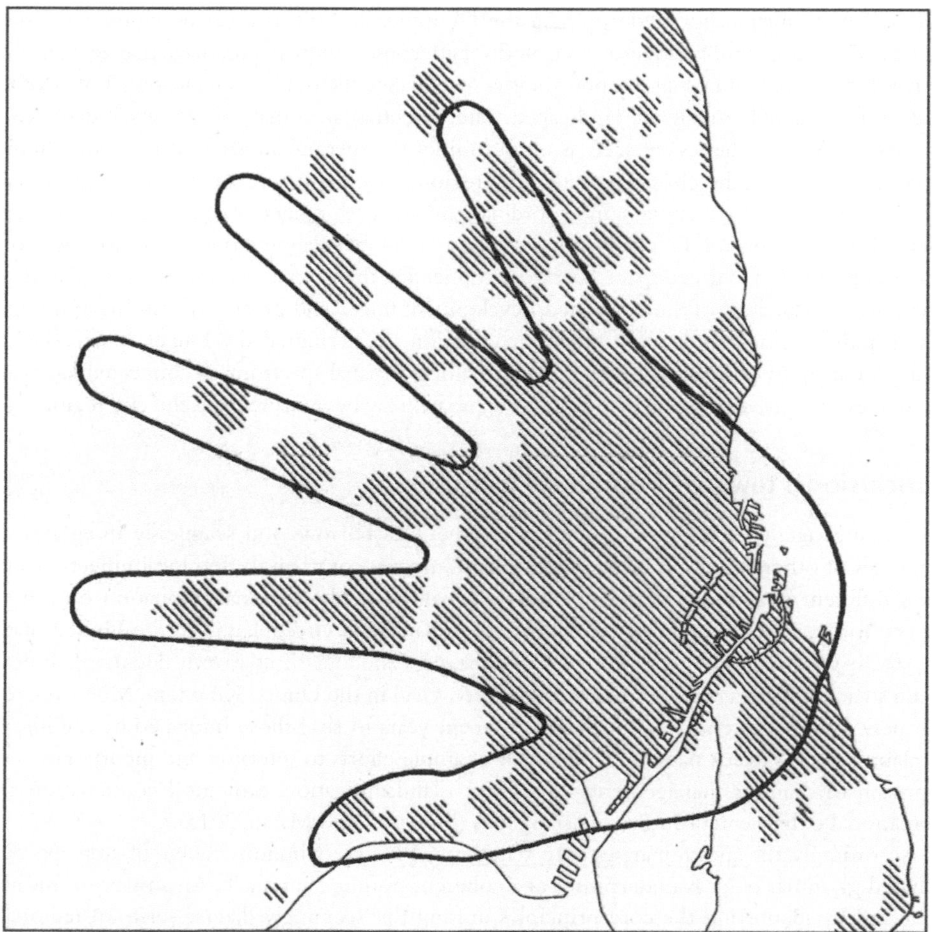

Figure 25.3 Cover of the 1947 Copenhagen Finger Plan

Source: The Danish Town Planning Institute (copyright permission obtained)

recreational connectivity between the city centre area and the agricultural areas within the city's wider hinterland (Vejre et al., 2007).

The Finger Plan encompasses 34 municipalities, thereby supplying the key strategic planning guidance structuring the overall morphology of the Copenhagen city region. It is in this context that planning in areas located within the ambit of the Finger Plan must be conducted relative to an assessment of development in the city region as a whole and must safeguard the continuity of the finger city structure. Accordingly, such development has to be evaluated relative to the 'green structure' comprising the various undeveloped wedges between the fingers. Within this green structure, a series of development types considered potentially erosive of the green character of the wedges are explicitly forbidden. These include energy infrastructure installations such as wind and solar farms, the construction of large facilities for commercial purposes and the establishment of intensive livestock farming, which it is believed would impair the landscape quality of the wedges. However, in keeping with the desire to maintain a high-quality landscape within these wedges, normal agricultural practices and forestry is encouraged (DNA, 2015). Recent iterations of the Finger Plan have expanded the range of ecosystems services perceived as emanating from the wedges and have thus encouraged an enlarged array of developments resonant with the principle of 'multifunctionality' underpinning the GI approach. This includes provision for climate change adaptation, carbon sequestration, biodiversity conservation and drainage management initiatives that concurrently supply opportunities to enhance the outdoors recreational offer of the wedges. For example, canals and landscaped water retention areas that are sensitively designed in response to local aesthetics are seen as opportunities to synergise an array of uses while simultaneously protecting the character of the city region's green structure. Hence, although initially formulated as a means to contain and coordinate urban development during a period of unprecedented urbanisation, the Finger Plan has served as a durable planning framework for the provision and protection of green space corridors connecting the central city area with its hinterland. The institutionalisation of these twinned 'development finger' and 'green structure' morphologies within politics, planning and the public imagination has permitted the longevity necessary to realise an array of ecosystems services benefits along a spatial spectrum of connected landscape typologies that give expression to how a GI approach may be interpreted at the city region scale.

Conclusions: towards a common language

Like a language, GI is a dynamic idea that sometimes borrows and seamlessly integrates the emphases of other dialects in local expressions. Thus, it is common to find local inflections that stress different dimensions of GI not fully resonant with the regional intonation one might expect; for example, the garden city informed Philadelphia Greenplan35 in the United States, the storm water management centred Copenhagen Cloudburst initiative in Denmark and the green structure orientation of the London Green Grid in the United Kingdom. Moreover, several new regional accents have emerged in recent years to rival those informed by the history of planning in Western nations. Witness for example efforts to interpret and incorporate a GI approach to drainage management, urban heat island mitigation, ecological conservation and recreational enhancement in many Asian cities (Newman and Matan, 2013).

Accordingly, the question arises as to whether GI is a cacophonous babble of concepts or if a shared grammar of ideas underpins it as a coherent planning approach. An answer to this may be found by identifying the core principles uniting this seemingly diverse series of regionally accented interpretations. Such scrutiny reveals how advocates of GI are unified in contending that planning for the protection and enhancement of green and blue spaces, and the natural processes they accommodate, should precede the allocation of lands for development. It is in this sense that those endorsing GI seek to advance a 'narrative of necessity' that reframes green and blue spaces as essential 'infrastructure' underpinning the successful functioning of society (Lennon and Scott, 2014). Advocates of GI are thereby united in their desire to work *with* rather than against natural processes.

Also threading the varieties of GI planning together is the centrality of 'multifunctionality'. Hence, advocates of the GI approach contend that by exploring and exploiting the variety of overlapping opportunities to work with natural processes, the varying attributes of green and blue spaces can be synergised to realise an array of benefits across different systems, such as hydrology, transportation, recreation, energy, economy and nature conservation (Rouse and Bunster-Ossa, 2013).

Furthermore, irrespective of its distinctive regional accent, the GI approach promotes spatial connectivity; albeit the emphasis may be variably placed on ecological permeability through a landscape, working with a natural and spatially dispersed hydrological hierarchy, and/or facilitating human mobility via active forms of transport such as walking and cycling. Thus, a shared characteristic of all endeavours operating beneath the GI banner is the promotion of spatial connectivity that seeks to enhance an assorted configuration of various functions, which at a minimum respect existing hydrological conditions and ecological integrity while concurrently facilitating anthropocentric utility. Hence, with a focus on natural processes, connectivity and multifunctionality, the various GI accents are united as an approach that seeks to reverse traditional planning practices wherein attention is directed at discrete land parcels for the provision of single functions (for example, drainage, conservation, recreation), often entailing resistance to natural processes (for example, the natural course and flooding characteristics of a river).

Consequent on the high profile allocated the benefits obtained from green and blue spaces, a key achievement of 'talking GI' has been to help embed the language and ideas of ecosystems services (MA, 2005) within environmental planning. This has been achieved by highlighting the benefits of working 'with' and enhancing natural processes through well planned green and blue spaces at all scales (Lennon and Scott, 2014). Realising this has involved collaboration with and across a variety of different disciplines to locate ways of synergising knowledges and efforts to consolidate and expand ecosystems services co-benefits. This attribute of GI planning underpins each of the illustrative case studies presented in this chapter. For example, operationalising the Seattle Green Factor necessitated extensive inter-departmental collaboration as stormwater engineers and city planners sought to harmonise their respective assessment methods and codes. This emphasis on partnerships and cooperation across administrative boundaries and organisational structures resonates with contemporary efforts in the theory and practice of environmental planning to promote 'joined-up thinking' and 'integrated governance' arrangements (Thomas and Littlewood, 2010), most notably as a means to spatially coordinate planning activities in confronting the challenges of climate change (Lennon et al., 2016). Hence, despite an array of regional accents, the desire to work collaboratively across disciplines and with natural processes, respect context, promote multifunctionality and foster connectivity, contour a shared understanding of what the GI concept represents. It is these shared attributes that enable advocates of the concept to communicate across contexts, boundaries and scales with a common language for environmental planning.

References

Brüel, M. (2012). 'Copenhagen, Denmark: Green City amid the finger metropolis', in: T. Beatley (ed.) *Green cities of Europe: global lessons on green urbanism*. Washington, DC: Island Press, pp 83–108.

DNA (2015). *The Finger Plan*. Copenhagen: Danish Nature Agency (DNA).

Fábos, J. G. (2004). 'Greenway planning in the United States: its origins and recent case studies'. *Landscape and Urban Planning*, 68(2): 321–42.

FCC (2013a). *Baldoyle-Stapolin local area plan*. Swords, Dublin: Fingal County Council.

FCC (2013b). *Portmarnock South local area plan*. Swords, Dublin: Fingal County Council.

Jongman, R. H. G., Külvik, M. and Kristiansen, I. (2004). 'European ecological networks and greenways'. *Landscape and Urban Planning*, 68: 305–19.

Lennon, M. and Scott, M. (2014). 'Delivering ecosystems services via spatial planning: reviewing the possibilities and implications of a green infrastructure approach'. *Town Planning Review*, 85(5): 563–87.

Lennon, M. and Scott, M. (2016). 'Re-naturing the city'. *Planning Theory and Practice*, 17(2): 270–6.

Lennon, M., Scott, M., Collier, M. and Foley, K. (2016). 'The emergence of green infrastructure as promoting the centralisation of a landscape perspective in spatial planning – the case of Ireland'. *Landscape Research*, 42(2): 146–63.

MA (2005). *Millennium ecosystems assessment (MA): ecosystems and human well-being: synthesis report*. Washington, DC: Island Press.

Mell, I. (2016). *Global green infrastructure: lessons for successful policy-making, investment and management*. London: Taylor and Francis.

Newman, P. and Matan, A. (2013). *Green urbanism in Asia: the emerging green tigers*. Singapore: World Scientific Publishing.

Rouse, D. C. and Bunster-Ossa, I. F. (2013). *Green infrastructure: a landscape approach*. Washington, DC: American Planning Association.

Seattle City (2017). *Seattle Green Factor*. Seattle Department of Construction and Inspections. Available at: www.seattle.gov/dpd/codesrules/codes/greenfactor/default.htm

Thomas, K. and Littlewood, S. (2010). 'From green belts to green infrastructure? The evolution of a new concept in the emerging soft governance of spatial strategies'. *Planning Practice and Research*, 25(2): 203–22.

Vejre, H., Primdahl, J. and Brandt, J. (2007). 'The Copenhagen Finger Plan: keeping a green space structure by a simple planning metaphor', in B. Pedroli, A. Van Doorn, G. De Blust, M. L. Paracchini, D. Wascher and F. Bunce (eds.) *Europe's living landscapes: essays exploring our identity in the countryside*. Zeist: KNNV Publishing, pp 311–28.

26

Marine planning and coastal management

Sue Kidd

Introduction

Half a century ago on Christmas Eve 1968, Apollo 8 astronaut Bill Anders took what has become one of the most influential environmental images of all time (Cosgrove, 2008). It was a picture of earthrise over the moon. Not only did it convey a sense of the Earth's fragility but from the perspective of this chapter, however, perhaps the most arresting feature of the image was that the Earth appeared as a bright blue disk streaked with immense cloud systems. It revealed in a new and compelling way that over 70% of the Earth's surface is covered by oceans.

It therefore seems surprising that most discussion regarding environmental planning tends to focus on the land. However, this chapter shows that attention is increasingly also extending to the sea and indicating that an integrated approach spanning both land and sea is needed if we are to respond in a meaningful way to the environmental and sustainability challenges facing the world in the 21st century.

The chapter starts by outlining the vital part that marine and coastal areas play in sustaining life on Earth and the complex web of ecosystem services they provide. It then presents an overview of the current state of these important ecosystems and the drivers of change that are leading to major threats to their health. Subsequently, key international treaties related to marine and coastal areas and the associated environmental commitments that guide policy development in most parts of the world are highlighted. The chapter then explores the suite of related policy responses that have emerged including Integrated Coastal Management (ICM), Marine Protected Areas (MPAs) and most recently the new systems of Marine Spatial Planning (MSP) that are being established all over the world. The chapter concludes by setting out some thoughts on the way ahead for environmental planning action more generally drawing inspiration derived from the ocean.

The importance of marine and coastal environments

Since the Apollo 8 moon landing, not only has our understanding of outer space increased significantly so too has our understanding of the sea, yet even today the ocean remains largely unexplored. Despite the embryonic state of much of our knowledge we do now understand that the world's oceans – their temperature, chemistry, currents and plant and animal life – are critical components determining the state of global ecology, increasingly referred to as the overall Earth System which makes the Earth habitable not only for humankind but for most, and perhaps all, life on Earth (Ocean Literacy Campaign, 2013).

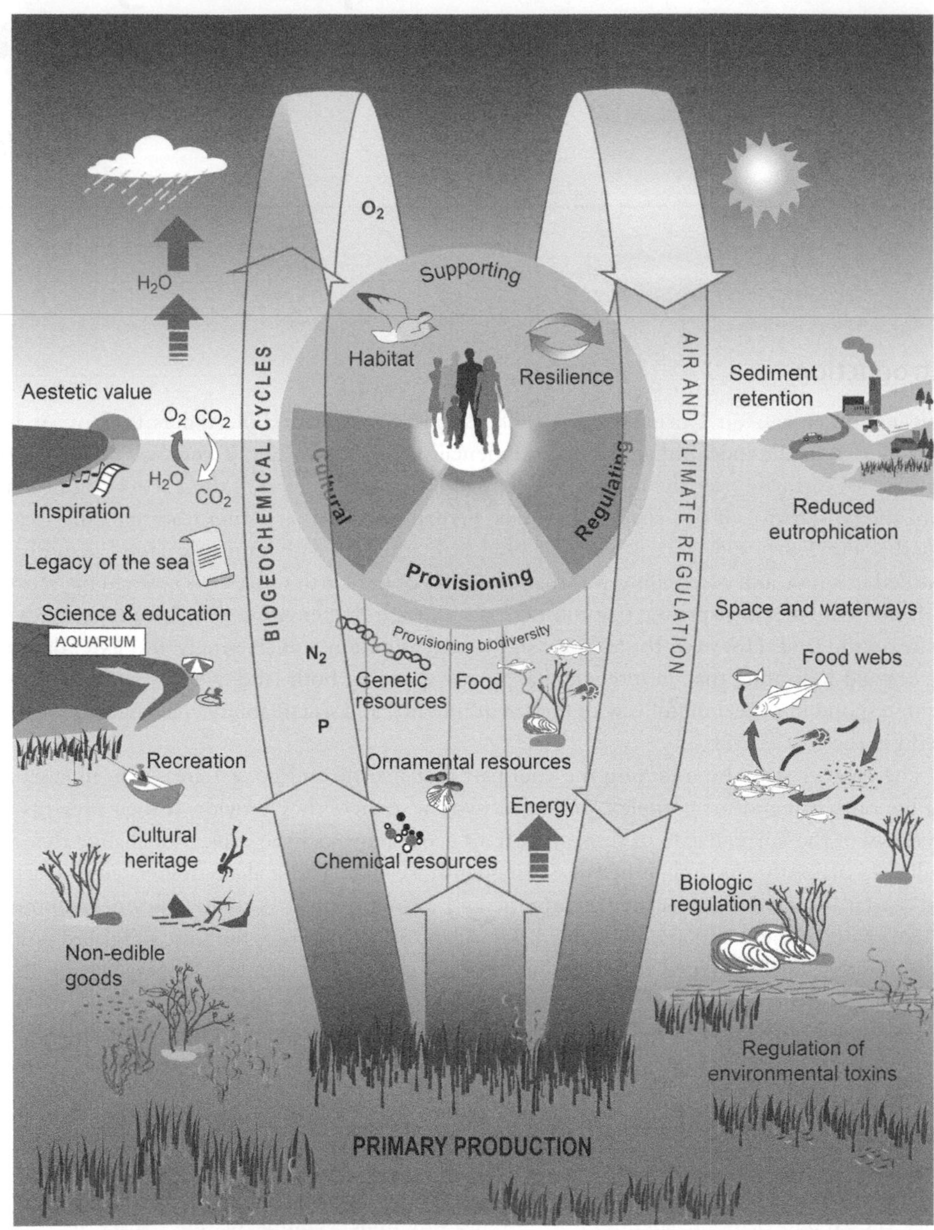

Figure 26.1 What's in the sea for me?

Source: adapted from Swedish Environmental Protection Agency (2009, p. 17)

The United Nations has played a significant role in collating and disseminating the latest scientific knowledge related to the marine and coastal environment particularly through work undertaken as part of the Millennium Ecosystem Assessment (UNEP, 2006) and more recently associated with its Sustainable Development Goal activities (United Nations, 2015, 2016). This work has highlighted, for example, that over half of the Earth's oxygen is produced by the photosynthesis of marine plants and that ocean ecosystems are one of the Earth's most important carbon sinks – absorbing around 30% of the carbon dioxide produced by humans – and through this process buffering the planet from the impact of climate change. At the same time, coastal habitats are now recognised as providing valuable natural defences against coastal flooding and the combined effects of sea level rise and increasing storminess associated with climate change. Beyond such fundamental supporting and regulating services, the sea also offers a wide array of provisioning and cultural services. For example, it is the world's largest source of protein with more than 3 billion people depending on the ocean as their primary source. At the same time, coastal areas continue to be favoured places for human habitation, tourism and recreation often because of their valued ocean-related cultural services. The complex web of ecosystem services provided by marine and coastal areas is shown in Figure 26.1.

The current state of the marine and coastal environment

The Millennium Ecosystem Assessment marked an important staging point in collating global understanding about ecosystem change and human wellbeing and the key messages related to marine and coastal ecosystems were significant. It concluded not only that people are dependent on the ocean and coasts and their resources for their survival and wellbeing but also that most services derived from marine and coastal ecosystems are being degraded and used unsustainably and are deteriorating faster than other ecosystems. In addition, it noted that the major drivers of change, degradation or loss of marine and coastal ecosystems and services are mainly anthropogenic and that many of these drivers are terrestrial in origin and related as much to human activity and development trends on land as in marine areas (UNEP, 2006). This is because the sea is the major sediment and nutrient sink from the land with pollutants and sediments from land-based activities released into water and air ultimately likely to find their way to the sea, creating pressures on the functioning of marine and coastal ecosystems (University of Liverpool, 2016).

These headline messages are reiterated in the most recent global review of the state of the oceans – the First Global Integrated Marine Assessment (United Nations, 2016). It concluded that today no part of the ocean has completely escaped the impact of human pressures, including even the most remote areas. The main drivers for change identified in the assessment are illustrated in Figure 26.2.

What is evident from these assessments is that, although humans have always looked to the sea for food, transport and trade, waste disposal and cultural and spiritual fulfilment, and coastal areas have long been favoured places for human settlement, there has been a dramatic uplift in the scale of human activity and associated land-sea interactions and environmental pressures in recent times. The assessments also indicate that in many parts of the world this trend of uplift is continuing. For example, today 16 of the world's 23 mega cities (with populations exceeding 10 million) are in coastal locations (Pelling and Blackburn, 2014), and with the prospect of the global population raising from 7.6 billion in 2017 to over 11 billion by 2100 (United Nations, 2017), ongoing urbanisation of coastal areas is anticipated.

Beyond general trends of globalisation and the growing importance of international connectivity, one of the factors driving contemporary coastal and marine development is that the sea is increasingly being seen as source of 'Blue Economy' or 'Blue Growth' opportunities and

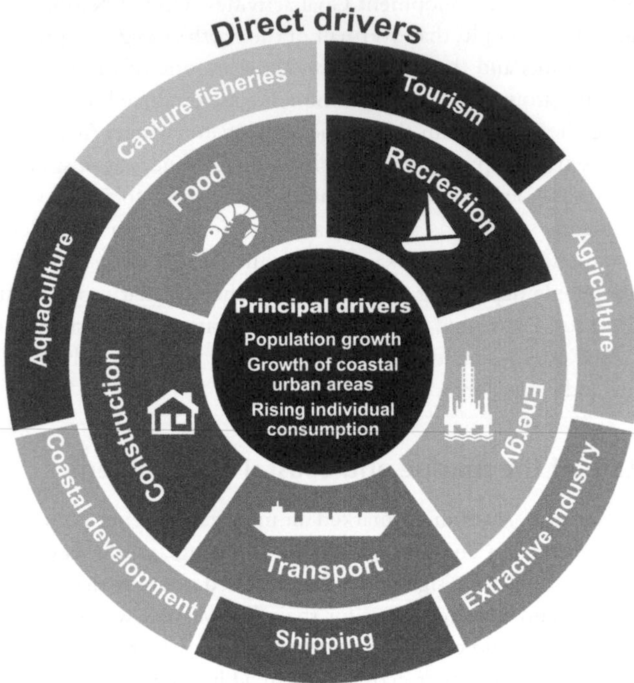

Figure 26.2 Principal direct and indirect drivers for change in the marine environment
Source: adapted from GRID-Arendal and UNEP (2016, p. 5)

these are the source of new human pressures on the ocean (European MSP Platform, 2017). As a consequence, mature maritime sectors such as fishing, shipping and offshore oil and gas production are now frequently accompanied by other offshore uses such as aquaculture and wind power developments which are bringing a new era of fixed built development into coastal and marine environments. Meanwhile, technological advances are opening new business possibilities in sectors such as blue biotechnology, ocean renewable energy and deep sea mineral extraction (Ecorys, 2012).

It is therefore evident that the combined effects of current human development trends on land and in the sea pose significant threats to the health of marine and coastal ecosystems and that new approaches to environmental management and sustainability are needed to address these threats.

Legal framework for the sea

Before considering some of the key policy responses to marine and coastal issues, it is important to understand that the legal regime for the sea is very different from that of the land. The reason for this is that together with the atmosphere, Antarctica and outer space, the High Seas are recognised as one of the world's global commons and historically the legal framework for all parts of the sea has been guided by the principle of *mare liberum* (free sea for everyone). While today, following the 1982 United Nations Convention on the Law of the Sea (UNCLOS), coastal states are able to gain sovereignty over adjacent coastal waters to hold in public trust for their inhabitants (including their territorial sea extending up to 12 nautical miles from the coast and Exclusive Economic Zone (EEZ) typically extending up to 200 nautical miles), this sovereignty must be exercised subject to UNCLOS and to other rules of international law (Osherenko, 2006). This means that international obligations and commitments are much more significant determinants of policies of all kinds in marine and coastal settings than is typically the case on the land where national legal and political regimes are much more influential (Rothwell and Stephens, 2016).

Table 26.1 lists some of the key international agreements related to marine and coastal areas and highlights associated environmental obligations and commitments that have guided

Table 26.1 Key international agreements related to marine and coastal ecosystems

Agreements	*Environmental considerations*
1971 Ramsar Convention on Wetlands of International Importance especially as Waterfowl Habitat	States have an obligation to ensure the conservation and wise use of wetlands designated to be of international importance including estuaries, deltas and tidal flats, mangroves and other coastal areas and coral reefs.
1982 United Nations Convention on the Law of the Sea (UNCLOS)	States have an obligation to protect and preserve the marine environment.
1992 Rio Declaration/Agenda 21 Chapter 17/Convention on Biological Diversity (CBD)	States have an obligation to protect ocean and coastal areas and ensure rational use and development of their living resources through activity at global, national and local levels. CBD establishes the Ecosystem Approach as the guiding paradigm for environmental planning.
1995 United Nations Fish Stocks Agreement	States have an obligation to ensure long-term sustainability and promote optimum utilisation of fish stocks.
2002 Johannesburg World Summit on Sustainable Development Plan of Implementation	Includes reiteration of UNCLOS commitments and application of the Ecosystem Approach, promotes integrated, multidisciplinary and multi-sectoral coastal and ocean management at the national level and strengthened regional sea cooperation.
2010 Aichi Biodiversity Targets	By 2020, at least 10% of coastal and marine areas, are conserved through protective designations and are effectively and equitably managed to produce ecologically representative and well-connected systems.
2012 Rio+20 World Summit on Sustainable Development Plan of Implementation	Promotes action to maintain or restore fish stocks, establish networks of Marine Protected Areas and to protect the marine environment from land-based activities.
2015 UN 2030 Agenda: Sustainable Development Goal 14	Calls for action to conserve and sustainably use the oceans, seas and marine resources for sustainable development and sets targets for implementation.
2017 Our Ocean, Our Future: Call for Action	Expresses a commitment to halting and reversing the decline in the health and productivity of the ocean and its ecosystems and to protecting and restoring its resilience and ecological integrity.

Source: compiled by Author from material in United Nations (2015)

the development of policy responses and action programmes. Safeguarding marine and coastal ecosystems are central concerns in international law related to the sea and the table charts the evolution of thinking on the type of measures required to achieve this. Two recurring themes are prominent. The first, now captured in Sustainable Development Goal 14 (United Nations, 2015), is an underlying international commitment to conserve and sustainably use the oceans, seas and marine resources. The second is a recognition that this requires the adoption of an integrated, multidisciplinary and multi-sectoral approach to coastal and ocean management and – as all oceans are connected – transnational cooperation, particularly at a regional sea scale.

Policy responses

Integrated Coastal Management

Early efforts to address environmental degradation of maritime environments focussed on the coast where human interactions with the sea are most concentrated and adverse environmental impacts most obvious. These efforts increased understanding of the complexity of both natural and legal/institutional systems at the land-sea interface and revealed that a lack of joined up working was a major reason for coastal ecosystem decline. As a result, Integrated Coastal Management (ICM) (or Integrated Coastal Zone Management [ICZM] as it is also known) emerged as a conscious management process that recognises the interrelationships between coastal and ocean uses and the environments they effect. ICM is designed to avoid the fragmentation inherent in single sector approaches to management by establishing processes for integrated and rational decision making related to the conservation and sustainable use of coastal resources and space (Cicin-Sain et al., 1998).

The 1992 United Nations Rio Earth Summit and the suite of international agreements associated with it, including the Convention on Biological Diversity (CBD) and Chapter 17 of Agenda 21, was influential in the development of ICM worldwide by requiring signatory parties to have suitable tools in place by 2002 to better manage coastal areas. Europe proved an important testbed through the EU ICZM Demonstration Programme which ran between 1996 and 1999 and reflection on this experience led to the distillation of a set of ICZM principles (see Box 26.1) that are still internationally recognised as a sound basis for environmental management and sustainability in coastal areas (Haines-Young and Potschin, 2011).

Box 26.1 Eight principles of good ICZM

Principle 1:

A broad thematic and geographic perspective

Principle 2:

A long-term perspective

Principle 3:

Adaptive management/sound knowledge basis

Principle 4:

Local specificity

Principle 5:

Working with natural processes/carrying capacity

Principle 6:

Involving all the parties concerned

Principle 7:

Involvement of all relevant administrative bodies at national, regional and local level

Principle 8:

Use of a combination of instruments to facilitate coherence between sectoral policy objectives and coherence between planning and management

Source: summarised from Ruprecht Consult and the International Ocean Institute (2006, p. 7).

Although ICM initiatives remain an important focus of activity in many parts of the world, they have not been without their challenges. For example, research concerning the implementation of ICM in Europe revealed that even where ICM arrangements were in place the complexity of responsibilities at the coast continued to prevent agencies from taking a 'joined-up' approach. In many countries there was ad hoc development of ICM initiatives and a lack of coherence in ICM coverage. Information obstacles were also significant in preventing coordination between science and policy makers and between different sectors. Equally, a democratic deficit was evident with limited opportunity for coastal stakeholders to contribute in a meaningful way to decision making. Perhaps the most significant finding however, was that ICM activity was mainly in the form of short-term 'pilot' initiatives, and it was considered that this project-based approach to ICM was unlikely to provide an adequate long-term response to coastal planning and management concerns (Shipman and Stojanovic, 2007).

Marine Protected Areas

Alongside the development of ICM initiatives, many countries also became active in designating coastal wetland areas for special protection in response to the ambitions of the 1971 Ramsar Convention (Matthews, 1993). Following the 1992 Convention on Biological Diversity, attention increasingly extended further offshore as nations began to make progress in establishing wider networks of MPAs drawing in particular upon the pioneering work undertaken for the Great Barrier Reef in Australia, which was designated in 1975 as one of the world's first and largest marine parks. However, MPA designation lagged well behand that of terrestrial designations (Green and Paine, 1997), and renewed effort was called for in the 2010 Aichi Targets which set the challenge of establishing protective designations covering at least 10% of the world's coastal and marine areas by 2020. By 2016, although only around 4% of the global ocean was covered by MPAs, many countries had reached the Aichi targets in their national waters (UNEP-WCMC and IUCN, 2016). In many respects, this can be regarded as a significant step forward as MPAs have increasingly been recognised as an important way of contributing to marine environment conservation goals by reducing anthropogenic impacts on marine areas of particular ecological value mainly through control of fishing activity (Halpern et al., 2009).

However, although MPAs are now regarded as an essential component in the suite of environmental planning and management responses in marine and coastal areas, their limitations have also become apparent (Pendleton et al., 2018). Indeed, as experience of MPA designation and management has grown the need for additional, more comprehensive policy responses has become evident. For example, it is now appreciated that MPAs by virtue of their size and management scope are insufficient on their own to address the environmental challenges facing marine and coastal ecosystems. Although increasing attention is being placed on developing coherent networks of MPAs to facilitate the movement of species between islands of protection, research has revealed that in many cases their conservation ambitions are seriously compromised due to the degradation of unprotected surrounding ecosystems. Indeed, some authors have argued that MPAs can do more harm than good as they may simply displace activities elsewhere and cause unintended adverse consequences for other marine areas. So, while MPAs cannot be regarded as a panacea for the declining health of marine and coastal ecosystems, many do see their experience as valuable by demonstrating what might be achieved by adopting more integrated planning and management approaches at a broader scale. Together with ICM, MPA experience has therefore informed the development of Marine Spatial Planning, which is emerging as a critical third element in the suite of policy responses to coast and ocean issues today.

Marine Spatial Planning

Marine Spatial Planning has been defined as 'a public process of analysing and allocating the spatial and temporal distribution of human activities in marine areas to achieve ecological, economic, and social objectives that are usually specified through a political process' (Ehler and Douvere, 2009, p. 18). The CBD is the most important global convention supporting the development of MSP through its requirement that states adopt the Ecosystem Approach in their efforts to promote conservation and sustainable use of the marine and coastal biological diversity (Maes, 2008). Indeed, the Conference of the Parties (COP) to the CBD have concluded that multi-objective MSP has the potential to transform the way the oceans are managed by building upon and supplementing ICM and MPA activity (Secretariat of the Convention on Biological Diversity and the Scientific and Technical Advisory Panel, 2012).

Other United Nations institutions have been equally enthusiastic in their support of MSP, notably UNESCO produced its influential guide, *MSP: a step by step approach towards ecosystem-based management*, which has informed the development of MSP world-wide (Ehler and Douvere, 2009). In 2017, UNESCO co-hosted an International Conference on MSP in Paris in association with the European Commission's DG Mare where it was reported that over 60 countries were now active in some way in the development of MSP and forecast that by 2030, 80 out of the 150 countries worldwide with marine waters would have engaged with MSP (Ehler, 2017).

Although the growth of MSP in the early years of the 21st century marks a significant step towards better management of the world's coasts and oceans, it must be recognised that environmental and sustainability concerns are not the only factors driving its uptake. For many, coastal states awareness of the 'Blue Economy' and its potential for growth has also been significant. Indeed, in some instances, particularly in those countries with offshore islands, the greater part of their territorial area is in fact sea and the scope for increasing ocean activity and associated state revenues has also been a key driver for MSP development (Jones et al., 2016). It is with such perspectives in mind that calls for a more critical analysis of MSP are growing, not least from an environmental management and sustainability viewpoint (Kidd and Ellis, 2012; Ellis and Flannery, 2016).

The review of emerging practice that was reported at the 2017 International MSP conference in Paris provides some insights into the challenges ahead. For example, concerns were expressed about the effectiveness of the new marine plans which were mostly advisory in nature rather than being regulatory and enforceable. Similarly, while the ambition is that MSP should be comprehensive and integrated, it appears that one of the key sectors of environmental concern – fishing – is frequently not included in the scope of MSP. Equally, although almost all MSP initiatives claim to be ecosystem-based, this is often difficult to validate. Another issue identified is that most MSP activity has tended to be at a national or subnational level and that very few transboundary MSP examples are evident. Perhaps of most significance though was the observation that 60% of the world's ocean lies in the High Seas or areas beyond national jurisdiction and as yet a way of developing MSP processes for this, the major part of the sea, has still to be found.

Charting a course ahead

This chapter has provided an overview of environmental issues and policy responses related to marine and coastal areas. It has highlighted the critical part these ecosystems play in the overall Earth System and in supporting human life and wellbeing. Yet, marine and coastal ecosystems are facing harmful and increasing pressure from human activity both from landward development and in the sea and are judged to be deteriorating faster than other ecosystems. It is therefore not surprising that an extensive international legal framework has been put in place to galvanise global efforts to conserve and sustainably use marine resources – a critical agenda that is now captured in the United Nations' Sustainable Development Goal 14. This framework is of particular significance as all parts of the sea have common property rights associated with them and action at all levels has to respect obligations and commitments set out in international treaties and agreements. For these reasons, a common pattern of coastal and marine policy responses is emerging in many parts of the world. This includes experimentation with Integrated Coastal Management, the formal designation of Marine Protected Areas, and most recently the development of new systems of Marine Spatial Planning.

Investigation of each of these policy responses reveals a mixed picture. On the one hand, it is evident that all are at an embryonic stage in comparison to environmental planning and sustainability practices on the land and that there is still a long way to go before material improvements in the health of the world's coasts and oceans are to be delivered. On the other hand, each area has attracted committed communities of interest which are pioneering innovative ways forward. In this way, marine and coastal areas are providing valuable new insights which may help chart the course ahead for environmental planning action not only for the sea but also more generally. Some key examples of this are provided as a conclusion to this chapter.

Developing an Earth System perspective: first, research and practice related to the protection and conservation of marine and coastal ecosystems have revealed the complex interconnectedness of land and marine ecosystems and the critical role that the oceans play in shaping the dynamics of the overall Earth System. This means that not only is it now increasingly acknowledged that efforts to address the deteriorating health of marine and coastal ecosystems need to have both land and sea dimensions but also that an overarching Earth System perspective could (and arguably should) inform all environmental planning activities.

Integrated action is key: second, and in many respects a reflection of the above, marine and coastal planning and management experience has revealed the inadequacies of sectoral approaches to governance and that this often lies at the heart of environmental concerns. It has also developed awareness of the need for more integrated/joined-up (multi-scale, multi-sectoral and multidisciplinary) action if meaningful progress towards more sustainable patterns of human development is to be made. ICM experience in particular has provided some valuable guiding principles which are worth re-examination in framing the future planning and management of marine and coastal areas and also more widely.

Governance Innovation: third, there seems little doubt that innovation in patterns of governance to deliver more sustainable patterns of human development will be one of the most important challenges of the 21st century. It is interesting therefore to observe that the special features of the legal framework for the sea may provide a useful source of inspiration in this regard beyond its emphasis on integration. This is because, unlike the land, the legal framework for the oceans has its starting point in common ocean property rights, combined with respect for freedom of access but also a shared concern and responsibility to conserve and protect marine ecosystems and living resources. This 'Global Commons' perspective with its social equity and environmental stewardship underpinnings seems particularly well suited to contemporary global grand challenges and there are signs that it is beginning to extend its influence and stimulate innovation in longstanding governance arrangements for the land. For example, in Europe, the development of MSP and associated consideration of land-sea interactions, has stimulated countries such as the Netherlands and Germany to combine planning for land and sea and revise at least some governance remits accordingly. In looking to the future, such cases are worth close observation to see how arrangements marry very different legal contexts and priorities over time and might assist the ongoing search for more effective ways to respond to the powerful environmental insights highlighted in Bill Anders' earthrise photograph taken half a century ago.

References

Cicin-Sain, B., Knecht, R. W., Jang, D. and Fisk, G. W. (1998). *Integrated coastal and ocean management: concepts and practices*. Washington, DC: Island Press.

Cosgrove, D. (2008). 'Images and imagination in 20th-century environmentalism: from the Sierras to the Poles'. *Environment and Planning A*, 40(8): 1862–80.

Ecorys (2012). *Blue growth scenarios and drivers for sustainable growth from the oceans, seas and coasts: final report*. Brussels: European Commission.

Ehler, C. (2017). *World-wide status and trends of marine/maritime spatial planning*. Keynote address 2nd International Conference on Marine/Maritime Spatial Planning. Paris: UNESCO. Available at: www.msp2017.paris/presentations

Ehler, C. and Douvere, F. (2009). *Marine spatial planning: a step-by-step approach toward ecosystem-based management*. Intergovernmental Oceanographic Commission and Man and the Biosphere Programme. IOC Manual and Guides No. 53, ICAM Dossier No. 6. Paris: UNESCO.

European MSP Platform (2017). *MSP for blue growth: how to plan for a sustainable blue economy?* Brussels: European MSP Platform.

Ellis, G. and Flannery, W. (2016). 'Marine spatial planning: Cui bono?' *Planning Theory and Practice*, 17(1): 122–8.

Green, M. and Paine, J. (1997). *State of the world's protected areas at the end of the twentieth century*. Cambridge, UK: World Conservation Monitoring Centre.

GRID-Arendal and UNEP (2016). *World ocean assessment overview*. Norway: GRID-Arendal.

Haines-Young, R. and Potschin, M. (2011). *Integrated coastal Zone management and the ecosystem approach*. Deliverable D2.1. PEGASO Grant Agreement No. 244170. CEM Working Paper No. 7.

Halpern, B. S., Lester, S. E. and Kellner, J. B. (2009). 'Spillover from marine reserves and the replenishment of fished stocks'. *Environmental Conservation*, 36(4): 268–76.

Jones, P., Lieberknecht, L. and Qiu, W. (2016). 'Marine spatial planning in reality: introduction to case studies and discussion of findings'. *Marine Policy*, 71: 256–64.

Kidd, S. and Ellis, G. (2012). 'From the land to sea and back again? Using terrestrial planning to understand the process of marine spatial planning'. *Journal of Environmental Policy and Planning*, 14(1): 49–66.

Maes, F. (2008). 'The international legal framework for marine spatial planning'. *Marine Policy*, 32(5): 797–810.

Matthews, G. (1993). *The Ramsar Convention on Wetlands: its history and development*. Gland: Ramsar Convention Bureau.

Ocean Literacy Campaign (2013). *Ocean literacy: the essential principles and fundamental concepts of ocean sciences for learners of all ages*. Version 2, a brochure resulting from the 2 week online workshop on 'Ocean Literacy through Science Standards'. National Oceanic and Atmospheric Administration, USA.

Osherenko, G. (2006). 'New discourses on ocean governance: understanding property rights and the public trust'. *Journal of Environmental Law and Litigation*, 21: 317–81.

Pelling, M. and Blackburn, S. (2014). *Megacities and the coast: risk, resilience and transformation*. Abingdon: Routledge.

Pendleton, L., Ahmadia, G., Browman, H., Thurstan, R., Kaplan, D. and Bartolino, V. (2018). 'Debating the effectiveness of marine protected areas'. *ICES Journal of Marine Science*, 75(3): 1156–9.

Rothwell, D. and Stephens, P. (2016). *International law of the sea* (2nd edn). London: Bloomsbury.

Ruprecht Consult and The International Ocean Institute (2006). *Evaluation of integrated coastal zone management in Europe: final report*. Brussels: European Commission.

Secretariat of the Convention on Biological Diversity and the Scientific and Technical Advisory Panel – GEF (2012). *Marine spatial planning in the context of the convention on biological diversity*. A study carried out in response to CBD COP 10 decision X/29, Montreal, Technical Series No. 68.

Shipman, B. and Stojanovic, T. (2007). 'Facts, fictions, and failures of integrated coastal zone management in Europe'. *Coastal Management*, 35(2–3): 375–98.

Swedish Environmental Protection Agency (2009). *What's in the sea for me? – ecosystem services provided by the Baltic Sea and Skagerrak*. Stockholm: Swedish Environmental Protection Agency.

UNEP (2006). *Marine and coastal ecosystems and human wellbeing: a synthesis report based on the findings of the Millennium Ecosystem Assessment*. Nairobi: UNEP.

UNEP-WCMC and IUCN (2016). *Protected planet report 2016*. Cambridge UK and Gland, Switzerland: UNEP-WCMC and IUCN.

United Nations (2015). *Sustainable development goals fact sheet*. New York: United Nations.

United Nations (2016). *The first global integrated marine assessment: world ocean assessment I*. New York: United Nations.

United Nations (2017). *World population prospects: the 2017 revision*. New York: United Nations.

University of Liverpool (2016). *Marine proofing for good environmental status of the sea: good practice guidelines for terrestrial planning*. Woking: Celtic Seas Partnership, WWF.

27

Transport and air pollution

Anil Namdeo and Saad Almutairi

Growth of private transport

Across the world, the transport sector is a very important component of the economy and an integral part of development. Its role is even more important within the global economy where the mobility of people, goods and information are intrinsically linked to economic opportunities. High-density transport infrastructure and connected networks are commonly associated with high levels of development. When transport systems are efficient, they provide economic and social opportunities and benefits that result in positive multiplier effects, such as better accessibility to markets, employment and additional investments. When transport systems are deficient in terms of capacity or reliability, they can have environmental and economic costs, often resulting in adverse environmental, health impacts and lower quality of life (Rodrigue, 2017).

The number of passenger cars has increased globally in recent decades. For example, in Europe, car manufacturing reached 18.5 million cars in 2015, 18% above 2010 figures (OCIA, 2018). 252 million cars were running on the roads of the EU in 2015 (ACEA, 2017). In 2017, there were 32 million passenger cars in the United Kingdom (Department for Transport, 2018). The number of vehicles is rapidly growing across the world, for example, China now has over 300 million registered vehicles. This value grew by 13.7% in 2016 alone and is only expected to continue growing rapidly (Zheng, 2017). Globally, the number of vehicles is set to vastly increase in the upcoming years, with the prediction that the number of cars on roads is expected to double by 2040 (Smith, 2016). A study into walkability in Asian cities by the Asian Development Bank found that Asian cities with previously high levels of pedestrians are now finding large decreases in walking and huge rises in vehicle numbers, particularly small motorbikes.

These issues provide a difficult context for environmental planning, not just with regard to managing the impacts or providing infrastructure but also with regard to transitioning to more sustainable cities. Overall, this chapter aims to provide a high-level overview of air pollution issues for planners before turning attention to some strategies that can address this growing concern.

Transport and air pollutant emissions

Up until the 1950s, the main air pollution problem in both developed and rapidly industrialising countries was typically high levels of smoke and sulphur dioxide emitted following the combustion of sulphur-containing fossil fuels such as coal, which were used for domestic and industrial purposes. However, today the major threat to clean air is posed by traffic emissions. All transport modes consume energy, and the most common source of energy is fossil fuels like petrol and diesel. The complex combination of chemicals discharged by exhaust fumes is created by the incomplete combustion of fuel within a vehicle's engine (Myung and Park, 2011). Petrol and diesel vehicles emit a wide variety of pollutants, principally carbon monoxide (CO), nitrogen oxides (NOx), volatile organic compounds (VOCs) (including chemicals such as methane, benzene, 1,3-butadiene, formaldehyde and polycyclic aromatic hydrocarbons). Additionally, particulate matter (PM) can comprise an array of chemicals including sodium chloride, black carbon, mineral dust and trace metals, VOCs and secondary particles (Hueglin et al., 2005; Vallero, 2008).

Most vehicular emissions are released by exhaust pipes; however, other pollutants (for example, PM) are emitted by braking systems, worn out and wearing tyres, abrasive road surfaces, leaks in tubing and engine casings and due to the process of evaporation (for example, VOCs). Moreover, indirect emissions released into the atmosphere by vehicles go through chemical transformations and become particles or gases that have a detrimental effect on the environment. Furthermore, the combination of exhaust gases is reliant on the type of engine (for example, four-stroke spark ignition, diesel) and vehicle operating conditions.

Transport is a significant and growing contributor to overall air pollution exposure. It is estimated to be responsible for up to 30% of PM emissions in European cities and up to 50% of PM emissions in OECD countries – mostly due to diesel traffic. However, the total contribution of transport to PM pollution can vary widely, from 12%–70% of the total pollution mix. Low- and middle-income countries suffer disproportionately from transport-generated pollution, particularly in Asia, Africa and the Middle East. In part, this is due to the use of old and inefficient diesel vehicles and a lack of public and active transport networks (WHO, 2018a). Emissions from vehicles also lead to high concentrations of air pollutants above EU standards in many of Europe's cities (European Environment Agency, 2017).

There is widespread acceptance that integrating decisions across land use planning, transport and environment policy sectors is crucial for sustainable development (Geerlings and Steed, 2003). However, relatively little work has been carried out on the issue of policy integration, particularly in relation to transport, land use planning and environment policies. Sustainability requires that policy making for urban travel be viewed in a holistic sense: that planning for transport, land use and the environment no longer be undertaken in isolation one from the other. Without adequate policy coordination, the effectiveness of the whole package of measures and their objectives is compromised.

Transport related air pollution and health

Clean air is essential for life, health, the environment and the economy. However, due to man-made and natural sources, clean air is contaminated with pollutants which have harmful effects on human health and ecosystems. A World Health Organization (WHO) study in 2005 critically assessed the health effects of transport related air pollution and determined that pollution from transport, specifically particulate matter in urban environments, was the cause of many health concerns (Krzyzanowski et al., 2005).

Exposure to air pollution is one the world's biggest public health challenges, shortening lifespans and damaging quality of life for many people. Air pollution is a major public health risk ranking alongside cancer, heart disease and obesity, accounting for over 16% of all deaths worldwide (Landrigan et al., 2018, p. 470). The detrimental health effects that air pollution has on the human body are well known globally. Numerous studies have shown that different forms of pollution are the direct and indirect causes of death and diseases on a global scale. The Lancet Commission on Pollution and Health (Landrigan et al., 2018, p. 471) stated that 'pollution is the largest environmental cause of disease and premature death in the world today'. A review by WHO concluded that long-term exposure to air pollution reduces life expectancy (WHO, 2018b). The negative health effects of exposure to air pollution build up over time and can include asthma, chronic obstructive pulmonary disease, cancer, dementia, stroke and heart disease, obesity and diabetes.

Several epidemiological studies have confirmed strong correlations between daily levels of exposure to poor air quality and deaths, particularly due to cardiovascular and respiratory diseases. Air quality has a considerable impact on peoples' health; approximately 6.5 to 9 million people die worldwide per annum due to the inferior quality of outdoor and indoor air (WHO, 2018b).

In the United Kingdom, due to air pollution, more than 50,000 people die prematurely with a reduction in life expectancy of on average 7 to 8 months, whilst exposure to $PM_{2.5}$ caused 29,000 deaths in 2008 (COMEAP, 2010). The Royal College of Physicians (2016) recently suggested that combined exposure to both $PM_{2.5}$ and NO_2 causes approximately 40,000 deaths per annum in the United Kingdom. Although complex to quantify, it is estimated that air pollution costs society nearly £20 billion annually in the United Kingdom due to the health-related consequences of people suffering diseases and early deaths.

The adverse effects of air pollution are also uneven. Those with pre-existing respiratory and heart conditions and children are most vulnerable to its effects. Furthermore, people from socially and economically disadvantaged backgrounds are more likely to be exposed to higher levels of air pollution and are more at risk of negative health effects. Poor air quality can affect health at all stages of life. Those most affected are the young and old. In the womb, maternal exposure to air pollution can result in low birth weight, premature birth, stillbirth or organ damage. In children, there is evidence of reduced lung capacity, while impacts in adulthood can include diabetes, heart disease and stroke. In old age, a lifetime of exposure to air pollution can result in reduced life expectancy and reduced wellbeing at end of life. There is also emerging evidence for a link between air pollution and an acceleration of the decline in cognitive function (Peters et al., 2019).

Air pollution is a global problem

With numerous more events occurring since the London Smog of 1952, it has become common for air pollution events to occur globally. While pollution levels have fallen in most developed nations since 1990 levels, as a whole they are still steadily increasing globally. While developed nations on a national scale are meeting the PM levels that the WHO has declared safe, these levels are often exceeded in urban and traffic dense environments. The WHO has predicted that by 2030, 60% of the global population will be living in urban environments, the areas most associated with heavy particulate matter and pollution from vehicles (WHO, 2019). The RAC Foundation published findings in 2014 that even the United Kingdom was failing to meet air quality levels for nitrogen oxide and particulate matter (Hitchcock et al., 2014). The EU has set out its own air quality limit values, which are much higher in comparison to the WHO's, which focus more on stricter PM levels as these pose a greater threat to human health. The report found that 91% of the EU's population are exposed to levels of $PM_{2.5}$ greater than what the WHO has deemed safe for human health. The report 'State of global air 2017' by the Health Effects Institution (2017) compared WHO targets with the global annual $PM_{2.5}$ concentrations. Very few nations met the WHO target, with 92% of the world's population living in areas exceeding 10 $\mu g/m^3$ for $PM_{2.5,}$ 50% of the global population resided in areas above the WHO

Interim Target 1 (of 35 μg/m^3); 64% lived in areas exceeding IT-2 (25 μg/m^3); 81% lived in areas exceeding IT-3 (15 μg/m^3). Most of the polluted countries in the world lie in Southeast Asia. According to the WHO database just 4 out of 632 locations in Asia were found to have air quality levels below the WHO standards for $PM_{2.5}$ (WHO, 2019).

An independent committee commissioned by the UK Department for Transport has shown that on-road emission levels of modern diesel cars are on average about seven times higher than the limit set by the Euro 6 emission standard (Department for Transport, 2016, pp. 22–3) and that although NOx emission limits for diesel cars in the EU were lowered by 85% between 2000 (Euro 3) and 2014 (Euro 6), over the same period on-road emission levels decreased only about 40% (ICEN, 2014).

Personal exposure to air pollution in various modes of transport

Exposure to air pollution refers to a particular time in a place where the pollution exists, which triggers interaction between air contaminants and the human body internally, through breathing and/or externally by means of irritation to the skin and eyes. Thus, exposure depends on other factors, such as physical activity that increases breathing rates and exposure duration and intensity, which has short and long-term consequences. Acute symptoms are attributed to short-term exposure to air pollution, whilst long-term exposure causes chronic symptoms.

Transport environments play a massive role in personal exposure. Due to the close vicinity to direct pollution from exhaust emissions, and particulate matter generated from braking and tyre wear, commuters and pedestrians on and near roads are exposed to much higher levels of pollution compared to someone in the countryside. There has been a great deal of research into exposure in transport microenvironments; however, most research has been conducted using data from monitoring stations at fixed points, such as Atkinson's et al. (2014) study in London or Shukla's (2015) study in Delhi. Yet despite this, all studies have shown that personal exposure to pollution is consistently higher in transport environments compared to others.

Personal exposure varies on different modes of transport, however. Studies have been conducted to measure the differences in personal exposure to particulate matter on different modes on transport on the same routes. A study by Chaney et al. (2017) in Salt Lake City found that commuters who cycled or walked were exposed to the greatest concentrations of $PM_{2.5}$, as they had no physical protection like the public transport commuters or drivers who had their windows closed. Driving with windows open increased the exposure concentrations by a significant amount, and the conclusion was that a car with windows closed was the most protective against pollution.

Levels of exposure vary from city to city, and studies have been conducted globally to compare the differences in pollution exposure. A study by Namdeo et al. (2014) compared the differences in commuter exposure between Newcastle upon Tyne and Mumbai. The results were conclusive in that commuters in Mumbai were exposed to much greater levels of particulate matter compared to the same types of commuting in Newcastle. Mumbai commuters were exposed to concentrations of particulate matter nearly ten times the amount compared to Newcastle. The study noted, however, that open windows and doors on buses can cause large variations in personal exposure, similar to combustion cars having different levels of exposure with open or closed windows as seen in Salt Lake City. Namdeo et al. (2014) noted that cyclists were exposed to mean lower amounts of particulate matter compared to other forms of transport.

A similar study, conducted by Nazelle et al. (2012) in Barcelona, concluded that car journeys were exposed to the highest levels of PM_{10} on average compared to other forms of commuting.

However, the study did not focus on the differences that active commuting can make to personal exposure due to increased inhaled doses of pollution. The study estimated inhaled doses that walking and cycling commuters would have been exposed to and stated that cyclists and pedestrians were exposed to similar/marginally higher concentrations of particulate matter compared to car drivers.

Among transport commuting modes, bus commuters were discovered to be more exposed to $PM_{2.5}$ than those using other modes of transport, such as travelling by car, walking and cycling. This exposure occurs not only during waiting times at bus stops but while seated in the cabin, with approximately 30% of pollutants inside being emitted by the bus itself (Beatty and Shimshack, 2011).

Actions to reduce transport-related air pollution

Cleaner air leads to increased productivity through improvements in public health, leading to reduced workplace absence, and through the creation of an environment that is appealing to businesses and the public alike. Emissions can be reduced significantly with good planning, such as enabling technological changes, behavioural changes and modal shifts to less polluting public transportation (DEFRA, 2018, p. 35; WHO, 2006, p. 89).

Reducing tailpipe emissions

Tailpipe emissions can be reduced by new abatement technologies, that could be promoted in regulation and legislation. For example, a Three-Way Catalyst (TWC) can achieve a reduction of approximately 95% of CO, NO_x, and hydrocarbon (HC) emissions. TWC uses the chemical processes of oxidation and reduction to turn destructive pollutants into undisruptive by-products. Similarly, Diesel Oxidation Catalysts (DOC) has the ability to promote oxidation of exhaust gas components by oxygen, present in ample quantities in diesel exhausts, thus reducing NOx emissions. However, DOC also oxidises some of the NO to NO_2, thus producing a higher percentage of primary NO_2. Another abatement technology, Exhaust Gas Recirculation (EGR), is deployed by diesel passenger cars Euro 5 and some later models to reduce NO_x formation. EGR lowers the temperature of combustion in order to reduce NO_x formation, which is done by taking a ratio of the exhaust gas and returning the same into the combustion chamber. Selective Catalytic Reduction (SCR) is another technology which has been used in recent years in diesel passenger cars to reduce NOx emissions. The primary reaction in the SCR process is the formation of ammonia (NH_3) and CO_2 when AdBlue, the most common diesel exhaust fluid (DEF), is mixed with exhaust gases and is rapidly hydrolysed. Since 2009, all diesel vehicles Euro 5 and after have been fitted with diesel particulate filters (DPF) to lower PM emissions. The primary concern relative to DPFs is the formation of high levels of ultrafine particulate emissions, which usually happens during regeneration.

Reducing emissions through vehicle technology

In contrast to petrol, diesel fuel possesses more energy and lasts longer in conventional vehicle (CV) engines; consequently, most European vehicle fleets have become more dependent on diesel vehicles. As a result of the lower carbon content of diesel fuel, diesel CVs have lower CO_2 emissions in comparison to petrol CVs. Nevertheless, diesel CVs are related to greater f-NO_2, NO_x, and PM_{10} emissions compared to petrol CVs. There is the potential to lower the emissions of various greenhouse gases (GHGs) and air pollution in the transport sector via the increasing adaptation to electric vehicles (EVs), which promises an improvement in air quality. Currently, several fuels are considered as alternatives to petrol and diesel. Fuels that can be employed in CVs with very few adjustments, including liquid petroleum gas (LPG), hydrogen and biofuels, are able to directly compete with petrol and diesel.

LPG vehicles have been extensively used in commercial vehicles. The higher butane content of LPG in relation to diesel fuels produces less NO_x emissions, likewise LPG comprises less carbon molecules than petrol or diesel, which results in a decrease in CO_2 emissions. Nevertheless, the efficiency of LPG engines is lower than diesel engines in term of producing energy which might influence the total GHG released by LPG engines fuel.

Biofuel vehicles comprise four principal types: biodiesel, which is prepared from the esterification of vegetable oils and methanol; alcohols (for example, methanol, ethanol), which are created from fermented sugar crops (grain crops); natural gas (methane), which is manufactured from the digestion of energy crops; and second-generation biofuels. Even though CO_2 emissions from vehicles which utilise alcohols or biodiesel as fuels are usually lower than those of standard vehicles, these liquid biofuels include less energy than petrol or diesel.

Electric vehicles (EVs): EVs are frequently mentioned as both a growth area and a simple means to reduce pollution. The reality is a little more nuanced, however. According to their technology, EVs can be categorised into four classes based on their technology, which are:

- The battery electric vehicle (BEV) relies purely on electricity for propulsion that is stored in the battery pack. The battery pack with an electric motor replaces the internal combustion engine (ICE) and fuel tank;
- The hybrid electric vehicle (HEV) uses both ICE and an electric powertrain, including a battery pack that can be recharged by converting kinetic energy (for example braking) to electricity;
- The plug-in-hybrid electric vehicle (PHEV) is driven by battery power until reaching depletion mode then ICE takes its place in propelling the vehicle. In addition, the battery pack can be recharged externally by connecting it to any outlet charging point whether at home, work or in public. Internal charging is possible through regenerative braking and decelerating;
- The extended-range electric vehicle (E-REV) is similar to the PHEV, but when the battery is depleted, an ICE run by fossil fuel is used to feed the electric motor with electricity and charges the battery pack simultaneously.

Overall, electric vehicles can play a significant role in reducing air pollutants and consequently improving air quality. In Taiwan, Li et al. (2016) pointed out that electrifying motorcycles and cars will influence air quality remarkably. They developed a scenario where the electricity is generated from thermal plants (mainly coal fired). It is estimated that emissions attributed will decrease. A dramatic decrease occurs for CO of 85%, VOCs of 79%, NO_x of 27% and $PM_{2.5}$ of 27%. It is worth mentioning that emissions of NO_x attributed to road transport is estimated to decrease by 33.9 Gg/year, although at power plants it will increase by 20.3 Gg/year. Similarly, $PM_{2.5}$ emissions from transport are estimated to decrease by 7.2 Gg/year, nevertheless, at the plant site they will increase by 0.8 Gg/year. In addition, the study estimates reductions in harmful pollutants in urban and rural areas. In urban areas, NO_x, VOC, CO, $PM_{2.5}$ and ozone, and SO_2 will be reduced by 18%, 21%, 65%, 6%, and 3%, respectively. Moreover, according to DeLuchi et al. (1989), the 100% replacement of conventional cars by EVs would result in a reduction in road transport emissions, the virtual elimination of CO and HC emissions and considerable NO_x reductions. Conversely, it should be noted that SO_2 emissions may escalate if the energy generation is from coal fired power stations which are the principal source of electricity in countries such as the United Kingdom.

Buekers et al. (2014), in their study conducted across 27 countries of the European Union, indicated that in 2010 an annual benefit of €30.3 million was expected in the United Kindgom alone from avoiding the external costs to tackle consequences in relation to health and climate impacts. This is for only a 5% EV market penetration rate, making an annual travelling distance of 10,000 km. The benefit was predicted to escalate to €46.6 million by 2030. Another study indicated that replacing 40% of ICE vehicles with EVs in Madrid, Spain will mitigate NO_x and CO emission concentrations by 17% and 22%, respectively (Soret et al., 2014).

Electrification of buses: the electrification of buses is likely to reduce passenger exposure to pollution. It has been demonstrated that doses of inhaled polluted air for passengers using electric buses are the lowest compared to passengers on diesel buses, petrol cars and diesel cars, who had also been exposed to pollutants (Zuurbier et al., 2010). An evaluation of the relationship between reducing bus emissions and morbidity was conducted in New York City by merging bus $PM_{2.5}$ and NO_x spatial concentrations at residential centroids with hospitalisation data for those residents who live near those centroids. It was found that stricter transport bus emissions standards are associated with reduced emergency department visits for respiratory illnesses, such as asthma and bronchitis (Ngo, 2015).

Reducing non-exhaust emissions

Particulate emissions from non-exhaust sources are a result of the friction required for braking and maintaining traction on the road, which are essential for road safety. However, these particles are also harmful to human health and the environment – and a source of microplastics in our oceans. There is a need to develop more evidence on non-exhaust emissions and to develop new international regulations for particulate emissions from tyres and brakes.

Reducing emissions by modal shift

Emissions savings and improved air quality will be associated with an increased modal share of sustainable transport options (walking, cycling, bus and metro usage). Modal shift to lower emission modes of travel plays a central role in reducing transport emissions. Within environmental planning, there is a core need to encourage more sustainable modes of transport like cycling, walking and public transport, and shifting freight from road to rail, even to inland waterways and sea.

Active travel: the 'Clean air strategy' developed by the UK government, for example, encouraging an increase in cycling and walking for short journeys delivers a reduction in traffic congestion and emissions from road transport, as well as health benefits from more active lifestyles (DEFRA, 2018). The 'Cycling and walking investment strategy' of DEFRA announced an investment of £1.2 billion for cycling and walking from 2016–21 to double the level of cycling by 2025 and to reverse the decline in walking (DEFRA, 2018). This included investing to improve and expand cycle routes between the city centres, local communities and key employment and retail sites in eight cities to get more people cycling. This fund also had provision to support local projects including training and resources to make cycling and walking safer and more convenient (DEFRA, 2018).

Public transport: congestion in major cities can be tackled through increased public transport, which will have an impact on exhaust and non-exhaust emissions. Modal shift to rail, particularly on electrified lines, can help to reduce road traffic congestion and emissions.

Planning and policy to reduce transport emissions

With the global shift towards a low carbon, circular economy already underway, there is a greater need to integrate land use planning, transport and environment policy to meet the increasing mobility needs of people and goods. An integrated framework of these policies will benefit citizens and consumers by delivering improvements in air quality, reductions in noise levels, lower congestion levels and improved safety. The public will benefit from less energy consuming cars, from better infrastructure for alternative fuels, better links between modes of transport and better safety and fewer delays due to the roll-out of digital technologies.

City planning policy could be developed such that cities generate a fewer number of trips, avoid shorter trips by cars and encourage the use of active modes of travel and a shift to public transport. An example is provided in the recommendations set out in the UK 'National planning policy framework', which states that:

> For larger scale residential developments in particular, planning policies should promote a mix of uses in order to provide opportunities to undertake day-to-day activities including work on site. Where practical, particularly within large-scale developments, key facilities such as primary schools and local shops should be located within walking distance of most properties.
>
> *(Department for Communities and Local Government, 2012, para 38)*

The UK's Committee on Climate Change have also recommended that 'One approach the government must take is to reduce the demand for travel by at least 5% below the baseline trajectory on a local and national scale' (Committee on Climate Change, 2017, p. 13). Additionally, the Committee has identified that local governments should implement policy to encourage 'better land-use planning to provide easy access to public transport infrastructure' (Committee on Climate Change, 2017, p. 128).

On an international scale, the EU Strategy for low-emission mobility highlights that cities and local authorities will play a crucial role in delivering the vision of low carbon clean air mobility. Local authorities in EU are already implementing incentives for low emission alternative energies and vehicles, encouraging active travel (cycling and walking), public transport and bicycle and car-sharing or pooling schemes to reduce congestion and pollution. While the UK Health Alliance on Climate Change recommend actions to improve air quality in general, with a particular focus on road transport given its position as a major contributor to air pollution. The Alliance recognised that action on road transport can unlock large health co-benefits through increases in cycling, walking and other active transport.

More broadly, Europe's answer to the emission reduction challenge in the transport sector is an irreversible shift to low emission mobility. By mid-century, greenhouse gas emissions from transport will need to be at least 60% lower than in 1990 and be firmly on the path towards zero. EU strategy states that emissions of air pollutants from transport that harm our health need to be drastically reduced without delay. The EU strategy integrates a broader set of measures to support Europe's transition to a low carbon economy and supports jobs, growth, investment and innovation. The strategy will benefit European citizens and consumers by delivering improvements in air quality.

References

ACEA (2017). *ACEA report, vehicles in use, Europe 2017*. Brussels: European Automobile Manufacturers' Association.

Atkinson, R. W., Kang, S., Anderson, H. R., Mills, I. C. and Walton, H. A. (2014). 'Epidemiological time series studies of $PM_{2.5}$ and daily mortality and hospital admissions: a systematic review and meta-analysis'. *Thorax*, 69(7): 660–5.

Beatty, T. K. M. and Shimshack, J. P. (2011). 'School buses, diesel emissions, and respiratory health'. *Journal of Health Economics*, 30(5): 987–99.

Buekers, J., Van Holderbeke, M., Bierkens, J. and Int Panis, L. (2014). 'Health and environmental benefits related to electric vehicle introduction in EU countries'. *Transportation Research Part D: Transport and Environment*, 33: 26–38.

Chaney, R. A., Sloan, C. D., Cooper, V. C., Robinson, D. R., Hendrickson, N. R., McCord, T. A. and Johnston, J. D. (2017). 'Personal exposure to fine particulate air pollution while commuting: an examination of six transport modes on an urban arterial roadway'. *PLoS ONE*, 12(11): e0188053.

COMEAP (2010). *The mortality effects of long-term exposure to particulate air pollution in the UK*. Chilton: Committee on the Medical Effects of Air Pollutants.

Committee on Climate Change (2017). *Meeting carbon budgets: closing the policy gap: 2017 report to the parliament*. Committee on Climate Change. Available at: www.theccc.org.uk/wp-content/uploads/2017/06/2017-Report-to-Parliament-Meeting-Carbon-Budgets-Closing-the-policy-gap.pdf

DEFRA (2018). *Clean air strategy 2018* (Draft: Citizen Space). Department for Environment, Food and Rural Affairs [Online]. Available at: https://consult.defra.gov.uk/environmental-quality/clean-air-strategy-consultation/user_uploads/clean-air-strategy-2018-consultation.pdf

DeLuchi, M., Wang, Q. and Sperling, D. (1989). 'Electric vehicles: performance, life-cycle costs, emissions, and recharging requirements'. *Transportation Research Part A: General*, 23(3): 255–78.

Department for Communities and Local Government (2012). *National planning policy framework*. London: Department for Communities and Local Government.

Department for Transport (2016). *Vehicle emissions testing programme, moving Britain ahead*. Department for Transport [Online]. Available at: https://assets.publishing.service.gov.uk/government/uploads/system/uploads/attachment_data/file/548148/vehicle-emissions-testing-programme-web.pdf

Department for Transport (2018). *Table VEH0101: Licensed vehicles at the end of the quarter by body type*, Great Britain from 1994 Q1; also United Kingdom from 2014 Q3. Vehicle Licensing Statistics, Department for Transport [Online]. Available at: www.gov.uk/government/collections/vehicles-statistics#published-in-2018

European Environment Agency (2017). *Air quality in Europe – 2017 report*. Copenhagen: European Environment Agency.

Geerlings, H. and Steed, D. (2003). 'The integration of land use planning, transport and environment in European policy and research'. *Transport Policy*, 10(3): 187–96.

Health Effects Institution (2017). *State of global air 2017*. Boston: Health Effects Institution.

Hitchcock, G., Conlan, B., Kay, D., Brannigan, C. and Newman, D. (2014). *Air quality and road transport*. London: RAC Foundation.

Hueglin, C., Gehrig, R., Baltensperger, U., Gysel, M., Monn, C. and Vonmont, H. (2005). 'Chemical characterisation of PM2.5, PM10 and coarse particles at urban, near-city and rural sites in Switzerland'. *Atmospheric Environment*, 39(4): 637–51.

ICEN (2014). *New ICCT study shows real-world exhaust emissions from modern diesel cars seven times higher than EU, US regulatory limits* [Press release]. The International Council on Clean Transportation. Available at: www.theicct.org/news/press-release-new-icct-study-shows-real-world-exhaust-emissions-modern-diesel-cars-seven-times

Krzyzanowski, M., Kuna-Dibbert, B. and Schneider, J. (eds.) (2005). *Health effects of transport-related air pollution*. Copenhagen: World Health Organization.

Landrigan, P. J., Fuller, R., Acosta, N. J., Adeyi, O., Arnold, R., Baldé, A. B., Bertollini, R., Bose-O'Reilly, S., Boufford, J. I., Breysse, P. N. and Chiles, T. (2018). 'The Lancet Commission on pollution and health'. *The Lancet*, 391(10119): 462–512.

Li, N., Chen, J.-P., Tsai, I. C., He, Q., Chi, S.-Y., Lin, Y.-C. and Fu, T.-M. (2016). 'Potential impacts of electric vehicles on air quality in Taiwan'. *Science of the Total Environment*, 566(Supplement C): 919–28.

Myung, C. L. and Park, S. (2011). 'Exhaust nanoparticle emissions from internal combustion engines: a review'. *International Journal of Automotive Technology*, 13(1): 9.

Namdeo, A., Ballare, S., Job, H. and Namdeo, D. (2014). 'Commuter exposure to air pollution in Newcastle, UK, and Mumbai, India'. *Journal of Hazardous, Toxic, and Radioactive Waste*, 20(4).

Nazelle, A. de, Fruin, S., Westerdahl, D., Martinez, D., Ripoll, A., Kubesch, N. and Nieuwenhuijsen, M. (2012). 'A travel mode comparison of commuters' exposures to air pollutants in Barcelona'. *Atmospheric Environment*, 59(1): 151–9.

Ngo, N. S. (2015). 'Analyzing the relationship between bus pollution policies and morbidity using a quasi-experiment'. *Medicine*, 94(37): e1499.

OCIA (2018). *Production statistics*. Paris: International Organization of Motor Vehicle Manufacturers.

Peters, R., Ee, N., Peters, J., Booth, A., Mudway, I. and Anstey, K. J. (2019). 'Air pollution and dementia: a systematic review'. *Journal of Alzheimer's Disease* (Preprint): 1–19.

Rodrigue, J.-P. (2017). *The geography of transport systems*. New York: Routledge.

Royal College of Physicians (2016). *Every breath we take: the lifelong impact of air pollution*. Report of a Working Party. London: Royal College of Physicians.

Shukla, A. (2015). 'Characterization of ambient PM2.5 at a pollution hotspot in New Delhi, India and inference of sources'. *Atmospheric Environment*, 109(1): 178–89.

Smith, M. N. (2016). 'The number of cars will double worldwide by 2040'. *Business Insider*, 20 April [Online]. Available at: http://uk.businessinsider.com/global-transport-use-will-double-by-2040-as-china-and-india-gdp-balloon

Soret, A., Guevara, M. and Baldasano, J. M. (2014). 'The potential impacts of electric vehicles on air quality in the urban areas of Barcelona and Madrid (Spain)'. *Atmospheric Environment*, 99(0): 51–63.

Vallero, D. A. (2008). 'Transport and dispersion of air pollutants', in D. A. Vallero (ed.) *Fundamentals of air pollution*. London: Elsevier, p 213.

WHO (2006). *Air quality guidelines. Global update 2005. Particulate matter, ozone, nitrogen dioxide and sulfur dioxide*. Germany: World Health Organization [Online]. Available at: www.euro.who.int/__data/assets/pdf_file/0005/78638/E90038.pdf?ua=1

WHO (2018a). *Health and sustainable development: air pollution*. Available at: www.who.int/sustainable-development/transport/health-risks/air-pollution/en/

WHO (2018b). *WHO news release: 9 out of 10 people worldwide breathe polluted air, but more countries are taking action*. Geneva: World Health Organization.

WHO (2019). *Global health observatory (GHO) data*. World Health Organization [Online]. Available at: www.who.int/gho/urban_health/en/

Zheng, S. (2017). 'China now has over 300 million vehicles . . . that's almost America's total population'. *South China Morning Post*, 19 April [Online]. Available at: www.scmp.com/news/china/economy/article/2088876/chinas-more-300-million-vehicles-drive-pollution-congestion

Zuurbier, M., Hoek, G., Oldenwening, M., Lenters, V., Meliefste, K., van den Hazel, P. and Brunekreef, B. (2010). 'Commuters' exposure to particulate matter air pollution is affected by mode of transport, fuel type, and route'. *Environmental Health Perspectives*, 118(6): 783–9.

[illegible]

Part 4

Methods and approaches to environmental planning

Hilda Blanco, Iain White, Richard Cowell and Simin Davoudi

Environmental planning is known as much for its decision-making methods and approaches as for its underlying philosophy and values. After the US passage of its National Environmental Protection Act in 1969 which established the practice of conducting environmental impact reviews/statements (EIS) of major developments funded by the federal government, many states in the United States adopted similar review processes for major developments. This was followed soon after by the adoption of similar practices around the world; early adopters included countries such as Australia, Canada and New Zealand (Morgan, 2012). It is important to note that EIS is a public process aimed at identifying environmental impacts and at reducing or mitigating such impacts. Environmental impact reviews have become incorporated into urban/spatial planning practice since major developments often require local permits for new construction. Historically, spatial planning itself has had a fundamentally environmental planning motivation and approach. Zoning, a major instrument of spatial planning, was aimed, although its history is intertwined in the United States with environmental injustices, at separating uses to reduce the pollution and nuisances of industrial cities at the turn of the 20th century.

Nicole Gurran, in **Chapter 28**, reviews current urban planning practice and its relation to environmental planning and, in addition to the continuing practice of environmental reviews of major new developments, notes urban planning's current emphasis on sustainable urban form, with its focus on regional and neighbourhood scale factors that emphasise place-making, including compact development and regulation of urban growth; on the protection of environmental sensitive areas and biodiversity conservation; and on sustainable design, for example, building orientation to maximise solar access, as well as the profession's recent explicit focus on climate change mitigation and adaptation. The chapter also discusses the increasing emphasis in urban planning on environmental justice and participation in the planning process.

While EISs focus on reviews of expert studies on the potential negative impacts of proposed development projects, **Cecilia Wong and Wei Zheng** in **Chapter 29** review how sustainable indicator systems operationalise policy problems to track progress on policy goals. In their chapter, the authors highlight the role of indicators in UN plans and reports, especially, indicators connected to sustainability, which were first tracked at the national level but are increasingly used at the urban and neighbourhood scale. Difficulties with interpretation and potential manipulation of indicators began to be addressed at the turn of this century through the development of the European Environment Agency (EEA) framework which provides causal links between driving forces, through pressures, states, and impacts, linking impacts to policy responses, in effect, providing the rationale for indicators. Wong and Zheng discuss the growing use of indicators by cities and major third sector national and international agencies, such as LEED and BREEAM and how their initial focus on the building scale has broadened to include neighbourhood scales. Indicator systems are easily grasped, which accounts for their popularity, and although they face several major methodological issues, for example, the weighting of factors, they will continue to be useful in monitoring environmental policy and planning.

Three chapters in this section focus on different ways to assess natural resource use through various types of environmental input-output analysis, i.e. different types and scales of materials flow and metabolic impact analyses. These types of analyses provide a basis for improving environmental performance at various scales, from national, to urban, to corporate, and even individual, by highlighting the environmental inputs on which such units depend upon, such as the mix of energy and water sources, and the environmental outputs, such as products, and services, as well as solid waste, air pollution and greenhouse gas (GHG) emissions. The emphasis in this type of analysis is the flow of resources, in terms of environmental inputs and outputs. **Robin Curry** and **Geraint Ellis** in **Chapter 30** review a range of such approaches at the national scale, for example, developed by the UN and Eurostat, and provide examples at different scales. As Curry and Ellis point out, materials flow analysis, applied at the national scale, measures materials and product inputs, production and consumption, exports, solid and liquid waste, as well as air emissions but does not include the environmental impacts of resource use, and is not necessarily aligned with policy areas or well understood by policy makers. The chapter also discusses several other metabolic types of analyses, such as the ecological footprint, which links resource use to total area of productive land and water (Global Footprint Network, 2019), and the carbon footprint, which limits the analysis to emissions with an impact on climate change. Both the analyses of ecological footprints and of carbon footprints have become popular, since they provide vivid ways of illustrating and comparing resource use or the emissions of individuals, organisations, cities, and nations. The chapter also briefly reviews life cycle analysis, a type of metabolic analysis focused on products that traces the environmental impacts of a product from extraction through use and disposal or recycling.

While Curry and Ellis discuss a broad range of material flow analyses, **Paulo Pinho** and **Ruben Fernandes**, in **Chapter 31**, focus on urban metabolism analysis. In their chapter, they discuss how the concept of urban metabolism has broadened beyond environmental inputs and outputs to include social, economic, and institutional aspects, as well as the role of power relations. Pinho and Fernandes further describe their concept of Metabolic Impact Analysis (MIA), aimed at enriching the EIS process and focused at the scale of urban development plans and large urban projects and infrastructures. The authors argue that, for plans and major projects, it is not sufficient to conduct typical EIS studies but that impact studies need to rely on a city-wide analysis of a city's urban metabolism and, based on such an analysis, the comparison of the urban metabolism inputs and outputs of alternative projects can be more holistically assessed. Such an approach would ensure that the overall impacts of a new project will improve a city's sustainability. This approach illustrates how metabolic analysis, which has remained, to a large extent, an expert type of analysis, can be incorporated into the standard practice of environmental impact review in spatial planning.

In **Chapter 32**, **Carlo Ingrao and associates** focus on life cycle assessment (LCA), another metabolic approach that seeks to encompass the entire process of production/consumption, including the extraction of primary materials, production, distribution, use, and disposal of all types of products, and elements in the human and urban environment including buildings and infrastructures. Chapter 32 applies this approach to buildings, which, as the authors point out, consume 40% of global energy demand. LCA methodology for buildings has been increasingly institutionalised, including the establishment of several international standards for conducting such analyses. Ingrao and associates discuss the comprehensive methodology that analyses each stage in the process, i.e. the production of construction materials, construction processes, building maintenance and operations, as well as the dismantling of the building at the end of its life to achieve greater efficiencies in the use of materials and energy. This methodology could provide the basis for green certification programmes and national and local buildings standards. It is applied to a sector essential to achieve significant reductions in GHG emissions.

Ken Willis, in **Chapter 33**, reviews how cost-benefit analysis, the major economic instrument for evaluating public projects, has developed over the past few decades to increasingly incorporate the more recent developments in environmental economics. Environmental economics has enriched public finance theory and practice through several methods of accounting for/quantifying environmental goods and bads, such as methods to estimate replacement costs, mitigation expenditures, revealed and stated preference methods, contingent valuation and choice experiments. In addition, the chapter discusses how climate change poses special challenges to environmental economics and cost-benefit analysis, including how to determine an appropriate discount rate for such studies, and whether to use a partial or general equilibrium approach.

In **Chapter 34**, **Tomas Badura**, **Kerry Turner** and **Silvia Ferrini** focus on natural capital and ecosystem valuation and their emerging role in policy making. The chapter provides an up-to-date, critical review of different environmental economics methods, monetary and non-monetary to value ecosystem services and biodiversity. It discusses how environmental accounting can be used to incorporate natural assets and emphasises the role that quantitative and qualitative non-monetary methods can play in the system of national accounts (SNA). This latter system is an exciting development since using such methodologies the World Bank has begun to include the value of natural capital in their accounting of the wealth of nations (Lange et al., 2018). The authors also briefly present their Balance Sheet Approach, a decision-support system, incorporating a combination of methods to aid environmental decision making.

Robert Paterson and **Frederick Steiner**, in Chapter 35, focus on suitability analysis, a methodology with historical origins in urban planning and landscape architecture, which has been enhanced by the use of geographic information systems (GIS). The methodology is increasingly used around the world to identify and compare sites for urban expansion, environmental protection, and for a wide range of facilities. Suitability analysis links places, through maps, to their environmental characteristics, such as type of soil, elevation, vegetation, native species, as well as other characteristics, including social and economic. It relies on a multi-criteria analysis methodology to identify areas suitable for proposed uses. While many of the environmental methods and approaches presented in this section are technical, based on expert methodologies, suitability analysis enables planners, decision makers and the public to take the materiality and location of the environment into account when planning new development, urban facilities or natural resource or preservation areas. The authors illustrate how this method was used in the development of the Travis County Greenprint Plan and the City of Austin's Comprehensive Plan in Texas to include a wide range of stakeholders in the planning process. The method provides a way to incorporate multiple dimensions of the environment, both resources and constraints, in a decision-making approach that makes clear the weights assigned to different factors and provides the visual display of spatial analysis layers.

In **Chapter 36**, **Alister Scott** begins with twin premises: (i) that the ecosystem services approach has become the major framework for incorporating the natural environment into the policy framework, for example, the Millennium Ecosystem Assessment, and (ii) that the environment is widely perceived as an obstacle to economic growth. The chapter then provides a powerful analysis of how the ecosystem services approach remains a specialised environmental field and how it has failed to become incorporated into economic development or spatial planning. Moving beyond this analysis, Scott argues for the need to integrate or mainstream the ecosystem services approach into spatial planning and broader policy making. Using Rogers' diffusion of innovations theory (2003), Scott illustrates how the ecosystem services approach has not moved beyond the first stage of the generation of new knowledge or innovation to the second stage, that of persuasion. Note that the more advanced stages for the diffusion of an innovation, according to Rogers, are decision and implementation stages. Rogers' diffusion theory provides a useful lens for analysis and for moving forward, and Scott uses the theory to identify barriers to mainstreaming, for example, 'complex and widespread use/misuse of ecosystem services', lack of exemplars, and illustrates how such mainstreaming can be achieved, for example, reconnection of paradigms, improved dialogues among different agencies. The chapter illustrates both the barriers and ways to overcome them with two recent planning processes, the Birmingham Green Living Spaces Plan and the South Downs National Park Local Plan.

This section on methods and approaches begins with a focus on spatial planning, its current interests and practices, and its continuing reliance on EIS processes for addressing environmental issues related to development. The focus on spatial planning returns in the last two chapters, with Paterson and Steiner's chapter on suitability analysis and Scott's chapter on the need for mainstreaming the ecosystem services approach into spatial planning and broader policy making. Suitability analysis combines a way of visualising different aspects of the natural environment or ecosystem services and the urban environment, as well as ways to value or weigh these different aspects. Its visual aspect and accessible multi-criteria analysis make it a useful method for engaging the public and decision makers in planning and policy. The mainstreaming approach that Scott argues for in Chapter 36 also emphasises the need for plans and policies that incorporate multiple objectives and interests and involve a wide range of stakeholders and decision makers. While the EIS approach and, more recently, Strategic Environmental Assessments (SEAs) of plans and programmes are focused on reducing the potential negative environmental impacts of projects, plans and programmes, the focus of both suitability analysis and efforts to mainstream ecosystem services into plans and programmes are proactive. They aim to more positively incorporate nature and its services into spatial planning and public policy. Pinho and Fernandes' Metabolic Impact Analysis (MIA) is also a novel, proactive way of using a city-wide urban metabolism analysis to frame individual environmental impact analyses, thus providing an urban-scale metabolism model against which individual projects or plans can be assessed. The popularity of environmental or sustainability indicator systems, as analysed by Wong and Zheng (Chapter 29) both for internal and external tracking of environmental characteristics/performance at different scales, from national and local government to the business firm, also demonstrates the interest on the part of policy makers and the public on these topics.

The methods and approaches profiled in this section document the advances in analytic methodologies, for example, materials flow analysis, life cycle analysis and ecosystem valuation. They also highlight the paucity of environmental methods that incorporate environmental systems or ecosystem services into spatial planning and policy making in proactive ways that facilitate mainstreaming. Moving forward Scott's mainstreaming agenda will also require moving beyond educational silos and incorporating knowledge and appreciation of nature's services in the education of planners, policy analysts, economic development experts, civil engineers and architects.

References

Global Footprint Network (2019). *Ecological footprint*. Available at: www.footprintnetwork.org/our-work/ecological-footprint/

Lange, G. M., Wodon, Q. and Carey, K. (eds.) (2018). *The changing wealth of nations 2018: building a sustainable future*. Washington, DC: World Bank.

Morgan, R. K. (2012). 'Environmental impact assessment: the state of the art'. *Impact Assessment and Project Appraisal*, 30(1): 5–14.

Rogers, E. M. (2003). *Diffusion of innovations* (5th edn). London: Simon and Schuster.

28

Addressing sustainability issues through land use regulation and zoning

Nicole Gurran

Introduction

Land use regulations and associated techniques such as zoning offer powerful methods for promoting environmental sustainability. Yet regulatory practices, which control development by separating land into separate 'zones' according to permitted activities, are not inherently sustainable. In fact, what we now know to be largely unsustainable forms of urban development – isolated and low-density suburban estates, sprawling business parks, car dependent 'big box' retail – reflect an unfortunate trend towards highly restrictive single use 'mono' zoning form of land use regulation which emerged in the post war era (Hirt, 2007). By the early 1990s, pressure to reform traditional modes of land use control in support of more compact growth and diverse housing emerged in nations such as the United States (US) and Australia, coinciding with a growing environmental agenda. Since that time, the diffusion of sustainability objectives, concepts, and models across urban policy and planning practice has transformed local land use regulation (Berke and Conroy, 2000; Jepson and Haines, 2014; Gurran et al., 2015). The range of land use regulations and tools now used by planners to implement sustainability goals is the focus of this chapter.

The chapter is structured as follows. First, the key environmental sustainability concepts are outlined and how they can be promoted through the institutionalised process of land use planning and development assessment. Next, specific techniques for incorporating sustainability within land use zoning and development control are identified. Third, the strengths and weaknesses of land use regulation and zoning are considered, including questions about the costs of regulatory planning and where these are incurred. Finally, the chapter canvases emerging information technologies and the potential for improved and more participatory planning and development assessment through e-planning initiatives.

Sustainability concepts and land use regulation

'Sustainability' has become an overarching normative goal for land use planning, now affirmed through most land use and environmental planning legislation and instruments throughout the world. For instance, the National Planning Framework (paragraph 6) for England states that: 'The purpose of the planning system is to contribute to the achievement of sustainable development' (Department for Communities and Local Government, 2012).

Originating with the Brundtland Commission's definition of 'sustainable development as development which meets the needs of the present without compromising the needs of future generations to meet their own needs' (WCED, 1987, p. 44), planning for sustainable development means protecting and enhancing environmental values, enabling and promoting economic objectives, while also delivering social benefits in an equitable way (Naess, 2001). In reality, achieving balance across the triangle of environmental, economic, and equity goals can be difficult to achieve (Campbell, 1996).

Subsequent international meetings of the United Nations Conference on Environment and Development (UNCED) confirmed and extended these core ideas, adding the 'precautionary principle' – that is, that a lack of scientific certainty should not be used to justify decisions with a risk of significant environmental harm; and emphasising transparency and participation in decision making (United Nations, 1997; Jordan, 2008).

These principles form a basis for sustainable community planning practice. This means recognising ecological limits in identifying development land, minimising the urban footprint and preserving farming and environmentally sensitive areas (Beatley, 1995; Beatley and Manning, 1997; Wheeler, 2013). Efficient transport networks promote walking and cycling; urban design nurtures a sense of place, while affordable housing policies support social inclusion (Wheeler, 2013). These principles are implemented through the institutionalised process of land use planning and development control.

Government and institutional context for land use regulation

Land use regulation is established by government within a wider policy and legislative system. These systems differ throughout the world but have common features, identified by the International Society of City and Regional Planners (ISOCARP). First, land use regulation depends on higher legislative power. This is typically an overarching environmental planning law or laws, which establish administrative responsibility and organisational arrangements for planning, including the basis for defining specific land use regulations. Depending on which level of government has responsibility for the environment and land use planning, this source of power will be articulated through national or state (and equivalent) bureaucracy and legislation.

Second, environmental planning systems involve strategic policies as well as controls/regulations. For instance, strategic plans or planning policies will typically contain objectives and indicative spatial arrangements for future development and change. In the United States, these strategic instruments are often comprehensive plans. More detailed land use controls operationalise these strategic policies, such as though land use zones or indicative site-based classifications.

Third, the planning system is triggered by the need to obtain permission to carry out development. Land use regulation is rarely proactive or retrospective – that is, regulatory planning is unable to compel actors to undertake a particular change in the use of land or buildings. Rather, land use regulation is activated when an individual or organisation wants to undertake a development and needs permission to do so. The process of considering the development application and issuing permission or refusal is the process of implementing the land use plan.

Fourth, planning systems enshrine legally enforceable rights for members of the public to be consulted about proposed new or changed land use regulation and certain types of development.

Typically, these consultation rights include the right to object to a plan or decision; and sometimes to appeal the outcome of this decision through the legal process.

Finally, land use planning systems include arrangements to finance things like public infrastructure on which private development depends (for example, local roads, footpaths, open space) and also to cover the costs of administering the planning process (Ryser and Franchini, 2015).

All of these elements of environmental and land use planning systems can operate in ways which further sustainability. For instance, the objectives of overarching environmental planning legislation set a framework for the ways in which the system will operate, including the ways in which land use plans are made and development applications determined.

These objectives should then carry through the strategic policies and land use planning controls defined to implement them. As will be discussed later, designing financial contribution requirements can directly support environmental objectives – such as requirements for developers to incorporate sustainability within local infrastructure (like green drainage systems), or by encouraging a compact urban form oriented around transit and existing urban services.

Objectives and risks being managed: towards sustainable land use zoning and development control

The techniques and methods of land use planning precede contemporary notions of environmental sustainability. Early 20th-century town and country planning efforts aimed to preserve rural landscapes, improve urban housing conditions and manage pollution from heavy industry (Hall, 1996). In practice, this came to mean siting activities to minimise risk to the environment; separating those activities which are incompatible with one another; and limiting sites for development to preserve existing environmental, cultural, economic and agricultural values. Thus, the primary way in which the land use planning system can promote environmental sustainability is by regulating the location and design of development.

Within this framework, land use zoning is a particular technique for allocating land in relation to forecasted need for housing, commercial, industrial and agricultural activities. A land use 'zone' is a spatial control which groups activities that may be carried out within a particular area, on the basis that these activities are likely to be compatible with one another and having regard to the environmental characteristics of the area. Importantly, our understanding of environmental conditions is continuing to change and in particular has become more sophisticated since the turn of the 21st century. This creates difficulties when new environmental knowledge suggests that a land use zone is no longer appropriate for the assigned use, as discussed later.

Land use zoning is not used in all jurisdictions. The practice was abandoned in England with the passage of the 'Town and Country Planning Act 1949', but local development plans continue to identify indicative sites for particular activities (Cullingworth and Nadin, 2014). In the United States, land use zoning is prevalent and has tended to be very rigid, often permitting a single land use only – such as detached residential houses (Hirt, 2012).

Once land has been assigned for particular types of activities or development, specific controls on building height, bulk and scale, site coverage, overshadowing, privacy, open space, landscaping and so on are typically defined within zoning ordinances or other regulatory schemes, codes, or guiding documents. Controls on the potential environmental impact of a development – for instance, operating hours for a business, traffic movements and parking, as well as the aesthetic appearance of a building, are often set.

Environmental impact assessment (EIA) is a particular methodology for regulating developments, which by their very nature present a significant potential environmental impact, wherever they are sited. These include forms of heavy industry, power plants or other types of infrastructure. Specific requirements for undertaking 'environmental impact assessment' are usually incorporated within planning and or environmental protection law. These provisions typically identify the types of development that will be subject to additional impact assessment requirements, the additional considerations and the process by which the assessment will be carried out (for example, see Thomas and Elliott, 2005). EIA has become an important mechanism for ensuring that large developments address environmental concerns, such as potential habitat loss, impacts on water catchments, or potential traffic congestion and resultant air pollution.

Environmental sustainability through land use planning and regulation

All of these land use planning and regulatory techniques – allocating land for preservation or development, defining development controls and establishing the parameters for assessing significant environmental impacts – can be used to promote, or undermine, sustainability outcomes. For instance, an over-abundance of land for housing development will usually result in piecemeal and sprawling patterns of growth, while highly restrictive land use zones which permit only a single use, are likely to increase distances and the need to travel between home, work, and shops. Development controls can leave decisions about design and material selection to 'the market' which usually results in the status quo being maintained or may even undermine sustainability objectives – for instance heavy parking requirements which increase the land area required for shops and services.

In the following sections, the approaches to sustainable land use regulation are divided into three dimensions: sustainable urban form (regional/neighbourhood scale), biodiversity conservation and enhancement (focusing on ecological systems and processes) and sustainable design (incorporating sustainability features within the design and construction of buildings).

Sustainable urban form

There are different ways of measuring and defining sustainable urban form, but general principles involve containing the spread of development through higher density and mixed-use configurations. Called 'urban containment' in the United Kingdom, 'urban consolidation' in Australia, and 'growth management' or 'smart growth' in the United States, the regulatory mechanisms for delivering more compact settlements include: designated 'green belt', or urban growth boundaries; 'upzoning' for higher density development within existing areas, particularly surrounding services and transport; and policies to enable former industrial sites to be rehabilitated and reused. Mixed-use zones, which promote a diversity of compatible uses – such as ground floor retail and upper storey commercial or residential development – are also techniques for achieving more compact growth.

The urban design movement known as 'new urbanism' incorporates many of these features – revolving around the notion of a traditional neighbourhood structure, interconnected street grids, diverse housing types, mixed uses, and active transport (Grant, 2009). To implement new urbanist forms of development it has often been necessary to adjust traditional land use regulations – for instance, to reduce minimum allotment sizes or to permit attached dwellings and mixed uses within residential areas (Knaap and Talen, 2005).

All of these strategies should reduce the need for car-based travel, reducing greenhouse gas emissions associated with car dependency. At the same time, containing the spread of urban development preserves rural land, wildlife habitat and biodiversity.

Biodiversity conservation and enhancement

Land use zoning is a primary method for protecting environmentally sensitive areas and promoting biodiversity conservation. Areas of high natural conservation area should be protected through restrictive zoning or equivalent control. Buffers for riparian areas and between core wildlife habitat and urban areas, can be designated through zones or 'overlay' controls (Jepson and Haines, 2014; Gurran et al., 2015). As well as restrictions on land use, specific criteria for considering the impacts of proposals, and for mitigating any negative impacts, can form part of the development control framework. It may be prudent to include requirements that development in certain areas be referred to an expert environmental agency.

Biodiversity values can be made explicit within subdivision design, for example, through requirements for housing to be clustered in less sensitive areas, or to maintain wildlife habitat (Jepson and Haines, 2014). Planning rules may permit more intense development of a site, in return for demonstrated improvements to overall conservation values – for example, by undertaking environmental rehabilitation work, or establishing a wildlife habitat corridor or refuge within the site.

Green 'offsets' are a controversial tool for managing the impacts of development on biodiversity values (Hamin and Gurran, 2009; McKenney and Kiesecker, 2010). The notion accepts that some environmental impacts are inevitable but seeks to establish an overall net environmental improvement by undertaking mitigation efforts on the site or elsewhere. Sophisticated markets for purchasing environmental offsets have emerged in jurisdictions where these are needed to support particular planning proposals because of requirements established in the local planning framework.

'Tradeable' or 'transferable' development rights operate on a similar basis. In an environmental context, transferable development right schemes seek to preserve sensitive lands by 'transferring' the hypothetical development 'right' from one site to another, less sensitive area, which may benefit from increased development potential (Pruetz and Pruetz, 2007; Linkous, 2016). Often the development rights are 'sold' as part of the transfer. The model compensates landholders who would otherwise suffer financial loss if not able to develop their land. Nevertheless, implementation can be complicated, and it is important to avoid any implication that landholders are owed compensation for longstanding environmental constraints affecting the development of their land.

Sustainable design controls

Development controls within local plans or ordinances can ensure that minimum levels of sustainability are incorporated within building design and construction. These could take the form of requirements for building orientation to maximise solar amenity and or ventilation and cooling through to requirements for energy efficient domestic appliances which reduce domestic greenhouse gas emissions (Hamin and Gurran, 2009). The practice of water sensitive urban design (WSUD) manages urban stormwater in ways that mimic the natural water cycle, limiting run off to waterways and reducing demand for new water supply and treatment facilities. Planning controls for new residential subdivisions or larger sites can include requirements for

WSUD such as rainwater tanks, rain gardens, sediment ponds and space for wetlands, the latter which are often able to be integrated within wider landscaping and open space.

Planning requirements can also consider requiring facilities such as bicycle parking and showers, which encourage active transportation, and provisions for waste sorting, recycling, and compost in residential and commercial buildings.

Climate change, land use zoning and development control

Land use planning should aim to reduce or eliminate urban and domestic sources of greenhouse gas emissions, while also helping communities adapt to the risks of climate change which are already underway. The principles for sustainable urban form and design, as well as biodiversity conservation (with natural vegetation operating as a 'carbon sink'), all help mitigate global warming. However, more specific considerations are needed to protect communities and infrastructure from increased risks arising from things such as hotter temperatures, increased rainfall, sea level rise, more intense storm events, flooding and fire. Existing risk models may need to be re-examined in the light of changed climate projections which may increase fire or flood risks.

Ideally, land use planning for climate change seeks to promote resilience across ecosystems and urban settlements. For example, land use zones should prevent inappropriate development in flood prone areas, and development controls should establish appropriate setbacks to accommodate sea level rise and storm surge. Special design codes can reduce vulnerability to fire risk, for instance through particular building materials, landscaping provisions, or provision for emergency access routes and so on. It is important to identify and avoid potential conflict between climate change responses designed to reduce greenhouse gas emissions (for instance, compact, higher density development) and the need for communities to adapt to future climate change (Hamin and Gurran, 2009). For example, in certain climatic zones, a concentration of high-density development may increase the risk of urban 'heat islands' (Corburn, 2009). This risk can be avoided through design, siting and landscaping strategies (including provision for green walls and roofs), but it is necessary to embed these requirements within planning controls if they are to be implemented in a systematic way.

Table 28.1 summarises the different ways in which land use zoning and other forms of development regulation can promote environmental sustainability.

Environmental justice and participation

The notion of 'environmental justice' means that no group of people – irrespective of their socio-economic status, ethnicity, or race– should unfairly shoulder negative impacts arising from particular policies, decisions or developments (Holifield, 2001). Within the urban planning literature, an ongoing concern has been whether land use zoning and development decisions operate to reinforce disadvantage by targeting particular neighbourhoods for industrial zones or high impact forms of infrastructure (Whittemore, 2017). Conversely, it is also possible to use land use zoning to protect lower income groups by preventing high impact land uses from occurring within residential areas. Similarly 'inclusive' zoning approaches ensure affordable housing opportunities are incorporated within all residential areas, thus preventing the spatial disadvantage that might occur through segregation of lower income earners or minority groups (Whittemore, 2013).

Embedding effective community consultation and making efforts to include diverse groups of people is an important way to promote environmental justice in the planning system. It is often argued that community involvement should occur during the strategic plan making

Table 28.1 Sustainable land use zoning and development control

Objective	*Zoning and techniques for development control*
Sustainable urban form (preserving and enhancing biodiversity, agricultural land, efficient transport networks and infrastructure)	Land use zones restrict or prevent development on environmentally sensitive and rural land
	Minimum allotment sizes in rural areas reflect agricultural needs; and in urban areas enable efficient growth
	Mixed use and higher density zones promote compact development, particularly around transit
	Height, setback, and density controls are aligned with land use zoning objectives
Conserving and enhancing biodiversity	Land use zones or overlays protect wildlife habitats and native vegetation; wetlands; water catchments, coastal features and processes
	Landscape requirements preserve and enhance urban biodiversity
	Environmental offsets
	Tradeable development rights
	Incentives for conservation agreements
	Clustering of development on less sensitive areas of environmentally significant sites
Energy, water and waste management (sustainable design)	Requirements for passive energy utilisation/energy saving in the design of buildings
	Requirement for water saving approaches
	Requirement for water sensitive urban design in new subdivisions/ redevelopment areas
	Requirement for retention/planting of endogenous species in sensitive areas
	Requirement for waste minimisation strategies in building materials as well as the construction and operation of new developments
	Green building criteria/points systems/performance targets in local planning controls
Climate change mitigation and adaptation	All of the techniques listed above, and these below
	Requirements for passive solar design/cooling; energy efficiency in buildings
	Appropriate set backs for sea level rise
	Specific criteria, referrals, development controls or regulations for buildings in fire prone/flood prone areas

Source: Author's own, derived from Gurran et al. (2015, p. 1887)

process (Burby, 2003). However, it is unrealistic to expect that all members of the community have the time or capacity to participate in planning processes. Planners have an ethical duty to consider the needs of all members of the community, whether or not these members are vocal participants in consultation processes, when making recommendations and decisions. For instance, the code of ethics for the American Institute of Chartered Planners (AICP) states that:

> We shall seek social justice by working to expand choice and opportunity for all persons, recognizing a special responsibility to plan for the needs of the disadvantaged and to

promote racial and economic integration. We shall urge the alteration of policies, institutions, and decisions that oppose such needs.

(AICP, 2016, p. 1)

Strengths and weaknesses

An important strength of land use regulation is that it operates to bind all actors in the development process – land owners, developers, public officials, elected representatives, ideally, providing a certain and transparent framework for decision making. In many communities, effective land use planning, through appropriate and enforced zoning and related controls, has also contributed to harmonious and efficient urban development. In other settings, the means to prevent land use conflict and protect the environment, by controlling and separating incompatible activities through systems such as zoning, has had the effect of preserving rural and scenic areas. Strong land use regulation is particularly important in situations where individual landholders and incremental development processes may otherwise result in piecemeal and fragmented forms of growth.

However, land use regulation is only one mechanism for implementing sustainability. Since land use regulation is triggered only when a landowner needs permission to carry out change on their land, zoning and development controls are generally not able to operate retrospectively, to require particular actions to occur or to cease. Land use planning has been described as 'reactive' rather than 'proactive' for this reason.

There are some specific weaknesses in land use regulation which are often identified in relation to EIA. First, in many jurisdictions, environmental studies required to properly assess the potential impact of a development, are usually provided by the developer or proponent themselves, which introduces a perceived conflict of interest since the proponent's consultants are unlikely to recommend against an activity going ahead (Thomas and Elliott, 2005). Another concern is the site by site basis of EIA. Cumulative impacts of major developments are not well addressed within individual EIAs which focus on the potential impact of the individual project rather than multiple similar developments. The approach known as 'strategic environmental assessment' addresses this concern by applying an EIA approach to a whole area, as part of a future planning exercise, considering the potential impact of a range of envisaged development types (Gunn and Noble, 2009). However, these processes are expensive and depend on very high quality data.

Costs and who pays

There is ongoing debate across research, practice and public policy about the potential costs of environmental and land use regulation. There are three broad ways in which land use regulation may impose specific costs on development which, in theory, might then be passed on to the wider community through a reduced range of services or through more expensive housing:

- Environmental and land use controls and requirements, ranging from requirements for specific studies to accompany proposals through to the use of particular building materials or installation of particular technology, designed to reduce or eliminate the environmental impact of a development;
- Constraints on the supply of land for urban development, intended to promote a more compact urban form, and or to protect rural land or areas of high ecosystem value;
- Requirements for developers to contribute towards green infrastructure or to pay impact fees which are used to mitigate environmental harm.

A systematic analysis of studies investigating the specific costs associated with environmental controls – including the need for additional studies and requirements for mitigation measures – found no evidence of significant cost effects (Nelson et al., 2009). This counter-intuitive finding may be explained with reference to the way that land is priced in the development process. According to the 'residual' method of land valuation, land values are determined by subtracting the total costs of development, including regulatory compliance costs, from potential sale price and the margin for profit. The remaining or 'residual' amount is what is offered to the landholder at the point of sale. In this way, environmental requirements, including any impact fees, can be absorbed into land costs, provided that they are able to be identified in advance – i.e. clearly specified in the land use plan and associated regulations. At the same time, this means that when the cost – including necessary costs of environmental mitigation – exceed the economic value of a potential development, the development will not proceed.

It is important to ensure that environmental concerns are not used as a basis for unnecessarily restricting diverse or affordable housing in communities where there are insufficient opportunities for lower income residents to rent or purchase a home. This dilemma is common in high amenity coastal or alpine communities where fragile ecosystem and scenic values represent a barrier to growth but also drive demand for tourism and residential development. This problem can be managed by ensuring that development that does proceed includes a component for affordable housing to accommodate lower income workers or others with ties to the location.

Overall, if land for new housing, commercial or other economic uses is limited during periods of high demand, land prices will increase, or developers will seek alternative locations. However, the evidence seems clear that growth management approaches which seek to accommodate growth and direct it to particular locations, rather than preventing development altogether, do not necessarily create affordability pressures (Pendall, 2000; Anthony, 2006; Landis, 2006). More widely, the increased certainty, environmental quality and coordinated infrastructure provision in areas which are comprehensively planned will often have a 'demand' effect – pushing up prices as the market values well designed and serviced areas (Mathur et al., 2004; Anthony, 2006).

Information technologies to improve regulations, assessment and decision-making tools

Internationally, the rise of geographic information systems (GIS) and similar tools for analysing spatial data has begun to affect the ways in which land use regulations are developed, implemented and communicated. In many jurisdictions, all land use regulations are available online, often in user friendly formats which locate and explain applicable regulations by address or development type. Proposals for specific developments are increasingly able to be lodged electronically and exhibited online as part of the public consultation process. Together, the range of software applications for planners include GISs for analysing spatial data, computer-aided design software, for undertaking and assessing building design and 3D modelling and visualisation software for design and modelling of potential development scenarios (Russo et al., 2018).

Many jurisdictions have also used information technology to facilitate public engagement processes, from online websites for consultation material (studies, artists renderings/flyovers of potential planning scenarios) through to interactive polls and social media campaigns (Afzalan and Muller, 2018).

This technology has enabled creative approaches for embedding sustainable design within the development assessment and control process (Siew et al., 2013). Powerful online systems for assessing the environmental performance of a particular building or even neighbourhood subdivision/precinct can provide a method for enforcing sustainability requirements while also educating designers, developers, and landowners (Retzlaff, 2008). For example, in the state of NSW, Australia, most development needs to comply with a planning policy known as the 'Building Assessment Sustainability Index – BASIX' (Gurran, 2011). A series of online questions, customised for the climatic location and development type, assess the sustainability of the building design, awarding points for the inclusion of energy and water efficient features such as solar orientation, shading, water tanks, glazing and so on. Similar mandatory or voluntary points or rating systems have emerged in many jurisdictions supporting a market for design features and appliances which support energy and water efficiency. More widely, sustainability metrics, such as 'BREEAM' (Building Research Establishment Environmental Assessment Method), which emerged in the United Kingdom in the early 1990s, is one of the most well-known systems for developments to measure and demonstrate their environmental performance and is used in over 70 nations worldwide. The Leadership in Energy and Environmental Design (LEED) green building rating system was developed by the US Green Building Council, also in the early 1990s. LEED rates aspects of building and neighbourhood design, construction, operation and maintenance and has been incorporated in building codes across the United States and worldwide. Overall, the efficacy of these instruments in promoting more sustainable building design depends on how effectively they are incorporated into land use regulations or, in the case of voluntary schemes, are endorsed by the industry and market.

Summary and conclusions

Land use regulation and zoning can set an important framework for promoting environmentally sustainable development. Traditional land use controls and zoning remain appropriate techniques for protecting sensitive land and maintaining urban boundaries in many jurisdictions. However, strictly segregated land uses can undermine sustainability outcomes in urban areas. Land use regulations which enable mixed activities and diverse housing provide a starting point for more sustainable community planning. Specific techniques to protect and enhance biodiversity values; reduce and eliminate carbon dioxide emissions in the built environment; and create climate resilience through design and green infrastructure while promoting social equity through affordable housing and participatory decision processes are critical elements of sustainable land use regulation.

References

Afzalan, N. and Muller, B. (2018). 'Online participatory technologies: opportunities and challenges for enriching participatory planning'. *Journal of the American Planning Association*, 84(2): 162–77.

American Institute of Chartered Planners (AICP) (2016). *AICP code of ethics and professional conduct*. Washington, DC: AICP.

Anthony, J. (2006). 'State growth management and housing prices'. *Social Science Quarterly*, 87(1): 122–41.

Beatley, T. (1995). 'Planning and sustainability: the elements of a new (improved?) paradigm'. *Journal of Planning Literature*, 9(4): 383–95.

Beatley, T. and Manning, K. (1997). *The ecology of place: planning for environment, economy, and community*. Washington, DC: Island Press.

Berke, P. R. and Conroy, M. M. (2000). 'Are we planning for sustainable development? An evaluation of 30 comprehensive plans'. *Journal of the American Planning Association*, 66(1): 21–33.

Burby, R. J. (2003). 'Making plans that matter – citizen involvement and government action'. *Journal of the American Planning Association*, 69(1): 33–49.

Campbell, S. (1996). 'Green cities, growing cities, just cities? Urban planning and the contradictions of sustainable development'. *Journal of the American Planning Association*, 62(3): 296–312.

Corburn, J. (2009). 'Cities, climate change and urban heat island mitigation: localising global environmental science'. *Urban Studies*, 46(2): 413–27.

Cullingworth, B. and Nadin, V. (eds.) (2014). *Town and country planning in the UK*. London: Routledge.

Department for Communities and Local Government (DCLG) (2012). *The national planning policy framework*. London: DCLG.

Grant, J. L. (2009). 'Theory and practice in planning the suburbs: challenges to implementing new urbanism, smart growth, and sustainability principles'. *Planning Theory and Practice*, 10(1): 11–33.

Gunn, J. H. and Noble, B. (2009). 'A conceptual basis and methodological framework for regional strategic environmental assessment (R-SEA)'. *Impact Assessment and Project Appraisal*, 27(4): 258–70.

Gurran, N. (2011). 'Thirty years of environmental planning in New South Wales', in N. Gurran (ed.) *Australian urban land use planning: principles, systems and practice*. Sydney: Sydney University Press, pp 143–83.

Gurran, N., Gilbert, C. and Phibbs, P. (2015). 'Sustainable development control? Zoning and land use regulations for urban form, biodiversity conservation and green design in Australia'. *Journal of Environmental Planning and Management*, 58(11): 1877–902.

Hall, P. (1996). *Cities of tomorrow*. Oxford: Blackwell.

Hamin, E. M. and Gurran, N. (2009). 'Urban form and climate change: balancing adaptation and mitigation in the US and Australia'. *Habitat International*, 33(3): 238–45.

Hirt, S. (2007). 'The devil is in the definitions'. *Journal of the American Planning Association*, 73(4): 436–50.

Hirt, S. (2012). 'Form follows function? How America zones'. *Planning Practice and Research*, 28(2): 204–30.

Holifield, R. (2001). 'Defining environmental justice and environmental racism'. *Urban Geography*, 22(1): 78–90.

Jepson, E. J. and Haines, A. L. (2014). 'Zoning for sustainability: a review and analysis of the zoning ordinances of 32 cities in the United States'. *Journal of the American Planning Association*, 80(3): 239–52.

Jordan, A. (2008). 'The governance of sustainable development: taking stock and looking forwards'. *Environment and Planning C: Government and Policy*, 26(1): 17–33.

Knaap, G. and Talen, E. (2005). 'New urbanism and smart growth: a few words from the academy'. *International Regional Science Review*, 28(2): 107–18.

Landis, J. D. (2006). 'Growth management revisited – efficacy, price effects, and displacement'. *Journal of the American Planning Association*, 72(4): 411–30.

Linkous, E. R. (2016). 'Transfer of development rights in theory and practice: the restructuring of TDR to incentivize development'. *Land Use Policy*, 51: 162–71.

Mathur, S., Waddel, P. and Blanco, H. (2004). 'The effect of impact fees on the price of new single family housing'. *Urban Studies*, 41(7): 1303–12.

McKenney, B. A. and Kiesecker, J. M. (2010). 'Policy development for biodiversity offsets: a review of offset frameworks'. *Environmental Management*, 45(1): 165–76.

Naess, P. (2001). 'Urban planning and sustainable development'. *European Planning Studies*, 9(4): 503–24.

Nelson, A. C., Randolph, J., McElfish, J. M., Schilling, J. M., Logan, J. and LLC Newport Partners (2009). *Environmental regulations and housing costs*. Washington, DC: Island Press.

Pendall, R. (2000). 'Local land use regulation and the chain of exclusion'. *Journal of the American Planning Association*, 66(2): 125–42.

Pruetz, R. and Pruetz, E. (2007). 'Transfer of development rights turns 40'. *Planning and Environmental Law*, 59(6): 3–11.

Retzlaff, R. C. (2008). 'Green building assessment systems: a framework and comparison for planners'. *Journal of the American Planning Association*, 74(4): 505–19.

Russo, P., Lanzilotti, R., Costabile, M. F. and Pettit, C. J. (2018). 'Adoption and use of software in land use planning practice: a multiple-country study'. *International Journal of Human – Computer Interaction*, 34(1): 57–72.

Ryser, J. and Franchini, T. (eds.) (2015). *International manual of planning practice* (6th edn). The Hague, Netherlands: International Society of City and Regional Planners (ISOCARP).

Siew, R. Y. J., Balatbat, M. C. A. and Carmichael, D. G. (2013). 'A review of building/infrastructure sustainability reporting tools (SRTs)'. *Smart and Sustainable Built Environment*, 2(2): 106–39.

Thomas, I. and Elliott, M. (2005). *Environmental impact assessment, theory and practice*. Sydney: Federation Press.

United Nations (1997). *Agenda 21: Programme of Action for Sustainable Development: Rio Declaration on Environment and Development: Statements of Forest Principles*; the Final Text of Agreements Negotiated by Governments at the United Nations Conference on Environment and Development (UNCED), 3–14 June 1992, Rio de Janeiro, Brazil: United Nations Department of Public Information.

World Commission on Environment and Development (WCED) (1987). *Our common future*. Oxford: Oxford University Press.
Wheeler, S. M. (2013). *Planning for sustainability: creating liveable equitable and ecological communities* (2nd edn). London: Routledge.
Whittemore, A. H. (2013). 'How the Federal Government zoned America: the Federal Housing Administration and zoning'. *Journal of Urban History*, 39(4): 620–42.
Whittemore, A. H. (2017). 'The experience of racial and ethnic minorities with zoning in the United States'. *Journal of Planning Literature*, 32(1): 16–27.

29

Indicator-based approaches to environmental planning[1]

Cecilia Wong and Wei Zheng

Introduction

> Salmon and trout, rather like the miner's canary, are sensitive indicators of pollution. Where they abound, water quality is good. Where populations have declined, or disappeared, all too often, the cause is pollution.
>
> *(Woodroofe, cited in Solbe, 1997, foreword)*

Indicators, in a simple way, can be defined as surrogates or proxy measures of some abstract, multi-dimensional concepts. The example of fish species illustrates well the dual qualities of a good indicator: the ability to communicate complex issues in simple terms as well as being underpinned by rigorous scientific knowledge. This means that when developing indicators to support policy planning, we should not just focus on complex modelling techniques but also take the user's perspective into account. Indeed, it is precisely such dual qualities that make indicators an important part of any policy toolkit. However, too often, the published indicators are either too complex to understand or too full of flaws to serve any meaningful purpose.

In the early 1990s, a new consensus emerged within planning circles to pay more attention to environmental protection and conservation. Following the publication of the Brundtland Commission Report (WECD, 1987) and the endorsement of national governments in the 1992 United Nations Conference in Rio de Janeiro, the international action plan 'Agenda 21' urged that 'indicators of sustainable development need to be developed to provide solid bases for decision-making at all levels' (UNCED, 1992, chapter 40). The 1996 Habitat II conference in Istanbul further reinforced the importance of community-based indicator projects to guide and track the progress towards achieving sustainability.

The global environmental agenda since the 1990s has brought with it a boom in using sustainability indicators for assessing environmental impact and capacity (Maclaren, 1996; Macnaghten et al., 1995) and spurred local action and broadened concern to encompass the wider community-based issues. In spite of the wide acceptance of the Brundtland definition (WECD, 1987, p. 8) that sustainable development is 'development that meets the needs of the present generation without compromising the ability of the future generations to meet their own needs', different organisations and actors defined the concept differently. Its interpretation is very much dependent on philosophical considerations that are influenced by political, ethical, religious and cultural factors (Schaller, 1993). Based on the experience of the last 25 years, this chapter aims to examine the nature, application and challenges of adopting indicator-based approaches to planning in the environmental field.

The global-urban nexus of sustainability indicators usage

The earlier use of indicators in the 1960s was very much grounded on the grand ideology of social reform and welfare at the national level to remove the gap in development between the developed and less-developed world (Taylor, 1981). Following the collapse of communism in the 1980s, the mood of top-down social reform became less enthusiastic. Instead, it was replaced by another form of ideology – the emphasis on global environmental concerns and the importance of inter- and intra-generation equality and equity of environmental resource consumption. It was the bottom-up nature of the sustainability agenda that triggered the so-called 'community indicators movement' (Innes and Booher, 2000). Whilst the approaches to the indicators movement in the 1960s and the 1990s were somewhat different, they both shared the value of improving the quality of living of people and places (Wong, 2006).

From the global perspective, the United Nations Commission on Sustainable Development (UNCSD, 1996) took a lead by publishing the 'Indicators of sustainable development: framework and methodologies' in August 1996. The publication contained methodology sheets for 134 sustainable development indicators under social, environmental, economic and institutional dimensions. Based on voluntary testing in 22 countries and expert group consultations, a revised set of 58 indicators were included in the 'Indicators of sustainable development: guidelines and methodologies' (September 2001). The testing of the indicators received support from Eurostat, which led to the publication of the UNCED indicators at the European level (Eurostat, 2001). The Millennium Development Goals (MDGs) were then proposed by the United Nations for poverty eradication though environmental sustainability was also emphasised. Indicators were proposed for tracking progress towards MDGs (United Nations, 2015a). Due to the watering down of the environmental focus in various indicator sets, the Environmental Sustainability Index (1999–2005), presented at the 2000 World Economic Forum, narrowed the scope down to only focus on environmental sustainability (Saisana and Tarantola, 2002).

The international agenda of developing indicators to monitor sustainable development progress has been very much centred at the national level (Sustainable Cities International, 2012). Building on the MDGs, 17 Sustainable Development Goals (SDGs) were agreed to by 193 world leaders in 2015 for the 2030 sustainable development agenda (United Nations, 2015b). In March 2016, the Interagency Expert Group agreed on a set of 232 indicators to monitor the 169 targets of the SDGs over the next 15 years (UNDESA, 2017). It is, however, important to note that the Expert Group opted for a set of indicators covering all 169 targets. This means that half of these indicators do not have acceptable country coverage or agreed upon methodologies, or both (Dunning, 2016).

Recent international policy discourse has shifted away from national and regional perspectives to focus on cities as drivers of the growth agenda. This shift is closely related to the unstoppable pace of urbanisation in developing countries (UNDESA, 2012). The global financial crisis and the challenges brought by climate change call for drastic solutions, and cities are seen as the key to tackling the crises and moving towards a sustainable and prosperous path (Wong, 2015). This has been affirmed in Habitat III's New Urban Agenda that cities will be the driver to end poverty, develop sustainable and inclusive urban economies and deliver environmental sustainability (UN-Habitat, 2016). As the coverage of the SDG indicators at the country level is not satisfactory, progress monitoring at the city level will be even more problematic. In 2004, UN-Habitat (UNHSP, 2004) introduced guidelines to collect 19 Global Urban Indicators (GUIs) to monitor the Habitat Agenda and the MDGs with respect to shelter; social development and eradication of poverty; environmental management; and economic development. GUIs have been regularly collected in a sample of cities worldwide and analysed in periodic 'State of the world's cities' reports. UN-Habitat (2012) introduced the City Prosperity Index (CPI) in its 'State of the world's cities 2012/13' report. The CPI is a comprehensive composite index that covers productivity; infrastructure; quality of life; equity and inclusion; environmental sustainability, and governance and legislation (Wong, 2015). Since then, UN-Habitat has adopted the CPI as a global monitoring framework for SDG 11 (Sustainable Cities and Communities) and the New Urban Agenda.

A consensus has emerged from the international community that it is at the urban level where political leaders, planners and local residents interact and benefit from policy co-production, and progress monitoring of goals and targets that fit with the local urban context (Wong and Watkins, 2009). Despite the fact that regional policy has been the orthodox spatial development approach of the European Commission, its 'Directorate-General for Regional Policy' has recently changed to 'Directorate-General for Regional and Urban Policy' to embrace cities in its policy agenda. The urban focus is also witnessed in China following its introduction of the 'National new urbanisation plan' (NUP) in March 2014. In order to accommodate China's unprecedented rate of rapid urbanisation, the concept of a 'new' type of urbanisation was introduced for coordinating urban and rural development; supporting an urban lifestyle for ecological production and consumption; making basic urban public services available to all permanent urban residents; optimising macro-level city layouts; and integrating 'ecological civilisation' into the urbanisation process. The Plan introduced 18 key indicators to monitor its urbanisation level; basic public services; infrastructure; and resources and environment.

Nature and function of indicators in environmental planning

Indicators are the result of operationalising abstract concepts of policy problems (Carley, 1981) and offering a guide indicating how a particular issue is structured or changing (Miles, 1985). This empirical definition, however, misses the normative dimension of indicators as a yardstick to measure progress and goal achievement (Bauer, 1966). The measurement of progress towards sustainable development presupposes certain innate knowledge and principles to benchmark against (Wong, 2006). Such a rationalist perspective means that value judgment would be involved in viewing some effects as better or worse. But, the norm of assessment is susceptible to change and interpretation. This highlights the underlying tension of indicators as policy instruments, which are subject to the politicisation of interpretation and the possibility of manipulation even at the measurement stage through the choice of indicators, data sources and methods. The value-laden aspect of indicators clashes with both the rationalist and the empirical ideology, as the foundation of securing objective knowledge from belief, opinion and even prejudice is somewhat less convincing.

Bell and Morse (1999, p. 9) thus queried the measurement of sustainability: 'how can something so vague be so popular?' Their findings show that it is exactly the vagueness and flexibility of such an open concept that makes it so attractive to different stakeholders and thus remains in the mainstream. Although there is a general consensus that sustainability should encompass social equity, economic growth and environmental protection, the emphasis attached to these different aspects of development varies from definition to definition. The development of indicators measuring sustainability adds an extra dimension of ambiguity, as sustainability indicators can be classified in many different ways according to their functions and roles in the decision-making process.

A framework of different categories of indicators has been recommended as the approach to evaluate different aspects of policy performance in the monitoring guidance issued by the United Nations, European Commission and different national government departments. The pressure-state-response (PSR) link model put forward by the Organisation for Economic Co-operation (OECD) and the United Nations in 1993 offers a new perspective on indicator-based approaches in environmental planning (Hammond et al., 1996). Some sustainability indicators are developed to provide a simple description of the current state of development (state indicators), others are used to diagnose and gauge the process that will influence the state of progress towards sustainability (pressure, process or control indicators), or to assess the impact brought by policy changes (target or performance indicators) (OECD, 2003). Based on the PSR model, the European Environment Agency put forward the widely adopted DPSIR framework (Smeets and Weterings, 1999). The DPSIR framework (see Figure 29.1) shows a chain of causal links starting with 'driving forces' (economic and human activities) through 'pressures' (pollutants, waste) to 'states' (of the natural and socio-economic system) and 'impacts' on ecosystems and environmental functions, and leading to policy 'responses' (setting priorities, targets). This framework describes the interactions between society and the environment by providing a more structured and integrated approach for indicators reporting, for example, the European Environment Agency (EEA) 'State of the environment report' (Kristensen, 2004).

The PSR and the DPSIR models provide a very neat and logical way of conceptualising the chain effect of human activities on the changing state of our environment and resources, which have attracted much attention on their role in environmental planning (for example, Spanò et al., 2017; Lin et al., 2013; Bidone and Lacerda, 2004). Ideally, monitoring frameworks should be guided by causal theories, but the complexity and inter-relations between different issues make it rather difficult to untangle the web of causalities in operation. When preparing for the 'Indicators of sustainable development for the UK' report (DoE, 1996), the Working Group abandoned the idea of adopting the PSR model and separated out indicators concerning the

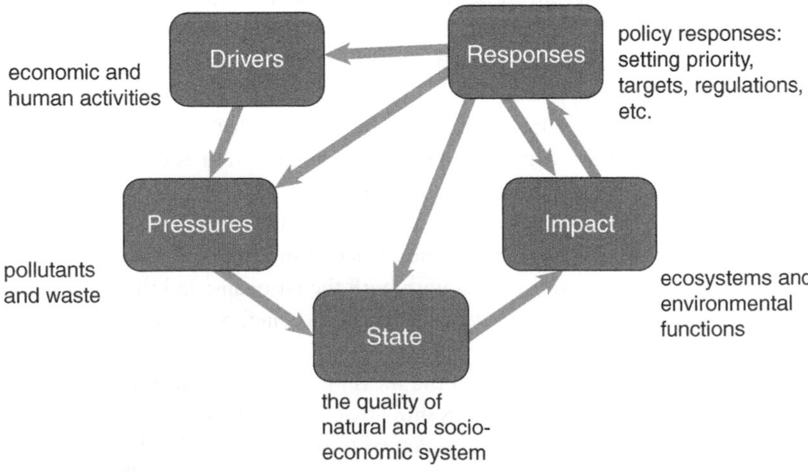

Figure 29.1 The DPSIR framework

Source: adapted from the figure in European Environment Agency (1997, p. 4)

economy, the environment and the actors involved (Cannell et al., 1999). For others, however, the linear relationship captured in the PSR model was seen as over-simplifying the complexity of real life and more complicated models were proposed (for example, Briggs et al., 1995; Wieringa, 1999). These extended models provide further sub-division of the process by chasing the driving forces outside the ecosystem to identify different sources of pressure and making a more fine-grained distinction between effects and impacts of change and the actions taken in response. However, the reality tends to be too complex to be captured by even these more elaborated models, as most effects also exert pressures on other variables (Dunn et al., 1998). Hoernig and Seasons (2004) thus argue that when deriving a monitoring framework, planners should take into account the model/conceptualisation of the interrelationship of different components, and the pragmatic policy making framework and policy needs.

Information overload and synthesis is a major concern when dealing with a large number of indicators. Researchers have increasingly recommended a tier structure to develop more efficient and coherent indicator systems. For instance, Innes and Booher (2000) proposed a three-tier indicator system to provide intelligence on city performance: system performance indicators to reflect the central values of concern and how the urban system is working; policy and programme indicators to reflect the activities and outcomes of various elements of the system; and rapid feedback indicators to help individuals, agencies and businesses to make day-to-day decisions. The European Commission (2000) also proposed a similar reference framework for the monitoring of its Structural Funds. Indicators were collected to monitor three tiers of programme objectives: global, specific and operational objectives. In the United Kingdom, there has been a tradition of collecting a small number of Sustainability Count Headline Indicators out of a larger number of Quality of Life core indicators (Wong, 2006). For example, there was a core set of 147 core indicators in 2004 (DEFRA, 2004), of which 15 were headline indicators, though the number was drastically reduced to 12 headline and 23 supplementary indicators in 2015 (ONS, 2015).

It is interesting to note that all these classifications rest upon a layered indicator structure of measurement by developing indicators from a general strategic level to gauge the overall health of the urban system through the measurement of policy outcomes to the more imminent/intermediate measures of policy feedback. The question to be asked is whether a tiered indicator structure should be adopted and, if so, how will the tiered framework operate?

There is an argument for developing more strategic and effective indicators to measure long-term and high order outcomes, that is, cross-cutting and overarching issues. These indicators will have to be easily understood by a whole range of stakeholders and organisations that need to be brought into action to deliver the visions and objectives of sustainable development. This tiered approach of structuring indicator sets has the obvious advantage of serving different analytical and policy purposes at different spatial levels and to avoid information overload.

Indicators at multiple scales

While sustainable development is very much an international initiative, it also has tremendous appeal at the local community level. However, the issues to be addressed are rather different under different spatial contexts. The spatial scale of indicators is very important as indicator values can change markedly with different spatial specifications (Sawicki and Flynn, 1996; Wong, 2015, pp. 6–7). Table 29.1 provides examples of indicator sets developed for different spatial scales, from international to city and community levels. While most indicator systems adopt a holistic perspective by encompassing the environmental, social and economic pillars of sustainability, a few just focus strongly on environmental issues.

Table 29.1 Frameworks of indicator-based approaches at different scales: some examples

Scale	*Framework/tool*	*Example indicators*	*Source*
International	Sustainable Development Goals SDG Indicators	✓ Number of deaths, missing persons and directly affected persons attributed to disasters per 100,000 population (in Goal 13) ✓ Proportion of land that is degraded over total land area (in Goal 15)	United Nations Department of Economics and Social Affairs (UNDESA) (2017)
	Millennium Development Goals (MDG) Indicators	✓ CO_2 emissions, total, per capita and per $1 GDP (Purchasing Power Parity (PPP)) (in Goal 7) ✓ Proportion of total water resources used (in Goal 7)	United Nations (2008)
	Commission on Sustainable Development (CSD) Indicators	✓ Carbon dioxide emissions ✓ Proportion of land area covered by forests	United Nations (2007)
	The pressure-state-response (PSR) model	✓ Pressure: urban air emissions (SOx, NOx, VOC) ✓ State: concentrations of air pollutants ✓ Response: green space (areas protected from urban development)	Organisation for Economic Cooperation and Development (OECD) (2003)
	Environmental Sustainability Index	✓ SO_2 emissions per land area ✓ Percentage of urban population with access to safe drinking water	The World Economic Forum (2000)
	The DPSIR (driving forces, pressures, state of the environment, impacts, and response) Framework	✓ Driving force: population growth ✓ Pressure: the amount of land used for roads ✓ State: the concentration of phosphorous and sulphur in lakes ✓ Impact: the loss of biodiversity ✓ Response: recycling rates of domestic waste	European Environment Agency (EEA) (1997)
	Common Monitoring and Evaluation Framework (CMEF)	✓ Increase in efficiency of water use in agriculture in Rural Development Programme (RDP) supported projects ✓ Increase in efficiency of energy use in agriculture and food-processing in RDP supported projects ✓ Total investment in renewable energy production (€)	European Commission (2015)

Scale	*Framework/tool*	*Example indicators*	*Source*
	Sustainable Development Indicator Framework for Africa	✓ Water abstraction by sector ✓ Energy use intensity per $1,000 (PPP) GDP (kg oil equivalent) ✓ Human and economic losses due to disasters	United Nations Economic Commission for Africa (2014)
	Eco-Efficiency Indicators (EEI)	✓ Water intensity [m^3/GDP] ✓ Land use intensity [km^2/GDP] ✓ Emission to air intensities [t/GDP]	United Nations Economic and Social Commission for Asia and the Pacific (2009)
National	Environmental performance indicators	✓ Amount of electricity purchased/consumed ($/kWh) ✓ Amount of waste going to landfills (tonnes) ✓ Amount of grey water recycled (L)	Department of the Environment, Water, Heritage and the Arts (DEWHA), Australia (1999)
	Sustainable Development Indicators (SDIs)	✓ UK greenhouse gases emissions ✓ Raw material consumption of non-construction materials ✓ Number of air pollution days classed as moderate or high – urban or rural	Department of Environment, Food, and Rural Affairs (DEFRA), UK (2013)
	Sustainable Development Indicator Framework	✓ Acres of major terrestrial ecosystems ✓ Conversion of cropland to other uses ✓ Soil erosion rates	The US Interagency Working Group on Sustainable Development Indicators (SDI Group), 1998
City-region/ City	City Prosperity Index	✓ Access To Improved Water ✓ Waste Water Treatment ✓ Share of Renewable Energy	UN Habitat (2015)
	A framework for common local indicators in regional sustainability assessment	✓ Nature conservation and management actions (Investment in nature conservation and management actions within regional protected areas[s]) ✓ Land use (Area distribution of Corine Land Cover classes [%])	Mascarenhas et al. (2010)
	The European Green City Index	✓ CO_2 emissions ✓ Energy consumption of residential buildings ✓ Use of non-car transport ✓ Waste reduction and policies	Economist Intelligence Unit (2009)
	Indicators for Sustainability	✓ Green spaces (e.g. Percentage of trees in the city in relation to city area and/or population size) ✓ Mobility (e.g. Average commute time and cost)	Sustainable Cities International (2012)

(*Continued*)

Table 29.1 (Continued)

Scale	*Framework/tool*	*Example indicators*	*Source*
	China Urban Sustainability Index	✓ Water supply (Water access rate) ✓ Industrial pollution (Industrial SO_2 discharged per GDP) ✓ Green jobs (number of environmental professionals per capita)	Urban China Initiative (2010)
	Sustainability index for Taipei	✓ Electricity consumption per person ✓ Urban population density ✓ Permeable rate in urban lands ✓ Government expenditure on pollution prevention and resource recycling	Lee and Huang (2007)
	The compass index of sustainability for Orlando	✓ Vehicle miles traveled ✓ Acres of land covered by buildings, roads, and other human-built structures ✓ Miles of streams passing State water quality standards ✓ Diversity of local job base by sector	Atkisson and Hatcher (2001)
	Urban Future toolkit in the UK	✓ Population growth (Percentage natural increase + net migration) ✓ Land recycling (Percentage of all new developments built on previously developed land) ✓ Access to public green space (Percentage of population with good access to public green space)	Boyko et al. (2012)
Community	LEED-ND	✓ Credit 1: certified green building (percentage of floor area certified) ✓ Credit 8: storm water management (percentile rainfall event) ✓ Credit 11: on-site renewable energy sources (percentage of annual electrical and thermal energy cost)	US Green Building Council (2009)
	BREEAM Communities	✓ Green infrastructure (in social and economic wellbeing issue) ✓ Resource efficiency (in resources and energy assessment issue) ✓ Access to public transport (in transport and movement assessment issue)	Building Research Establishment (BRE) (2012)

Scale	*Framework/tool*	*Example indicators*	*Source*
	CASBEE(-UD)	✓ Environmentally friendly buildings (in environmental quality of urban development) ✓ Development of traffic facilities (in environmental quality of urban development) ✓ CO_2 emissions from traffic sector or building sector (in environmental load of urban development)	Institute for Building Environment and Energy Conservation (IBEC) (2014)
	The HQE2R approach towards sustainable neighbourhoods	✓ Percentage of buildings with a standard of heating/cooling or insulation above legal norms ✓ Consumption of drinking water in the residential sector of the neighbourhood ✓ Useful surface area of open public spaces per resident	Charlot-Valdieu et al. (2004)

As discussed earlier, international institutions such as the United Nations, EEA and the OECD have focussed on developing indicator systems to guide national monitoring of sustainable development. Responding to these international agendas, different countries have formulated their own national objectives and associated indicator frameworks to track the progress of development towards sustainability. Relevant government departments in countries such as the United States and the United Kingdom have developed their own comprehensive sustainable development indicator sets. The Australian Government places emphasis on ecological sustainability development via legislation and the use of environmental performance indicators (DEWHA, 1999). In academia, efforts have been made to develop different indicator frameworks. For instance, environmental vulnerability indicators were developed to support planning decisions and were applied in a comparative evaluation of four South Pacific countries (Villa and McLeod, 2002).

The global sustainable development agenda requires local efforts and capacity development, especially efforts at a city scale. Cities are seen as the clusters of socio-economic activities and directly interact with the natural environment. The recent shift of international discourse to urban sustainability means that a bottom-up approach to environmental planning is necessary as the drivers of development are at the local level. It is thus unsurprising that an enormous number of indicator frameworks and toolkits have been developed for environmental planning at the city scale. UN-Habitat's GUIs have been used as the benchmark of cities worldwide until the recent introduction of the CPI to monitor SDG 11 and the New Urban Agenda. As shown in Table 29.2, the two indicator frameworks overlap a lot in terms of the core issues. However, there are also some differences between the two, such as the inclusion of green space and sustainable building in the CPI but not in the GUIs.

Under different city contexts, various indicator sets have been developed such as the sustainability index for Taipei (Lee and Huang, 2007), the compass index of sustainability for Orlando of Florida (Atkisson and Hatcher, 2001) and the UK Urban Future toolkit (Boyko et al., 2012). Some indicator frameworks are city-region based. For instance, a conceptual framework for

Table 29.2 Themes and issues of global urban indicators and city prosperity index

Themes	*Issues*	*Global Urban Indicators*	*City Prosperity Index*
Environment and resources	Water management	√	√
	Energy use	√	√
	Waste management	√	√
	Land use		√
	Disaster reduction	√	√
	Air quality		√
	Pollution reduction	√	√
	Green space		√
	Sustainable and resilient building		√
Social development	Urbanisation	√	√
	Access to housing	√	
	Access to public services	√	√
	Access to infrastructure	√	√
	Social equity and inclusion	√	√
	Poverty	√	√
	Health	√	
	Security	√	√
	Transportation and communication	√	√
Economic development	Productivity	√	√
	Employment	√	√
Governance	Public participation	√	
	Transparency and efficiency of governance	√	
	Municipal finance	√	√
	Public-private partnership	√	√
	Promoting decentralisation and strengthening local authorities	√	

common local sustainability indicators, with a participatory approach, has been developed for the Algarve region of Portugal (Mascarenhas et al., 2010). Similarly, the Swedish Association of Local Authorities and Regions (SALAR) has proposed a common set of 25 local indicators to enable municipalities to evaluate their environmental performance and make comparisons (Mineur, 2007). The Sustainability Solution Space for Decision-making (SSP) is an indicator-based decision toolkit with systemic and procedural features for sustainability assessment of city-regions (Wiek and Binder, 2005).

Community, the spatial scale at which land development takes place, is a linchpin of upper-level policy making and local sustainability delivery (Sharifi and Murayama, 2013). A pledge for context-specific indicators can be found from various community/neighbourhood assessment frameworks worldwide to support environmental planning. Examples include the Leadership in Energy and Environmental Design for Neighbourhood Development (LEED-ND), Building Research Establishment's Environmental Assessment Method Communities (BREEAM Communities), Comprehensive Assessment System for Building Environmental Efficiency for Urban Development (CASBEE-UD) and the HQE^{2}R methodology. Similar to the higher spatial levels, the three pillars of sustainability constitute the backbone of community sustainability which covers issues in relation to water, energy, natural resources, housing, social inclusion, community wellbeing, employment, business, economy, mixed land use, accessibility and infrastructure (Zheng et al., 2017). Since community is the basic spatial unit of development, urban design principles are often included in the indicator system for community planning (Porta and Renne, 2005).

Challenges of applying indicator-based approach in environmental planning

The crux of many challenges associated with the use of indicators to inform environmental planning is our expectation of a simple statistical measure to perform the complex policy functions of both communicating the message in simple terms but at the same time standing the test of scientific methodology.

The analysis of different indicator systems at various spatial scales shows that their core principles and underlying values of sustainable development are very similar. It is good to see that concerted global effort has been made to achieve the SDGs. The reliance on top-down guidance from international organisations and national governments, nonetheless, may not be plausible or relevant to address complex local issues. However, the number of SDG indicators proposed is too large to be fully implemented at the local level, given that different countries are under various development phases, face different challenges and have different data infrastructure and analytical capabilities. As discussed in implementation theory, whether a party recognises a distinction between policy formulation and implementation has provoked debates over the top-down and bottom-up approaches (Hill and Hupe, 2002). Proponents of a top-down approach argue that it helps to manage the implementation process on a step by step basis (Gao et al., 2014). However, from a bottom-up perspective, the need to engage with different voices means that policy formulation and implementation process cannot be easily separated (Hill and Hupe, 2002).

With the diversity of issues at different spatial scales under different spatial contexts, there has been a continuous process of delegation of power to local and regional actors to carry out indicators collection under the centrally designed monitoring regime (Innes and Booher, 2000; Wong, 2000). There are obvious merits of combining both the top-down and bottom-up approach in developing indicator frameworks (Wong and Watkins, 2009). The critical local issues, identified from the bottom-up process, can be set against the top-down strategic goals and overarching guidance framework to test the comprehensiveness and validity of the issues included. This will allow flexibility for local communities to identify issues that reflect their particular concerns and circumstances and facilitate the implementation of strategic goals and plans at the local level. More importantly, this will encourage cities and their associated communities to play a proactive role in making use of indicators as a collective way of 'seeing' to develop their own vision of sustainable development (Wong et al., 2015).

Given that there is not a definitive definition of sustainable development and there are many untested assumptions of cause and effect of different aspects of sustainability, it is difficult to find a prudently proved causal framework to guide indicator analysis for environmental planning. As far as current knowledge stands, it is fair to say that the development of environmental indicators is more advanced than that of socio-economic indicators in terms of moving towards a causal framework of analysis, though it is not without difficulties and reservations (Wong, 2015). Although frameworks like the PSR and DPSIR were proposed to explore the interrelationships between different variables, they actually separated indicators into different groups and ignored the dynamic interactive nature of different types of indicators. Niemeijer and De Groot (2008) criticised DPSIR's focus on the causal chain rather than the complicated interrelationships, which could simplify the complexity of the processes behind environmental indicators.

Indicators can be aggregated to produce a single aggregated index value and examples include UN-Habitat's CPI and Eurostat's 'Environmental Pressure Indices'. Composite indices, on the whole, tend to be more appropriate to provide a synoptic overview of issues at a higher spatial scale but are less responsive to pinpoint issues at the lower rungs of the spatial hierarchy (Sawicki and Flynn, 1996). Hence sustainability indicator sets, comprising a broad range of indicators, are rapidly emerging in local communities across the world. The other concern over the use of composite indices is that they conceal detailed information on different aspects of sustainability and are heavily reliant upon the weighting schemes used to combine the indicators (Wong, 2006). This is especially problematic when the relationship between the indicator and the phenomenon concerned is ambiguous. Some composite indices are sensitive to the weighting attached, which will require strong justification for their application. The arbitrary weighting system may thus undermine the effectiveness of these composite indices (Chowdhury, 1991).

The choice between different approaches to develop indicators is a difficult one as technical methodological issues are only part of the story. It is the wider politics and policy context that help to shape the decision. It is a fine balancing act to develop a methodologically robust framework that is widely accepted by key stakeholders. In order to appeal to a wide spectrum of stakeholders, there is a tendency to propose more and more indicators, as witnessed in the recent set of SDG indicators. The pursuit of methodological excellence of some indicator sets, through major statistical processing and manipulation exercises, could undermine the transparency of the indicator creation process and stifle debate and discussion. On the other hand, the bottom-up consensus building process could erode the rigour of indicator development, as mediocre solutions tend to be adopted to satisfy the interest of all stakeholders.

The inherent tensions between complexity and transparency will always exist and be a politically contested area of debate. As Sawicki (2002) observed, community indicator projects aiming at consensus building tended to lean on using simplistic methodologies to gain a wider spectrum of communication. There is thus an articulated fear that the failure of local communities to grasp the abstract concept of indicators and certain aspects of sustainability will lead to side

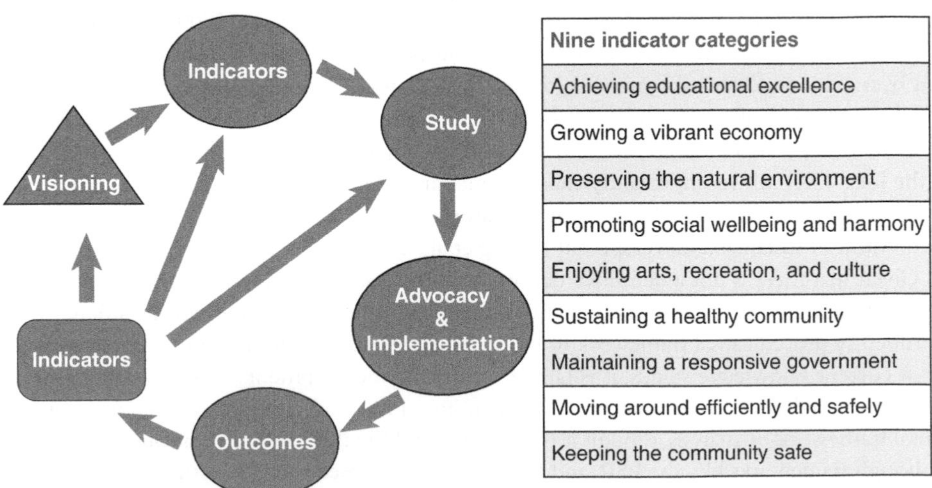

Figure 29.2 Jacksonville Council on Citizen Involvement Community Indicators

Source: adapted from 'Indicators and benchmarks in successful regional visions and plans' (Seven-50, 2012, p. 2)

lining of these components in the sustainable development agenda. However, it is interesting to find that the more successful projects, such as the Jacksonville Community Indicators Project in the United States, managed to adopt a more complex causation model in their indicator framework (see Figure 29.2).

With the constant bombardment of all sorts of sustainability-related indicators over the last two decades, the sentiment is to rationalise these indicator sets and keep the methodology simple and transparent for wider understanding. Hence, tiered indicator structures emerged to provide a framework to integrate the analysis and interpretation of indicator values with policy process, objectives and the wider policy operation environment. However, there is a concern that practitioners on the ground may find the distinction between terminologies such as context, drivers, inputs, outputs, responses, outcomes and impacts too difficult to grasp. In some cases, such as the development of the Sustainability Count Headline Indicators in the United Kingdom, the architects themselves seemed to be somewhat confused with the nature and purpose of the indicators; however, it is more likely that they were subject to all sorts of political pressure to water down the original conceptual framework by taking into account the 'bottom-up' representations (Wong, 2000).

Conclusion

This chapter provides a critical discussion of the nature and function of indicator-based approaches to environmental planning by examining the experiences and lessons learnt internationally over the last 25 years. We argue that the complex issues of sustainable development are best monitored at the local level, though with guidance from national and international actors to provide a robust conceptual framework and technical methodological support. The discussion also highlights some inherent tensions and challenges of developing indicators for environmental planning. To end our discussion, we put forward a number of principles to underpin the development of a sustainable indicator framework for local development:

- The proposed set of indicators should serve to provide a strong platform for stakeholders to develop their own indicator framework;
- The framework should be transferable, and form a strong backbone, to link up with other monitoring frameworks at different spatial scales;
- Analysis should no longer focus on single indicator values but more on how to flexibly combine indicators to yield meaningful policy intelligence;
- Analysis should include 'spatiality' by emphasising the importance of spatial linkages and connections;
- Focus on making use of indicators to help planners and stakeholders to question the values, assumptions, and core strategies that can lead to policy actions which, in turn, can be used to modify existing policy and actions to address new issues;
- Indicator frameworks should help provide a communicative and iterative learning approach to monitoring and should embed monitoring right at the heart of the policy-making process.

Note

1 The material in this chapter is based in part on research funded by the UK Economic and Social Research Council (ES/N010698/1) on 'Eco-urbanisation: promoting sustainable development in metropolitan regions of China'.

References

Atkisson, A. and Hatcher, R. L. (2001). 'The compass index of sustainability: prototype for a comprehensive sustainability information system'. *Journal of Environmental Assessment Policy and Management*, 3(4): 509–32.

Bauer, R. A. (1966). *Social indicators*. Cambridge, MA: MIT Press.

Bell, S. and Morse, S. (1999). *Sustainability indicators*. London: Earthscan.

Bidone, E. D. and Lacerda, L. D. (2004). 'The use of DPSIR framework to evaluate sustainability in coastal areas. Case study: Guanabara Bay basin, Rio de Janeiro, Brazil'. *Regional Environmental Change*, 4(1): 5–16.

Boyko, C. T., Gaterell, M. R., Barber, A. R., Brown, J., Bryson, J. R., Butler, D., Caputo, S., Caserio, M., Coles, R., Cooper, R. and Davies, G. (2012). 'Benchmarking sustainability in cities: the role of indicators and future scenarios'. *Global Environmental Change*, 22(1): 245–54.

BRE (Building Research Establishment) (2012). *BREEAM communities: technical manual*. Watford, UK: BRE Global. Available at: www.breeam.com/communitiesmanual/content/resources/otherformats/output/bre_printoutput/breeam_communities.pdf

Briggs, D., Kerrell, E., Stansfield, M. and Tantrum, D. (1995). *State of the countryside environment indicators*, a final report to the Countryside Commission. Northampton: Nene Centre for Research.

Cannell, M. G. R., Palutikof, J. P. and Sparks, T. H. (1999). *Indicators of climate change in the UK*. London: DETR.

Carley, M. (1981). *Social measurement and social indicators*. London: George Allen and Unwin.

Charlot-Valdieu, C., Outrequin, P. and Robbins, C. (2004). *Brochure HQE²R No2: the HQE²R toolkit for sustainable neighbourhood regeneration projects*. The European Commission Community Research. Available at: https://app.box.com/s/bc3sptnucj

Chowdhury, O. H. (1991). 'Human development index: a critique'. *The Bangladesh Development Studies*, 19(3): 125–7.

DEFRA (Department for Environment, Food and Rural Affairs) (2004). *UK Government quality of life counts: update 2004*. London: DEFRA.

DEFRA (Department for Environment, Food and Rural Affairs) (2013). *Sustainable development indicators*. National Statistics. Available at: https://assets.publishing.service.gov.uk/government/uploads/system/uploads/attachment_data/file/223992/0_SDIs_final__2_.pdf

DEWHA (Department of the Environment, Water, Heritage and the Arts) (1999). *Guidelines for section 516A reporting – Environment Protection and Biodiversity Conservation Act 1999*. Australian Government. Available at: www.environment.gov.au/system/files/resources/4f0f2a97-22a8-40a8-85d0-5d453842b51b/files/reporting-guidelines.pdf

DoE (Department of Environment) (1996). *Indicators of sustainable development for the United Kingdom*. London: HMSO.

Dunn, J., Hodge, I., Monk, S. and Kiddle, C. (1998). *Developing indicators of rural disadvantage*, a final report to the Rural Development Commission. Cambridge: Department of Land Economy, University of Cambridge.

Dunning, C. (2016). '230 indicators approved for SDG Agenda'. *Center for Global Development* [Blog posts], 15 March. Available at: www.cgdev.org/blog/230-indicators-approved-sdg-agenda

Economist Intelligence Unit (2009). *European green city index: assessing the environmental impact of Europe's major cities*. Munich, Germany: Siemens AG. Available at: www.siemens.com/entry/cc/features/greencityindex_international/all/en/pdf/report_en.pdf

EEA (European Environment Agency) (1997). *Air pollution in Europe 1997: executive summary*. Copenhagen: European Environment Agency. Available at: www.eea.europa.eu/publications/92-9167-059-6-sum

European Commission (2000). *Indicators for monitoring and evaluation: an indicative methodology*. The New Programming Period 2000–2006: Methodological Working Papers (Working Paper No. 3). Brussels: European Commission.

European Commission (2015). *Working paper: common evaluation questions for rural development programmes, 2014–2020*. Brussels: European Union. Available at: https://enrd.ec.europa.eu/sites/enrd/files/uploaded-files/wp_evaluation_questions_2015.pdf

Eurostat (Statistical Office of the European Union) (2001). *Environmental pressure indicators for the EU*. Luxembourg: Eurostat.

Gao, J., Christensen, P. and Kørnøv, L. (2014). 'The changing Chinese SEA indicator guidelines: top-down or bottom-up?' *Environmental Impact Assessment Review*, 44: 22–30.

Hammond, A., Adriaanse, A., Rodenburg, E., Bryant, D. and Woodward, R. (1996). *Environmental indicators: a systematic approach to measuring and reporting on environmental policy performance in the context of sustainable development.* Washington DC: World Resources Institute.

Hill, M. J. and Hupe, P. L. (2002). *Implementing public policy: governance in theory and practice.* London: Sage.

Hoernig, H. and Seasons, M. (2004). 'Monitoring of indicators in local and regional planning practice: concepts and issues'. *Planning Practice and Research*, 19(1): 81–99.

IBEC (Institute for Building Environment and Energy Conservation) (2014). *CASBEE for urban development: technical manual.* Institute for Building Environment and Energy Conservation. Available at: www.ibec.or.jp/CASBEE/english/download.htm

Innes, J. E. and Booher, D. E. (2000). 'Indicators for sustainable communities: a strategy for building on complexity theory and distributed intelligence'. *Planning Theory and Practice*, 1(2): 173–86.

Kristensen, P. (2004). *The DPSIR framework.* Paper presented at the 27–29 September workshop on a comprehensive/detailed assessment of the vulnerability of water resources to environmental change in Africa using river basin approach. UNEP Headquarters, Nairobi, Kenya. Available at: http://wwz.ifremer.fr/dce/content/download/69291/913220/file/DPSIR.pdf

Lee, Y. J. and Huang, C. M. (2007). 'Sustainability index for Taipei'. *Environmental Impact Assessment Review*, 27(6): 505–21.

Lin, Y. C., Huang, S. L. and Budd, W. W. (2013). 'Assessing the environmental impacts of high-altitude agriculture in Taiwan: a driver-pressure-state-impact-response (DPSIR) framework and spatial emergy synthesis'. *Ecological indicators*, 32: 42–50.

Maclaren, V. W. (1996). 'Urban sustainability reporting'. *American Planning Association Journal*, 62(2): 184–201.

Macnaghten, P., Grove-White, R., Jacobs, M. and Wynne, B. (1995). *Public perceptions and sustainability in Lancashire: indicators, institutions, participation.* Preston: Lancashire County Council.

Mascarenhas, A., Coelho, P., Subtil, E. and Ramos, T. B. (2010). 'The role of common local indicators in regional sustainability assessment'. *Ecological Indicator*, 10(3): 646–56.

Miles, I. (1985). *Social indicators for human development.* London: Frances Pinter.

Mineur, E. (2007). *Towards sustainable development: indicators as a tool of local governance.* PhD. Umeå University.

Niemeijer, D. and de Groot, R. S. (2008). 'Framing environmental indicators: moving from causal chains to causal networks'. *Environment, Development and Sustainability*, 10(1): 89–106.

OECD (Organisation for Economic Co-operation and Development) (2003). *OECD environmental indicators: development, measurement, and use.* Reference Paper. Paris: OECD Environment Performance and Information Division.

ONS (Office for National Statistics) (2015). *Sustainable development indicators.* July. London: ONS.

Porta, S. and Renne, J. L. (2005). 'Linking urban design to sustainability: formal indicators of social urban sustainability field research in Perth, Western Australia'. *Urban Design International*, 10(1): 51–64.

Saisana, M. and Tarantola, S. (2002). *State-of-the-art report on current methodologies and practices for composite indicator development.* Ispra, Italy: European Commission, Joint Research Centre.

Sawicki, D. and Flynn, P. (1996). 'Neighbourhood indicators: a review of the literature and an assessment of conceptual and methodological issues'. *Journal of the American Planning Association*, 62(2): 165–83.

Sawicki, D. S. (2002). 'Improving community indicator systems: injecting more social science into the folk movement'. *Planning Theory and Practice*, 3(1): 13–32.

Schaller, N. (1993). 'The concept of agricultural sustainability'. *Agriculture, Ecosystems and Environment*, 46(1–4): 89–97.

Seven-50 (2012). *Indicators and benchmarks in successful regional visions and plans.* Southeast Florida Regional Partnership. Available at: http://seven50.org/wp-content/uploads/2012/05/Indicators-and-Benchmarks-in-Successful-Regional-Visions-and-Plans.pdf

Sharifi, A. and Murayama, A. (2013). 'A critical review of seven selected neighborhood sustainability assessment tools'. *Environmental Impact Assessment Review*, 38: 73–87.

Smeets, E. and Weterings, R. (1999). *Environmental indicators: typology and overview.* Copenhagen: European Environment Agency.

Solbe, J. (1997). *Water quality for salmon and trout* (3rd edn). Perthshire: Atlantic Salmon Trust.

Spanò, M., Gentile, F., Davies, C. and Lafortezza, R. (2017). 'The DPSIR framework in support of green infrastructure planning: a case study in Southern Italy'. *Land Use Policy*, 61: 242–50.

Sustainable Cities International (2012). *Indicators for sustainability: how cities are monitoring and evaluating their success.* Vancouver: Canadian International Development Agency.

Taylor, C. L. (1981). 'Progress towards indicator systems: an overview', in C. L. Taylor (ed.) *Indicator systems for political, economic and social analysis*. Cambridge, MA: Gunn and Hain, pp 1–10.

UNCED (United Nations Commission on Environment and Development) (1992). *Agenda 21*. Conches, Switzerland: UNCED.

UNCSD (United Nations Commission on Sustainable Development) (1996). *Indicators of sustainable development: framework and methodologies*. New York: UNCSD.

UNDESA (United Nations Department of Economics and Social Affairs) (2012). *World urbanization prospects – the 2011 revision*. New York: UNDESA.

UNDESA (United Nations Department of Economics and Social Affairs) (2017). *Inter-agency and expert group on SDG indicators*. New York: UNDESA. Available at: https://unstats.un.org/sdgs/iaeg-sdgs/

UN-Habitat (United Nations Human Settlements Programme) (2012). *State of the world's cities 2012/13: prosperity of cities*. Nairobi, Kenya: UN-HABITAT.

UN-Habitat (United Nations Human Settlements Programme) (2015). *The city prosperity initiative: 2015 global city report*. Available at: http://cpi.unhabitat.org/sites/default/files/resources/CPI_2015%20 Global%20City%20Report._0.pdf

UN-Habitat (United Nations Human Settlements Programme) (2016). *Habitat III new urban agenda: draft outcome document for adoption in Quito, October 2016*. Nairobi, Kenya: UN-HABITAT.

UNHSP (United Nations Human Settlements Programme) (2004). *Urban indicators guidelines: monitoring the Habitat Agenda and the millennium development goals*. Nairobi: UNHSP.

United Nations (2007). *Indicators of sustainable development: guidelines and methodologies*. New York: United Nations. Available at: www.un.org/esa/sustdev/natlinfo/indicators/guidelines.pdf

United Nations (2008). *Official list of MDG indicators*. New York: United Nations. Available at: http://mdgs.un.org/unsd/mdg/Host.aspx?Content=Indicators/OfficialList.htm

United Nations (2015a). *The millennium development goals report*. New York: United Nations.

United Nations (2015b). *Transforming our world: the 2030 agenda for sustainable development*. New York: United Nations.

United Nations Economic Commission for Africa (2014). *Sustainable development indicator framework for Africa and initial compendium of indicators*. Addis Ababa, Ethiopia: United Nations Economic Commission for Africa. Available at: www.uneca.org/sites/default/files/PublicationFiles/sdra_sustainable-develop ment-indicator-framework-sdif_15-00581.pdf

United Nations ESCAP (United Nations Economic and Social Commission for Asia and Pacific) (2009). *Eco-efficiency indicators: measuring resource-use efficiency and the impact of economic activities on the environment*. New York: United Nations Publication. Available at: https://sustainabledevelopment.un.org/content/ documents/785eco.pdf

US Green Building Council (2009). *LEED 2009 for neighborhood development rating system*. Washington, DC: US Green Building Council.

US Interagency Working Group on Sustainable Development Indicators (1998). *Sustainable development in the United States: an experimental set of indicators*. Washington, DC: US Interagency Working Group on Sustainable Development Indicators. Available at: http://teclim.ufba.br/jsf/indicadores/SDI%20 US%20SUST%20DEVEL%20INDICAT.PDF

Urban China Initiative (2010). *The urban sustainability index: a new tool for measuring China's cities*. Available at: http://urbanchinainitiative.typepad.com/files/usi.pdf

Villa, F. and McLeod, H. (2002). 'Environmental vulnerability indicators for environmental planning and decision-making: guidelines and applications'. *Environmental Management*, 29(3): 335–48.

WECD (World Commission on Environment and Development) (1987). *Our common future*. Oxford: Oxford University Press.

Wiek, A. and Binder, C. (2005). 'Solution spaces for decision-making – a sustainability assessment tool for city-regions'. *Environmental Impact Assessment Review*, 25(6): 589–608.

Wieringa, K. (1999). 'Towards integrated environmental assessment support for the European Community's environmental action programme process'. *International Journal of Environment and Pollution*, 11(4): 525–41.

Wong, C. (2000). 'Indicators in use: challenges to urban and environmental planning in Britain'. *Town Planning Review*, 71(2): 213–39.

Wong, C. (2006). *Quantitative indicators for urban and regional planning: the interplay of policy and methods*. London: Royal Town Planning Institute Library Book Series, Routledge.

Wong, C. (2015). 'A framework for "city prosperity index": linking indicators, analysis and policy'. *Habitat International*, 45: 3–9.

Wong, C. and Watkins, C. (2009). 'Conceptualising spatial planning outcomes: towards an integrative measurement framework'. *Town Planning Review*, 80(4/5): 481–516.

Wong, C., Baker, M., Webb, B., Hincks, S. and Schulze-Baing, A. (2015). 'Mapping policies and programmes: the use of GIS to communicate spatial relationships in England'. *Environment and Planning B: Planning and Design*, 42(6): 1020–39.

World Economic Forum (2000). *Pilot environmental sustainability index*. Yale Center for Environmental Law and Policy. Available at: http://sedac.ciesin.columbia.edu/downloads/data/esi/esi-pilot-environmental-sustainability-index-2000/2000-esi-full-report.pdf

Zheng, H. W., Shen, G. Q., Song, Y., Sun, B. and Hong, J. (2017). 'Neighborhood sustainability in urban renewal: an assessment framework'. *Environment and Planning B: Urban Analytics and City Science*, 44(5): 903–24.

30

Metabolic impact assessment

A review of approaches and methods

Robin Curry and Geraint Ellis

Introduction

Previous chapters in this book have described the complexity of the issues, concepts and processes that need to be addressed in any effective system of environmental planning. Despite this, the effectiveness of environmental planning systems can be judged by the alluringly simple question of 'How do we assess the extent of the resources being used by human settlements and how can we reduce the impact of this resource use on broader socio-ecological systems?' The answer is, of course, far from simple, and many of the approaches covered in this section of the book provide insights into how we understand different types of impact and resource use. More specifically, this chapter focuses on an idea that aims to assess the entire resource and ecological impact of human societies: that of 'metabolism'. The term is most commonly used to describe processes that occur within a living organism to maintain life, but it can be used as a powerful analogy and method for assessing resource consumption and waste production by human societies. This is often referred to as 'urban metabolism' and defined as: 'the sum total of the technical and socio-economic processes that occur in cities, resulting in growth, production of energy and elimination of waste' (Kennedy et al., 2007, p. 44).

This emphasises that cities are embedded in nature and, like organisms or ecosystems, also consume resources and need to excrete wastes, thus transforming raw materials in to toxins, by-products, human biomass and enduring attributes of the built environment. These processes define the essential state of sustainability of human settlements and by thinking of them as processing flows of resources, it helps identify priority areas for intervention. While natural ecosystems are usually self-sustaining and exhibit circular and relatively closed metabolisms, including the recycling of waste through the action of detrivores, modern cities tend to be unsustainable in that they have open, linear metabolisms that import resources from all over the world, accumulate material and produce enormous quantities of waste that presents substantial challenges in its management. By applying the analogy of metabolism to different types of human settlements, it facilitates understanding of system-wide resource use and allows universal benchmarks to be developed for comparing performance of different settlements or other geographic units. It can also stimulate wider reflections on the wider social and economic arrangements that reproduce existing forms of unsustainability. However, to be able to do this, we need to quantify all the inputs, outputs and storage of energy that sustain the biophysical and socioeconomic life of a settlement, and thus have data on the use of water, energy, wastes, nutrients and other materials.

The purpose of this chapter is to briefly describe some of the key tools and concepts that have been used to assess flows of these resources using different types of metabolic impact assessments. Many of these are based on core principles of 'Material Flow Analysis' which feed into a wide range of methods and assessments, including the 'Ecological Footprint', and its derivatives, which convert resource and carbon flows into a land equivalent to create a single indicator of sustainability. The chapter reviews a range of these approaches, providing examples of where they have been applied to environmental planning issues.

Material flow analysis

One of the key concepts of metabolic impact assessment is that of material flows, from which most environmental impacts originate. The term 'material flow analysis' (MFA) covers a range of approaches which aim to quantify the flows into, within and out of a specific unit. The most common use is an economy-wide MFA (Daniels and Moore, 2001), which depends on the use of statistics on the production of industrial (mainly manufactured) goods or products, which are usually only collected at the level of a national economy. This provides invaluable insights for strategic aspects of environmental planning as it seeks to understand how an economy relies on a globalised interdependence of resources – thus defying many principles of self-sufficiency and sustainability – and helps identify issues requiring priority action at a national level – such as a displacement of CO_2 emissions through the importation of carbon-intensive foods, energy insecurity or the erosion of critical natural capital, such as by importing tropical hardwoods. Techniques are being developed to apply similar approaches to city, regional and sub-regional levels (for example, Roy et al., 2015), allowing quantification of self-sufficiency at smaller geographic scales; however, there are significant data and methodological challenges when applied to sub-national spatial levels.

The theoretical basis of the MFA combines systems analysis with the mass balance/material balance principal of the conservation of matter, derived from the first law of thermodynamics (Hinterberger et al., 2003). The first law of thermodynamics states that matter (i.e. mass or energy) is neither created nor destroyed by any physical transformation (production or consumption) process. This basic principle is applied to understanding how material flows through a socio-ecological system, such as an economy, and hence help identify full environmental impacts, areas of inefficiency or other aspect of environmental performance. An important starting point in any analysis is to identify a boundary for a system being analysed and for an economy-wide material flow account; these are defined as (Eurostat, 2001):

- The extraction of primary (i.e. raw, crude or virgin) materials from the national environment and the discharge of materials to the national environment;
- The political (administrative) borders that determine material flows to and from the rest of the world (imports and exports). Natural flows into and out of the geographical territory are excluded.

This allows a mass balance account based on the following flows:

[Resource Flows In] (material imports + product imports) + **[Material Production]** + **[Resource Flows Out]** (material exports + product exports + waste production + emissions to air + dissipative outputs of products) = **[Net Addition to Stock]**

This relatively simple formula facilitates an analysis of the entire flow of materials through an economy. This began to be applied at a national level in the early 1990s, with a rapid growth of applications and a standardisation of methods (Fischer-Kowalski and Hüttler, 1998). A strength of MFA, and a reason for its popularity amongst national statistical agencies (for example, Eurostat, 2001), is that it links to standardised economic accounting via the UN System of Environmental and Economic Accounting (United Nations et al., 2003). This allows policy makers to align environmental and resource management with economic policy and, as such, forms the basis of the sustainable development indicators used by Eurostat (Eurostat, 2006). This allows a standardised means of benchmarking performance on resource use of different economies using MFA-derived indicators, examples of which are set out in Table 30.1.

An example of an economy-wide MFA is shown in Figure 30.1, taken from an MFA analysis of Ireland, carried out for the Irish Environmental Protection Agency (EPA) (Curry and Maguire, 2008a).

MFA provides a standardised method and indicators for resource use at a national level; it allows for the development of annual time series to monitor performance over time, facilitates comparative analysis, while using data routinely collected by national agencies. There have also been criticisms, including that while measuring material flows can highlight resource use, it

Table 30.1 Key MFA indicators

Input indicators	*Output indicators*
• Direct Material Input (DMI) – measures the direct input of materials for use into the economy, i.e. all materials which are of economic value and are used in production and consumption activities. • Total Material Requirement (TMR) – includes, in addition to TMI, the (indirect) material flows that are associated to imports but that take place in other countries. • Consumption indicators. • Domestic material consumption (DMC) – measures the total amount of material directly used in an economy (i.e. excluding indirect flows). • Total material consumption (TMC) – measures the total material use associated with domestic production and consumption activities. • Net Additions to Stock (NAS) – measures the 'physical growth of the economy', i.e. the quantity (weight) of new construction materials used in buildings and other infrastructure, and materials incorporated into new durable goods such as cars, industrial machinery, and household appliances. • Physical Trade Balance (PTB) – measures the physical trade surplus or deficit of an economy. PTB equals imports minus exports.	• Domestic Processed Output (DPO) – the total weight of materials, extracted from the domestic environment or imported, which have been used in the domestic economy, before flowing to the environment. • Total Domestic Output (TDO) – the sum of DPO, and disposal of unused extraction. This indicator represents the total quantity of material outputs to the environment caused by economic activity. • Direct Material Output (DMO) – the sum of DPO, and exports. This indicator represents the total quantity of material leaving the economy after use either towards the environment or towards the rest of the world. • Total material output (TMO) – measures the total of material that leaves the economy. TMO equals TDO plus exports.

Source: adapted from Eurostat (2001, p. 25)

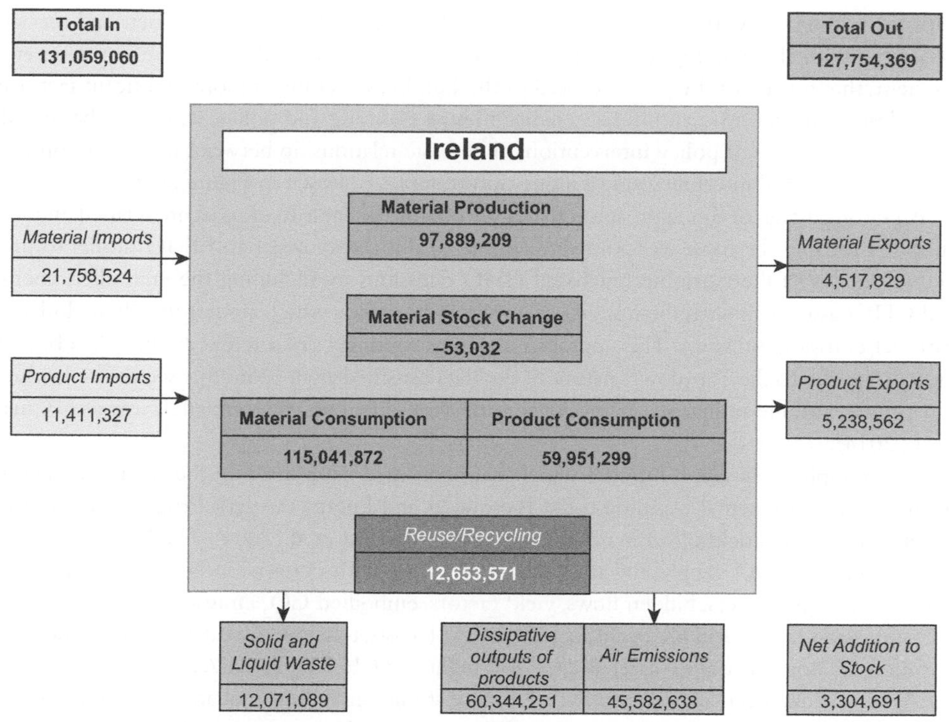

Figure 30.1 Material flow model of Ireland (2003)
Source: Curry and Maguire (2008a, p. 5)

cannot provide a measurement of the environmental *impacts* of this resource use. Additionally, the MFA-derived indicators set out in Table 30.1 may well be standardised to the satisfaction of statistical agencies, but they do not neatly align with policy areas and thus are often poorly understood by policy makers (Curry et al., 2011), and there is a need for further methodological development in order to derive direct insights into actual environmental impacts and its application to specific human settlements.

Environmentally extended input-output analysis (EE-IO)

Another metabolic approach, 'environmentally extended input-output analysis' (EE-IO), is also based on standardised economic reporting which describe the sale and purchase relationships between producers and consumers within an economy. EE-IO responds to some of the weaknesses of MFA by facilitating the use of an existing economic tool, input-output tables, to directly assess environmental factors such as resource use or greenhouse gas (GHG) emissions. Input-output analysis quantifies inter-industry or sector monetary transactions between the sectors of an economy to capture the interactions among all of the industries comprising an economic system and allows resource flows and emissions to be assigned to final consumption. EE-IO enables the capture of the entire supply chain impacts of any product or process, for example, in addition to estimating the emissions and resource use associated with a building over its lifetime of use (largely energy and water consumption), the EE-IO can estimate the emissions associated with the construction of the building and materials, products and services associated with the buildings use. This is important because, as buildings become more energy efficient, the impacts of the products used in the buildings becomes a more significant factor in their design and this in turn informs environmental planning and policy, while also being able to model the effects of policy interventions. The basic relationship between inputs and outputs of an economy and interpretation of input-output tables is shown in Figure 30.2.

A key advantage of this approach is that it allows for the capture of what are termed 'indirect effects'. Probably the most well-known example of this is the use of an EE-IO model to provide estimates of the consumption-based GHG emissions by including the embodied energy and GHGs associated with the imports of goods and services, which are not usually included in national emission estimates. This approach also uses routinely collected economic data but has limitations due to the 'top-down' nature of the data classified at an economic sector level, which can often combine industries or products with very different environmental impacts (Curry et al., 2014).

An example of an EE-IO model which has been used extensively in the United Kingdom to inform environmental planning is the Resources and Energy Analysis Programme (REAP), developed by the Stockholm Environment Institute (Barrett et al., 2004). The REAP methodology is based on IO supply and use tables, combined with conversion factors for embodied energies, energy carriers, hidden flows, yield factors, embodied CO_2 emission factors, transport, CO_2 emission factors and life cycle analysis (LCA) factors, to allow the calculation of a range of output indicators, including the ecological footprint and GHG emissions/carbon footprint. The REAP model has been applied in a wide range of studies including Scotland's Global Footprint project[1] and the One Planet Wales project.[2] The Northern Visions project (Curry and Maguire, 2008b) used the REAP model in Northern Ireland to calculate regional environmental impacts using different foot printing techniques (to be discussed later) by different policy areas, including housing. The study showed that housing was the largest regional environmental burden and could then be modelled to test the effectiveness of different policy options to reducing its environmental impact. It demonstrated that while improvements in new buildings will deliver a reduction in the footprint of the housing stock, this is smaller than improving existing houses due to the small number of new dwellings built each year compared to the overall stock (Curry and Maguire, 2011). This was an important finding in terms of environmental planning, as

INPUT	**Input-Output Tables**	**OUTPUT**
Producer Extraction of resources Domestic production plus Imports	• Provide the link between production and consumption • Describe the structure of the economy • And its interaction with the environment • Provide mass balance • Provide trade balance • Assign responsibilities for **material flows**	Consumer Waste emissions Final consumption plus Exports

Figure 30.2 Basic relationship between inputs and outputs of an economy
Source: adapted from Wiedmann and Barrett (2005, p. 8)

policy has focused on improving the energy efficiency of new buildings, rather than the more challenging process of retrofitting.

Life Cycle Analysis (LCA)

While both MFA and EE-IO take as their starting point material flows, they do so using a 'top-down' approach, that is, using data derived at an economic sector level, life cycle analysis (LCA) gathers data 'bottom-up' at a product or process level and combines this with an inventory of environmental releases, which allows the environmental impacts to be evaluated across an entire life cycle, from material acquisition and production to use and final disposal. LCAs have been carried out on numerous products and processes over many years and an ISO (International Organisation for Standardisation) standard for LCA was published in 2006. LCA is generally considered to have four main stages:

- Goal and scope;
- Inventory analysis;
- Impact assessment;
- Interpretation.

There is a wide range of published guidance on how to conduct an LCA, such as that published by the US EPA (Science Applications International Corporation, 2006), the European Union (European Commission Joint Research Centre, 2010) and the United Nations Environment Programme (UNEP). LCA is important as it is the only method that can fully incorporate all of the environmental impacts of a product, process or service, from extraction of raw materials, through production and use and reuse/recycling, to final disposal. It also forms the basis of what is commonly referred to as the 'carbon footprint' and the methodologies for evaluating the sustainability of renewable energy and biofuels (to be discussed later). The chapter by Ingrao et al. in this volume provides a full description and discussion of LCA and its application of metabolic impact assessment to buildings.

The carbon footprint

Carbon footprints (CF) can provide a vivid way of explaining the impact for individuals, organisations, regions and countries of products, processes or services on levels of the GHGs that cause climate change. While CO_2 is the largest contributor to 'Global Warming', other GHGs are significant for particular industries or countries (for example, methane from agri-food production) and the CF encompasses these, so a *carbon* footprint is essentially shorthand for the estimate of the climate impacts of any activity, with the impacts of all GHGs reported in carbon equivalents – CO_2eq (Weidema et al., 2008). There are a number of different types of 'carbon footprints', often classified as:

- CF of products, processes and services;
- CF of organisations;
- Regional or national CF.

The CF of products and processes uses the same methodology as LCA but limits the analysis to one impact area; climate change. The following definition is from the European Commission

Joint Research Centre on LCA: 'A carbon footprint is a life cycle assessment with the analysis limited to emissions that have an effect on climate change' (EC JRC, 2007, p. 1).

The most widely used standard for CF is published by the British Standards Institute (BSI, 2008), which describes the steps for calculating product CFs as being:

- Step 1: Building a process map;
- Step 2: Checking boundaries and prioritisation;
- Step 3: Collecting data;
- Step 4: Calculating the footprint;
- Step 5: Checking uncertainty (optional).

The relevance of the CF of products, processes and services to environmental planning may appear unclear, but this CF forms the basis of the more detailed estimates of the GHG emissions associated with major planning and policy areas such as energy and transport (Minx et al., 2009). For example, using more conventional methods, MacKay and Stone (2013) attempted to estimate the potential GHG emissions associated with shale gas extraction and use, concluding that:

> The carbon footprint (emissions intensity) of shale gas extraction and use is likely to be in the range 200–253g CO_2e per kWh of chemical energy, which makes shale gas's overall carbon footprint comparable to gas extracted from conventional sources (199–207g CO_2e/kWh(th)), and lower than the carbon footprint of Liquefied Natural Gas (233–270g CO_2e/kWh(th)).
>
> *(MacKay and Stone, 2013, p. 3)*

However, Howarth et al. (2011) derived CF using the LCA approach and established a more comprehensive picture of the impacts, noting that:

> Natural gas is composed largely of methane, and 3.6% to 7.9% of the methane from shale-gas production escapes to the atmosphere in venting and leaks over the lifetime of a well. These methane emissions are at least 30% more than and perhaps more than twice as great as those from conventional gas.
>
> *(Howarth et al., 2011, p. 679)*

The main difference between these assessments are how they treat the impacts of methane, which is a much more powerful GHG than carbon dioxide, so when methane leaks from wells during gas production (rather than being combusted to water and CO_2), it has an amplifying effect on the GHG-equivalent footprint of shale gas. Howarth et al. (2011) concluded that over 20 years, the GHG footprint for shale gas is 22% to 43% greater than that for conventional gas, while over 100 years the GHG footprint for shale gas is 14% to 19% greater than that for conventional gas, if these consequences of methane leakage are fully accounted for.

The relevance of CFs of organisations is clearly highly relevant to individual companies and institutions formulating their own internal environmental policies but can also provide insights into the collective impact of particular sectors or the range of economic activity taking place in a particular geographic unit. Although there is no international standard for organisational carbon reporting, the 'de facto' international standard are the guidance and standards developed by the GHG Protocol, a partnership between the World Resources Institute (WRI) and the World Business Council for Sustainable Development (WBCSD) (WRI and WBCSD, 2014), which provides guidance for emission reporting for organisations. This classifies operational boundaries for GHG reporting based on the scope of the emissions as summarised in Figure 30.3, covering:

- Scope 1 (Direct emissions): activities owned or controlled by your organisation that release emissions straight into the atmosphere;
- Scope 2 (Energy indirect): emissions being released into the atmosphere associated with your consumption of purchased electricity, heat, steam and cooling. These are indirect emissions that are a consequence of your organisation's activities but which occur at sources you do not own or control;
- Scope 3 (Other indirect): emissions that are a consequence of your actions, which occur at sources which you do not own or control and which are not classed as scope 2 emissions (business travel by means not owned or controlled by your organisation, waste disposal, or purchased materials or fuels).

Scope 3 are classified as discretionary, not because they are unimportant but due to challenges associated with estimating them. For example, estimating emissions from business travel by plane requires information on distance travelled, plane type and class of travel.

Finally, a key application in environmental planning is through the development of regional or national CFs – otherwise known as territorial GHG reporting – which is how GHG emissions are accounted for the purpose of reporting to national governments, or the United Nations

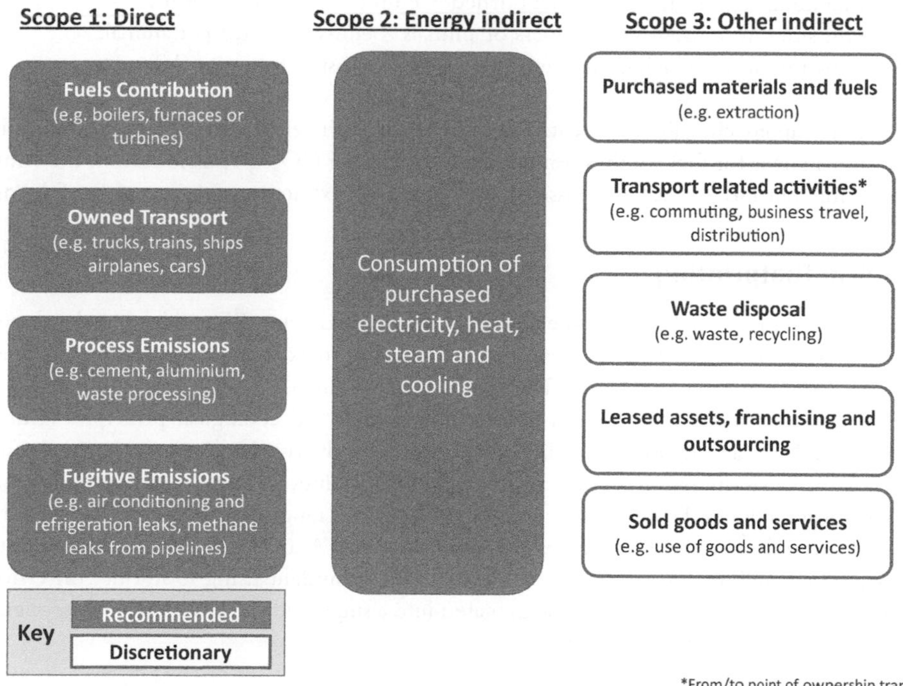

Figure 30.3 Operational boundaries for GHG reporting for carbon footprints of organisations
Source: adapted from WRI and WBCSD (2014, p. 26)

Framework Convention on Climate Change (UNFCCC). In the United Kingdom, this is set out in the UK Greenhouse Gas Inventory (DECC, 2014), which reports emissions in nine National Communication (NC) sectors, which are listed below:

- Agriculture;
- Business;
- Energy supply;
- Industrial processes;
- Land use, land use change and forestry;
- Public;
- Residential;
- Transport;
- Waste management.

All countries who report to the UNFCCC are required to use estimation methods from agreed guidelines and good practice to ensure that the GHG emissions for each country are complete and comparable. The basic equation used for calculating most emissions is:

$$\text{Emission} = \text{Emission factor} \times \text{Activity}$$

For example:

- For fuel combustion: amount of fuel burned × emission factor for that fuel;
- For emissions from livestock: numbers of animals × emission factor per animal;
- For emissions from vehicles: kilometres travelled × emission factor per kilometre.

As with other approaches, life cycle data forms the basis of the emissions factors. For example, the emissions factor for fuel combustion includes not just the CO_2eq emissions from the combustion of fuel but also the CO_2eq emissions associated with extraction and processing of the oil.

Ecological footprinting

A final approach to metabolism assessment is the ecological footprint (EF), which was developed by Wackernagel and Rees (1998) as a sustainability indicator to express the relationship between humans and the natural environment. The EF calculates human impacts (or use of ecological services) in a way that is consistent with thermodynamics and ecological principles (Curry and Maguire, 2011) and translates this into total area of productive land and water ecosystems required to produce the resources that the population consumes and assimilate the wastes that the population produces. To derive the equivalent land area, a range of energy and material flow data are derived from a range of sources including trade, MFA and LCA. This is then used to transform biotic and abiotic consumption into an area-equivalent using footprint conversion factors (see Figure 30.4), which is then aggregated into a single indicator. The area components of the footprint are (Borucke et al., 2013):

- Carbon;
- Fish;
- Crops;
- Built-up land;

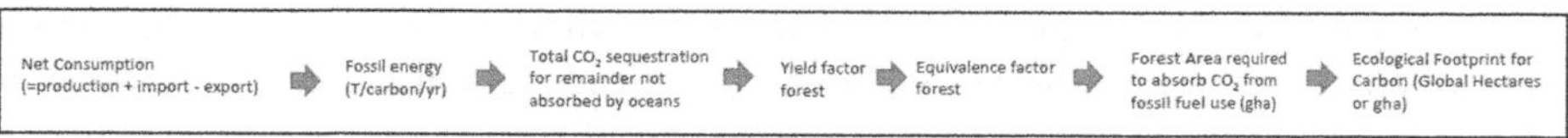

Figure 30.4 Calculation of the ecological footprint for carbon

Source: adapted from Lin et al. (2016, p. 6)

- Forest products;
- Grazing products.

The Global Footprint Network (GFN) has developed and published standardised guidance on the calculation and use of the EF (Borucke et al., 2013). An example of the calculation method for the carbon component is provided in Figure 30.4.

The EF provides a clear illustration of demand (in land area equivalent), which can then be directly compared to the global availability of resources (supply), referred to as biocapacity (BC). The biocapacity of a nation is defined as the sum of its bio-productive areas, which is converted into global hectares by multiplying the area by country-specific equivalence and yield factors. The combination of EF and BC enables the sustainability assessment and forms the basis of EF's effectiveness, as it places consumption in the context of globally available resources.

The clear link between capacity and consumption makes EF a very powerful tool in portraying (un)sustainability. It has therefore been popular within policy and research communities but subject to ongoing debate, for example, Fiala (2008) describes it as 'bad economics'. Nevertheless, it has been widely adopted and its ability to calculate the number of planets required to sustain current levels of consumption gives rise to the principle of 'One Planet Living'. This leads to the concept of a fair 'Earthshare': the amount of land everyone would get if all the ecologically productive land on Earth were divided evenly among the present population.

Conclusion

This chapter has reviewed some of the key approaches to metabolic assessments that are used in environmental planning. Each of these methods have their own strengths and weaknesses and no single method captures all of the material and energy flows and emissions in a way which meets all the needs of policy makers or is able to portray environmental impacts to the public. Here, there is a difficult balance to strike between the rigour and comparability of the standardised accounting methods which are required by government and international institutions to track progress of different countries against agreed targets (most notably, GHG), against those methods around which there may be less methodological agreement (particularly the EF), which evocatively raise awareness of the scale of our sustainability challenges.

While the concept of metabolism has been applied to human settlements for over five decades, it has attracted more interest in the last ten years, both because of our looming environmental crises but also because recent developments in data collection and analysis enable a more accurate modelling of metabolic processes that operate at different spatial and organisational scales. Future development in the collection of big data and the growing complexity of mapping and locational science will inevitably lead to further exciting developments in the field over the coming decade, with the expectation that virtually all environmental planning decisions could be underpinned by accurate metabolic data on the potential environmental impacts of future scenarios and options. Nevertheless, however sophisticated and adaptable our analytical methods become, the real strength lies in the fact that metabolic approaches are based on the relatively simple idea that if we regard our human settlements as if they were functioning organisms, we can begin to understand how these should relate to the wider global ecosystems in which they belong and, in so doing, uphold core principles of sustainability.

Notes

1 See www.gov.scot/Topics/Environment/SustainableDevelopment/funding/SAGprojects2003/footprint project.
2 See https://gov.wales/topics/environmentcountryside/climatechange/publications/ecological-footprint-of-wales-report/?lang=en.

References

Barrett, J., Wiedmann, T. and Ravetz, J. (2004). *Report No. 1. Development of physical accounts for the UK and evaluating policy scenarios*. Ecological Budget UK, Resources and Energy Analysis Programme, Stockholm Environment Institute at York, Centre for Urban and Regional Ecology, Manchester, Cambridge Econometrics.

Borucke, M., Moore, D., Cranston, G., Gracey, K., Iha, K., Larson, J., Larson, J., Lazarus, E., Morales, J. C., Wackernagel, M. and Galli, A. (2013). 'Accounting for demand and supply of the biosphere's regenerative capacity: The National Footprint Accounts' underlying methodology and framework'. *Ecological Indicators*, 24: 518–33.

British Standards Institution (2008). *Guide to PAS 2050. How to assess the carbon footprint of goods and services*. London: BSI.

Curry, R. and Maguire, C. (2008a). *Island limits. A material flow analysis and ecological footprint of Ireland*. County Wexford, Ireland: Environmental Protection Agency.

Curry, R. and Maguire, C. (2008b). *Northern visions. Footpaths to sustainability*. Belfast: SRI.

Curry, R. and Maguire, C. (2011). 'The use of ecological and carbon footprint analysis in regional policy making: application and insights using the REAP model'. *Local Environment*, 16(9): 917–36.

Curry, R., Maguire, C., Simmons, C. and Lewis, K. (2011). 'The use of material flow analysis and the ecological footprint in regional policy making: application and insights from Northern Ireland'. *Local Environment*, 16(2): 165–79.

Curry, R., Maguire, C. and Baird, J. (2014). 'Measuring the sustainability of a national economy: the application of three measures of sustainability to Ireland'. *Irish Geography*, 47(1): 1–32.

Daniels, P. L. and Moore, S. (2001). 'Approaches for quantifying the metabolism of physical economies. Part I: methodological overview'. *Journal of Industrial Ecology*, 5(4): 69–93.

Department of Energy and Climate Change (DECC) (2014). *An introduction to the UK's greenhouse gas inventory*: London: DECC. Available at: https://assets.publishing.service.gov.uk/government/uploads/system/uploads/attachment_data/file/349618/IntroToTheGHGI_2014_Final.pdf

European Commission Joint Research Centre (2007). *Carbon footprint – what it is and how to measure it*. Brussels: European Commission.

European Commission Joint Research Centre (2010). *International reference life cycle data system (ILCD) handbook – general guide for life cycle assessment – detailed guidance*: Brussels: European Commission.

Eurostat (2001). *Economy-wide material flow accounts and derived indicators. A methodological guide*. Luxembourg: Office for Official Publications of the European Communities.

Eurostat (2006). *Measuring progress towards a more sustainable Europe. Sustainable development indicators for the European Union. Data 1990–2005*. Luxembourg: Office for Official Publications of the European Communities.

Fiala, N. (2008). 'Measuring sustainability: why the ecological footprint is bad economics and bad environmental science'. *Ecological Economics*, 67(4): 519–25.

Fischer-Kowalski, M. and Hüttler, W. (1998). 'Society's metabolism. The intellectual history of materials flow analysis, part II, 1970–1998'. *Journal of Industrial Ecology*, 2(4): 107–36.

Hinterberger, F., Giljum, S. and Hammer, M. (2003). 'Material flow accounting and analysis (MFA). A valuable tool for analyses of society-nature interrelationships'. *Encyclopaedia of the International Society for Ecological Economics (ISEE)*.

Howarth, R., Santoro, R. and Ingraffea, A. (2011). 'Methane and the greenhouse-gas footprint of natural gas from shale formations'. *Climatic Change*, 106: 679–90.

Kennedy, C., Cuddihy, J. and Engel-Yan, J. (2007). 'The changing metabolism of cities'. *Journal of Industrial Ecology*, 11(2): 43–59.

Lin, D., Hanscom, L., Martindill, J., Borucke, M., Cohen, L., Galli, A., Lazarus, E., Zokai, G., Iha, K., Eaton, D. and Wackernagel, M. (2016). *Working guidebook to the national footprint accounts: 2016 edition*. Oakland: Global Footprint Network.

MacKay, D. and Stone, D. (2013). *Potential greenhouse gas emissions associated with shale gas extraction and use*. London: DECC. Available at: www.gov.uk/government/uploads/system/uploads/attachment_data/file/237330/MacKay_Stone_shale_study_report_09092013.pdf

Minx, J. C., Wiedmann, T., Wood, R., Peters, G. P., Lenzen, M., Owen, A., Scott, K., Barrett, J., Hubacek, K., Baiocchi, G. and Paul, A. (2009). 'Input-output analysis and carbon footprinting: an overview of applications'. *Economic Systems Research*, 21(3): 187–216.

Roy, M., Curry, R. and Ellis, G. (2015). 'Spatial allocation of material flow analysis in residential developments: a case study of Kildare County, Ireland'. *Journal of Environmental Planning and Management*, 58(10): 1749–69.

Science Applications International Corporation (2006). *Life cycle assessment: principles and practice*. Washington, DC: USEPA. Available at: https://cfpub.epa.gov/si/si_public_record_report.cfm?dirEntryId=155087

United Nations, European Commission, International Monetary Fund, Organisation for Economic Co-operation and Development and World Bank (2003). *Integrated environmental and economic accounting 2003 (SEEA)*. Washington, DC: UN, EC, IMF, OECD, WB.

Wackernagel, M. and Rees, W. (1998). *Our ecological footprint: reducing human impact on the earth, volume 9*. Gabriola Island, BC and Philadelphia, PA: New Society Publishers.

Weidema, B. P., Thrane, M., Christensen, P., Schmidt, J. and Løkke, S. (2008). 'Carbon footprint'. *Journal of industrial Ecology*, 12(1): 3–6.

Wiedmann, T. and Barrett, J. (2005). *Report no.2. The use of input-output analysis in REAP to allocate ecological footprints and material flows to final consumption categories*. York: Stockholm Environment Institute – York.

World Resources Institute (WRI) and World Business Council for Sustainable Development (WBCSD) (2014). *GHG protocol corporate accounting and reporting standard*. Geneva: WBCSD. Available at: https://ghgprotocol.org/corporate-standard

31

Urban metabolic impact assessment

From concept to practice

Paulo Pinho and Ruben Fernandes

Introducing the concept of urban metabolism

Over the last 20 years, both European Union (EU) and national environmental policies have been increasingly targeting urban areas as the focus of the most challenging environmental problems in Europe and elsewhere. This trend has been accompanied by a general perception of the need for a closer and more efficient articulation between environmental policies and urban planning policies. New paradigms emphasising the importance of this articulation, such as the compact city, the resilient city, the eco-city, the low carbon or the post-carbon city, as well as new policy instruments such as 'sustainability assessment' or 'strategic environmental assessment' (SEA), have been developed in an attempt to respond to the rapidly changing economic and social development processes and environmental conflicts that characterise present times.

This chapter begins with a discussion of the broad history of the concept of urban metabolism (UM) and an account of the range of its practical applications. An innovative methodology – 'Metabolic Impact Assessment' – designed to assess the likely impacts of urban plans and large urban projects in the overall metabolism of cities and metropolis is then presented. Finally, a case study of an urban development plan is presented to illustrate how this methodology can support decision making by enabling the comparison of the energy consumption requirements associated to different alternatives.

From a physical point of view, cities and metropolitan areas are complex systems that require a wide range of resources to support all social, economic and cultural processes that take place inside their boundaries (however difficult they are to define in many cases), and all the multifaceted interactions they generate or receive from the outside world. These resources include different forms of energy, water, soil, materials, food or nutrients, to name just a few of the most important ones.

This perception of how a city (or metropolis) works calls for the notion of 'metabolism', a central concept of biochemistry science. Inspired in this concept, and by pure analogy, 'urban metabolism' attempts to encompass the particular way in which cities capture and process the wide range of resources they need for functioning and the way they, subsequently, eliminate the generated residuals.

Considering the overall levels of urban environmental quality, social and economic development and quality of life offered by a particular city or metropolis, 'urban metabolism performance' or 'urban metabolism efficiency' can be simply defined as the ratio between a macro indicator of those urban qualities by a synthetic indicator of the amount of resources and residuals necessary, on an annual basis, to support the normal workings of that city or metropolis.

Surely some cities exhibit better metabolic performances than others and for very different reasons. The patterns of production and consumption of their inhabitants may differ significantly, as well as the particular urban structure and morphology, the quality and the energy efficiency of the built stock, the local climate or the natural shelter provided by the local environmental conditions, the local network of the green-blue spaces and corridors, the quality of the different urban infrastructures or the overall efficiency of the urban transport systems. To a larger or lesser extent all these factors are directly or indirectly influenced by local planning policies. For the influence of urban form policies on sustainable urban metabolism see the recent paper of Davoudi and Sturzaker (2017). Thus, the in-depth analysis of urban metabolism can constitute a powerful design and evaluation tool to guide the planning and management of cities and metropolis.

The evolution of urban metabolism

The concept of urban metabolism (UM) has been evolving since the 19th century, as schematically presented in Figure 31.1. The timeline includes the most important conceptual contributions found in the literature and the corresponding key dates.

Indeed, concerns regarding UM are not new (Barles, 2010). Although UM in its widest sense has almost a bi-secular past, its epistemology remains the subject of ongoing refinement. The notion of 'metabolism' per se was introduced in the early 19th century in the field of biology to characterise all chemical processes that occur in living organisms, resulting in growth, production of energy and elimination of wastes (Chen and Chen, 2012; Golubiewski, 2012).

In the following 50 years, it was broadly applied in what would become to be known as biochemistry science to represent processes of organic breakdown and combination not only within individual organisms, at a cellular scale, but also between organisms and their environment, as organisms convert raw materials from the environment to grow, reproduce and maintain themselves (Bancheva, 2014).

According to any biochemistry manual, metabolism is the set of life sustaining chemical transformations within the cells of organisms. Metabolism includes two processes: catabolism and anabolism. 'Catabolism' refers to the breaking down of organic matter, while 'anabolism' refers to the building up of essential cellular components. Usually, breaking down releases energy and building up consumes energy. The word metabolism can also refer to the sum of all chemical reactions that occur in living organisms, including digestion and the transport of substances into and between different cells.

Back to the history of the concept, in the late 19th century, metabolism was transposed to the field of agrochemistry. At the time, the European urban chemists concerned with population growth, the limitations of rural organic fertilisers and the resulting fear of soil depletion and food shortages used the concept of metabolism to understand and promote the cycle of organic wastes and nutrients between urban and agricultural lands. With growing population concentrations, cities began to be considered not only as centres of consumption but also through their abundant output of excreta, as important sources of fertilizers (Barles, 2010, p. 441).

Inspired by the previously mentioned studies, and especially by the work of the German chemist Justus von Liebig, Karl Marx introduced, in 1883, the concept of metabolism in social

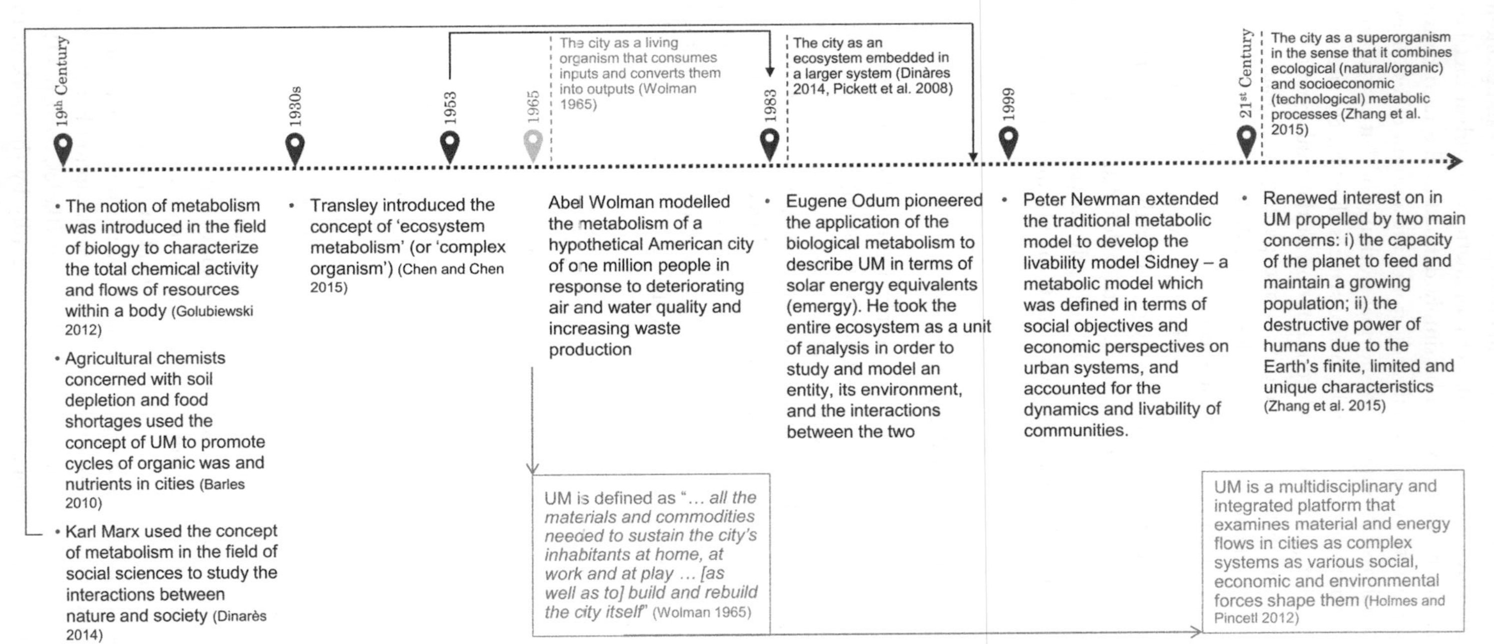

Figure 31.1 Urban metabolism through time

Source: adapted from Fernandes and Pinho (2017, p. 5)

sciences in order to describe the material and energy exchanges between nature and society that occurred through labour processes (see Dinarès, 2014; Zhang, 2013; Zhang et al., 2015). By proposing the concept of 'metabolic rift' to describe the socioecological impacts of large-scale industrial agriculture and capital accumulation, Marx argued that urbanisation processes and industrial agriculture led to rural-to-urban migration, creating a 'rift' in the metabolism of capitalist societies as humans steadily lost their ancestral relationships with nature (Bancheva, 2014; Rapoport, 2011).

In the 1930s, Arthur Transley introduced the concept of 'ecosystem metabolism'. He distinguished 'organism' from 'ecosystem' or, alternatively, 'complex organism' by arguing that the latter was a conceptual unification of organisms (biotic components) and contextual physical factors (abiotic components) into a single system (Chen and Chen, 2015).

In 1965, the concept of UM was explicitly mentioned for the first time in the pioneer article 'Metabolism of cities', authored by Abel Wolman and published in the journal 'Scientific American'. Concerned with the problems posed by rapid urbanisation in the 1960s, Wolman used the bio-inspired concept of metabolism in order to understand the effects of urban development on the environment of American cities, namely on air and water quality and waste production (Chen and Chen, 2015; Zhang, 2013). While Marx emphasised the social organisation of the consumption of Earth's resources, Wolman focused, in turn, on the patterns of resource utilisation and the resulting physical limits to the planet's carrying capacity (Pincetl et al., 2012).

More specifically, Wolman (1965) modelled the metabolism of a hypothetical American city of one million people by quantifying, in the form of an input-output chart, all input-output transactions (in addition to stocks) while tracking their respective transformations and flows within the city boundaries. In his metabolic model, Wolman regarded the city as analogous to a living organism, and divided the metabolic flows into *resource* flows (water, food and fuel) and *waste* flows (sewage, oil refuse and air pollutants). The latter are also referred to as 'metabolites'.

In the same way that an organism consumes resources to satisfy its needs (the *anabolism* referred to before) and transforms it into products and wastes, the so-called *catabolism*, so does a city consume raw materials, fuel and water in the urbanisation process, converts them into products, such as the built environment, and pushes away the metabolites that result from urban activity (Golubiewski, 2012; Costa et al., 2004; Zhang et al., 2015). Whenever a system cannot internally obtain all the resources needed, it must get them from the surrounding environment, which acts as the supporting system; the same principle applies to removal of wastes from the system (Zhang, 2013).

Up until the 1980s, industrial ecologists remained mainly focused on industrial metabolism. Therefore, urban metabolism has been mainly a normative idea where quantitative and accounting approaches to metabolism were favoured with the goal to improve the metabolic efficiency of cities or, alternatively, to reduce the amount of resources used per unit of economic output – a process that came to be known as 'dematerialisation' or 'decoupling' (Carpinteiro, 2005, cited in Dinarès, 2014). Though necessary, these approaches have been considered insufficient, not only because the consumption-driven flows have been neglected but also because the role played by stakeholders with different profiles and objectives in shaping the flows of energy and materials have not been taken into account (Barles, 2010).

During the 1970s, Odum proposed an alternative metabolic theory based on systems ecology, which supported the developing of a new method for the quantitative analysis of UM by taking the entire ecosystem as the unit of analysis (Kennedy et al., 2007). Since then, 'ecosystem metabolism' has developed into a research area with its own conceptual framework, specific theory, and set of definitions and methods (Golubiewski, 2012).

Building on his previous work from the 1950s in which the metabolic energy was used to represent the production of organic matter – 'photosynthesis', and the metabolic processes of ecological systems to represent its consumption – 'respiration', Odum modelled the metabolism of an urban system, of its environment, and of the interactions between both (Castán-Broto et al., 2012). The resulting method – the 'emergy method', in which 'emergy' stands for all the embodied energy consumed, directly and indirectly, in the processes in which resources are transformed into products or services – represents an attempt to apply biophysical value theory to both ecological and economic systems in order to study the flows of energy in the metabolism of socioeconomic systems (Dinarès, 2014, p. 555).

Based on the second law of thermodynamics (entropy), the emergy method considers that energy flows are hierarchically organised in accordance to their *transformability* potential – i.e. energy quality – to accomplish the same amount of work (Holmes and Pincetl, 2012). Such recognition has led Odum to propose the use of a single metric to compare the different kinds of flows in and out of a city – the solar energy equivalents or, alternatively, the emergy unit (Zhang et al., 2015; Liu et al., 2011). However, and so far, only very few empirical studies have looked at the metabolism of cities in terms of their embodied energy, following Odum's conceptualisation of UM through the emergy method (Kennedy et al., 2011).

The treatment of the city as a peculiar kind of 'ecosystem embedded in a larger system' (Dinarès, 2014, p. 558) is not, however, without critique. Recent research questions the use of such metaphors in contemporary urban discourses, arguing that they usually converge around the conceptualisation of the city as a metabolic system which can be examined in isolation from neo-Marxian interpretations of the processes of urban transformation, with emphasis on phenomena such as commodity chains, historical context, or urban form characteristics. As Gandy writes:

> The use of biological analogies may serve some heuristic or imaginative value in the context of architectural design for individual buildings but when applied to an entire city or region these essentially arbitrary combinations of scientific metaphors quickly become untenable and lose any analytical utility.
>
> (Gandy, 2004, p. 373)

In order to overcome these problems, the Marxist ecologists Erik Swyngedouw and John B. Foster extended, by the late 1990s, the view of cities as ecosystems beyond the original biophysical or functionalist processes of urban transformation to include the complex interactions between social and natural processes (Newell and Cousins, 2014). Since the 2000s, urban ecologists such as Steward Pickett and Nancy Grimm, drawing on complex systems theory, developed, in turn, conceptual frameworks to understand cities as 'urban ecosystems', i.e. 'complex, dynamic biological-physical-social entities, in which spatial heterogeneity and spatially focalized feedbacks play a large role' (Pickett et al., 2008, p. 148).

If any consensus can be derived from the literature, it would seem to be that the disputed nature of the concept of UM, whether used in relation to material and energy processes or as a societal metaphor, is related to the fact that it has developed from unrelated, and sometimes contradictory intellectual traditions – such as industrial ecology, ecological economics, political ecology, urban ecology and urban political ecology (Newell and Cousins, 2014; Castán-Broto et al., 2012) – which have, over time, ascribed different meanings and methods to the metaphor. Moreover, even if at some point the metaphor that cities are like organisms, or even like real ecosystems with respect to metabolism breaks down, one should note that the concept of UM is far more than a metaphor; it is a way to emphasise the material and energy flows and the related dynamic, interconnected and mutually transformative socioeconomic (technological) and natural (organic) processes of urban systems (Clift et al., 2015), providing a basis upon which sustainability lessons for cities can be drawn (Pincetl et al., 2012; Zhang et al., 2015). This is the perspective adopted in this chapter.

In sum, the concept of UM, and the approaches to UM, have evolved greatly over the course of the last 50 years. Whereas the first definitions of UM were developed within the biological and physical sciences and encompassed only the quantification of physical flows (such as energy, materials and substances) in and out of cities (Weisz and Steinberger, 2010), the UM concept has been recently broadened to include social, economic and institutional changes that affect or result from metabolic transactions (Liu et al., 2011), the role of power relations, institutions and neoliberal reforms in governing the metabolic flows. As Gandy (2004, p. 373) argued:

> [I]f the idea of UM can be disentangled from its organicist and functionalist antecedents . . . it can serve as a useful point of entry for a critical reformulation of the relationship between social and bio-physical processes . . . that produced new forms of urban and metropolitan nature.

Applications of the urban metabolism concept

Following the pivotal work of Wolman back in 1965, three studies were conducted in the 1970s on the cities of Tokyo, Brussels and Hong Kong by chemical engineers, ecologists and civil engineers, respectively (Zhang et al., 2015). These studies, which used the traditional methods of material and energy-flow analysis, recognised, nonetheless, the interdisciplinary nature of the UM concept (Kennedy et al., 2011).

Throughout the 1980s, the progress in the study of the metabolism of cities was quite modest (Bancheva, 2014; Rapoport, 2011). Even so, about 20 comprehensive empirical studies have been reported during this period, including cities such as Brisbane, Prague, Stockholm, Taipei and Vienna, in addition to numerous sectorial studies addressing particular components of the UM concept, for example, water, nutrients, and metals (see Mostafavi et al., 2014; Kennedy and Hoornweg, 2012; Zhang et al., 2015).

Some key contributions to the literature were published in the 1990s. Herbert Girardet used, in 1992, linear versus circular metabolism as a heuristic device to take the first steps towards the establishment of the link between UM and cities' sustainable development (Kennedy et al., 2011). Two years later, Bohle (1994) emphasised the potential of UM approaches to examine urban food systems in developing countries. In 1996, Rees and Wackernagel (1996) used the concept of UM to characterise the ecological footprint of cities. Three years later, Newman (1999) extended the metabolism approach to include liveability and health issues.

The turn of the millennium witnessed a renewed interest in the study of UM that has been mainly driven by concerns over the capacity of the planet to sustain the projected growth in human population. As a result, a second wave of empirical studies based on more consistent approaches to UM has followed. Even though the primary focus of such approaches was on quantitative methods, some works have taken UM in a political science context or even in a qualitative historical context (Kennedy et al., 2011).

More recently, literature reviews on the topic of UM have been provided by Kennedy et al. (2011) and Zhang (2013). Despite the vast amount of research on UM that has been undertaken to date, some problems remain to be solved from a planning standpoint. As argued by Newell and Cousins (2014), the UM metaphor has not been yet infused with the same theoretical or methodological rigour in planning literature as it was in other branches of the scientific literature. Furthermore, the lack of an explicit and disaggregated spatial dimension in most UM models developed or implemented to date have been limiting the application of this concept in urban planning practice and, in particular, in the analysis of the metabolic impacts of major urban development projects and plans on the whole city or metropolis. The methodology introduced in the following section and applied to a real case study intends to overcome this fundamental shortcoming of UM approaches.

Indeed, this shortcoming is not exclusive of the literature on UM methodologies. It can also be found in the literature on methodologies to support the preparation of 'sustainability assessments' and 'strategic environmental assessments' (see, for instance, a comprehensive review carried out by Noble et al., 2012). This may not be a surprise if one bears in mind that these environmental policy instruments emerged within the natural and environmental sciences and related engineering fields and not within urban and spatial planning. As such, they tend to privilege sectorial approaches, modelling air and water pollution or solid wastes management disregarding the spatial dimension and the complexities of how urban environments really work.

Metabolic Impact Assessment (MIA): concept and challenges

Based on the understanding of how a city works from a metabolic perspective, it is possible to move on to attempt to further understand the metabolic consequences of urban development processes. These processes are constantly affecting the present and shaping the future of our cities and metropolis. They are made of millions of decisions. Most of these decisions are associated with small and localised interventions on the built environment with, often times, minor impacts at the city scale (not always indeed!). However, from time to time, major decisions are necessary, associated with large and strategic interventions with, potentially, far reaching consequences on the cities' overall metabolic performances. In these cases, it would be important to count on a methodological tool that could anticipate the metabolic impact of such major interventions. Such a methodological tool has been developed and it is called Metabolic Impact Assessment, MIA for short (Pinho et al., 2011; Pinho et al., 2013).

As the name suggests, MIA has been designed to assess the impact of a particular development proposal on the existing UM performance of a city, metropolis or city region. Given the amount of resources and expertise involved, MIA targets, in particular, urban development plans and detailed local plans as well as large urban projects and infrastructures.

The application of MIA, and the scale at which MIA is applied, depends on the previous availability of an adequate and updated UM model for the reference area (the city, the metropolis or the city-region). One of the fundamental requisites of that model is its capacity to incorporate a sufficiently disaggregated spatial dimension when describing the most important energy and materials stocks and flows which take place within the reference area, the so-called throughputs, and in its interactions with the outside world, the inputs and the outputs of the model. Needless to say, developing such a model for a city is not easy, and differs from the more elementary 'black box' models of UM, which have been developed in the past to characterise the simple balance of the inputs and outputs of energy and materials.

However, when such a model is available, the application of MIA produces far better results when compared to traditional methods and techniques applied in strategic environmental assessment (SEA), (see, for example, Pinho et al., 2011). Through the systematic analysis of the changes in, and interactions among, stocks and flows of energy and materials, the UM rationale behind MIA offers, in comparison to those traditional approaches, a far more comprehensive and integrated framework to understand and integrate the complex web of interdependencies between the natural and the built environments, generated by a proposed large urban intervention.

Figure 31.2 represents the MIA process. The process is divided into six stages:

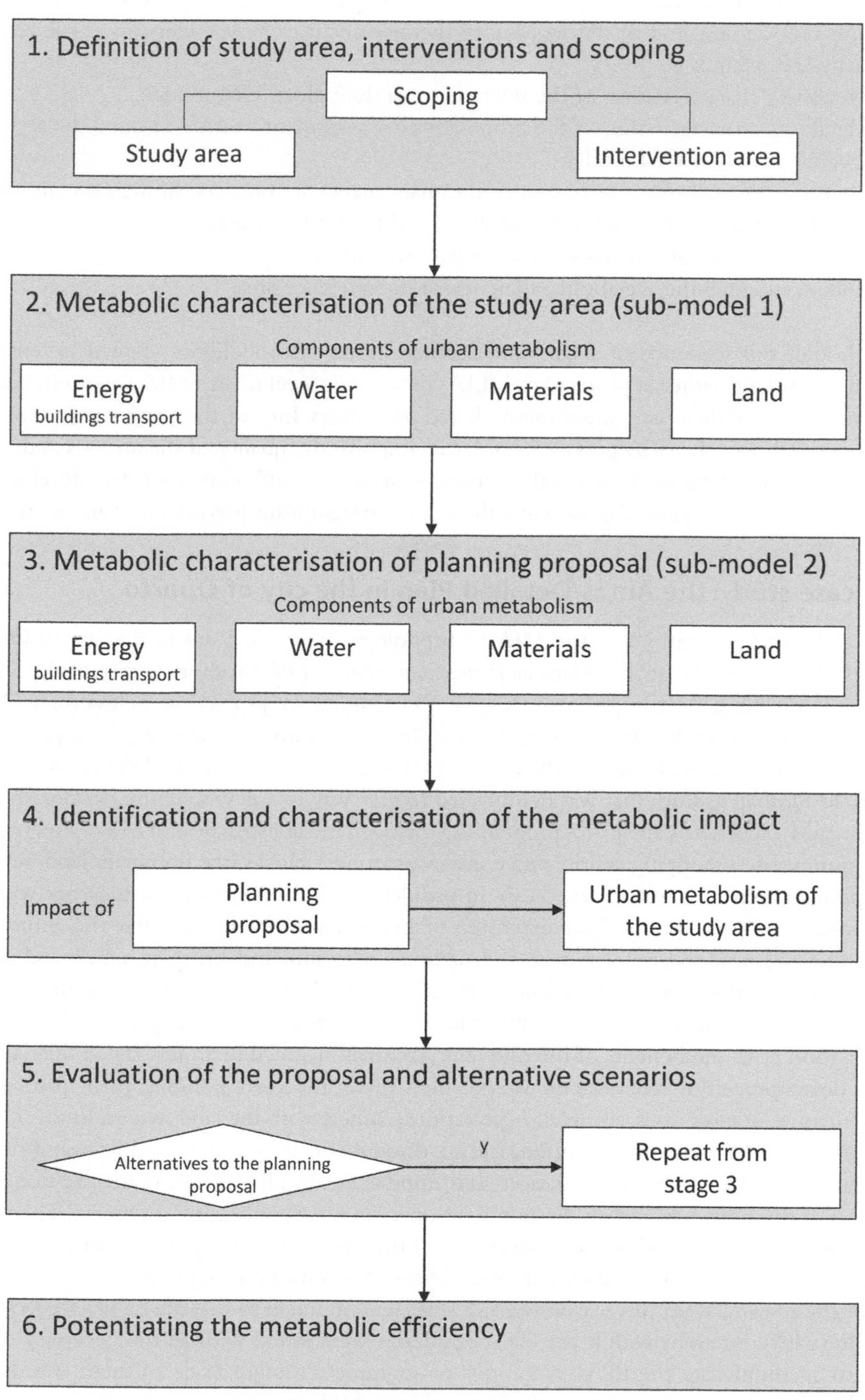

Figure 31.2 The six main stages of the MIA methodology

Source: adapted from Pinho et al. (2013, p. 188)

1 Definition of the reference (or study) area; of the proposed urban project (or urban plan) intervention area; and of the scoping of the assessment exercise (selection of the relevant analysis domains);
2 Metabolic characterisation of the reference area through an UM model;
3 Metabolic characterisation of the proposal under assessment bearing in mind the technical specificities of the UM model;
4 Identification and characterisation of the main metabolic impacts through the analysis of the likely interaction between the proposal and the reference area;
5 Evaluation of the proposal and of alternative scenarios;
6 Enhancement of the metabolic efficiency of the reference area.

Although this sequence of stages can resemble some methodologies applied in 'environmental assessment', either at project level (EIA) or strategic level (SEA), MIA is not intended to substitute either of these assessment methods and procedures. Instead, the aim is to introduce the MIA approach into the SEA process in order to improve the quality of the analysis and of the assessment results, particularly when these assessments are focused on major urban development projects and/or urban plans able to affect the whole or significant parts of cities and metropolis.

MIA case study: the Antas Detailed Plan in the city of Oporto

One of the first applications of the MIA methodology was carried out in the city of Oporto on an Urban Detailed Plan, the *Plano de Pormenor das Antas*, PPA for short, as part of the SUME project. For a detailed description of this 7th EU Framework project see Schremmer (2010). This plan, carried out by the Municipality of Oporto and now already largely implemented, covers an area of 41.3 ha located in the eastern part of the city, on and around the lands occupied by an old football stadium that was demolished to give way to a new stadium. Besides this new stadium and a sports hall, the plan's programme included the construction of a new street system to accommodate a shopping centre, office spaces, apartment blocks (the dominant land use), and new public green areas (see Figure 31.3). In parallel, the Metropolitan Junta together with the local metro company, approved the extension of an existing metro line to serve the entire area.

Besides the need to conform with the national and municipal urban planning and design regulations, the Portuguese legislation contains the so-called 'per equation' system, which is a regulatory mechanism intended to promote a fair redistribution among stakeholders of the urbanisation costs and benefits of the different proposals included in an integrated operation of urban development (or redevelopment). Needless to say, the corresponding participatory processes involve, in most cases, complex negotiations, namely with the landowners. In this respect, the case of the PPA was no exception, and so, throughout the planning and licensing process, the plan suffered a number of alterations and improvements. However, no one questioned the location of this fairly large urban development operation within the city of Oporto. The reason could have been the fact that the main driver of the operation was the construction of a new stadium expected to be located close to the old one it was meant to replace.

For the research team this reason was not sufficient, so much so that after a brief GIS search, two alternative locations within the city boundaries were found with all the necessary conditions to accommodate the PPA's ambitious programme, although none of them was directly served by the metro system (Figure 31.4). Could the MIA methodology provide a sound basis to decide, from an UM perspective, what would be the best alternative location for an urban development operation similar to the PPA? In other words, and independently of the individual merits of the different proposals contained in the PPA, would the chosen location be able to

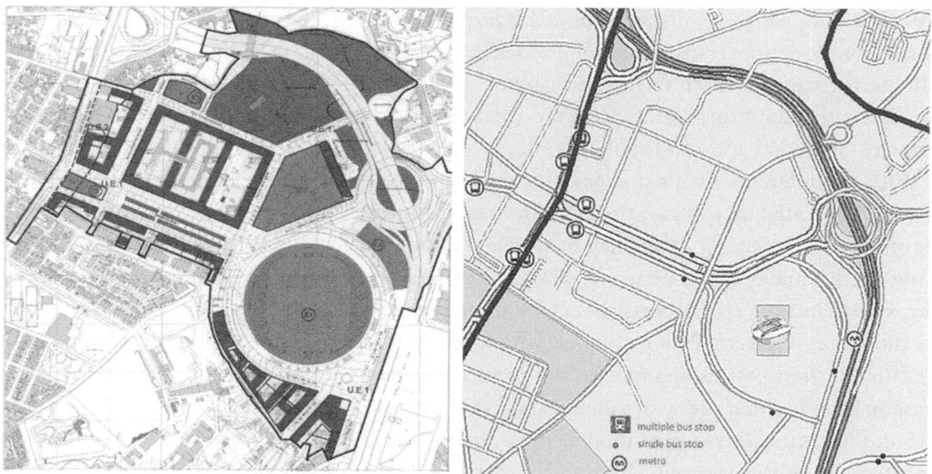

Figure 31.3 The urban detailed plan of Antas: groundfloor map on the left and street system and public transport systems on the right

Source: adapted from Pinho et al. (2011, pp. 32 and 34)

Figure 31.4 PPA (Antas) on the right and the two alternative locations, Prelada and Campo Alegre

Source: adapted from Pinho et al. (2011, p. 74)

potentiate or, at least, minimise the (metabolic) impact of the operation on the overall UM performance of the city?

To address this question and test the capabilities of the MIA methodology, an UM model for the city of Oporto, previously developed in the research centre during the first part of the SUME project, was revisited (Pinho et al., 2010). The model was constituted by a number of sub-models, namely buildings energy, transport energy, water and (construction) materials. Considering that the three alternative locations – Antas, Prelada and Campo Alegre – had fairly similar physical and urban infrastructural characteristics, and assuming that the functions and features of the new buildings and open spaces would observe the same programme and so would not change significantly from location to location, the scoping exercise selected the energy component of transport as the fundamental factor able in itself to differentiate the metabolic performance of each of the alternatives. To isolate the effect of the metro line, since the two alternative locations were not served by metro, the Antas location was analysed under two scenarios, with and without the metro line.

With the reference area made to coincide with the administrative boundaries of the city of Oporto, the initial option and the two alternative locations identified, and the scoping process electing transport energy as the key metabolic factor of analysis, a move on to the second stage of the methodology could be made. The UM general model devised included the well-known 'four steps' land use – transports model for the entire city (Figure 31.5) (for a critical review of this model see Cervero, 2006), coupled with a transport energy consumption matrix, including the different transport modes (public and private) operating in the city. Since the city of Oporto is not an island within the wider metropolitan area, the in and out traffic flows had to be considered and fed into the transport model. Subsequently, the different types of motorised flows were allocated to the street system through the classical Saturn model (see Van Vliet, 2010), which enabled the identification of the most congested streets at peak hours.

The stage three of MIA involved the forecasting, for the three alternative locations, of all new trips generated by the activities associated with the PPA proposals and their distribution by the different modes of transport. For this forecasting exercise, the reference metrics for trip generation by built square meter for each type of land use (for example, housing, shopping, offices and sports facilities) were taken into consideration. The corresponding modal split was calculated by analogy with other areas of the city with similar socio-economic characteristics and similar transport facilities.

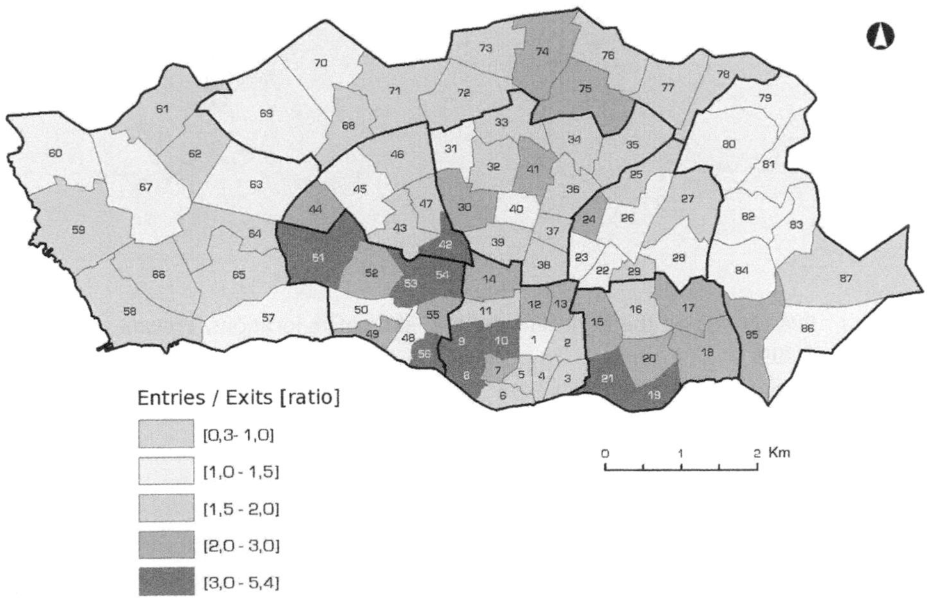

Figure 31.5 'Four steps' land use – transports model: step 1, trip generation in the city of Oporto: ratio between entries and exits between 7:30 and 9:30 am per area unit

Source: adapted from Pinho et al. (2011, p. 40)

In stage four, the extra volume of trips distributed according to the modal split calculated in the previous stage was plugged into the 'four steps' model for each alternative location under analysis. Through the transports energy matrix referred to previously it was possible to arrive at the expected metabolic impact of each alternative location on the city as a whole, expressed in energy units or, for an easier understanding, in the percentage increase of the total energy currently consumed by the city's different transport systems. The final results pointed to an overall increase in the annual energy consumed in transport in the city of Oporto of:

- +1.00% if the PPA stayed, as planned, in the Antas location but without metro line, and +0.46% with the metro line;
- +0.75% if the PPA moved to the Campo Alegre location;
- +0.60% if the PPA moved to the Prelada location.

Given the experimental nature of this exercise, stages five and six of the MIA methodology were not properly developed as they would need the active involvement of the different stakeholders and the public. Actually, in a real case, and given the obvious technical complexities of this methodology, both stakeholders and the public must be involved right from the beginning of the process.

Nevertheless, the final results seemed to point out that, disregarding the metro effect, the two hypothetical alternative locations for the PPA, Campo Alegre and Prelada in particular, would be able to perform better than the actual chosen location at Antas. However, with the metro line, which actually corresponds to the existing conditions, the metabolic performance of the Antas location comes out as, by far, the best alternative.

This application of MIA reinforced the local consensual view that the introduction of the metro system in the Oporto metropolitan area significantly changed local urban mobility patterns and modal choices, which brought about very positive impacts on the overall metabolic performance of this city and metropolitan area.

Conclusion

Over the last couple of years urban metabolism evolved from an organicist and functionalist formulation into a richer and more critical conceptualisation able to encompass the complex web of relationships between social and biogeophysical processes, which take place in cities and metropolitan areas and are, all together, responsible for their form, structure, dynamics and endless change.

This wider vision of the urban metabolism concept offers a coherent rationale to develop the Metabolic Impact Assessment (MIA) concept as an innovative decision support tool geared towards the analysis and understanding of the complex metabolic consequences associated with the implementation of new urban projects and plans in a particular city or metropolitan area. In such cases, the MIA approach attempts to go beyond current methods applied in 'strategic environmental assessment' and in 'sustainability assessment' by providing a comprehensive view of energy use changes (and greenhouse gas emissions).

The MIA methodology introduces the spatial dimension into the urban metabolism field, an aspect that is crucial to urban planning, and is seldom seen in the urban metabolism modelling literature. MIA is able to deal with and generate alternative projects and plans, in terms of nature, size and location (as seen in the Oporto case study) that better fit the specificities of the metabolic performances of cities and metropolitan areas. In this way, MIA is not just another assessment tool but also, and foremost, a design tool able to contribute to the final shaping of an urban project or plan.

Although the application of MIA to an urban development plan (the PPA) in the city of Oporto was brief and partial, it illustrates how applying the method adds a spatially explicit metabolic dimension to environmental impact assessments. Although the method is technologically complex and resource demanding, we expect further research can reduce this complexity and enable a more generalised use of MIA in municipal or metropolitan planning practice.

References

Bancheva, S. (2014). *Integrating the concept of urban metabolism into planning of sustainable cities*. Working Paper No. 168. London: Bartlett Development Planning Unit, UCL.

Barles, S. (2010). 'Society, energy and materials: the contribution of urban metabolism studies to sustainable urban development issues'. *Journal of Environmental Planning and Management*, 53(4): 439–55.

Bohle, H.-G. (1994). 'Metropolitan food systems in developing countries: the perspective of "urban metabolism"'. *GeoJournal*, 34(3): 245–51.

Castán-Broto, V., Allen, A. and Rapoport, E. (2012). 'Interdisciplinary perspectives on urban metabolism'. *Journal of Industrial Ecology*, 16(6): 851–61.

Cervero, R. (2006). 'Alternative approaches to modeling the travel-demand impacts of smart growth'. *Journal of the American Planning Association*, 72(3): 285–95.

Chen, S. and Chen, B. (2012). 'Network environ perspective for urban metabolism and carbon emissions: a case study of Vienna, Austria'. *Environmental and Science and Technology*, 46(8): 4498–506.

Chen, S. and Chen, B. (2015). 'Sustainable urban metabolism'. *Encyclopedia of Environmental Management*: 1–8.

Clift, R., Druckman, A., Christie, I., Kennedy, C. and Keirstead, J. (2015). *Urban metabolism: a review in the UK context*. Future of Cities: Working Paper. London: Government Office for Science.

Costa, A., Marchettini, N. and Facchini, A. (2004). 'Developing the urban metabolism approach into a new urban metabolic model'. *WIT Transactions on Ecology and the Environment*, 72: 31–40.

Davoudi, S. and Sturzaker, J. (2017). 'Urban form, policy packaging and sustainable urban metabolism'. *Resources, Conservation and Recycling*, 120: 55–64.

Dinarès, M. (2014). 'Urban metabolism: a review of recent literature on the subject'. *Documents d'Anàlisi Geogràfica*, 60(3): 551–71.

Fernandes, R. and Pinho, P. (2017). *Urban metabolism and planning: existing gaps and future research*. Paper presented at the AESOP Conference 'Spaces of Dialog for Places of Dignity', Instituto Superior Técnico, Lisbon.

Gandy, M. (2004). 'Rethinking urban metabolism: water, space and the modern city'. *City*, 8(3): 363–79.

Golubiewski, N. (2012). 'Is there a metabolism of an urban ecosystem? An ecological critique'. *Ambio*, 41(7): 751–64.

Holmes, T. and Pincetl, S. (2012). *Urban metabolism literature review*. Edited by the Institute of the Environment Center for Sustainable Urban Systems. Los Angeles, CA: UCLA.

Kennedy, C. and Hoornweg, D. (2012). 'Mainstreaming urban metabolism'. *Journal of Industrial Ecology*, 16(6): 780–2.

Kennedy, C., Cuddihy, J. and Engel-Yan, J. (2007). 'The changing metabolism of cities'. *Journal of Industrial Ecology*, 11(2): 43–59.

Kennedy, C., Pincetl, S. and Bunje, P. (2011). 'The study of urban metabolism and its applications to urban planning and design'. *Environmental Pollution*, 159(8–9): 1–8.

Liu, G., Yang, Z. and Chen, B. (2011). 'Comparison of different urban development strategy options to the urban metabolism optimal path'. *Procedia Environmental Sciences*, 5: 178–83.

Mostafavi, N., Farzinmoghadam, M., Hoque, S. and Weil, B. (2014). 'Integrated urban metabolism analysis tool (IUMAT)'. *Urban Policy and Research*, 32(1): 53–69.

Newell, J. P. and Cousins, J. J. (2014). 'The boundaries of urban metabolism: towards a political – industrial ecology'. *Progress in Human Geography*, 39(6): 1–27.

Newman, P. W. G. (1999). 'Sustainability and cities: extending the metabolism model'. *Landscape and Urban Planning*, 44: 219–26.

Noble, B. F., Gunn, J. and Martin, J. (2012). 'Survey of current methods and guidance for strategic environmental assessment'. *Impact Assessment and Project Appraisal*, 30(3): 139–47.

Pickett, S. T. A., Cadenasso, M. L., Grove, J. M., Groffman, P. M., Band, L. E., Boone, C. G., Burch, W. R., Grimmond, C. S. B., Hom, J., Jenkins, J. C., Law, N. L., Nilon, C. H., Pouyat, R. V., Szlavecz, K., Warren, P. S. and Wilson, M. A. (2008). 'Beyond urban legends: an emerging framework of urban ecology, as illustrated by the Baltimore ecosystem study'. *BioScience*, 58(2): 139–50.

Pincetl, S., Bunje, P. and Holmes, T. (2012). 'An expanded urban metabolism method: towards a systems approach for assessing urban energy processes and causes'. *Landscape and Urban Planning*, 107: 193–202.

Pinho, P., Oliveira, V., Cruz, S., Barbosa, M. and Silva, M. (2010). *Manual of evaluation methodologies*. Unpublished SUME Report D 3.1, OIR/CITTA, Vienna.

Pinho, P., Oliveira, V., Cruz, S., Barbosa, M. and Silva, M. (2011). *Good practice guide on assessing impacts of urban structures on urban metabolism*. Unpublished SUME Report D 3.2, OIR/CITTA, Vienna.

Pinho, P., Oliveira, V., Cruz, S. and Barbosa, M. (2013). 'Metabólic impact assessment for urban planning'. *Journal of Environmental Planning and Management*, 56(2): 178–93.

Rapoport, E. (2011). *Interdisciplinary perspectives on urban metabolism*. Working Paper. London: UCL Environmental Institute.

Rees, W. and Wackernagel, M. (1996). 'Urban ecological footprints: why cities cannot be sustainable – and why they are a key to sustainability'. *Environmental Impact Assessment Review*, 16(4–6): 223–48.

Schremmer, C. (2010). 'SUME – sustainable urban metabolism for Europe', in P. Pinho and V. Oliveira (eds.) *Planning in times of uncertainty*. CITTA – FEUP Edições, pp 33–48.

Van Vliet, D. (2010). *SATURN v10.9, Saturn manual & appendices*. Leeds: The Institute for Transport Studies, The University of Leeds.

Weisz, H. and Steinberger, J. K. (2010). 'Reducing energy and material flows in cities'. *Current Opinion in Environmental Sustainability*, 2: 185–92.

Wolman, A. (1965). 'The metabolism of cities'. *Scientific American*, 213(3): 179–90.

Zhang, Y. (2013). 'Urban metabolism: a review of research methodologies'. *Environmental Pollution*, 178: 463–73.

Zhang, Y., Yang, Z. and Yu, X. (2015). 'Urban metabolism: a review of current knowledge and directions for future study'. *Environmental and Science Technology*, 49(19): 11247–63.

32

Application of Life Cycle Assessment in buildings

An overview of theoretical and practical information

Carlo Ingrao, Antonio Messineo, Riccardo Beltramo, Tan Yigitcanlar and Giuseppe Ioppolo

Introduction

The urban system is a territorial ecological system being strongly affected by human activities and needs (Ioppolo et al., 2014). Urban areas offer a series of benefits to citizens because they allow them to: enjoy living spaces and services; improve their quality of life, mainly thanks to greater proximity to health facilities and better access to job markets; and take advantage of economies of scale (Ioppolo et al., 2016). In this context, the built environment is a material, spatial and cultural product of human labour that combines physical elements and energy to shape living, working and playing spaces for people. It represents the human surroundings that provide the setting for human activity and so encompasses places and spaces that were created or modified by people including buildings, parks and transportation systems (Roof and Oleru, 2008). The built environment provides people with most direct, frequent and unavoidable images and experiences of everyday life. It plays multiple key roles in the society of today and can be considered the result of many socio-economic factors and processes that are central to achieving the aims of sustainable development (Forsberg and von Malmborg, 2004). Sustainable development, in turn, can be described as enhancing the quality of life, thus allowing people to live in a healthy environment and improve social, economic and environmental conditions for present and future generations (Ortiz et al., 2009; Ioppolo et al., 2016).

However, built environments of today are created by utilising large amounts of energy and materials, affecting the health of humans and the natural environment in negative ways (Forsberg and von Malmborg, 2004). These negative impacts of the built environment are amplified by the expansion of urban areas to meet the needs and requirements of an increasing number of people. Current trends predict that the number of people living in urban areas will keep rising, reaching almost 5 billion by 2030 out of an 8.1 billion world population (Castanheira and Bragança, 2014).

There are increasing global concerns related to climate change and overall sustainability of the built environment, putting urban development on the top of the agenda in almost every city across the world (Dizdaroglu and Yigitcanlar, 2014; Yigitcanlar and Dizdaroglu, 2015). The built environment is, indeed, the biggest contributor to the emissions of greenhouse gases (GHGs), accounting for up to 50% of global carbon dioxide emissions (Monkiz Khasreen et al., 2009; Arbolino et al., 2017). Buildings contribute some of the largest impacts because of huge consumption volumes of materials, energy and fuels (Monkiz Khasreen et al., 2009; Ingrao et al., 2016).

Energy consumption associated with the life cycle of buildings accounts for around 40% of the global demand and can be split into the embodied and operational energy. In Dixit's et al. (2012), embodied energy is defined as the energy associated with the construction of a building and building materials during all processes of production, on-site construction, use and maintenance, and final demolition and disposal. The operational energy, on the other hand, is the energy spent to maintain the inside environment of a building through various processes and activities, such as heating and cooling, lighting and operating buildin%g appliances (Dixit et al., 2012). Literature shows that the operational energy can vary from 70%–90% during the whole life cycle energy consumption of a building (Asdrubali et al., 2013), while the embodied energy generally ranges 10%–30% based upon the building system considered.

The embodied energy and environmental impacts associated with the building sector have been raising serious concerns, as well as generating increasing attention from the scientific community at the international level. To obtain quality and environmentally sound buildings, the application of Life Cycle Thinking (LCT) for both development and improvement of their life cycles is essential. According to the LCT approach, the consequences that the current systems of the construction of buildings and infrastructure have on their quality, safety and sustainability are considered and evaluated during the design stage. Additionally, application of LCT allows us to understand how each downstream phase is affected by the technological choices that are made for implementation and development of the upstream or previous phase, and how these can be addressed and improved, if needed. Such an approach represents the foundation for assessing the improvements that can be made all along the life cycle of a building for its enhanced quality and sustainability.

The application of the LCT approach could consider, from the early stage of the study, the regulations of the Urban General Plans (UGP). This is useful when UGPs establish specific criteria for construction related to the characteristics of the subareas of the inner city, new residential areas, industrial or commercial areas, agricultural areas, or environmentally valuable areas. Such regulations sometimes not only set design requirements for buildings but also provide guidelines on material and techniques to be used, with different levels of requirements (Beltramo, 2017).

The significance of energy consumption and GHG emissions in the building operation stage is a global concern (Cellura et al., 2014), and is addressed by diverse research groups worldwide with the final objective of finding room for improvements. It is well known by now that buildings require energy over their lifespan and, so, comprehensive environmental assessments should consider energy consumption, natural and primary energy resource exploitation and pollutant emissions, from a life cycle perspective (Cellura et al., 2014). Improving the performance of buildings in the operational phase should remain as a primary goal in the design stage, but this could increase the embodied energy share. Some measures to reduce the operating energy demand are: improving the overall thermal insulation performance of the building envelope, operating upon both the opaque and transparent areas; and installing electric appliances, lighting, water-heating, cooking, and heating and cooling plants with high energy efficiency rates. However, those measures could cause an increase in the embodied energy of the building due to the life cycle of involved raw materials, auxiliaries and equipment, and to the related installations (Cellura et al., 2014).

According to Tola (2014), sustainable construction can be achieved through application and proper combination of the following solutions: utilisation of renewable energies; saving of primary energy resources; definition of novel construction technologies; usage of eco-compatible building material commodities; and installation of heating and cooling systems that are well integrated into, and form a whole high energy performing system within the building. In this context, Life Cycle Assessment (LCA) is globally recognised as a valid and relevant method to make complete and reliable evaluations and considerations on the energy and environmental behaviour of buildings throughout their life cycles (Cellura et al., 2014).

In buildings, LCA can serve as a tool for evaluation of important aspects like: embodied and operational energy; production of building materials; logistics involved in supply of those materials to construction yards; consumption of water; emissions of GHGs and other pollutants; and generation of wastes and related management and treatment systems (Ingrao et al., 2016). It also enables ways to investigate how those aspects can be interconnected, as well as solutions to improve the entire building supply chain. LCA can also be used to compare materials and construction techniques; to show how alternative solutions could make real improvements to the system; and to conduct sensitivity analyzes.

There is increasing literature to deepen the knowledge and enrich the current literature on the energy, CO_2 emission related, and environmental issues in buildings and building components – such as, Ramesh et al. (2010), Sharma et al. (2011), Cabeza et al. (2014) and Chau et al. (2015).

This chapter builds upon these studies and other related studies and aims to provide theoretical and practical information regarding the application of LCA for the estimation of energy and environmental issues associated with the life cycle of buildings. In this way, the chapter synthesises up-to-date information on these topics for building designers and LCA practitioners, as well as for graduate and undergraduate students.

Theoretical and practical information on the application of Life Cycle Assessment in buildings

LCA is acknowledged worldwide to be a technique for assessing the environmental aspects and potential impacts associated with a product's life cycle by compiling an inventory of the most relevant inputs and outputs; evaluating the potential environmental impacts associated with those input and output flows; and interpreting the results of the inventory and impact phases in relation to the objectives of the study. It is based upon a clear methodology that is articulated in the following phases: goal and scope definition; Life Cycle Inventory (LCI); Life Cycle Impact Assessment (LCIA); and life cycle interpretation. Elaboration of LCA is not a linear process, starting with the first phase and ending with the last one; instead, it follows an iterative procedure which leads to increasing levels of detail (Udo de Haes, 2002).

This section contains theoretical and practical aspects related to application of LCA in buildings, based upon requirements from ISO 14040:2006 and ISO 14044:2006 (ISO, 2006a, 2006b) and EN 15978:2011 (EN, 2011) to make clear the importance of LCA as an advanced environmental assessment tool. It specifies the calculation method for assessment of the environmental performance of a building and provides the means for the reporting and communication of the results obtained (EN, 2011).

Goal and scope definition

The first phase is about defining the main objectives of the study and the background that warrants LCA, such as a gap in the literature or the innovation aspects that are associated with a given product, technology, or system. Also, the way the study will be used is determined during this phase, namely whether it will contribute to an ecolabel and so will become public or will remain private (López-Mesa et al., 2009). This stage identifies the level of detail required for the application at hand, and establishes a procedure aimed at ensuring the overall quality of the study (Udo de Haes, 2002). Furthermore, an important part of this phase is the definition of the Functional Unit (FU) and system boundaries consistent with the aim and scope of the study that best represent the system under investigation to facilitate data collection and to enable comparisons with similar products and systems. The FU is a key element in any LCA and measures the unit of product that is going to be studied. It also provides the reference point to which the inputs and outputs are linked to the resulting environmental impacts (Klöppfer and Grahl, 2014).

System boundaries include the phases of production of the resources, materials, and energies that are utilised within the system, as well as the processes and stages through which the system is developed and modelled. The system boundaries include all the unit operations which constitute the system in order to facilitate its modelling and the development of the next phases of Life Cycle Inventory (LCI) and Life Cycle Impact Assessment (LCIA) (Klöppfer and Grahl, 2014). When modelling complex systems like buildings or civil infrastructures, the system boundaries are usually defined based upon the organisational, managerial and technological aspects related to the supporting firm or agency. This is done to best understand all the processes and stages of the system, and the way the input and output inventories, as well as the environmental impacts, should be quantified and analysed (Clasadonte et al., 2013).

FUs and system boundaries are usable for assessing buildings and building materials, depending upon the surrounding conditions and the system investigated, whether it is a material, a product, an element, a portion of the building or the whole building frame. In the latter case, the FU could be 1 m^2, or 1 m^3 of finished building, or the building in its entirety, with specific dimensions and construction features. Many LCAs have been developed thus far. By scanning the literature, it is clear that the choice of both FUs and system boundaries is an important aspect in any LCA and needs to be conducted properly. However, the literature also makes clear that LCAs differ in the choice of FUs and system boundaries, even for the same type of product and system. This emphasises the absence of agreement on the system FU and border to be considered to facilitate comparisons between similar products and systems. It also makes clear the need for more careful selection of FUs and system boundaries to avoid affecting negatively the findings of a study and, thus, its scientific value and reliability (Ingrao et al., 2015).

Another step that should be developed as part of this phase is defining the service life of both the building and the overall building operation system that is used within it, as this influences their maintenance and the energy consumption associated with its operation. The next section provides a brief discussion of each phase within the building's life cycle boundary.

Building's life cycle phases: a brief overview

CONSTRUCTION OF THE BUILDING AND OF THE ENTIRE BUILDING OPERATING SYSTEM

In this phase, types and amounts of building materials are defined and estimated as the result of the design phase. As part of this modelling phase, the inclusion of all Building Operating Systems

(BOSs) is essential for correctly computing the input and output inventories and the resulting environmental impacts associated with the life cycle of the building. Those systems are for heating and cooling, sanitary-hot-water production and lighting, as well as for water and electricity supply. Their construction is realised through the assembly of semi-finished products and components that are manufactured using raw materials produced from resources extraction.

The installation of the building material commodities, as well as of the building systems, is modelled considering the usage of auxiliaries and building equipment, as well as their consumption of fuels and energies. The building-BOS construction border is depicted in Figure 32.1.

BUILDING MANAGEMENT

This phase estimates the environmental impacts resulting from maintaining the building over its lifespan. In this regard, based upon building service life, maintenance cycles are defined for building material commodities and building systems to make sure that the overall quality performances established in the design phase are attained. In this phase, it is necessary to identify the materials, products and components to be replaced when their service life expires. For each of these components, the environmental flows associated with their production and manufacturing, transport to the building site and installation should be accounted for.

Finally, this phase provides the definition and assessment of the disposal scenarios for the replaced materials and components.

OPERATION

In this phase, the environmental impacts associated with the consumption of operational energy and water are estimated, together with the impacts of the indoor heating and cooling systems and the sanitary-hot-water production. This phase is expressed as kWh/m^{2}year and is modelled through the calculation of annual primary energy demand for indoor heating and cooling, lighting and hot water production for internal sanitary use. However, detailing the related calculation goes beyond the scope of this chapter, and so the reader is referred to the subject literature for specific details.

END-OF-LIFE OF THE BUILDING MATERIALS AND PRODUCTS AND OF THE BUILDING OPERATION SYSTEM

This phase is about the end-of-life scenarios to be considered for the building materials and components that are utilised for construction of both the building and its building operating systems. The related assessment is strictly dependent upon the usage and installation of those materials and components, as well as upon their connections with other materials and products. Therefore, this phase is generally modelled as the end-of-life of the entire building and of the building operating systems based upon their relevant service lifespans.

This phase is modelled through computation of resources, energies and fuels, as well as of auxiliaries and equipment for building dismantling. In addition, transport of the resulting wastes to the treatment plants and treatment of those wastes are accounted for in the assessment as part of building end-of-life.

LIFE CYCLE INVENTORY

LCI involves the compilation and quantification of both inputs and outputs and includes data collection and calculation. This is recognised as the most challenging part of preparing an LCA

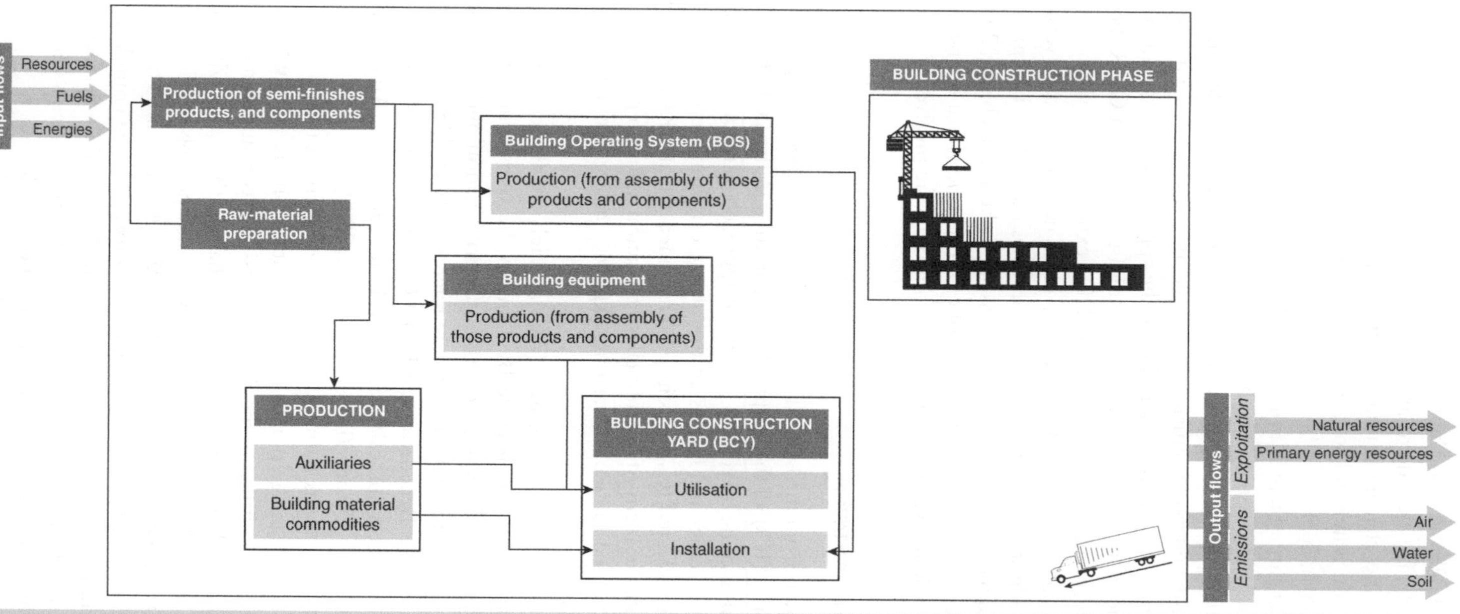

Legenda

The building material commodities are those installed within the building with different functions (i.e. load bearing, filling, insulating, cladding, covering, finishing, and so on) and, actually, form the building.

The auxiliaries are those materials, products, systems and equipment that are used for construction of the building but do not remain installed in it all along its service life

Both of them are strictly dependent upon the definitive metric computation and the construction techniques established in the design phase. For this reason, they have been just defined here and sort of categorised, but not reported in details, as there exist huge variability in them.

The Building Operating System (BOS) encompassed all systems for heating and cooling, sanitary-hot-water production, lighting, as well as for water and electricity supply

All interconnections represent both input/out process and transport flows for material supply.

Figure 32.1 Building construction border

Source: Authors' own

(Ingrao and Siracusa, 2018). In buildings, all data inventoried are related to the system FU and its border/boundaries and are then allocated to the building's life cycle phases for elaboration and assessment (Udo de Haes, 2002; De Benedetto and Klemeš, 2009). Inputs are represented by the resources, materials and energies that are consumed within the system investigated, while outputs are material emissions in air, water and soil, as well as the use of natural and primary energy resources (Ingrao and Siracusa, 2018).

Site-specific data or primary data are obtained from the building design phase and represent the material and energy flows through the system. The building material commodities and plants as well as the auxiliaries that are used for building construction are modelled by collecting primary data related to their production and manufacturing and whole life cycle. The data quality is crucial and, ideally, all the data required should be obtained from the appropriate sources. In practice, this is seldom possible and, therefore, generic data are used, at least for part of LCI (Robertson, 2013). Primary data are, indeed, combined with background data extrapolated from databases of acknowledged scientific value and relevance, like Ecoinvent in its most up-to-date version (Ecoinvent 3, 2018), and properly adjusted if needed. Ecoinvent provides most of the background materials and processes often required in LCA case-studies (Frischknecht and Rebitzer, 2005). Difficulties are sometimes experienced in obtaining quality data for each individual operational unit (Robertson, 2013), especially when several manufacturing companies are involved.

LIFE CYCLE IMPACT ASSESSMENT AND INTERPRETATION

This phase is carried out using results from the LCI. The assessment is performed by aggregating in a limited set of Impact Categories (ICs) all the output flows quantified in the LCI phase (De Benedetto and Klemeš, 2009). This is done based on a mid-point approach using predetermined classification and characterisation schemes and factors. In the 'characterisation' step, the contribution of each input and output to selected ICs are assigned (De Benedetto and Klemeš, 2009). Mid-point results are expressed in the form of specific characterisation values represented by equivalent indicators such as, $kgCO_2$ eq for 'global warming', $kgPM_{2.5}$ eq for 'respiratory inorganics' and kgC_2H_3Cl eq for 'carcinogens'.

ICs can, then, be grouped into Damage Categories (DCs), which are the environmental aspects negatively impacted due to a product's life cycle. By doing so, the assessment is extended to the end-point approach, which means including the phases of 'normalisation' and 'weighting' in the assessment. Normalisation is done to establish common references to enable comparisons of different environmental impacts (De Benedetto and Klemeš, 2009). Normalisation relates each environmental effect to an average value calculated on a European, North American or Asian scale, obtaining dimensionless effect scores (López-Mesa et al., 2009).

Weighting is developed to estimate results by means of equivalent numerical parameters to express them as 'weighting points' or 'damage points' or 'eco-points' or, more simply, 'points'. Weighting points can be obtained by multiplying the dimensionless results from normalisation by weighting factors, that are calculated using different methods having their own advantages and disadvantages (López-Mesa et al., 2009). This allows ranking and possibly to define the importance of different results (De Benedetto and Klemeš, 2009). Weighting enables highlighting processes and outputs with the greatest impacts that represent priorities to be considered in the search for, and assessment of, improvement potentials. It is particularly useful when assessing improvements or conducting sensitivity analyses for comparison of different systems because it enables global evaluation and rating of those systems on the same point scale.

Finally, interpretation is a systematic procedure to evaluate the information from the inventory analysis and impact assessment of the product system and to draw conclusions from all the foregoing results of the study (De Benedetto and Klemeš, 2009). In this phase, as part of the drawing of conclusions, improvements are identified and assessed for enhanced energy efficiency and sustainability of buildings.

Discussion and conclusion

This chapter aims to address the relevance and the main methodological aspects associated with the application of LCA to buildings. The procedural method discussed above leads to performing a total LCA (LCA_{tot}) which results from summing up contributions from:

- Construction and assembly of both building and BOS, starting from the production of the related component materials and products;
- End-of-life of the whole building structure and the whole BOS, which includes the disposal of the aforementioned materials, components and products;
- LCA associated with management of the building, including the BOSs;
- Phases of using both the building and the whole BOS, which includes computation of the consumption of water and energy.

The chapter highlights the importance and usefulness of life cycle analysis as applied to buildings, the need for minimising resource and energy consumption, as well as buildings' environmental impacts. These objectives can be achieved by making sure that through the measures and actions discussed earlier, LCA_{tot} is the minimum possible. This highlights the iterative aspect of the LCA process, according to Udo de Haes (2002), that enables greater and greater analysis of details, quality and sustainability.

In conclusion, through this iterative method, LCA can support the search for and the identification of innovation pathways in buildings, through the adoption of technological solutions that limit operational energy consumption, and related economic costs. It also provides the basis for guidelines on the evaluation and certification of buildings, as well as providing operators in the sector with a better understanding of the management costs and the direct and indirect benefits associated with sustainable buildings. Furthermore, it could motivate the production and manufacturing industry connected with buildings to improve quality, innovation and sustainability in their sector.

Through its application to buildings, LCA can support the creation of initiatives for environmental planning in urban and rural areas. As part of this, LCA results could be used to implement and promote Environmental Product Declarations (EPDs) for buildings as a whole in all their complexity, which could stimulate greater sensitivity and knowledge from stakeholders such as designers, producers, users and policy and decision makers on the life cycle environmental issues associated with buildings. The EN 15804:2014 norm defines the category rules for LCAs of all types of components and services for buildings (EN, 2014). However, when it comes to application of LCA at the whole building scale, limiting the FU to a material or component could lead to incomplete consideration of the functions that a building is designed to provide. Decisions based on isolated LCA for materials or components might lead to erroneous conclusions. Therefore, the value of an EPD can only be assessed properly in the context of the performance of a whole building or construction work (Rivela and Neila, 2013).

This chapter provided theoretical information on the lifecycle assessment of buildings and discussed several important methodological issues. As indicated at the beginning of the chapter, the built environment is the largest contributor to carbon emissions, and this sector will be crucial to reduce the unavoidable impacts of climate change. Understanding LCA methodology will be increasingly important for practitioners, designers and producers in this sector, as well as for graduate students entering related jobs and research.

References

Arbolino, R., Carlucci, F., Cirá, A., Ioppolo, G. and Yigitcanlar, T. (2017). 'Efficiency of the EU regulation on greenhouse gas emissions in Italy: the hierarchical cluster analysis approach'. *Ecological Indicators*, 81: 115–23.

Asdrubali, F., Baldassari, C. and Fthenakis, V. (2013). 'Life cycle analysis in the construction sector: guiding the optimization of conventional Italian buildings'. *Energy and Buildings*, 64: 73–89.

Beltramo, R. (2017). 'Environmental and landscape management system', in S. Massari, G. Sonnemann and F. Balkau (eds.) *Life cycle approaches to sustainable regional development* Abingdon: Routledge/Taylor & Francis Group, pp 149–55.

Cabeza, L. F., Rincón, L., Vilariño, V., Pérez, G. and Castell, A. (2014). 'Life cycle assessment (LCA) and life cycle energy analysis (LCEA) of buildings and the building sector: a review'. *Renewable and Sustainable Energy Reviews*, 29(C): 394–416.

Castanheira, G. and Bragança, L. (2014). 'The evolution of the sustainability assessment tool SBTool^{PT}: from buildings to the built environment'. *The Scientific World Journal*, 2014: 1–10.

Cellura, M., Fontana, M., Longo, S., Milone, D. and Mistretta, M. (2014). 'Energy and environmental assessment of retrofit actions on a residential building', in R. Salomone and G. Saija (eds.) *Pathways to environmental sustainability – methodologies and experiences* (1st edn). Cham: Springer International Publishing, pp 127–35.

Chau, C. K., Leung, T. M. and Ng, W. Y. (2015). 'A review on life cycle assessment, life cycle energy assessment and life cycle carbon emissions assessment on buildings'. *Applied Energy*, 143(1): 395–413.

Clasadonte, M. T., Lo Giudice, A. and Matarazzo, A. (2013). 'Guidelines for environmental labels in the agri-food SMEs', in R. Salomone, M. T. Clasadonte, M. Proto and A. Raggi (eds.) *Product-oriented environmental management systems (POEMS)*. Dordrecht: Springer Science+Business Media, pp 203–43.

De Benedetto, L. and Klemeš, J. (2009). 'The environmental performance strategy map: an integrated LCA approach to support the strategic decision-making process'. *Journal of Cleaner Production*, 17(10): 900–6.

Dixit, M. K., Fernández-Solís, J. L., Lavy, S. and Culp, C. H. (2012). 'Need for an embodied energy measurement protocol for buildings: a review paper'. *Renewable & Sustainable Energy Reviews*, 16: 3730–43.

Dizdaroglu, D. and Yigitcanlar, T. (2014). 'A parcel-scale assessment tool to measure sustainability through urban ecosystem components: the MUSIX model'. *Ecological Indicators*, 41: 115–30.

Ecoinvent (2018). *Ecoinvent v.3*. The Swiss Centre for Life Cycle Inventories.

EN (European Norm) (2011). 15978. *Sustainability of construction works – assessment of environmental performance of buildings – calculation method*. European Committee for Standardization (CEN).

EN (European Norm) (2014). 15804. *Sustainability of construction works – environmental product declarations – core rules for the product category of construction products*. European Committee for Standardization (CEN).

Forsberg, A. and von Malmborg, F. (2004). 'Tools for environmental assessment of the build environment'. *Building and Environment*, 39(2): 223–8.

Frischknecht, R. and Rebitzer, G. (2005). 'The ecoinvent database system: a comprehensive web-based LCA database'. *Journal of Cleaner Production*, 13(13): 1337–43.

Ingrao, C., Lo Giudice, A., Bacenetti, J., Tricase, C., Dotelli, G., Fiala, M., Siracusa, V. and Mbohwa, C. (2015). 'Energy and environmental assessment of industrial hemp for building applications: a review'. *Renewable and Sustainable Energy Reviews*, 51: 29–42.

Ingrao, C., Scrucca, F., Tricase, C. and Asdrubali, F. (2016). 'A comparative life cycle assessment of external wall-compositions for cleaner construction solutions in buildings'. *Journal of Cleaner Production*, 124: 283–98.

Ingrao, C. and Siracusa, V. (2018). 'Quality – and sustainability – related issues associated with biopolymers for food packaging applications: a comprehensive review', in N. G. Shimpi (ed.) *Biodegradable and biocompatible polymer composites – processing, properties and applications*. Duxford: Woodhead Publishing, Elsevier Ltd, pp 401–18.

Ioppolo, G., Heijungs, R., Cucurachi, S., Salomone, R. and Kleijn, R. (2014). 'Urban metabolism: many open questions for future answers', in R. Salomone and G. Saija (eds.) *Pathways to environmental sustainability – methodologies and experiences* (1st edn). Cham: Springer International Publishing, pp 23–32.

Ioppolo, G., Cucurachi, S., Salomone, R., Saija, G. and Shi, L. (2016). 'Sustainable local development and environmental governance: a strategic planning experience'. *Sustainability*, 8(2): 180–96.

ISO (International Organization for Standardization) (2006a). *14040 – environmental management – life cycle assessment – principles and framework*. Geneva: ISO.

ISO (International Organization for Standardization) (2006b). *14044 – environmental management – life cycle assessment – requirements and guidelines*. Geneva: ISO.

Klöppfer, W. and Grahl, B. (2014). *Life cycle assessment (LCA): a guide to best practice*. Weinheim: Wiley-VCH Verlag GmbH & Co. KGaA.

López-Mesa, B., Pitarch, Á., Tomás, A. and Gallego, T. (2009). 'Comparison of environmental impacts of building structures with in situ cast floors and with precast concrete floors'. *Building and Environment*, 44(4): 699–712.

Monkiz Khasreen, M., Banfill, P. F. G. and Menzies, G. F. (2009). 'Life-cycle assessment and the environmental impact of buildings: a review'. *Sustainability*, 1(3): 674–701.

Ortiz, O., Castells, F. and Sonnemann, G. (2009). 'Sustainability in the construction industry: a review of recent developments based on LCA'. *Construction and Building Materials*, 23: 28–39.

Ramesh, T., Prakash, R. and Shukla, K. K. (2010). 'Life cycle energy analysis of buildings: an overview'. *Energy and Buildings*, 42(10): 1592–1600.

Rivela, B. and Neila, F. J. (2013). 'Methodological issues of LCA application to building sector: challenges, risks and opportunities', in *23rd Annual Meeting of the Society of Environmental Toxicology and Chemistry (SETAC Europe)*, 12–16 May 2013, Glasgow, pp 68–77. Available at: http://oa.upm.es/29896/1/INVE_MEM_2013_166989.pdf

Robertson, G. L. (2013). *Food packaging – principles and practice* (3rd edn). Boca Raton: Taylor & Francis Group LLC.

Roof, K. and Oleru, N. (2008). 'Public health: Seattle and King County's push for the built environment'. *Journal Environmental Health*, 71(1): 24–7.

Sharma, A., Saxena, A., Sethi, M., Shree, V. and Goel, V. (2011). 'Life cycle assessment of buildings: a review'. *Renewable and Sustainable Energy Reviews*, 15(1): 871–5.

Tola, A. (2014). 'Bio-construction and renewable raw materials: the case of cork', in R. Salomone and G. Saija (eds.) *Pathways to environmental sustainability – methodologies and experiences* (1st edn). Cham: Springer International Publishing, pp 137–46.

Udo de Haes, H. A. (2002). 'Industrial ecology and life cycle assessment', in R. U. Ayres and L. W. Ayres (eds.) *A handbook of industrial ecology* (1st edn). Cheltenham: Edward Elgar Publishing Ltd, pp 138–48.

Yigitcanlar, T. and Dizdaroglu, D. (2015). 'Ecological approaches in planning for sustainable cities: a review of the literature'. *Global Journal of Environmental Science and Management*, 1(2): 159–88.

33

Environmental economics and cost–benefit analysis

Ken Willis

Introduction

Planning approval for any project typically entails a fairly intuitive judgement about whether the perceived benefits of a project exceed its perceived costs. Economics by contrast uses a quantitative numeraire, typically a monetary unit, to appraise projects, through cost-benefit analysis (CBA). CBA determines if a project, policy or plan is sound, that is, whether benefits exceed costs and by how much and, as a basis of comparing projects, to maximise the welfare of society by selecting projects where benefits are maximised relative to costs. Many projects produce environmental benefits, and environmental dis-benefits, which are not priced in the market place. Environmental economic valuation can estimate values for these non-priced effects. These non-market benefits are included in a CBA appraisal. In project appraisal, no method other than CBA has such a sound theoretical economic base, and widespread application.

Cost and benefit analysis

Introduction

Early applications of CBA were in water resource planning and flood control in the United States and in transport planning (see Foster and Beesley, 1963; Beesley et al., 1983). CBA records benefits as financial revenue plus consumer surplus on intra-marginal units of a good and cost as the resource or opportunity cost of using resources to produce the good, i.e. the social value foregone when resources are moved away from another economic activity into the project.

The need for CBA arises precisely because environmental projects are affected by so many distortions and non-marketed outcomes: the 'public good' nature of many environmental goods; environmental externalities such as noise and air pollution from development; the fact that items such as non-working journey time and the value of preventable fatalities and injuries are not explicitly priced in the market; and the fact that some goods (for example, car transport) are highly taxed (through fuel taxes) whilst other goods (for example, public transport) are often subsidised by the state.

A CBA of a project is undertaken relative to some alternative, which is usually the 'do nothing' scenario. The 'do nothing' is a realistic view of what is likely to happen in the absence of the scheme.

In CBA, all relevant costs and benefits associated with the project are enumerated over the life of the project. For example, in evaluating a motorway scheme, the relevant costs would include land, construction, maintenance costs of the road. The benefits would include savings in journey time, fuel, accidents, noise and any other *technological* externalities (i.e. any externality that affects production or satisfaction (utility) of consumers). However, pecuniary externalities or spill-over effects are omitted in CBA since under complete markets pecuniary externalities offset one another.

Market prices and 'shadow prices'

Optimality in an economy requires marginal rates of substitution (MRS) (marginal values in the demand for goods) to be equal between consumers (if this is not the case then the exchange of products between consumers will make one or more consumers better off and leave no consumer worse off). Optimality also requires marginal rates of transformation (MRT) (marginal costs of factors of production) to be equal in the production of goods (if this is not the case then by substituting factors of production between industries it would be possible to reduce aggregate inputs with no change in aggregate output). Finally, optimality requires MRS = MRT to ensure consumers preferences and welfare are maximised (Winch, 1971; Layard and Walters, 1978). Under these conditions, market prices reflect appropriate MRS and MRT and are used to measure costs and benefits in CBA.

However, market prices do not reflect appropriate MRS and MRT where there are (i) monopolistic prices, (ii) unemployed resources, (iii) subsidies, (iv) taxes, (v) constraints on use, (vi) public goods and (vii) externalities. In such situations, implicit or 'shadow prices' are used to derive appropriate MRS and MRT in CBA, and these shadow prices normally replace observed 'market prices' in a CBA.

'Economic rent' is any (transfer) payment to an owner or a factor of production in excess of the cost to bring that factor into use. Economic rent is not counted in CBA. Compulsory purchase can be used in planning to prevent monopolistic position and 'hold-out' by landowners demanding prices (economic rent) in excess of the existing use value or opportunity cost of the land. Both compulsory purchase (eminent domain in the United States) and the absence of compulsory purchase powers can lead to inefficiency (Munch, 1976).

The guiding rule for estimating the shadow price of a resource, whether land, labour or capital, is the value it produces in the use from which it is being transferred plus any additional amount above its existing use value required to induce the transfer. Where labour is drawn from the unemployment pool, the true cost of the unemployed labour to society is zero, apart from any loss to the black economy. However, an unemployed worker will not remain unemployed for ever, so the social opportunity cost of labour needs to be estimated over time, which is not an easy task.

Resources subject to subsidies are measured not at their market price but at their resource or factor cost (market price minus subsidy element). For example, where agricultural land is used in a project, its opportunity cost is not the lost agricultural output minus agricultural inputs, nor its market price since this is artificially raised by price guarantees and tariffs at the EU frontier, but what it would cost to replace this lost agricultural output on world markets.

Taxed inputs are measured at their factor cost, not their market value. In calculating the operational benefits of a new road, fuel savings are valued at the market price of fuel minus the tax on fuel. Tax is merely a transfer payment, not the resource value of the fuel saving.

Land use planning creates constraints on land use. Resources subject to constraints should be valued at their opportunity cost rather than their current use value since the former is what society is losing by transferring the land, for example, to housing rather than designating it as green belt land and restricting its use to agriculture and amenity purposes. Cheshire and Sheppard (1989) used a hedonic price model to estimate the opportunity cost of the planning system, by comparing the house price consequences of planning restrictions in two towns with different degrees of development restrictions.

Public goods and externalities, such as air pollution, noise, visual amenity, etc., are non-market goods. They can be valued by environmental economic techniques. These techniques include production function approaches (expenditure to mitigate or avert an externality), revealed preference (how much people spend on related goods and services, such as housing, to acquire or avoid the public good or externality) and stated preference approaches (how much people 'say' they would be willing to pay to acquire the public good, or would require in compensation to avoid its loss or endure the detrimental externality).

Decision rules

Since people have time preferences, benefits and costs are discounted using a social time preference rate. Project appraisal should also include an adjustment for risk (HM Treasury, 2016). Projects are deemed economically feasible if the net present value (NPV) of benefits (*B*) and costs (*C*) discounted over each year (*t*) of the life of the project from year 0 to the end of the life of the plan or project (*T*) is ≥ 0.

$$\text{NPV} = \sum_{t=0}^{t=T} B_t(1+i)^{-t} - \sum_{t=0}^{t=T} C_t(1+i)^{-t}$$

where i = discount or rate. The discount rate is either an opportunity cost rate or a social time preference rate. Any project is socially profitable if NPV > 0 or if projects are mutually exclusive the project with the highest NPV is implemented. A project which has a negative NPV now may have a positive NPV in the future, if costs are declining over time and/or benefits are increasing.

In the presence of a budget constraint, projects should be implemented in descending order of their NPVs or benefit/cost (B/C) ratios. However, where projects are implemented over more than one time period, in the presence of budget constraints, welfare is maximised not by taking the absolute advantage of one project over another but by the comparative advantage between periods. The objective is to maximise the combined NPV of the chosen projects over time (Willis, 2005).

Social welfare

The Pareto criterion in economics approves a project if one person in society is made better off by a change and no one is made worse off. However, planning projects and policies invariably have both gainers and losers; and planning rarely compensates losers. The Kaldor-Hicks criterion sanctions projects where gainers could hypothetically compensate losers and still be better off, thus leading to a Pareto improving outcome. But unless compensation takes place, the project will lead to someone losing.

Lichfield (1964) was an early advocate of CBA in planning, but he simply suggested a planning balance sheet which tabulated who gained and who lost as a result of a policy. This left the decision to politicians as to which variant of a plan was acceptable economically and politically. Since planning decisions are primarily political decisions, this led to the use of public choice theory to analyse planning policies. But as Evans (2003) points out, while public choice theory can explain which planning policy is selected, this does not mean that this planning policy maximises economic welfare; it may only maximise the politician's chance of being re-elected. Economic analysis has more general applications to planning than merely CBA (see Willis, 1980; Evans, 2004).

Distributional issues can be incorporated more formally in CBA. Economic welfare can be maximised by making choices through a social welfare function (SWF), for example:

$$W = (1/\alpha)\mu r^{\alpha} + (1/\alpha)\mu p^{\alpha},\ \alpha < 1\,\alpha \neq 0$$

where W is welfare, μ is net social benefit to the rich (r) and the poor (p), respectively, and α is a distributional weight, for example, based on marginal rates of income tax, or more theoretically and correctly the marginal utility of money. The utilitarian position assigns everyone a distributional weight of 1.0; so if the NPV of a project is positive, then the project should go ahead, irrespective of who gains. The smaller α, for example, -1 or less, the more egalitarian the criterion (Layard and Walters, 1978). As $\alpha \rightarrow -\infty$ then W comes to depend on the income of the poorest person (the position advocated by Rawls, 1971). Adler (2016), in a review of benefit-cost analysis and distributional weights, shows that a SWF framework depends on an interpersonally comparable utility function but also requires moral judgement and an estimate of the appropriate degree of inequality aversion.

However, redistribution is not without benefit and efficiency losses. Consider a simple SWF:

$$W = \left(\mu_r^a,\ \mu_p^b\right)$$

where μ is net social benefit to the rich (r) and the poor (p), respectively, and a and b are distributional weights. If money or utility changes from a project were: rich +200 and poor – 100, and the distributional weights were 0.5 and 1.5 respectively, the project would be deemed undesirable since:

$$\Delta W = 0.5\,(200) + 1.5\,(-100) = -50 < 0$$

The distributional anomaly here is that if the project went ahead, the rich could compensate the poor and leave them no worse off, with the rich still gaining 100. Indeed, if the entire 200 were redistributed from the rich to the poor, the poor would gain by 1.5 ★ (100) = 150, which would be better than not undertaking the project at all! The marginal cost of redistribution can be high (Browning, 1996).

Environmental economics

Public goods and externalities

Central to environmental economics is the concept of market failure. Market failure comprises 'public goods' and externalities (negative and positive). A 'public good' is one that is non-rival in consumption (its consumption by one person does not lessen its availability to others) and non-excludable (property rights cannot be extended over the good, so a price cannot be charged). Winch (1971) suggests three criteria to define public goods: opportunity cost, property rights of the producer and property rights of the consumer. These criteria can be permutated to produce different classes of public goods and externalities.

Non-marketed benefits are often associated with town planning and sustainable development projects. Environmental economics estimates 'shadow prices' for public goods and environmental externalities in a number of ways: in terms of effect on production (EOP), opportunity cost (OC), replacement cost (RC), mitigation expenditure (ME), revealed preference (RP) and stated preference (SP) methods (Garrod and Willis, 1999).

Effect on production approaches

The EOP approach is widely used to assess compensation to farmers for loss of agricultural output, due to restrictions prohibiting damaging operations to agricultural land, in the interests of protecting the environment and wildlife. Compensation is based on net annual profit foregone, for example:

$$FC = (c - a) - (b - d) + k$$

where FC = financial compensation payable, c = revenue from farming more intensively, a = variable costs incurred for this output level, b = revenue from restricted agricultural output, d = variable costs for this less intensive agricultural output level and k = annuitized value of any capital (barns, machinery, etc.) made redundant by farming at the less intensive position.

There are limitations to the EOP approach. Some physical relationships between activities affecting the environment, and the resulting effect on output, may not be well understood. It may also be difficult to separate the physical effects from multiple pollution sources, in trying to assign consequential losses in productivity. And the opportunity cost may differ from the financial cost. For example, trade restrictions and tariffs on agricultural goods entering the European Union (EU) result in the social value of agricultural output foregone being less than financial compensation paid to farmers. The shadow price of agricultural output can be estimated by calculating producer subsidy equivalents (PSE). PSE measures the social value of output foregone by revaluing domestic output, minus direct subsidies and taxes, at world market prices:

$$PSE = [V_m + D - V_w] / V_w$$

where V_m = output at domestic market prices, D = direct subsidies, and V_w = world market prices. PSE only concentrate on outputs. Where the effect of protection on inputs is significant then protection needs to be measured through the value added element in production, for example, by calculating an effective protection rate (EPR).

$$EPR = [VA_m - VA_w] / -VA_w$$

where VA_m = value added at domestic prices, VA_w = value added at world prices, where $VA_w = P_j(1 - A_{ij})$, $VA_m = P_j\,[(1 - \mathrm{T}_j) - A_{ij}\,(1 + T_j)]$, and P_j = nominal price of commodity j in free trade, A_{ij} share of input i in cost j at free trade prices, and T = nominal tariff on i or j as the case may be. A social OC measure can result in the opportunity cost of environmental protection being much less than its financial cost (Willis et al., 1988).

Some EOP policy impacts, such as a restriction on agriculture to promote natural habitats, may seem relatively easy to specify. But some policy effects on production may not be obvious for a CBA. For example, replacement of traffic lights with LED lights, in Sweden, reduces electricity consumption. But from what source is the electricity saved: coal fired or hydro-electric stations? This assumption makes quite a difference to carbon emission savings, and ultimately to whether CBA judges a project to be sustainable and socially beneficial (see Willis, 2010).

The RC approach assesses the value of a natural resource by how much it costs to replace or restore the resource after it has been damaged. This is the position adopted by the Comprehensive Environmental Response, Compensation and Liability Act (CERCLA) 1980, or 'Superfund' in the US, which requires full restoration or replacement of natural resources damaged by pollution incidents. Liability for damages rests with the polluter, who also has to pay for the clean-up of contaminated sites. Damages are assessed as the cost of restoration or replacement, plus compensable value of the services lost to the public for the time period of the discharge until the attainment of restoration or replacement. CERCLA is concerned with the use and non-use value of natural resources and recovering any loss of consumer surplus associated with the resource. The Deepwater oil spill in 2010 in the United States resulted in an estimated cost to British Petroleum of US$62 billion in natural resource restoration; criminal liabilities; and civil damages for loss of fisheries, tourism, etc. (see Wikipedia, 2018; NOAA, 2017).

Restoring resources to their pre-damaged state ensures that lost consumer surplus and non-use values are recovered. But restoration and replacement costs will over-estimate society's loss if these costs are greater than the value of use and non-use benefits lost. In such a case, it is rational to compensate individuals for their lost uses rather than impose the cost of restoration or replacement. The environmental loss, under sustainability or the need to maintain the stock of natural capital, can be valued in terms of the monetary requirement for (i) full restoration; (ii) economic efficiency, where the difference between marginal costs and marginal benefits is maximised; or (iii) some intermediate restoration level, where restoration costs do not exceed the use and non-use value of the environmental good (Ward and Duffield, 1992).

Preventative expenditure (PE) and mitigation expenditure (ME) are sometimes used to value non-marketed commodities or mitigate against an environmental externality. ME expenditure, for example, on double or triple glazing to mitigate road noise in houses will not mitigate road noise in gardens, so ME will not necessarily restore an individual to his pre-externality welfare position. Dickie (2017) provides examples and discusses the issues in the use of averting behaviour methods to value non-market goods, such as air pollution.

Revealed preference methods

Travel cost (TC) and hedonic price (HP) methods infer values for non-market goods and externalities from related expenditures on market goods.

TC uses the amount people are willing to spend travelling to gain access to an open access non-priced recreation site, as a means of inferring the recreational benefits that the site provides. TC is weakly complementary to on-site recreation. The individual TC method takes the form:

$$Vij = \mathrm{f}(Pij,\ Tij, \boldsymbol{Q_i}, \boldsymbol{S_k}, Yi)$$

where V_{ij} is the number of visits made by individual i to site j; P_{ij} is the travel cost incurred by individual i when visiting site j; T_{ij} is the time cost incurred by individual i when visiting site j; $\mathbf{Q}$ is a vector of the perceived qualities of the recreation site j; $\boldsymbol{S_k}$ is a vector of the characteristics of available substitute sites (this may refer only to sites of a similar nature or may be extended to include other recreational attractions); and Y_i is the household income of individual *i*.

The key issues surrounding the application of TC methods involve travel cost per mile or kilometre (do individuals use the marginal fuel cost of operating a car, or full cost including depreciation in judging travel cost), the value of non-working time to apply for the time cost involved, the allocation of costs between car occupants, the allocation of utility between the journey to the site and recreation on the site, multi-site journeys, the presence of substitute sites, as well as the particular model to use, for example, censored regression model, or a count data model such as a Poisson or a Negative Binomial model. Different assumptions about these key issues can give rise to different economic estimates for the value of the public good. Hanley et al. (2003) used a count data negative binomial TC model, based on actual trips made, trip costs, and perceived bathing water quality, to assess the value of improvements to bathing water quality across different beaches in south-west Scotland.

Hedonic price models (HPMs) infer values for externalities from expenditure on housing purchased to 'consume' these public goods. HPMs have been used to estimate WTP to avoid noise and air pollution; to gain access to good schools and commuting rail stations; and provide access to pleasant landscape features such as open space, woodland and trees, lakes, rivers and the coast, etc.

Problems encountered in the application of HPMs include multicollinearity between variables, omitted variable bias, and non-linearity in relationships. Some attributes are quite difficult to encapsulate in an HPM even with GIS and view-sheds, for example, the spatial configuration of landscape attributes in relation to each other.

HPMs were used by Cheshire and Sheppard (1989) to estimate the cost of the planning system, principally in terms of house prices, by comparing two towns at polar extremes of planning restrictiveness on development. Planning was estimated to have a significant impact on house prices. They later extended the application of the HPMs to assess the welfare benefits of town planning in terms of creating, preserving and maintaining public goods (Cheshire and Sheppard, 2002). However, it is quite difficult to identify which public goods are solely attributable to planning and which public goods are provided privately through non-development and charitable ownership and protection. HPMs can therefore over-estimate the value of the planning system in terms of welfare benefits.

HPMs of separate housing market segments, and the use of semiparametric smoothing estimators rather than standard specifications of HP functions, have been used to derive more accurate estimates of externalities such as noise (Day et al., 2007). GIS has also been used to provide additional information on the house characteristics to derive more accurate estimates of the value of views of parks, etc.

In some cases, it is impractical to use RP methods to value public goods. For example, it may not be possible to observe visits to a cultural heritage site if the good is inaccessible to the public (for example, in the case of undersea or underground archaeological remains, or paintings and books held in storage in galleries and libraries). RP requires data with variance in demand to allow variables to be used to explain the variation in demand and to estimate a model. This can make RP data quite difficult, time consuming and expensive to collect. RP methods may be subject to confounding or multicollinearity problems. But in SP, experimental design can ensure orthogonality between variables in the experiment. RP cannot be used to estimate non-use values, although charitable giving partly reveals altruistic, existence and bequest values to preserve some public goods. Stated preference (SP) methods can overcome these problems encountered in RP methods. SP asks consumers to state their willingness-to-pay (WTP) for a good. SP methods embrace both contingent valuation (CV) and choice experiment (CE) methods.

Stated preference methods: contingent valuation

CV asks respondents their WTP for a specific change in a good. This question can either be asked as an open-ended (OE) question (how much are you willing to pay for X rather than go without it?) or as a single bounded (SB) discrete choice (DC) question (are you willing to pay \$$A$ for good X?: 'yes' or 'no'). After the first DC question has been asked, this can be followed by a second DC question with a (i) higher bid if the response was a 'yes' to the first DC question or (ii) lower bid if the response was a 'no' to the first DC bid question, giving a double bounded DC CV. Iterative bidding provides yet another estimation method.

WTP values can be calculated non-parametrically as a distribution free model, or as a 'smoothing' Turnbull model, or alternatively as a parametric model. For a SB DC parametric model:

$$\text{WTP} = 1/\beta \ln\left[1 + e^{\alpha}\right]$$

where β is the coefficient associated with the bid amount, and α is the constant term in the model.

There have been many CV applications valuing public goods and environmental externalities. One early application of CV, within a CBA, in Britain was to assess the amenity value of green belt land (Willis and Whitby, 1985). CV has been widely employed in the United States to assess environmental losses under CERCLA (1980).

Some CV studies have investigated whether CV estimates are subject to biases, such as part-whole bias, starting point bias, bid ranges, the probability mass assignment within the interval bid range and the individual valuation summation issue in CBA (Hoehn and Randall, 1989). There has also been substantial research to assess under what conditions CV is incentive compatible (will respondents answer truthfully) and whether CV is applicable to derive accurate willingness-to-accept (WTA) compensation values for the loss of an environmental good (Vossler et al., 2012). Auction mechanisms such as a Vickrey auction or a Becker-DeGroot-Marschak auction may generate more accurate and reliable estimates.

Stated preference methods: choice experiment

The value of a good is dependent on its attributes. A choice experiment (CE) allows respondents to trade off attributes against each other, and against price. A CE is thus more flexible in determining the value of different combinations of environmental attributes and attribute levels. CEs are now widely applied in environmental economics to investigate the value of environmental goods and policy components.

The experimental design of a CE can ensure orthogonality, balance and efficiency. The design allows for main order effects of attributes to be estimated but can also include second order, or higher-order effects, if these need to be estimated. First order or main effects typically account for most of the variance in WTP values.

In a CE, the individual i is assumed to choose alternative attribute bundle j over alternative k if the utility provided by attribute bundle j is greater than that provided by attribute bundle k:

$$U_{ij} > U_{ik}$$

Choice is modelled using random utility theory: individuals will make choices based on the attributes and attribute levels (an objective component) along with some degree of randomness (a random component). This random component arises either because of randomness in the preferences of the individual or because the researcher does not have the complete set of information available to the individual (Train, 2003).

The conditional logit (CL) models the probability of choosing *j* over the *k* other attribute bundles, depending on differences amongst alternative attributes. Attributes that do not vary by alternative do not affect probabilities. Where $\mathbf{X}_{ij}$ is a vector of attributes of good *j*, individual *i*, then the probability that individual *i* choosing alternative *j* is (Haab and McConnell, 2003):

$$\Pr_i(j) = e^{\mathbf{X}_{ij}\beta} / \Sigma_{k=1}^{J} e^{\mathbf{X}_{ik}\beta}$$

The CL model differs from the multinomial logit (MNL) model. The MNL uses individual characteristics to explain the choice of an alternative (Haab and McConnell, 2003).

The CL model assumes preference homogeneity across individuals, the independence of irrelevant alternatives (IIA), with error terms which are independent, have zero correlation and are homoscedastic. Other models relax these assumptions. The nested logit, heteroscedastic extreme value, mixed logit, error component, multinomial probit, latent class and other models relax the CL assumptions and allow for heterogeneity in preferences and different structures for the error term (Train, 2003).

CE research in valuing public goods has explored the effects of choice complexity, attribute non-attendance, WTP versus WTA asymmetry, interaction effects between attributes, scale versus taste heterogeneity, utility maximisation versus regret minimisation and the impact of respondent attitude and socio-economic characteristics on value estimation (Willis, 2014). Other extensions include the reparameterization of utility in WTP space, rather than in preference space. Scarpa and Willis (2010) provide an example in relation to renewable energy.

The 'true' effect on welfare of a planning or environmental change is measured by the amount of money (Δ£) households are willing to pay for (or accept) a positive (or negative) change in environmental or public good. This is the amount that sets the two utility levels equal to each other. The general formula is:

$$\Delta\pounds = [\Sigma_j \beta_i (x_j^0 - x_j^1)] / \alpha = [\Sigma_j \beta_j (\Delta x_j)] / \alpha = \Sigma_j (\beta_j / \alpha)(\Delta x_j)$$

where β_j is the parameter estimate for the attribute *j*; *x* is the numerical level for a particular environmental attribute; α is the parameter estimate for the utility of money; the superscripts 0 and 1 refer respectively to the initial and final state; and the subscript *j* refers to the generic *j*-th attribute. Thus, a change in value is given by the sum of the products between part-worths of each attribute and the respective change in provision Δx_j. This type of analysis can also be used to estimate market share, or the proportion of respondents who sanction a project or policy with a particular combination of attribute levels (Willis, 2006).

Environmental economic values in CBA

Atkinson et al. (2018) provide an excellent review of the use of revealed and stated preference methods to CBA applications. There are many techniques in environmental economics which can be used to estimate the value of public goods and positive and negative environmental externalities and, from labour economics and other areas of economics, to estimate other shadow prices for a CBA. Often assumptions have to be made about the 'with-without' project positions. Such assumptions can ultimately determine whether CBA judges a project to be sustainable or socially beneficial (see Willis, 2010).

Climate change

Climate change and resource depletion pose special challenges to environmental economics and the cost-benefit analysis of sustainability because of issues on the extent to which depletion in natural capital can be offset by additions to human capital (Pearce et al., 2006), uncertainty about the effects on production of climate change, the specification of the 'with-without' positions, the extreme long run life of the effects, the likely non-marginal nature of the changes and how the future impacts should be discounted.

The atmosphere is a global commons, into which pollution is emitted by firms and individuals, creating environmental externality effects, or a 'public bad'. Greenhouse gas emissions and the increasing levels of these gases in the atmosphere have wide-ranging effects on freshwater supplies, food and agriculture, human health, coastal areas and ecosystems, as documented by the Stern Review (Stern, 2007) and others (for example, Harris et al., 2017). The analysis by Stern (2007) suggested the benefits of strong early action on climate change outweigh the costs of not acting.

The Stern Review has been criticised on the grounds that the science of climate change, the impact of greenhouse gas emissions, the absorption rates of greenhouse gas by oceans, trees, etc, as well as human adaptation, is not well known and uncertain, as well as making pessimistic assumptions about the effects of climate change on production.

There have also been a number of economic criticisms. Stern used a social discount rate based on the pure time preference rate and the Ramsey formula:

$$s = \gamma + \eta g$$

where s = the social discount rate; γ = pure time preference rate; η = the marginal elasticity of utility; and g = rate of growth of per capita consumption. Assumptions about γ, η and g gave rise to a low discount rate. Nordhaus (2007) thought this low interest rate untenable and inconsistent with market place real interest rates. Weitzman (2007) thought the Stern Review adopted extreme assumptions about the impact of climate change and discount rates which were too low.

Conventional CBA assumes that any project or policy is marginal: it will not significantly change relative prices and can be analysed using a partial-equilibrium approach. Dietz and Hepburn (2010) point out that this assumption may be inappropriate for large development projects in small economies, or in the analysis of mitigation of global climate change. Where changes are not marginal a general equilibrium approach should be adopted. A partial equilibrium approach is more likely to give rise to errors where η is high, with the possibility of counter-intuitive results arising when projects are evaluated in the context of a growing population.

Conclusion

The principles of CBA are well established (Brent, 1996), but environmental economic values need to be carefully applied in CBA. Often environmental values are transferred through meta-analysis from other studies. However a ceteris paribus assumption can rarely be applied. Values cannot readily be transferred (Florax et al., 2002). Individual valuation and summation (IVS) will result in valuation bias if programmes have substitute or complementary projects (Hoehn and Randall, 1989). IVS over estimates the true value of environmental benefits if programme schemes are substitutes. The values or benefits of a programme of sustainability projects may need to be either sequentially estimated, simultaneously estimated, or individually estimated depending on the circumstances.

Assumptions often have to be made in applying environmental values in CBA. For an energy consumption saving project, will energy savings be attributable to renewable or non-renewable sources (see Willis, 2010, for an example). And if a non-renewable source, what price should be attached to the reduced carbon emissions, given the wide disparity in the implied and estimated social price of carbon (Pearce, 2003).

Sustainable development often involves additional government expenditure on projects, raised by public taxation. Additional taxation causes a surplus loss or 'excess burden', which ought to be included in a CBA. Excess burden is the loss of utility incurred by the private sector from giving up resources to the public sector. The marginal cost of public funds (MCPF) is >1, assuming a negative substitution effect between work and leisure (Browning, 1987). Because taxes are not generally hypothecated, most public expenditure cannot be ascribed to a tax source. If it is assumed that public expenditure is derived from income tax, the marginal excess burden (MEB) depends upon the marginal tax rate (m), the elasticity of labour supply (η) and the progressivity of the tax system. Estimates of the MCPF have been made for different European countries (see Kleven and Kreiner, 2006).

Continued development and refinement of environmental economic techniques will result in more accurate, reliable and robust value estimates to include in a CBA. For example, allowing consumers to express uncertainty about intended behaviour would allow stated choice methodology to more accurately mirror actual behavior, and applying incentive menu regulation mechanisms should ensure increased welfare for consumers where goods, including environmental goods, are supplied under monopoly and oligopoly frameworks.

References

Adler, M. D. (2016). 'Benefit-cost analysis and distributional weights: an overview'. *Review of Environmental Economics and Policy*, 10(2): 264–85.

Atkinson, G., Braathen, N. A., Groom, B. and Mourato, S. (2018). *Cost benefit analysis and the environment: further developments and policy use*. Paris: OECD.

Beesley, M. E., Gist, P. and Glaister, S. (1983). 'Cost benefit analysis of London's transport policies'. *Progress in Planning*, 19(3): 169–269.

Brent, R. J. (1996). *Applied cost – benefit analysis*. Cheltenham: Edward Elgar.

Browning, E. K. (1987). 'On the marginal cost of taxation'. *American Economic Review*, 77: 11–23.

Browning, E. K. (1996). 'The marginal cost of redistribution: a reply'. *Public Finance Quarterly*, 24: 63–74.

Cheshire, P. and Sheppard, S. (1989). 'British planning policies and access to housing: some empirical estimates'. *Urban Studies*, 26(5): 469–85.

Cheshire, P. and Sheppard, S. (2002). 'The welfare economics of land use planning'. *Journal of Urban Economics*, 52(2): 242–69.

Day, B., Bateman, I. and Lake, I. (2007). 'Beyond implicit prices: recovering theoretically consistent and transferable values for noise avoidance from a hedonic price model'. *Environmental and Resource Economics*, 37(1): 211–32.

Dickie, M. (2017). 'Averting behaviour methods', in P. A. Champ, K. J. Boyle and T. C. Brown (eds.) *A primer on non-market valuation* (2nd edn). Dordrecht: Springer, pp 293–346.

Dietz, S. and Hepburn, C. (2010). 'On non-marginal cost – benefit analysis'. *Centre for Climate Change Economics and Policy, Working Paper No. 20: Grantham Research Institute on Climate Change and the Environment, Working Paper No. 18*. London: London School of Economics (LSE).

Evans, A. W. (2003). 'Shouting very loudly: economics, planning and politics'. *Town Planning Review*, 74(2): 195–212.

Evans, A. W. (2004). *Economics and land use planning*. Oxford: Blackwell.

Florax, R. J. G. M., Nijkamp, P. and Willis, K. G. (eds.) (2002). *Comparative environmental economic assessment*. Cheltenham: Edward Elgar.

Foster, C. D. and Beesley, M. E. (1963). 'Estimating the social cost of constructing an underground railway in London'. *Journal of the Royal Statistical Society Series A*, 126: 46–92.

Garrod, G. D. and Willis, K. G. (1999). *Economic valuation of the environment.* Cheltenham: Edward Elgar.

Haab, T. C. and McConnell, K. E. (2003). *Valuing environmental and natural resources: the econometrics of non-market valuation.* Cheltenham: Edward Elgar.

Hanley, N., Bell, D. and Alvarez-Farizo, B. (2003). 'Valuing the benefits of coastal water quality improvements using contingent and real behaviour'. *Environmental and Resource Economics*, 24(3): 273–85.

Harris, J. M., Roach, B. and Codur, A.-M (2017). *The economics of global climate change.* Global Development and Environment Institute, Tufts University, Medford, Massachusetts.

HM Treasury (2016). *The green book: appraisal and evaluation in central government.* London: HM Treasury.

Hoehn, J. P. and Randall, A. (1989). 'Too many proposals pass the benefit cost test'. *American Economic Review*, 79: 544–51.

Kleven, H. J. and Kreiner, C. T. (2006). 'The marginal cost of public funds: hours of work versus labor force participation'. *Journal of Public Economics*, 90(10–11): 1955–73.

Layard, P. R. G. and Walters, A. A. (1978). *Microeconomic theory.* New York: McGraw-Hill.

Lichfield, N. (1964). 'Cost benefit analysis in plan evaluation'. *Town Planning Review*, 35(2): 159–69.

Munch, P. (1976). 'An economic analysis of eminent domain'. *Journal of Political Economy*, 84(3): 473–97.

National Oceanic and Atmospheric Administration (NOAA) (2017). *Deepwater Horizon oil spill settlements: where the money went.* Washington, DC: NOAA, US Department of Commerce. Available at: www.noaa.gov/explainers/deepwater-horizon-oil-spill-settlements-where-money-went

Nordhaus, W. D. (2007). 'A review of the Stern review on the economics of climate change'. *Journal of Economic Literature*, 45(3): 686–702.

Pearce, D. W. (2003). 'The social cost of carbon and its policy implications'. *Oxford Economic Policy*, 19(3): 349–61.

Pearce, D., Atkinson, G. and Mourato, S. (2006). *Cost benefit analysis and the environment.* Paris: OECD.

Rawls, J. (1971). *A theory of justice.* Oxford: Clarendon Press.

Scarpa, R. and Willis, K. (2010). 'Willingness-to-pay for renewable energy: primary and discretionary choice of British households' for micro-generation technologies'. *Energy Economics*, 32(1): 129–36.

Stern, N. (2007). *The economics of climate change: the Stern review.* Cambridge: Cambridge University Press. Available at: http://webarchive.nationalarchives.gov.uk/+/www.hm-treasury.gov.uk/sternreview_index.htm

Train, K. E. (2003). *Discrete choice methods with simulation.* Cambridge: Cambridge University Press.

Vossler, C. A., Doyon, M. and Rondeau, D. (2012). 'Truth in consequentiality: theory and field evidence on discrete choice experiments'. *American Economic Journal: Microeconomics*, 4(4): 145–71.

Ward, K. M. and Duffield, J. W. (1992). *Natural resource damages: law and economics.* New York: John Wiley.

Weitzman, M. L. (2007). 'A review of the Stern review on the economics of climate change'. *Journal of Economic Literature*, 45(3): 703–24.

Wikipedia (2018). *Deepwater Horizon oil spill.* Available at: https://en.wikipedia.org/wiki/Economic_effects_of_the_Deepwater_Horizon_oil_spill

Willis, K. G. (1980). *The economics of town and country planning.* London: Collins (Granada).

Willis, K. G. (2005). 'Cost benefit analysis', in K. J. Button and D. A. Hensher (eds.) *Handbook of transport strategy, policy and institutions, volume 6: handbooks in transport.* Amsterdam: Elsevier, pp 491–506.

Willis, K. G. (2006). 'Assessing public preferences: the use of stated preference experiments to assess the impact of varying planning conditions'. *Town Planning Review*, 77(4): 485–505.

Willis, K. G. (2010). 'Is all sustainable development sustainable? A cost – benefit analysis of some procurement projects'. *Journal of Environmental Assessment Policy & Management*, 12(3): 1–21.

Willis, K. G. (2014). 'Cultural heritage: economic analysis and the evaluation of heritage projects', in D. Throsby and V. Ginsburgh (eds.) *Handbook of the economics of art and culture, volume 2.* Amsterdam: Elsevier-North Holland, pp 145–81.

Willis, K. G., Benson, J. F. and Saunders, C. M. (1988). 'The impact of agricultural policy on the costs of nature conservation'. *Land Economics*, 64: 147–57.

Willis, K. G. and Whitby, M. C. (1985). 'The value of Green Belt land'. *Journal of Rural Studies*, 1(2): 147–62.

Winch, D. M. (1971). *Analytical welfare economics.* Harmondsworth, London: Penguin Books.

34

Natural capital and ecosystem services valuation

Assisting policy making

Tomas Badura, Kerry Turner and Silvia Ferrini

Introduction

Across the globe, human populations and nature are deeply intertwined. Humans rely on the natural environment for production of fish, food, timber and fibre. We also utilise nature's ecological processes such as water and air purification, climate stabilisation, soil protection and pollination of nut and fruit trees. A number of wetland ecosystems protect communities against floods, storm surges and hurricanes. Nature provides cultural inspiration for arts and research and contributes to a sense of place and wonder. It is in these multiple ways that nature, through its stock of natural capital assets, provides a flow of tangible and intangible benefits, known as ecosystem services, that improve human wellbeing and contribute to economic development.

Some of these nature related benefits are, however, not exchanged through market mechanisms and assigned prices. This can, in some circumstances, make them less visible in current decision-making processes, leading to suboptimal choices. The full societal value of nature-related benefits is, therefore, difficult to capture in both public and private decision making, which often leads to increasing degradation of the natural environment with negative effects on human wellbeing (Millennium Ecosystem Assessment, 2005). To take just one example, the full value of mangrove forests over the long run lies in their ability to protect coastal communities from storms, harbour young fish and local biodiversity and contribute to cultural heritage. Nevertheless, mangroves continue to be removed to make way for shrimp farms which provide immediate financial returns from sales to international markets. This case illustrates the implications when the benefits of ecosystem conservation provide gains in welfare for the public at large (and in the case of, for example, ecosystems that sequester and store carbon, for the global population) but costs are distributed more locally. Many of the ecosystem goods and services are not traded in any markets, and information and knowledge about their values are less widely dispersed culturally through society. Nature's value is a multidimensional concept and not restricted to just monetary expression. Both private and public decision-making systems, hence, often do not have sufficient information on the multiple trade-offs that are associated with environmental management.

Assigning more priority to nature conservation and management in our decision making is socially desirable and an urgent requirement (TEEB, 2010). This could be done in multiple ways. One set of approaches relies on regulation, such as laws through which society decides, for example, to protect certain species and ecosystems (for example, biodiversity protection directives in the EU) or ban trade with others (for example, Convention on International Trade in Endangered Species). Another approach is to more fully value ecosystem related goods and services in monetary terms. This information can help to better inform cost-benefit decisions that balance the returns from investment in our natural capital against other investment decisions. It can also underpin other types of policy instruments such as tradable permits, payments for ecosystem services, offsets or better product labelling. This chapter focuses on the multiple ways through which nature related goods and services can be valued in monetary or non-monetary terms depending on the resources available and the given decision-making context. The chapter outlines the different valuation methods and techniques that might support better decision making related to nature and examines future prospects for valuation research. In conclusion, the chapter discusses the role of these methods in policy making and their increasing importance in efforts to incorporate the value of the natural environment into a system of national accounts.

Natural capital, ecosystem services and biodiversity

From an economic point of view, nature can be seen as a form of capital. In this view, natural capital, alongside other forms of capital (human, produced/manufactured and social capital), is seen as a stock of biotic and abiotic assets that has the ability to produce further goods and services that benefit human societies. We follow the Natural Capital Committee (NCC, 2014, p. 26) in defining natural capital as '[t]he elements of nature that directly and indirectly provide value or benefits to people, including ecosystems, species, freshwater, land, minerals, the air and oceans, as well as natural processes and functions'. The condition (or health) of individual ecosystem assets underpins their ability to produce flows of services that, combined with other capital inputs, produce goods. Such goods provide benefits to people that can be sometimes quantified and valued (this might be, but does not necessarily need to be, in monetary terms).

An important distinction between intermediate and final ecosystem services needs to be made for economic and monetary valuation purposes in particular (Fisher et al., 2009). This requires that valuation focuses on the very final contribution of the ecosystems to benefits that enhance welfare. The focus on 'only' final ecosystem services ensures that the analysis of the contribution of nature to human wellbeing is not overstated as it avoids the problem of double counting. For example, valuing both pollination services as well as crop production of a given agricultural ecosystem leads to valuing pollination twice – once by itself and second embodied in the crop production values – this overstates the contribution that nature makes to crop production and human wellbeing. A relevant distinction also needs to be made, particularly in the context of national accounting statistics, between the pollination related benefits that are reflected in national statistics (for example, crop pollination) and others which are not (for example, wildflowers that contribute to the attractiveness of landscape for the arts). While in the former the goal of valuation is to identify the contribution of the ecosystem to an accounted-for good (crop production), the latter aims to value benefits that are not included in the national statistics at all.

Biodiversity[1] is understood to underpin ecosystems' functioning and the services they provide; however, this relationship is complex and requires further research (Balvanera et al., 2006; Díaz et al., 2006; Mace et al., 2012; Naeem et al., 2012). It influences ecosystems' condition and health. Biodiversity regulates underpinning ecosystem processes, but it can also be a final ecosystem service and a good that can be subject to valuation (Mace et al., 2012). This underlines an essential component of any attempts to assess the contribution of nature to human wellbeing – the need for sufficient ecological understanding/knowledge of the processes that generate these contributions.

Valuation of ecosystem related goods and services builds on and is conditioned by the natural science understanding of how environmental change impacts the provision of these goods and services. Conventional economic valuation assesses the value at the margin – that is, it aims to value the results of a given (environmental) change, which is limited in its extent, and its implication on human wellbeing. It does not aim to quantify the total value of all ecosystems, nor, for example, the global value of a specific ecosystem service. Instead, economic valuation usually aims to provide evidence enabling comparison of alternative investment options, resulting from a small change in provision of ecosystem related goods and services. For example, valuation can help to compare natural versus man-made options for flood protection and water purification for a given community or evaluate different land use options for an area and the associated changes in bundles of ecosystem services. Some ecological economists take a different approach to the valuation of a stock of ecosystem services and have attempted to put monetary values on the stock of global ecosystems (Costanza et al., 1997, 2014). They argue that such numbers are analogous to national income accounting measures, such as GDP, with a constant set of weights. Despite the debate on this practice of ascribing monetary value to the natural capital, it is still the case that an understanding of the 'ecological production function' is necessary for the valuation of ecosystems. Biophysical assessment is the first step in most of valuation work and the results can be only as robust as the underpinning natural science.

An important dimension of the valuation of ecosystem related goods and services is the threshold effects some ecosystems can exhibit (Bateman et al., 2011). The use of economic valuation approaches is limited as systems approach such non-linearities, as the unit value is likely to change significantly. In these cases, an alternative approach might be chosen to avoid irreversible risks to provided benefits. This might be accomplished, for example, through laws and regulations and/or by imposing 'safe minimum standards' which ensure that the thresholds are not crossed and ecosystem resilience is maintained. However, this (again) requires a detailed ecological understanding of the ecosystem in question and where and when the risk of reaching a tipping point might be approached. A useful start might be a 'risk register' approach, a tool commonly used by organisations to identify and quantify potential risks, their likely impacts, mitigation measures and responsibilities to address the risks, which was applied on the case of UK natural capital by Mace et al. (2015).

Methods and techniques for the valuation of ecosystem related goods and services

Any attempts to quantify or qualify a relationship between nature and societies requires articulation of the type of value that the valuation process tries to elicit. This is often defined by the policy and decision-making context that the valuation is aiming to inform. The choice of the value type does not need to be mutually exclusive – for example, it is possible to consider the economic value in monetary terms of the benefits of a given forest's ability to purify water and regulate flows compared to man-made solutions, while also recognising in a non-monetary manner the significant role that forests play in shaping the lives of the community of people living in the area and their arts and other cultural assets. In fact, in many cases such complementary assessments are more likely to be generally acceptable and provide a more complete evidence base for decisions.

The choice of valuation method(s) relevant for a given assessment is dependent upon resources and data available as well as the purpose of the valuation research. Barton and Harrison (2017) and Harrison et al. (2018) provide some guidance on selection and combination of different assessment and valuation methods, building on experience from 27 case studies distributed across the EU. Figure 34.1 provides a simplified range of the methods that can be used for valuing ecosystem related goods and services. Broad categorisation can be made between non-monetary methods that are qualitative and quantitative and monetary methods that are based on either exchange value concept or those that aim to measure welfare values. A hybrid set of methods combine non-monetary and monetary approaches. The expanding attempts to develop and implement national accounting practices adds another dimension to how we use evidence in quantifying the relationships between people and nature. The rest of this chapter will provide an overview of monetary and non-monetary methods to value ecosystem related goods and services (see Figure 34.1) for policy appraisal and in the context of environmental accounting.[2]

Monetary approaches

Economists categorise the instrumental values associated with ecosystem related goods and services into use and non-use values. Use and non-use values are not mutually exclusive as people can hold both types of value for the same good. The use values relate to an act of engagement with nature and are further distinguished into direct use, indirect use and option values. Direct use values arise from a direct interaction with the environment and include extractive (for example, timber, food) and non-extractive activities (for example, recreation). Indirect use values are associated with benefits that are indirect – for example, values associated with natural hazards protection or carbon sequestration and storage services but also with welfare generated by photos or TV programmes concerning the natural environment. Option value relates to possible use of ecosystem related goods and services that might happen in the future, for example, through possible future biomedical research from the genes and species present in the tropical rainforests. Non-use (known as existence and bequest) values do not concern any active interaction with nature but relate to the welfare gained from knowledge of its continued existence or of its components.

Monetary values can be derived from market information, through observing consumer/producer behaviour, or through surveys. A range of detailed guidance material on the application of valuation methods has been produced in recent years and the reader is suggested to consult these for further information – for example, Freeman et al. (2014), Bockstael and McConnell (2007) and Johnston et al. (2015, 2017).[3] We examine these approaches in turn in the following paragraphs.

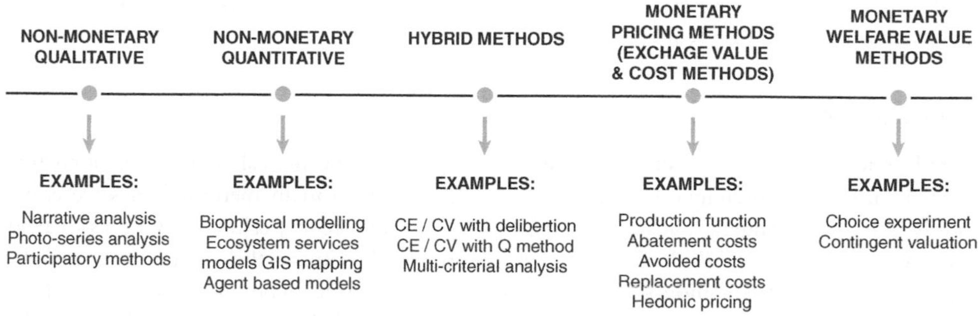

Figure 34.1 Spectrum of valuation methods

Source: Authors' own

Pricing methods

The market price of certain goods provides useful information as it signals the scarcity of the resource. Before using such information for valuing ecosystem related goods and services, prices need to be adjusted for any market distortions such as subsidies and taxes and for non-competitive practices. The production function approach (see for example, Fezzi et al., 2014) is an example of a pricing method frequently used for valuing the ecosystem provisioning services provided to agriculture. However, in this case, other inputs to production, including labour, expertise or manufactured and social capital, need to be disentangled in order to identify the contribution of natural capital to the value of a given market good. This approach could be applied to a variety of ecosystem services including crop pollination or protection from tropical storms.

Other pricing methods rely on costs as an approximation of the value of ecosystem inputs. One approach is the defensive expenditures technique that is based on the costs incurred by people to avoid damage caused by degradation of an ecosystem service. Another is to look at the damage costs avoided when any such service would be lost or be degraded, as for example, the loss of land and property related to storm surge due to mangrove forest removal. Costs of replacement or restoration of an ecosystem service are also sometimes used, but as they do not take into account the preferences of the users or beneficiaries of the services, their use is controversial (see Barbier, 2007).

The main limitation of the pricing approaches is that they rely on existing markets for capturing the value of ecosystem services. For many ecosystem-related benefits no markets exist, so other methods are required.

Revealed preferences methods

Revealed preference methods (see Bockstael and McConnell, 2007) can be useful when ecosystem goods or services are strongly associated with some existing market priced goods. Two methods are often used: the travel cost technique and hedonic pricing. The travel cost method (see for example, Egan et al., 2009) relies on the premise that the value of ecosystem related recreation benefits is implicitly shown in peoples' travel expenses. Analysis of actual travel, time costs and admittance fees to a particular site can reveal the value people hold for the recreation benefits of a given site. The hedonic pricing method (see for example, Day et al., 2007) captures the price of an ecosystem associated with an associated market priced good. For example, analysing the property markets and disentangling the house prices into components such as structural characteristics but also ecosystem related benefits, it is possible to value aesthetic views, air quality, flood risk or noise reduction.

Stated preferences

Stated preference valuation methods use surveys to portray hypothetical environmental change to elicit individuals' preferences regarding these changes. Two main methodologies are commonly used – contingent valuation method (CV) and choice experiments (CE; also called discrete choice models). While the two methods are broadly similar, the CV (see for example, Bishop et al., 2017) is a mono-attribute technique where one hypothetical scenario which encompasses all the characteristics of a portrayed environmental change is used for eliciting individuals' willingness to pay (WTP) for it. CE (see for example, Hanley et al., 2001) offers more flexibility as the bundle of environmental characteristics are represented by multiple attributes and willingness to pay is just one of these. Respondents are presented with a number of alternatives with varying attribute levels to choose from. CE results provide implicit willingness to pay for a change in each attribute, which, in turn, forms a richer set of information for decision making.

A major advantage of stated preference methods is that they can be used to value all ecosystem related goods and services and are the only methods that can be used to elicit non-use values. They can be particularly useful for situations that have not yet happened, for example, in terms of potential policy change. The disadvantage (and frequent criticism) of these methods lies in the hypothetical nature of the valuation exercise and the reliability of the results due to what is known as hypothetical bias (i.e. the fact that hypothetical WTP values are often higher than actual values). Improvements in survey design are possible following recent guidelines (for example, Johnston et al., 2017).

Value transfer

In cases when undertaking a new valuation study is not feasible due to time or resource constraints, a value transfer approach is adopted. Value transfer (see for example, Ferrini et al., 2014, 2015), sometimes also termed benefit transfer, consists of taking a value elicited in one context (study site) and applying this value to a site that is the objective of the study (policy site). The value(s) can be transferred with or without adjustments, aiming to reflect the differences between the study and policy sites (for example, income adjustments). Another approach is to estimate a value transfer function that establishes a statistical relationship between the values estimated and the site characteristics that can be then used for deriving values for another context. Recent approaches derive value functions that incorporate spatial context (see for example, Liekens et al., 2013; Ferrini et al., 2015). The advantage of the value transfer approaches lies in the growing availability of primary studies and in the lower time and budget requirements of the method. However, the saving in resource inputs comes with the costs of high-value transfer errors unless there is a reasonable degree of similarity between study and policy sites.

Integrated modelling approaches

Pricing and valuation methods are frequently integrated in decision support tools that combine elements of biophysical and economic methods coupled with geographical tools to assess ecosystem services and natural capital trade-offs. These tools vary widely in applicability and scope but do try to incorporate the complex trade-offs that are associated with ecosystem management in order to provide spatially explicit decision support advice. In light of increased data availability and methodological developments, it is likely that such integrated assessments will become more common place. For examples of Integrated modelling tools see for example, InVEST, Sharp et al. (2016); ARIES, Villa et al. (2014); or TIM, Bateman et al. (2014).

Non-monetary approaches

A range of non-monetary approaches are available for valuing ecosystem related goods and services (for example, Kenter, 2016; Barton and Harrison, 2017; Harrison et al., 2018; van Zanten et al., 2016). These could be broadly divided into biophysical and socio-cultural methods. The biophysical methods include, but are not limited to, biophysical modelling (for example, 'species distribution modelling', hydrological models), ecosystem services and integrated modelling, GIS mapping, emergy index or agent-based modelling. These methods can be used to, for example, quantify the benefits in non-monetary terms (for example, amount of water purified or number of pollutant particles removed from air for a given area and time period) but are also often key inputs for other approaches, such as most monetary methods, to provide an understanding of the 'ecological production function' supplying benefits. Socio-cultural valuation methods form a heterogeneous family of methods whose common characteristic is that they do not rely on either economic or biophysical methods. This family of methods includes, for example, narrative analysis, deliberative mapping, participatory scenario analysis or photo-series analysis. Socio-cultural methods are particularly suitable, for example, for understanding spiritual and sense of place related values – but can also complement other valuation methods.

Non-monetary approaches might be used alone or in parallel with monetary approaches (hybrid approaches in Figure 34.1) to provide more nuanced understanding of the relationship between people and nature. In fact, some of the approaches employed in socio-cultural valuation methods are often used in the first stage of monetary valuation methods – particularly in focus group deliberations.

Following this brief review of valuation methods, we now turn to the question of how to incorporate such values more fully into decision support indicators that measure human progress.

Environmental and ecosystem accounting

The past three decades have seen attempts to broaden the scope of national accounting to include in its remit environment, ecosystems and natural capital. Environmental accounting aims to construct a coherent and compatible system of statistics related to the natural environment that is fully compatible with the 'System of National Accounts' (SNA). SNA lies at the heart of current decision making and its headline indicator, gross national product, is a major driver of policy worldwide. A new statistical standard has been adopted that allows accounting for natural assets, such as forests and minerals, in the system of national accounts and an experimental ecosystem accounting process has been also introduced (for recent developments in ecosystem accounting see an overview given in Hein et al. (2015) and a potential application in Remme et al. (2015) and Vallecillo et al. (2018)).[4] Both aim to complement the SNA system with biophysical and monetary data on the links between the economy and the environment. The aim here is to provide a coherent statistical system that helps to monitor the condition and value of environmental and ecosystem assets and how these change over the years, as well as monitoring the annual flows of (ecosystem) services from these assets. Ecosystem accounting in particular poses numerous measurement, conceptual and theoretical challenges, particularly related to the spatial dimension of ecosystem accounting and the monetary methods that can be used within the accounting context. This formidable task is, however, worthwhile – as building environmental accounts will represent a much more comprehensive evidence base for sustainable progress as well as the integration of sustainability into economic policies.

Valuation is a particular challenge when it comes to ecosystem accounting (for further details see Atkinson and Obst, 2017; Obst et al., 2015). The major reason for this is that the SNA is based on exchange values (via pricing methods) which are not available for non-market goods and services related to nature. Welfare values, which in contrast to exchange values include consumer surplus,[5] can be estimated for some of the ecosystem services and related goods using some of the methods discussed in the previous section (most notably stated preference approaches). However, welfare values are not strictly compatible with the SNA system (for example, Obst et al., 2015).

This is the case given the assumption that the immediate and only aim of environmental accounts is full and consistent integration with the SNA for which only exchange value methods can be used. However, as is argued elsewhere (Badura et al., 2017), another more pragmatic approach might be chosen – one that allows experimentation and development of so-called satellite accounts. This approach allows more flexibility in the use of multiple valuation methods and types of values (including socio-cultural), as each account would be collated with a specific policy question in mind. The last approach (see for example, UNU-IHDP and UNEP, 2014; Lange et al., 2018) to environmental accounting aims to develop a comprehensive 'wealth' account that relies on valuation methods that aim to estimate the shadow prices of capital assets – which translate as the contribution of a marginal unit of the asset to human wellbeing. This is the most data-intensive approach and due to pragmatic reasons often employs pricing methods but is theoretically compatible with other valuation methods such as stated preference approaches.

The role of natural capital and ecosystem services accounting in supporting policy making

This chapter has summarised the different methods developed over the past decades which aim to incorporate the value of the benefits that people receive from natural capital and ecosystem services into decision making. The proliferation of the valuation research reflects a growing policy demand and related commitments for the incorporation of the different environmental values into international, national and local decision-making support systems (see Box 34.1), reporting and accounting frameworks. Valuation research has sought to fulfil a number of different policy support needs, including the evaluation of local policy alternatives (especially through monetary cost-benefit analysis (CBA) but also through non-monetary valuation research); national-scale assessments and scenario assessments (for example, UK NEA, 2014); and in some cases the provision of an evidence base for litigation (for example, Bishop et al., 2017). More recently, valuation research has been also discussed in the context of monetary environmental accounts (natural capital accounts) to extend or supplement national income accounts.

The diversity of valuation research reflects the plurality of values and views that is often at stake in environmental decision making. The range of valuation methods can be broadly distinguished into (pricing and welfare value based) monetary methods (qualitative and quantitative), non-monetary methods and the hybrid forms of the two. Each group of valuation methods can be used to inform different stages in the environmental policy-making process and also to reflect the plurality of perspectives and values involved. This can help to form a comprehensive evidence base for a wider decision support system that provides both data assessments as well as actual policy implementation pathways (see a generic example in Box 34.1). Monetary valuation research is generally used to support policy evaluation and formulation, either for systematic assessment of the use and supply of ecosystem services benefits (for example, via natural capital accounting) or for evaluation of different options for future policy or past projects evaluation (for example, via CBA, see for example, Bateman et al., 2016). The non-monetary methods on the other hand are well suited to quantify or qualify the relationships between nature and humans that are not easily captured in market-based terms. The quantitative non-monetary methods can provide essential biophysical and socio-economic indicators that can be used directly to monitor specific policy targets (for example, biodiversity and subjective wellbeing indicators) and/or are often used with qualitative methods as inputs to monetary methods. Qualitative non-monetary methods are particularly well placed to capture the intrinsic characteristic of some values including relational and socio-cultural values. They can help, through deliberative processes, with the understanding of the wider views and potential conflicts regarding policy implementation. Finally, hybrid methods can be deployed for specific policy questions (for example, Q methodology and multi-criteria analysis).

Box 34.1 Decision support systems – the balance sheet approach

A key requirement for policy making is as comprehensive an evidence base as possible. Construction of the evidence base has to be enabled through a decision support system. This system should encompass, among other things, basic data collection, scoping and agenda setting frameworks, data interpretation and modelling and the monitoring of option choice success. The 'Balance Sheet Approach (BSA)' is one such support system which has general applicability, and is as much about investigating the policy issues and questions as it is about finding answers (Turner, 2016). The BSA is both a process and a set of tools to aid environmental policy making and it allows for the collation, interrogation, analysis and presentation of the evidence base in any given policy context (UK NEA, 2014; see Figure 34.2). It is composed of three interlinked and overlapping sets of analyses. In the first 'balance sheet' a strategic level analysis, based on conventional environmental and economic efficiency analyses (e.g. CBA, environmental impact assessments) and often employing monetary valuation methods, considers the given policy change from a macroeconomic economic efficiency point of view. Further, two 'balance sheets' allow for different spatial scales and in particular focus, through trade-off analysis, on distributional impacts and fairness and equity concerns, i.e. who gains and who loses from policy option choices. This involves broadening of the conception of values considered and the identification of conflicting issues both of which are well addressed by non-monetary valuation approaches. The focus on gains and losses and compensation often requires a downscaling to regional/local spatial scales (i.e. the move from balance sheet one to two in Figure 34.2). The transition from the second to

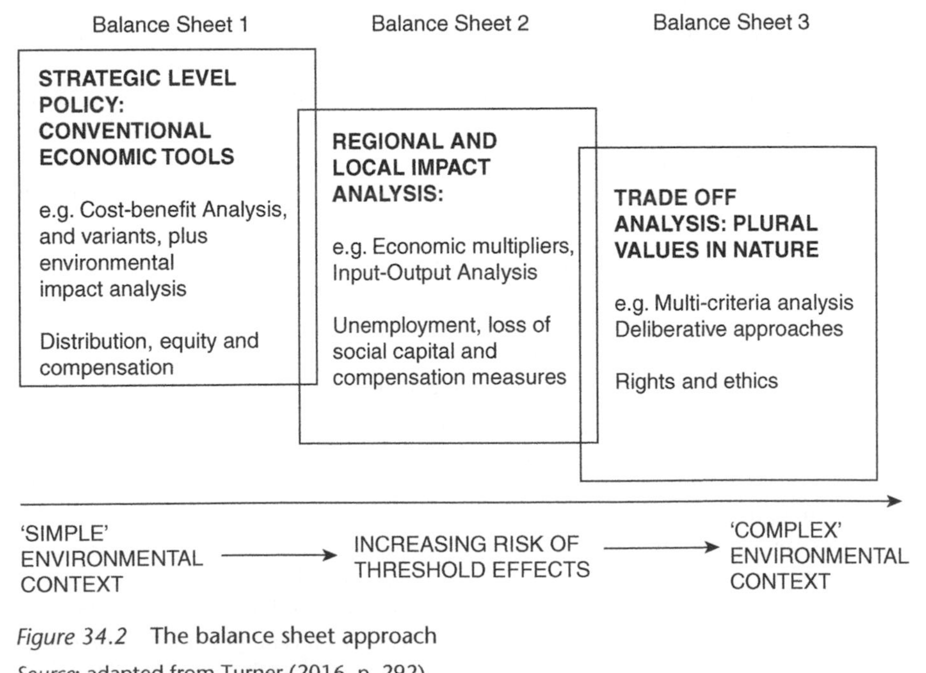

Figure 34.2 The balance sheet approach

Source: adapted from Turner (2016, p. 292)

the third 'balance sheet' specifically focuses on the identification of actual compensation measures (for example, financial or in-kind) and how such measures can be implemented by facilitating negotiation between the stakeholder groups involved. In this context, qualitative monetary methods, such as deliberative techniques and processes, are particularly useful.

Source: Authors' own text, based on Turner (2016)

Notes

1 For a simple article about the role of biodiversity in the world see this very well written summary: www.theguardian.com/news/2018/mar/12/what-is-biodiversity-and-why-does-it-matter-to-us.
2 We will use the term environmental accounting interchangeably for the family of approaches related to incorporation of the natural environment in the national accounts.
3 See also, for example, http://teebweb.org/.
4 For official accounting documents see https://seea.un.org/ and www.wavespartnership.org.
5 Consumer surplus represents an economic gain from the difference between the maximum price that a consumer is willing to pay for a given good and the market price that she needs to pay (or is paying) for it, aggregated across the relevant population.

References

Atkinson, G. and Obst, C. (2017). *Prices for ecosystem accounting*. Technical Report. Prepared as part of the work program of the World Bank WAVES project. Available at: www.wavespartnership.org/en/knowledge-center/prices-ecosystem-accounting

Badura, T., Ferrini, S., Agarwala, M. and Turner, K. (2017). *Valuation for natural capital and ecosystem accounting*. Synthesis report for the European Commission. Norwich: Centre for Social and Economic Research on the Global Environment, University of East Anglia.

Balvanera, P., Pfisterer, A. B., Buchmann, N., He, J. S., Nakashizuka, T., Raffaelli, D. and Schmid, B. (2006). 'Quantifying the evidence for biodiversity effects on ecosystem functioning and services'. *Ecology Letters*, 9(10): 1146–56.

Barbier, E. B. (2007). 'Valuing ecosystem services as productive inputs'. *Economic Policy*, 22(9): 177–229.

Barton, D. N. and Harrison, P. A. (eds.) (2017). *Integrated valuation of ecosystem services. Guidelines and experiences*. European Commission FP7, 2017. EU FP7 OpenNESS Project Deliverable 33–44.

Bateman, I. J., Mace, G. M., Fezzi, C., Atkinson, G. and Turner, K. (2011). 'Economic analysis for ecosystem service assessments'. *Environmental and Resource Economics*, 48(2): 177–218.

Bateman, I., Day, B., Agarwala, M., Bacon, P., Badura, T., Binner, A., De-Gol, A. J., Ditchburn, B., Dugale, S., Emmett, B., Ferrini, S., Fezzi, C., Harwood, A., Hillier, J., Hiscock, K., Hulme, M., Jackson, B., Lovett, A., Mackie, E., Matthews, R., Sen, A., Siriwardena, G., Smith, P., Snowdon, P., Sünnenberg, G., Vetter, S. and Vinjili, S. (2014). *UK national ecosystem assessment follow-on. Work package report 3: economic value of ecosystem services*. London, UK: UNEP-WCMC, LWEC.

Bateman, I., Agarwala, M., Binner, A., Coombes, E., Day, B., Ferrini, S., Fezzi, C., Hutchins, M., Lovett, A. and Posen, P. (2016). 'Spatially explicit integrated modeling and economic valuation of climate driven land use change and its indirect effects'. *Journal of Environmental Management*, 181: 172–84.

Bishop, R. C., Boyle, K. J., Carson, R. T., Chapman, D., Hanemann, W. M., Kanninen, B., Kopp, R. J., Krosnick, J. A., List, J., Meade, N., Paterson, R., Presser, S., Smith, V. K., Tourangeau, R., Welsh, M., Wooldridge, J. M., DeBell, M., Donovan, C., Konopka, M. and Scherer, N. (2017). 'Putting a value on injuries to natural assets: the BP oil spill'. *Science*, 356(6335): 253–4.

Bockstael, N. E. and McConnell, K. E. (2007). *Environmental and resource valuation with revealed preferences*. Dordrecht: Springer Netherlands.

Costanza, R., d'Arge, R., de Groot, R., Farber, S., Grasso, M., Hannon, B., Limburg, K., Naeem, S., O'Neil, R. V., Paruelo, J., Raskin, R. G., Sutton, P. and van den Belt, M. (1997). 'The value of the world's ecosystem services and natural capital'. *Nature*, 387: 253–60.

Costanza, R., de Groot, R., Sutton, P., van der Ploeg, S., Anderson, S. J., Kubiszewski, I., Farber, S. and Turner, R. K. (2014). 'Changes in the global value of ecosystem services'. *Global Environmental Change*, 26(1): 152–8.

Day, B., Bateman, I. and Lake, I. (2007). 'Beyond implicit prices: recovering theoretically consistent and transferable values for noise avoidance from a hedonic property price model'. *Environmental and Resource Economics*, 37(1): 211–32.

Díaz, S., Fargione, J., Chapin, F. S. and Tilman, D. (2006). 'Biodiversity loss threatens human well-being'. *PLoS Biology*, 4(8): e277.

Egan, K. J., Herriges, J. A., Kling, C. L. and Downing, J. A. (2009). 'Valuing water quality as a function of water quality measures'. *American Journal of Agricultural Economics*, 91(1): 106–23.

Ferrini, S., Schaafsma, M. and Bateman, I. (2014). 'Revealed and stated preference valuation and transfer: a within-sample comparison of water quality improvement values'. *Water Resources Research*, 50(6): 4746–59.

Ferrini, S., Schaafsma, M. and Bateman, I. J. (2015). 'Ecosystem services assessment and benefit transfer', in R. J. Johnston, J. Rolfe, R. S. Rosenberger and R. Brouwer (eds.) *Benefit transfer of environmental and resource values: a guide for researchers and practitioners*. Dordrecht: Springer, pp 275–305.

Fezzi, C., Bateman, I., Askew, T., Munday, P., Pascual, U., Sen, A. and Harwood, A. (2014). 'Valuing provisioning ecosystem services in agriculture: the impact of climate change on food production in the United Kingdom'. *Environmental and Resource Economics*, 57(2): 197–214.

Fisher, B., Turner, R. K. and Morling, P. (2009). 'Defining and classifying ecosystem services for decision making'. *Ecological Economics*, 68(3): 643–53.

Freeman III, A. M., Herriges, J. A. and Kling, C. L. (2014). *The measurement of environmental and resource values: theory and methods* (3rd edn). Abingdon: RFF Press.

Hanley, N., Mourato, S. and Wright, R. E. (2001). 'Choice modelling approaches: a superior alternative for environmental valuation?' *Journal of Economic Surveys*, 15(3): 435–62.

Harrison, P. A., Dunford, R., Barton, D. N., Kelemen, E., Martín-López, B., Norton, L., Termansen, M., Saarikoski, H., Hendriks, K., Gómez-Baggethun, E., Czúcz, B., Llorente, M. G., Howard, D., Jacobs, S., Karlsen, M., Kopperoinen, L., Madsen, A., Rusch, G. M., Eupen, M., Verweij, P., Smith, R., Tuomasjukka, D. and Zulian, G. (2018). 'Selecting methods for ecosystem service assessment: a decision tree approach'. *Ecosystem Services*, 29(Part C): 481–98.

Hein, L., Obst, C., Edens, B. and Remme, R. P. (2015). 'Progress and challenges in the development of ecosystem accounting as a tool to analyse ecosystem capital'. *Current Opinion in Environmental Sustainability*, 14: 86–92.

Johnston, R. J., Rolfe, J., Rosenberger, R. S. and Brouwer, R. (eds.) (2015). *Benefit transfer of environmental and resource values: a guide for researchers and practitioners*. Dordrecht: Springer.

Johnston, R. J., Boyle, K. J., Adamowicz, W., Bennett, J., Brouwer, R., Cameron, T. A., Hanemann W. M., Hanley, N., Ryan M., Scarpa, R., Tourangeau, R. and Vossler, C. A. (2017). 'Contemporary guidance for stated preference studies'. *Journal of the Association of Environmental and Resource Economists*, 4(2): 319–405.

Kenter, J. O. (2016). 'Deliberative and non-monetary valuation', in M. Potschin, R. Haines-Young, R. Fish and R. K. Turner (eds.) *Routledge handbook of ecosystem services*. London: Routledge, pp 271–88.

Lange, G. M., Wodon, Q. and Carey, K. (eds.) (2018). *The changing wealth of nations 2018: building a sustainable future*. Washington, DC: World Bank.

Liekens, I., Schaafsma, M., De Nocker, L., Broekx, S., Staes, J., Aertsens, J. and Brouwer, R. (2013). 'Developing a value function for nature development and land use policy in Flanders, Belgium'. *Land Use Policy*, 30(1): 549–59.

Mace, G. M., Norris, K. and Fitter, A. H. (2012). 'Biodiversity and ecosystem services: a multilayered relationship'. *Trends in Ecology and Evolution*, 27(1): 19–26.

Mace, G. M., Hails, R. S., Cryle, P., Harlow, J. and Clarke, S. J. (2015). 'REVIEW: towards a risk register for natural capital'. *Journal of Applied Ecology*, 52(3): 641–53.

Millennium Ecosystem Assessment (2005). *Ecosystems and human well-being: synthesis*. Washington, DC: Island Press.

Naeem, S., Duffy, J. E. and Zavaleta, E. (2012). 'The functions of biological diversity in an age of extinction'. *Science*, 336(6087): 1401–6.

Natural Capital Committee (NCC) (2014). *State of natural capital: restoring our natural assets*. London: Natural Capital Committee.

Obst, C., Hein, L. and Edens, B. (2015). 'National accounting and the valuation of ecosystem assets and their services'. *Environmental and Resource Economics*, 64(1): 1–23.

Remme, R. P., Edens, B., Schröter, M. and Hein, L. (2015). 'Monetary accounting of ecosystem services: a test case for Limburg province, the Netherlands'. *Ecological Economics*, 112: 116–28.

Sharp, R., Tallis, H. T., Ricketts, T., Guerry, A. D., Wood, S. A., Chaplin-Kramer, R. and Bierbower, W. (2016). *InVEST +VERSION+ User's Guide*. The Natural Capital Project, Stanford University, University of Minnesota, The Nature Conservancy, and World Wildlife Fund.

TEEB (2010). *The economics of ecosystems and biodiversity. Mainstreaming the economics of nature: a synthesis of the approach, conclusions and recommendations of TEEB*. Geneva: TEEB.

Turner, R. K. (2016). 'The balance sheet approach', in M. Potschin, R. Haines-Young, R. Fish and R. K. Turner (eds.) *Routledge handbook of ecosystem services*. London: Routledge, pp 289–98.

UK National Ecosystem Assessment (UK NEA) (2014). *The UK National Ecosystem Assessment follow-on: synthesis of the key findings*. Cambridge: UNEP-WCMC, LWEC, UK.

UNU-IHDP and UNEP (2014). *Inclusive wealth report 2014. Measuring progress toward sustainability*. Cambridge: Cambridge University Press.

Vallecillo, S., La Notte, A., Polce, C., Zulian, G., Alexandris, N., Ferrini, S. and Maes, J. (2018). *Ecosystem services accounting: Part I – outdoor recreation and crop pollination*, EUR 29024 EN. Luxembourg Publications Office of the European Union.

van Zanten, B. T., Van Berkel, D. B., Meentemeyer, R. K., Smith, J. W., Tieskens, K. F. and Verburg, P. H. (2016). 'Continental-scale quantification of landscape values using social media data'. *Proceedings of the National Academy of Sciences of the United States of America*, 113(46): 12974–9.

Villa, F., Bagstad, K. J., Voigt, B., Johnson, G. W., Portela, R., Honzák, M. and Batker, D. (2014). 'A methodology for adaptable and robust ecosystem services assessment'. *PLoS ONE*, 9(3): e91001.

35

Suitability analysis

A fundamental environmental planning tool

Robert Paterson and Frederick Steiner

Introduction

Within the field of city and regional planning, suitability analysis refers at its most fundamental level to simply considering where certain human activities (i.e. uses of land – such as, residential, agricultural, industrial, waste management, energy production and conservation) make the most sense within a given landscape area (i.e. would logically work well and create minimal harm to human or natural systems at a given location). Using this rather basic definition, it is clear that humans and collectives of humans have been engaged in suitability analysis, at least implicitly, since the very first civilizations formed. For example, in North America, the ancient Anastasi civilization selected cliffs in the Four Corners region of the southwest as locations for their villages because they provided security; proximity to wood, water, and food; and access to the sun (the focus of their religion) among other factors. The Native American people of the Pacific Northwest region often identified locations for their community houses near rivers where the pink or chum salmon ran as this species had low fat content and therefore could be dried and stored for long periods of time for the winter months. Colonial settlers of the New World often sited their communities close to deep water ports and along navigable rivers to ship and receive goods and people and because floodplains typically provided relatively flat, fertile soils that were suitable for agricultural cultivation. Fall lines were also important, initially as a barrier for navigation and eventually as a source of energy.

In the current Information Age, where spatial data about many facets of the built and natural environment are ubiquitous, we find that suitability analysis is foundational to virtually every level of planning and type of planning undertaken whether it's land use plans for regions, counties, cities, towns, districts and neighbourhoods or all the varieties of functional and special purpose plans that are needed to plan and operate public facilities, utilities and services in villages, towns and cities (for example, open space plans, school siting, infrastructure plans, energy facility plans and others) (Berke et al., 2015). This is not surprising, for so long as there are human settlements, we will always want to know the best place to locate certain activities and uses in ways that pose few or minimal risks as well as the best opportunities to function well.

While there are several related areas of suitability analysis in the resource management fields (such as the US Department of Agriculture's land capability analysis and Land Evaluation and Site Assessment systems), this chapter explains the evolution and application of suitability planning, largely from a North American city planning perspective, noting early exemplars and important methodological and technological developments (see Steiner, 2000 for these other systems). The chapter is divided into five more parts. The next part briefly outlines the historical development of the suitability mapping from the earliest efforts at the turn of the 20th century to the seminal work of Ian McHarg (1969) in applying and documenting a detailed methodology that is foundational to current methods, applications and technologies. The following two parts describe a five-step outline of the suitability analysis as currently practiced in mainstream US planning and elaborates on suitability practice using examples from Central Texas planning efforts. This is followed by a brief description of the use of land suitability analysis beyond the North American context and the conclusion discusses the challenges and opportunities for further refinements of suitability analysis as part of planning support systems and participatory planning more generally.

The beginnings of suitability analysis in city planning

Systematic consideration of land suitability for a variety of rural and urban planning purposes in the United States has been traced to the late 19th century when landscape architects from the firm Olmsted, Olmsted and Eliot used overlays of soils and topography on maps pasted onto windows so that sunlight penetrating the surfaces revealed the combined information of the two maps as a single composite layer to inform the planning and design of the Boston Metropolitan Park as well as the town plan for Billerica, Massachusetts (Collins et al., 2001; Steiner et al., 2000; Hopkins, 1977). According to Steiner (2000), a variety of plans created between 1910 and 1950 employed overlay mapping techniques; notable among these was the Regional Plan of New York (1929) which addresses suitability of locations for residential development but largely from an economic over an environmental perspective. It was not until 1950 that a planning text described a method or theory of overlay mapping for suitability analysis when an article written by British planning educator and town planner Jacqueline Tyrwhitt explains the use of four maps (elevation, hydrology, geology and soils) drawn on transparent paper to then combine to get a composite understanding of landscape features to inform planning and design (Collins et al., 2001). That description proved to be important for the diffusion of the method, as overlay mapping based suitability analysis became widely used in British town planning and other post World War II planning efforts.

In the 1960s, in North America, George Angus Hill, a natural resources planner in Ontario, Canada, and Philip Lewis, a landscape architect and planner working for the Wisconsin State Department of Land and Forests, independently mapped large areas of Ontario and Wisconsin using suitability overlay techniques. Hill focused on combining environmental and physiographic information to evaluate specific geographic subareas called 'landscape units' and ranked each unit's suitability for major land use classes (Collins et al., 2001), while Lewis focused on combined maps of water, wetlands, vegetation and view sheds to spatially delimit areas of prime environmental importance throughout the state (Steinitz et al., 1976). However, arguably the most important development of this era was the work of Ian McHarg and colleagues at the University of Pennsylvania. McHarg is credited with being the first to link suitability analysis with an ecological theory and environmental inventory methodological underpinning. McHarg argued that 'a region is understood as a biophysical and social process comprehensible through the operation of [natural] laws and time. This can be reinterpreted as having explicit opportunities and constraints for any particular human use' (1997, p. 321). In the seminal text 'Design with Nature', McHarg explained suitability analysis:

> [C]onsists of identifying the area of concern as consisting of certain processes, in land, water, and air – which represent values. These can be ranked the most valuable land and the least, the most valuable water resources and the least, the most and least productive agricultural land, the richest wildlife habitats and those of no value, the areas of great or little scenic beauty, historic buildings and their absence, and so on.
>
> *(McHarg, 1969, p. 34)*

At its most basic level suitability analysis asks, for a given land unit, what uses are more or less compatible and appropriate given the intrinsic properties of the land itself and its surroundings. Each land unit (for example, parcel, drainage area, or other geography) has specific identifiable degrees of vulnerability as it relates to natural systems (for example, natural hazards, wetlands, slope stability and habitat value) and specific attractiveness attributes as it relates to human systems (for example, road or highway access, water and wastewater serviceable and market potential for development). The combining of these layers of mapped information both environmental and human systems based provides insights to the most suitable locations for a variety of land use needs, where it is possible to rate all land units suitabilities for all land uses under consideration. In McHarg's pioneering work, environmental and human system information was mapped on clear Mylar sheets or transparent paper where shades of grey represented the most intense degree of preferred feature (for example, susceptible to erosion or high scenic value might have three rankings lesser to greater with three gradients of grey mapped for a landscape unit). When a series of maps with differing layers of environmental information are then overlaid with all features sharing the same light grey (avoid) to dark grey (use appropriate) gradient, the darkest areas would be the most suitable.

Fast forward to today, and it is fair to say suitability analysis has become much easier, with the advent of computer based mapping and spatial analysis calculations available through geographic information system (GIS) software. There is literally a plethora of choices of layers of environmental, social, economic, infrastructure, scenic, housing and other information available to the typical planner or at least possible access to such data from national, state, local, private and non-profit spatial data clearinghouses and information sources. One of the main challenges seen as land use and environmental planning educators and practitioners is that land use and environmental planners today seem to struggle with knowing when enough is enough and truly focusing on what is centrally important to the suitability analysis task at hand. Today's practitioners have so much data availability that, in some ways, the task in front of them is akin to drinking from a fire hose, when a simple faucet worth of data is all one may really need. Adding more and more data is not necessarily a better thing in suitability analysis, and in many cases it is noted that excessive layers leads to redundancy of data, may overwhelm participants to a planning process, providing little or no added value to the analysis. This observation is expanded on with some illustrations of suitability analyses in the following section.

The suitability analysis method and examples

Figures 35.1 and 35.2 represent the typical range of plans found in most metropolitan regions of the United States to guide development as well as suitability analysis foci typically found in the analysis to inform each plan type. Berke et al. (2015) refer to these as the 'web of plans' that make

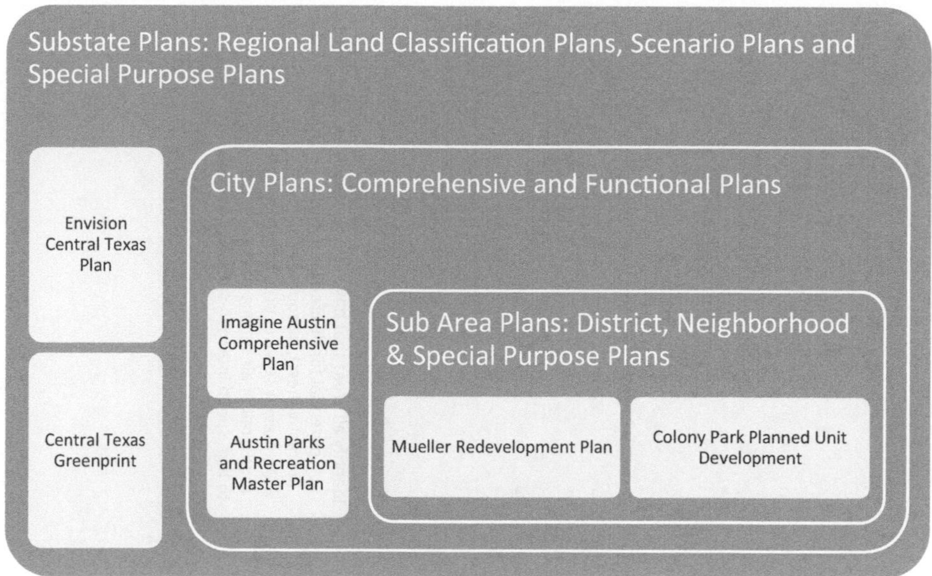

Figure 35.1 The web of plans: examples from Central Texas

Source: Authors' own

up planning in most regions of the United States reflecting differing scales, purposes and tasks. Sub-state plans would include regional plans that typically are focused on broad landscape scale classifications of land use for urban and rural uses such those used as part of regional scenario plans, regional urban growth boundary or service boundary designation, and regional greenspace or habitat conservation plans. The primary task of suitability analysis for these plans is typically defining the urban, rural, conservation, or protected area divide. For example, suitability analysis strongly informed the decisions for urban growth boundary designations following the creation of the Portland Metro 2040 growth concept plan in Oregon by identifying areas that were important to agricultural (for example, prime agricultural soils and areas with high agricultural productivity) and forestry industries (sustainable yield forested areas) while concentrating urban growth near existing urbanised areas with adequate (or planned) central water and sewer service to reduce speculative land pressure on rural uses.

City comprehensive plans (also known as general plans in California and city master plans in New Jersey) nest within the larger regional plans and provide more nuanced guidance to land development and conservation activities within and at the edge of the city through future land use maps or growth concept maps that are informed by multiple versions of suitability mapping performed for each land use type under consideration (for example, single family residential, multi-family residential, heavy and light industrial, general retail and office, and conservation oriented land uses). In addition, most cities and towns in the United States will have a fair number of functional plans that inform the operation and expansion of the city's public facilities and services on a citywide basis (for example, parks and recreation, water and wastewater, transportation, housing, storm water and flood control, police and fire and others). These functional plans are expected to nest within the city comprehensive plan and the larger regional and/or county plan(s) and often use suitability analysis to inform decisions about future service zone expansion/phasing, expansion of service capacity and the timing/location of new facilities and service.

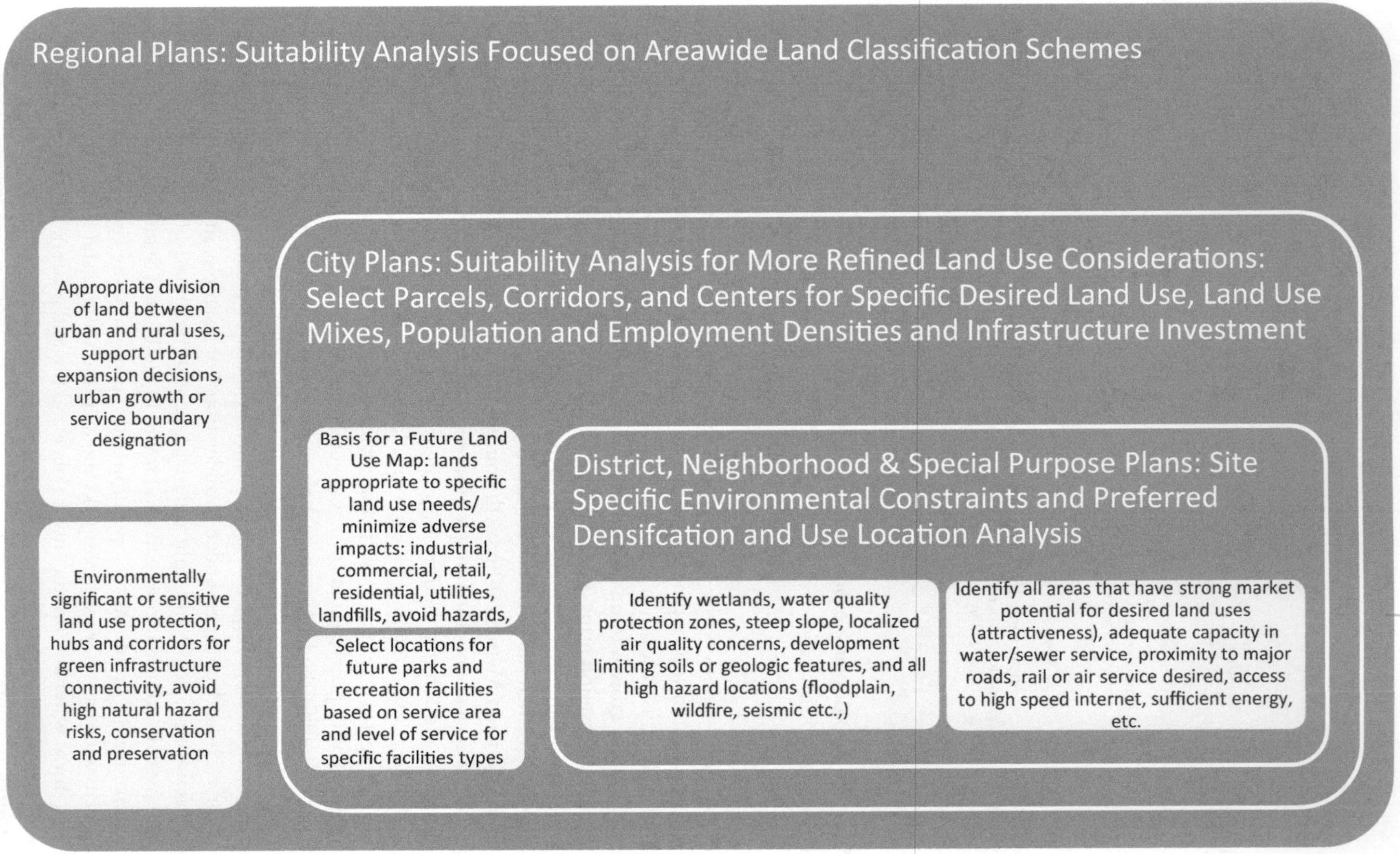

Regional Plans: Suitability Analysis Focused on Areawide Land Classification Schemes

Appropriate division of land between urban and rural uses, support urban expansion decisions, urban growth or service boundary designation

Environmentally significant or sensitive land use protection, hubs and corridors for green infrastructure connectivity, avoid high natural hazard risks, conservation and preservation

City Plans: Suitability Analysis for More Refined Land Use Considerations: Select Parcels, Corridors, and Centers for Specific Desired Land Use, Land Use Mixes, Population and Employment Densities and Infrastructure Investment

Basis for a Future Land Use Map: lands appropriate to specific land use needs/ minimize adverse impacts: industrial, commercial, retail, residential, utilities, landfills, avoid hazards,

Select locations for future parks and recreation facilities based on service area and level of service for specific facilities types

District, Neighborhood & Special Purpose Plans: Site Specific Environmental Constraints and Preferred Densifcation and Use Location Analysis

Identify wetlands, water quality protection zones, steep slope, localized air quality concerns, development limiting soils or geologic features, and all high hazard locations (floodplain, wildfire, seismic etc.,)

Identify all areas that have strong market potential for desired land uses (attractiveness), adequate capacity in water/sewer service, proximity to major roads, rail or air service desired, access to high speed internet, sufficient energy, etc.

Figure 35.2 Suitability foci for scales of plans

Source: Authors' own

For example, many city park and recreation plans have not only projections of service needs for things like baseball diamond or trails but also proximity standards for access to different types of parks, which logically lend themselves to suitability mapping for parcels that would best fit that need and serve the greatest population with the minimum amount of waste.

Beneath the city scale plans, in scale, are district, special area and neighbourhood plans which once again should nest within (be compatible with and informed by) but may also use an even more refined scale of analysis than found at the city scale (for example, drainage areas may be delineated down to a 5-acre area at a district level while at a city scale 64 acres may be as deep as drainage feature mapping goes). There is a wide array of district plan types found in most US cities including downtown plans, historic districts, hospital districts, airport districts, port plans, corridor plans (linear districts), entertainment districts, neighbourhood districts, cultural and arts districts and many more. Land use and environmental considerations often become very fine grained, down to a parcel level in district plans. In many cases, where large undeveloped parcels are present, district plans may suggest options to subdivide the areas using proposed streets and land use designations to reflect urban design scale intentions (for example, massing and scaling of development in relation to streets and surrounding development). For example, suitability mapping in a corridor district plan may be used to identify prime spots to locate Light Rail or Bus Rapid Transit stops based on a variety of factors including projected ridership along the corridor based on expected job and residential densities, proximity to large tracts suitable for densification at a stop, logical locations for intermodal transfers, and other factors.

In all of the types of plans described, the planner typically has a five-step process to complete a suitability analysis (keeping in mind that this often will require repeating numerous times for various planning needs). The first step is identifying the land use or activity that needs to be analysed (for example, siting a large windfarm or identifying parcels that are needed to create a regional open space network with strong ecological service potential). The second step is identifying the environmental and human systems spatial data that would be a constraint to or a facilitator of that intended use (for example, for a windfarm a constraint would be proximity to residential and for an open space plan a facilitator might be parcels with high habitat value and floodplains that border on other conservation-oriented parcels).

The third step involves converting all spatial data into compatible and combinable data formats so that maps the planner wishes to overlay can be combined in an accurate manner in GIS. This can be the most laborious aspect of suitability mapping today simply because we do not have uniform standards for creating and compiling spatial information in the United States, EU, and many other parts of the world. This means mapped information may be based on different projection systems, differing categories and measurement approaches across city departments, and adjoining cities, and using different metadata documentation approaches. Some environmental and human system data are spatially recorded in GIS in vector format (for example, polygons, lines and points such as floodplains, streams and soil classes) while other data are recorded in raster format (grid cells such as digital elevation data) and thus planners must convert all data used to a common format that is combinable in a meaningful and accurate manner. Moreover, it often is necessary to rescale, reclassify and change the valence of environmental and human system data so when maps (data) are combined, constraints and facilitators in the correct direction are added. For example, greater slope is generally a constraint to most human land uses (for example, industrial and office land uses is less likely as slope increases), but proximity to a wetland is inverse to this and also a constraint (i.e. as proximity to a wetland is less favourable for industrial use).

The fourth step is deciding the 'decision rules' that will be applied in combining the spatial data of constraints and facilitators either sequentially or in a single computational step for the land use under consideration. For example, it is very common practice to first remove certain land areas as absolutely not suitable for some land use purposes (for example, high natural hazard zones like floodways, earthquake surface fault rupture or liquefaction zones, and ridge lines for wildfire risks) and then do additional suitability analysis using different decision rules such as either straight additive combining of constraints and facilitators data using a common scale such as low, medium, high (ordinal scaling) or use of weights to reflect either participatory process preferences or expert judgement of the relative importance of certain constraints and facilitators (for example, an open space plan may place greater double weight on a map layer of known endangered species habitat over soils types or scenic resources values when combining layers of map information).

The fifth and final step is to combine the environmental and human system map layer data using the decision rules to create the composite map that shows where the land is more and less suitable for the land use under study. It is often necessary to repeat the mapping using differing assumptions and decision rules among stakeholders as a form of social learning before you may settle on a single combined map that all feel reflects the needs of the study.

In the next section, two examples are provided of suitability analysis from plans completed in Central Texas, not because these are believed to be exemplars in any sense but rather to show a range of variation in data used and approach as often found in practice. (Plus, it should be added, that the authors were involved in various capacities in these undertakings.) It is suggested that good suitability analysis for land use and environmental planning should be treated the same way as all forms of well executed planning analysis: with transparency to data, assumptions and methods used; with sensitivity analysis to explore uncertainties and the possible range of outcomes from differing approaches; and with full participation of stakeholders to provide authenticity and shared understanding of analysis.

Two Central Texas examples: the 'Travis County Greenprint' and the 'Imagine Austin' plan

In this section, the role that suitability analysis played in two Central Texas plans is briefly described, and these plans are used to illuminate some of the strengths and challenges confronting suitability analysis in environmental and land use planning practice.

Greenprint plans, created by the Trust for Public Land as both an analytical and participatory planning enterprise, identify prime lands worthy of conservation and garner buy-in from local officials and organisations to fund a desired regional open space or 'greenprint' network. The 'Travis County Greenprint' was a collaborative enterprise among the Trust for Public Lands, Travis County, the City of Austin and the University of Texas at Austin. Representatives from federal, state, regional, local, neighbourhood, environmental justice, land conservation and special districts were convened for a two-year suitability mapping project. The key purpose of the suitability mapping was to identify currently unprotected areas that offered the 'highest conservation benefit' based on locally identified goals and criteria, and to mobilise public support for conservation efforts. A technical advisory team developed a suitability rating matrix with weights assigned to specific land use features or attributes important to four conservation goals: protect water quality and quantity, improve recreational opportunities, protect sensitive and rare environmental features, and protect cultural resources. Table 35.1 details the spatial data that were identified as important to the four goals and the relative weighting of importance of each attribute to the associated goal. A total of four maps were created and modified with input from community organisations for each goal before the final four maps were adopted by the group. These four maps, reflecting the four conservation goals, were then combined into a final master

Table 35.1 Underlying suitability maps (goals and weights) that informed final composite Greenprint suitability map

Goals and supporting criteria	*Weight assigned (percent)*	*Data source*
Suitability Map 1 Goal: Protect water quality and quantity		
Criteria 1: Streams/Riparian Corridor	10	US Geological Survey Hydrologic Units
Criteria 2: Waterbody	10	City of Austin GIS Data
Criteria 3: Floodplain	25	City of Austin GIS Data
Criteria 4: Aquifer Down Dip Zone	3	City of Austin Data
Criteria 5: Steep Slopes	5	USDA Geospatial Data
Criteria 6: Native Prairies	3	City of Austin GIS Data
Criteria 7: High Quality Woodlands	5	University of Texas GIS Data
Criteria 8: Recharge Zones	20	City of Austin GIS Data
Criteria 9: Recharge Contributing Zones	9	City of Austin GIS Data
Criteria 10: Alluvial Soils	10	NRCS, Trust for Public Lands
Suitability Map 2 Goal: Improve Recreational Opportunities		
Criteria 1: Greenspace	10	Capital Area COG (CAPCOG) GIS Data
Criteria 2: Water Access	10	City of Austin GIS Data
Criteria 3: Adjacent to Existing Parks	10	City of Austin GIS Data
Criteria 4: Community Gardens	5	Trust for Public Lands GIS Data
Criteria 5: Park Equity	15	City of Austin GIS Data
Criteria 6: Riparian Corridors	5	USGS NHD GIS, Lower Colorado River Authority (LCRA)
Criteria 7: Wildlife Corridors	5	USGS GIS, LCRA
Criteria 8: Trail Connectivity	5	USGS GIS, City of Austin
Criteria 9: Trail Corridors	10	City of Austin
Criteria 10: Floodplain	25	City of Austin GIS, FEMA
Suitability Map 3 Goal: Protect Rare/Sensitive Environmental Features		
Criteria 1: High Quality Woodlands	15	University of Texas and Travis County
Criteria 2: Migratory Bird Habitat	10	TPL National GIS Database
Criteria 3: Habitat Connectivity	15	COA, LCRA, USGS NHD GIS
Criteria 4: Geologic Features	10	City of Austin GIS
Criteria 5: Senstitive Environmental Features	10	USGS NHD, CAPCOG GIS
Criteria 6: Native Prairies	10	City of Austin GIS
Criteria 7: Threatened and Endangered Species	20	Texas Parks and Wildlife
Criteria 8: Alluvial Soils	10	City of Austin, TPL National GIS, NRCS
Suitability Map 4 Goal: Protect Cultural Resources		
Criteria 1: Working Lands	20	CAPCOG GIS
Criteria 2: Viewsheds	20	COA GIS
Criteria 3: Federal/State Historical Sites	20	Texas Historical Commission
Criteria 4: Scenic Corridor	25	TPL National GIS
Criteria 5: Adjacent to a Conservation Easement	15	COA GIS

Source: Adapted from The Trust for Public Land (2008, p. 11)

suitability map, with each map having equal weight (25% weighting). The final composite suitability map became the final priority conservation lands greenprint map as adopted by the stakeholders (see Figure 35.3).

The greenprint suitability analysis included many environmental objectives and a large number of data layers and, in some ways, perhaps employed questionable analysis. For example, two of the suitability maps had over ten layers of data with one having three layers that were only given 3% weights relative to the other layers in the analysis, suggesting negligible importance in the final mapping (which also begs the question of why they were included at all). As more and more layers of data get piled into a suitability map, it becomes less and less clear what you are really seeing spatially. From experience, it is a very good idea to look carefully at each layer in isolation to fully understand the spatial distribution of the attribute, the environmental or human processes or concerns it represents and only then decide what it adds before incorporating it

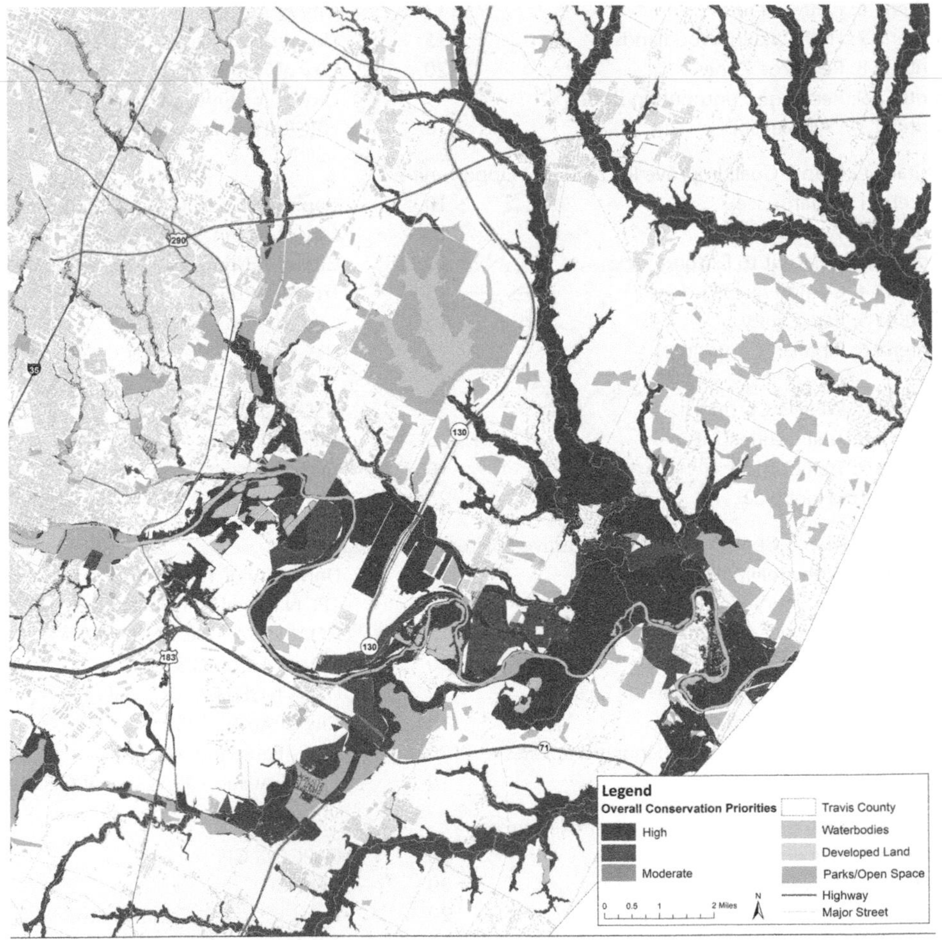

Figure 35.3 The final Greenprint map based on the combination of four different environmental and socio-cultural suitability maps to delineate priority lands for land conservation efforts

Source: The Trust for Public Land (2008, p. 19)

into the study. Moreover, in several cases some layers seem to be double counting, although presented as individually unique in the data matrix. For example, floodplains and alluvial soils are likely to overlap considerably, and yet both carry considerable weight in the analysis and are presented as distinct. Disagreements over weights chosen were considerable in the mapping process and the decision to give equal weighting to the four goal layers for the final suitability map had less to do with technical agreement and more to do with political compromise to come to closure. That said, the final greenprint map is largely considered a local success. It was used extensively by Travis County to guide conservation lands acquisitions, as well as smaller cities in the county, following its completion.

A second example is the suitability mapping that was used to inform the City of Austin's comprehensive plan entitled 'Imagine Austin'. This plan created a growth concept map instead of a traditional future land use map, largely to avoid getting bogged down in the politics that can arise with parcel level land use classification in traditional comprehensive plans (see Figure 35.4).

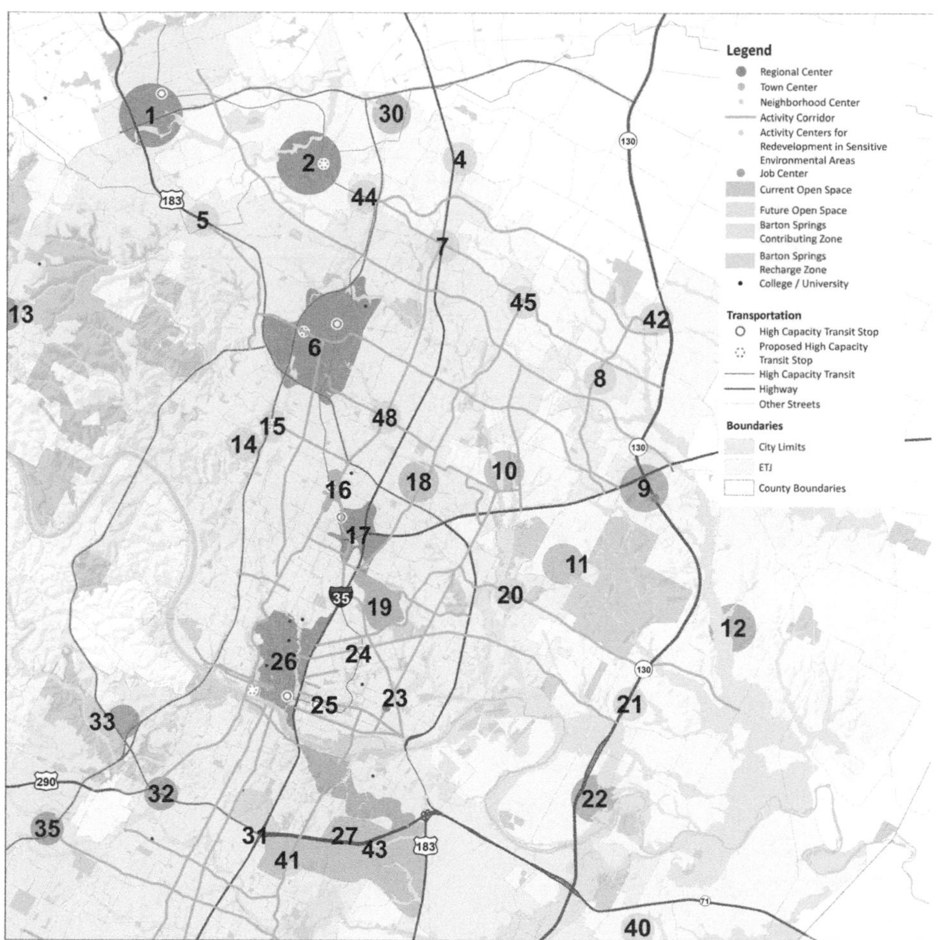

Figure 35.4 Final activity centre locations and centre acreage changes informed by suitability mapping

Source: City of Austin (2012, Appendix D, p. A-31)

At the heart of 'Imagine Plan' is the 'Centres and Corridors' map (see Figure 35.4) that had two levels of suitability mapping applied. At the first level was suitability mapping of where certain kinds of activity centres should be located to absorb future growth as either greenfield development or redevelopment (greyfields) centres. Second, those centres were reviewed and re-evaluated in light of constraints mapping, and modified to avoid conflicts where they arose. Factors viewed as favourable to 'Activity Centre' designation (facilitators) that informed the suitability mapping included proximity to highways and major arterial intersections, proximity to existing or planned high capacity transit, centres already identified in existing plans, available developable lands and already completed development agreements. Human system-based constraints to centres before evaluating them against environmental systems constraints included proximity to heavy industrial or hazardous materials land uses and landfills. The location and area of the 'Growth Concept' centres were then modified in light of many sequentially reviewed environmental suitability constraint maps that included consideration of conflicts with

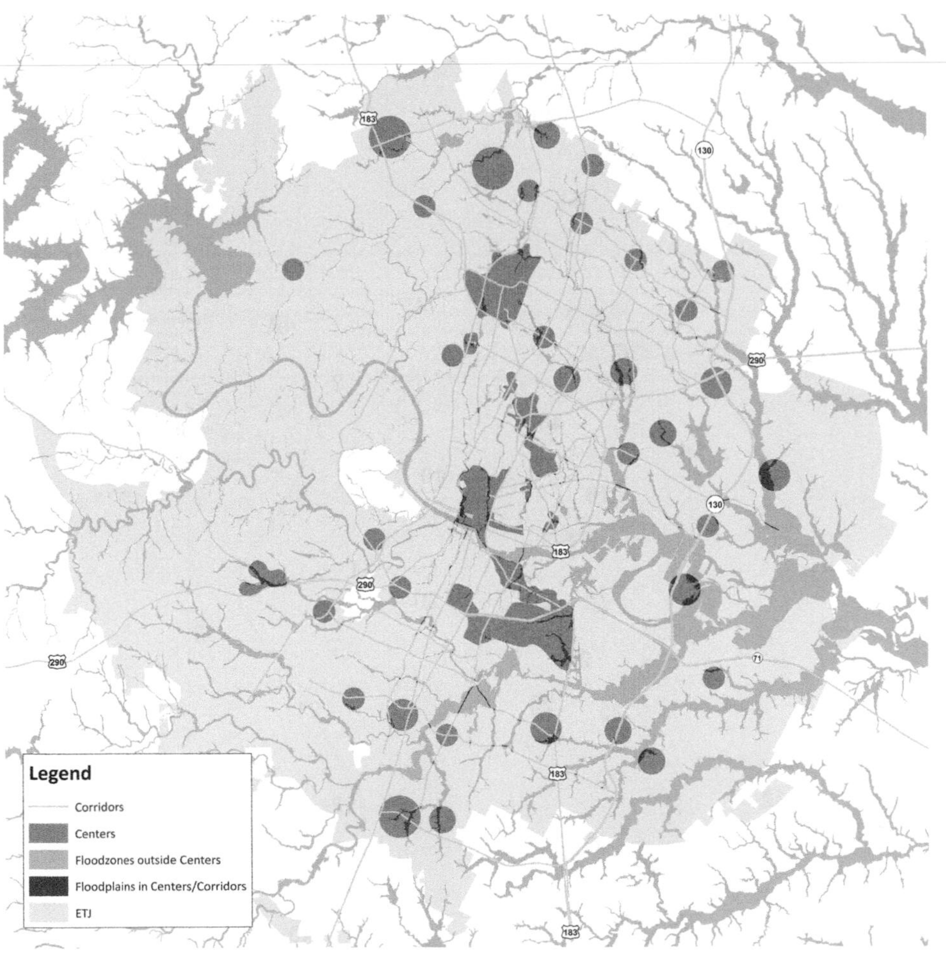

Figure 35.5 Example of floodplain constraint mapping which was used to adjust activity centre sizes and locations

Source: City of Austin (2012, Appendix D, p. A-43)

floodplains, wildfire hazard risk, stream water quality buffers, headwaters, steep slopes, prime farmlands and expansive soils (see, for example, Figure 35.5). The Growth Concept map's activity centres were moved in some instances due to incompatibility or reduced in area in other cases in light of the suitability mapping exercise.

In both examples, it is clear suitability analysis played an important role in shaping the planning process for important environmental and human system considerations. In 'Travis County Greenprint', over thirty different environmental, economic and cultural attributes were considered in combination through the suitability analysis to identify priority lands for conservation purposes. In the 'Imagine Austin' comprehensive plan, suitability analysis was used to both locate and then modify, for environmental purposes, citywide activity centre land use designations. The next section briefly discusses the use of land suitability analysis in the EU and Asia, and the chapter concludes with some closing thoughts on challenges and opportunities for suitability mapping.

Suitability mapping in a global context

A recent OECD study (2017) reported on land use planning practice for 32 OECD countries based on planning faculty survey responses from the member countries. Specific details on planning process are largely not covered by the report. However, the report does ask about challenges for land use planning systems and the number one most cited problem was land development in environmentally valuable places and loss of biodiversity (23 of 32 countries). While this does not necessarily mean suitability analysis is not being used, it may be indicative of the need for greater use. Suitability analysis is commonplace in many EU member states' regional and local spatial plans in response to various EU directives and as well as national spatial planning guidance. For example, the EU Habitat and Bird Directives call for use of spatial overlay techniques in both the selection process of parcels suitable for inclusion in the Natura 2000 conservation system as well as part of spatial planning practices that could adversely impact conservation objectives (Simeonova et al., 2017). At the national level, a good example is Great Britain's National Planning Policy Framework which notes:

> The planning system also provides a key role in ensuring the suitability of a proposed development for its location in terms of risks from existing pollution or any polluting effect the new development might have on its surroundings, and in mitigating those effects.
>
> *(DCLG, 2012, p. 65)*

British local land use plans are also required to undertake sustainability appraisals which can be informed by land use suitability analysis (DCLG, 2012). On the other side of the world, the Land Administration Law of the People's Republic of China requires land use master plans to be updated every ten years and explicitly calls for use of land suitability analysis as part of the five step planning process (Metternicht, 2017). In fact, use of land use suitability analysis can be found in planning documents on every continent of the world; however, the actual extent of the practice has not been systematically studied to date.

Challenges and opportunities

If we peg modern-day suitability analysis to McHarg's publication of 'Design with Nature', ecologically grounded suitability analysis will soon be entering its 50th anniversary in 2019. Reflecting on this, what are the most significant challenges and opportunities for this analytic planning process as it reaches middle age? Based on the authors combined 50 plus years of teaching, research, and practice, 2 key challenges and opportunities are identified for the application of the technique going forward. The first challenge, alluded to in the proceeding sections, is making sure planners 'do not conflate quantity with quality', and that they recall the underlying ecological theoretical premise articulated by McHarg (1969) some 50 years ago. Suitability mapping should elucidate ecological processes of natural and human systems, making connections between systems and what makes sense within systems as we locate land uses and activities in a sensible fashion that works, not obfuscate processes or obscure the importance of key features in the mounds of data that are simply readily available. The second major challenge is continuing to view suitability analysis as an inherent and indispensable component of all forms of planning. While it was noted in the proceeding sections that suitability mapping is foundational to all forms of planning – that does not mean it is as widespread in usage as it ought to be. For example, the ever-escalating losses from natural hazards in the United States provide clear evidence that many cities and regions still are not paying attention to such fundamental constraints as avoiding high hazard locations as a starting point for land use and environmental planning.

In terms of opportunity, the continued growth of 'GeoDesign' as both a field and technological development is viewed as further cementing the importance of suitability analytic techniques into advance computing and environmental analysis for the foreseeable future. The promise of integrating data systems from building information systems to neighbourhood to cities to regions suggests a computational scalability and sophistication that would enable far more nuanced multi-scalar suitability analysis than is currently possible. Second, the advances in planning support systems software are also viewed as a promising opportunity for growth of the technique. For example, the 'Community Viz' software system integrates interactive suitability analysis capabilities that allow rapid on the fly sensitivity analysis in a fairly easy to use interface. The authors believe that further advance in these software technologies may help make what appears to be a fairly technical and, in some ways, unapproachable technique far more user friendly to lay audiences engaged in planning processes.

References

Berke, P., Newman, G., Lee, J., Combs, T., Kolosna, C. and Salvesen, D. (2015). 'Evaluation of networks of plans and vulnerability to hazards and climate change: a resilience scorecard'. *Journal of the American Planning Association*, 81(4): 287–302.

City of Austin (2012). *Imagine Austin*. City of Austin, Austin. Available at: www.austintexas.gov/sites/default/files/files/Planning/ImagineAustin/webiacpreduced.pdf

Collins, M. G., Steiner, F. and Rushman, M. (2001). 'Land-use suitability analysis in the United States: historical development and promising technological achievements'. *Environmental Management*, 28(5): 611–21.

Department for Communities and Local Government (DCLG) (2012). *The national planning policy framework*. London: DCLG.

Hopkins, L. D. (1977). 'Methods for generating land suitability maps: a comparative evaluation'. *Journal of the American Institute of Planners*, 43(4): 386–400.

McHarg, I. L. (1969). *Design with nature*. Garden City, NY: Natural History Press/Doubleday.

McHarg, I. L. (1997). *A quest for life*. New York: John Wiley and Sons.

Metternicht, G. (2017). 'Land use planning'. United Nations Global Lands Outlook Working Paper. United Nations Convention to Combat Desertification (UNCCD).

OECD (2017). *Land-use planning systems in the OECD: country fact sheets*. Paris: OECD Publishing. Available at: http://dx.doi.org/10.1787/9789264268579-en

Simeonova, V., Bouwma, I., van der Grift, E., Sunyer, C., Manteiga, L., Külvik, M., Suškevičs, M., Dimitrov, S. and Dimitrova, A. (2017). *Natura 2000 and Spatial Planning*. Final report for the European Commission (DG ENV) (Project 07.0202/2015/716477/ETU/ENV. B.3).

Steiner, F. (2000). *The living landscape: an ecological approach to landscape planning* (2nd edn). New York: McGraw-Hill.

Steiner, F., McSherry, L. and Cohen, J. (2000). 'Land suitability analysis for the Upper Gila River watershed'. *Landscape and Urban Planning*, 50(4): 199–214.

Steinitz, C., Parker, P. and Jordan, L. (1976). 'Hand drawn overlays: their history and prospective uses'. *Landscape Architecture*, 66(5): 444–55.

The Trust for Public Land (2008). *The Travis County Greenprint for growth*. Austin: The Trust for Public Land. Available at: http://data.capcog.org/Information_Clearinghouse/presentations/2006-10-01_Travis_County_Greenprint_for_Growth_Final_Report.pdf

36

Mainstreaming the environment in planning policy and decision making[1]

Alister Scott

Introduction

This chapter assesses theoretically and practically how we can improve environmental mainstreaming in policy and decision making from its relatively weak position at the present time. Significant research evidence via the Millennium Ecosystem Assessment (MEA) (2003), UKNEA (2011), TEEB (2010), ESPA (2018), WWF (2016) and IPBES (2018) highlights serious and ongoing environmental decline at multiple scales impacting across multiple habitats and species as a direct consequence of human interventions and actions. Despite increased knowledge and awareness of the multiple benefits that the environment provides for society; for example, in the natural capital value of the Great Barrier Reef estimated at £56 billion (Deloitte, 2017, p. 7), there still remains a prevalent narrative that the environment is an impediment to economic growth. This highlights the tension and potential incompatibility between short-term political and economic priorities versus longer-term environmental goals.

Mainstreaming is about reconciling these tensions and 'moving environmental issues from the periphery to the centre of decision-making, whereby environmental issues are reflected in the very design and substance of sectoral policies' (EEA, 2005, p. 12). Thus, mainstreaming processes necessarily involve a change in culture and behaviours across multiple audiences if they are to be successful. This requires attention to both the process and outcomes from mainstreaming activities, recognising that the term suffers from uncritical and excessive usage in research and practice initiatives which obfuscates its complexity and importance (Scott et al., 2018). Indeed, too many environmental challenges are identified, assessed, diagnosed and treated within environmental silos without the involvement of other stakeholders, often leading to conflict, rejection or alienation within economic, political and social arenas (Scott et al., 2013).

Furthermore, decision makers rarely understand or have access to sufficient scientific data to guide them or access to tools that take full account of the value(s) of the environment. Indeed, in reviewing 17 ecosystem service decision-support tools, Bagstad et al. (2013) found that most were too resource intense for routine use in public and private sector decision making.

This chapter draws on emerging ideas from environmental and spatial planning theory and practice to offer an improved conceptualisation and operationalisation of mainstreaming concomitant with the development and enhancement of mechanisms and tools to facilitate this. However, first we need to identify and understand the barriers that hinder current mainstreaming efforts.

Exposing disintegrated development in environmental planning theory and practice

Disintegrated development arguably represents the key cultural barrier to overcome (Scott et al., 2013). It captures the way much research, policy and decision making occurs within separate silos, each with their favoured paradigms, agency champions, preferred tools, approaches and stakeholders. This creates separate sectoral pathways with associated processes and outcomes that rarely intersect. Furthermore, the shift from government to governance has dramatically increased the number and complexity of actors, agencies and partnerships operating within the same contested spaces, resulting in increased participatory conflict with attendant problems of accountability and transparency to overcome (Lockwood, 2010).

Within the built and natural environment arena there are two 'competing' paradigms: spatial planning and the ecosystem approach. Allmendinger and Haughton (2010 , p. 83) define spatial planning as 'shaping economic, social, cultural, and ecological dimensions of society through "place making" with a shift towards more positive, integrated and resource-based contexts'. Whilst the the UN Convention on Biological Diversity (CBD, 2010 , p. 12) defines the ecosystem approach (EcA) as 'a strategy for the integrated management of land, water and living resources that promotes conservation and sustainable use in an equitable way'. Despite their similarities, few researchers have assessed their intersection or interdependencies (Scott et al., 2013) which has led to significant problems in built environment practice as illuminated within the three following case studies.

The fallacy of ecosystem services (ES) mainstreaming

Today the ES concept is believed to be the dominant policy-making framework for the natural environment sector (Reed et al., 2017). However, it has yet to be mainstreamed successfully across economic development and planning sectors, despite spurious claims to have done so (UKNEAFO, 2014). So whilst there have been dedicated national ecosystem assessments (for example, Schröter et al., 2016), new voluntary environmental markets from payments for ecosystem services programmes (for example, Reed et al., 2017), green accounting methods (for example, World Bank, 2010), habitat banking and trading development rights schemes (Santos et al., 2015) and improved communication on the importance of ecosystems and biodiversity to human wellbeing (Luck et al., 2012), the ES concept struggles to gain traction outside the natural environment arena (Posner et al., 2016).

The reasons for this failure are based on a number of co-related factors. First, the rapid pace of advancement of ecosystem science itself within the environmental sector, complete with its own separate sub-discourses and environmental critiques. Second, the limited attempts to engage with other sectors within research with a predilection for working with the usual environmental suspects as partners. For example, the planning profession consistently has been absent in ecosystem services research; yet the majority of policy and decision making is made using statutory land use planning documents and processes. Third, the economic and planning arenas have developed and championed their own guiding paradigms, narratives and tools (for example, Spatial Planning; Building Information Modelling; and SMART cities) reinforcing disciplinary and professional divides with the environment. Finally, the technocentric vocabulary of ES and natural capital requires significant investment in time to understand and apply to non-environmental work practices with few attempts in evidence to translate these ideas for other audiences (see TEEB, 2010 for a notable exception).

The value(s) and cost(s) of parks

There has been significant research valuing public parks and green infrastructure with evidence of multiple environmental benefits flowing from effective management regimes; for example, water regulation, climate change regulation, pollination, biodiversity, recreation, mental and physical health and wellbeing and food growing, albeit with differential spatial and socio-economic impacts (Wolch et al., 2014). Vivid Economics (2017) have recently calculated the value of London's parks as £91 billion using natural capital accounting methods.

However, decision makers primarily view parks in neoclassical economic cost and profit terms which currently does not generate sufficient tangible financial revenue via taxes and charges to offset long-term management liabilities in increasing maintenance budgets (Mackrodt and Helbrecht, 2013). Thus, cutting resources for parks is common, as the benefits of such investments are not easy to capture or to transfer into existing financing models or market-based instruments (Hanley and Barbier, 2013). Furthermore, through the lens of traditional economic cost benefit analysis, concepts such as increased Gross Value Added (GVA) and Gross Domestic Product (GDP) provide tangible and powerful political indicators that may lead to further erosion of parks and green spaces under the guise of economic development.

Low-impact development and scalar disintegration

The scalar dimension of policy disintegration is evidenced within development at Brithdir Mawr in Pembrokeshire national park (Adams et al., 2013). Here, conflict with a permaculture development in a farm setting in open countryside resulted in a complex planning case lasting over 10 years. Buildings were constructed in open countryside without planning permission and when spotted led to enforcement action being taken by Pembrokeshire Coast National Park Authority. The subsequent planning decision required several low-impact buildings to be demolished, including a roundhouse, as they contravened approved housing policy relating to development refusals in open countryside. A subsequent public inquiry further confirmed the buildings be demolished resulting in a global and national campaign to save the roundhouse. According to the planning inspector, the development could not be sustainable as it contravened development plan policies which, under plan-led legislation, provided the legal definition of sustainability (Scott, 2001). Simultaneously, the Welsh Assembly Government launched their new policy on low impact development where they included the condemned Brithdir Mawr roundhouse as an exemplar of sustainable development.

Here, the relative flexibility of national planning guidance collided with the more bureaucratic statutory and dated local plan processes that dealt with housing in isolation from the wider environmental and societal benefits that low impact developments might offer. The issue also exposed the problem of a risk averse planning system dealing with something new and innovative where, at that time, there was limited policy responses or case law to draw upon in relation to determining low impact developments, although planning agreements with temporary planning permission had been used previously as part of an adaptive management process; for example, Tinkers Bubble Somerset (Scott, 2001). However, the planning system did make provision for houses in open countryside for agricultural workers based on functionality and viability assessments, where workers were required to live onsite. Yet this was rooted in the provisions of the 1947 Agriculture Act where profitability was a key consideration. This in no way captures the diverse nature of contemporary farming/permaculture activities and the production of multiple environmental benefits that could have helped prove a case for such a development.

Environmental mainstreaming towards a mainstreaming typology

The previous case studies highlight the need for improved theoretical and practice pathways to make the transition from policy conflict and 'disintegration' towards more effective environmental mainstreaming.

Drawing from examples in practice undertaken by the author in developmental mainstreaming projects, the UKNEAFO (2014) and Scott et al. (2018), a mainstreaming continuum has been developed for this chapter to capture and characterise different modes of environmental mainstreaming activity that were observed (Figure 36.1). The different stages reflect increasing capacity, capability and culture/behaviour change within a given case study setting but crucially should not be seen as a normative framework. Rather, they reflect the art and science of what is possible and pragmatic at that time.

Thus, some initiatives will start mainstreaming activity through simply retrofitting environmental benefits/services into to a plan, policy, project or programme (PPPP) retrospectively without influencing the rest of the plan process or document (Retrofit). Usually this takes the form of an action plan which can act as an evaluation tool for future progress. A more common approach is to create a dedicated environmental chapter/section/project as part of a PPPP but with limited integration or cross referencing across the other chapters/sections/areas, reflecting the different individuals who developed them in isolation (Incremental). Arguably this is the most common stage of environmental mainstreaming that was observed, but it is vulnerable without integration. This mode occurs when a PPPP identifies specific multidisciplinary challenges that require the integration of multiple policy areas and staff working together within planned assessments and interventions. Often this involves task and finish groups which are assembled for such activities (Challenge-led). Finally, there is a mode that champions a systems perspective. Here the PPPP challenges are identified collectively through an understanding of the drivers, interactions and interdependencies affecting the PPPP leading to co-developed interventions across different interdisciplinary teams with shared values and goals identified from the outset (Systemic).

The continuum is dynamic with movement either way possible due to improved knowledge exchange/transfer (for example, successful case studies) and/or particular planning decisions and case law. Movement can also be dramatic when a tipping point or crisis emerges which transforms actions. Good examples of this are the recent issues in plastic pollution and climate emergency where the Blue Planet 2 TV Series and campaigning by Greta Thunberg respectively have mobilized global, national, and local action and legislation.

Rogers (2003) classic diffusion model provides a useful way of conceptualising this and mainstreaming processes more generally as different environmental innovations move through stages of knowledge generation, persuasion, decision (adoption/rejection), implementation and

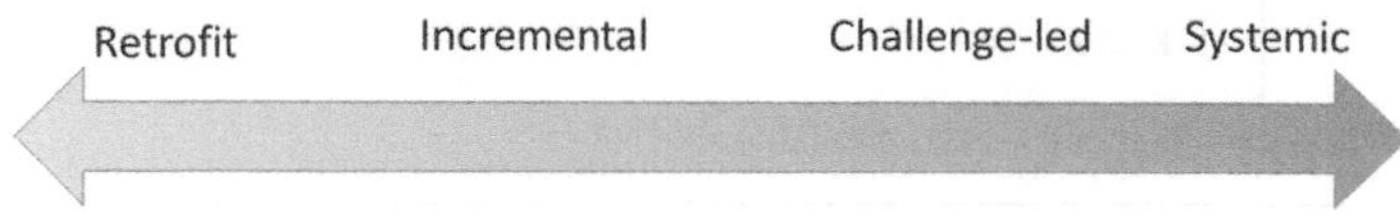

Figure 36.1 A mainstreaming continuum

Source: Author's own

confirmation (Figure 36.2). Here the efficacy of the communication channels, the nature and power of those stakeholders and gatekeepers involved, and the nature of the idea or innovation itself, become key features in determining progression. Overcoming the persuasion stage is perhaps the key gateway to reaching the challenge and systemic stages.

The current barriers that prevent the persuasion stage being breached are summarised below:

- The environmental narrative has been traditionally constructed and viewed in policy and decision making as a constraint to overcome;
- The complex and widespread use/misuse of ES, natural capital and biodiversity language has alienated other interests who struggle to keep up with the evolution of ecosystem science;
- The lack of exemplars and social learning platforms that can demonstrate environmental mainstreaming inhibits further uptake;
- The widespread use of policy-based evidence approaches by key gatekeepers hinders acceptance of different ideas to the accepted policy narrative. Thus, evidence-based policy is often rejected;
- The micropolitics within agencies and partnerships can block meaningful progress and change, with the status quo preferred.

We can flip these barriers into opportunities within a stepped process:

- Reconnecting disparate paradigms (for example, the ecosystem approach and spatial planning) to help integrate different theoretical viewpoints within shared values and vocabularies;
- Reframing and translating core environment concepts within the vocabulary and everyday terms and concepts that are used and prioritised by economic and social interests and audiences:
  - identifying 'hooks' reflecting key policy or legislation, duties or priorities that relate to a particular user group to build initial traction;
  - identifying 'bridges' reflecting a term, concept or policy priority that is used and readily understood across multiple groups and publics to build traction;
- Translating the environmental concepts within these hooks and bridges amongst policy and decision makers to start inclusive and safe discussion spaces generating new pathways.

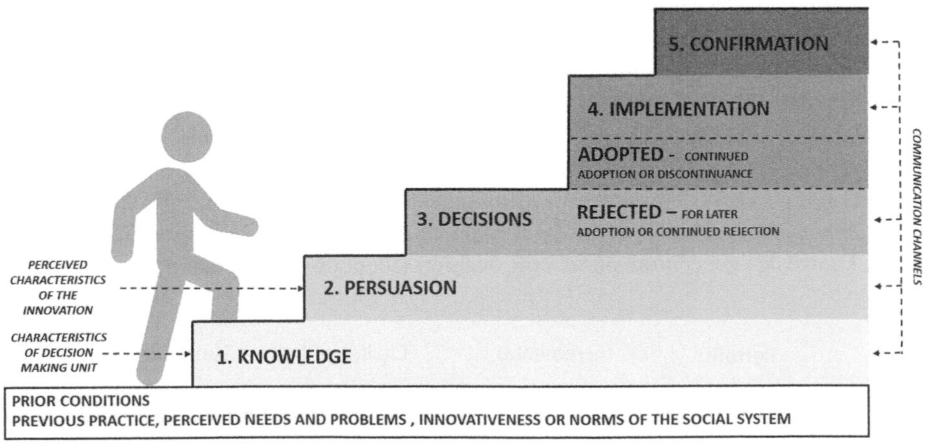

Figure 36.2 A model of five stages in the innovation decision process

Source: adapted from Rogers (2003, p. 170)

All the above points highlight the need for developing more inclusive processes and improved dialogues based on mutual respect and understandings of different groups' world views within safe social learning and knowledge exchange spaces which can readily incorporate and integrate competing theories, ideas and cultures and thus break down silos (Cowell and Lennon, 2014). The connecting of paradigms from their hitherto separate pathways provides the conceptual glue to join up the disparate and disintegrated natural and built environments between the ecosystem approach and spatial planning.

Table 36.1 captures these synergies within a mapping exercise of the 12 Malawi principles (Ecosystem Approach (EcA)) against six spatial planning principles advanced by the UNECE (2008). Both EcA and Spatial Planning are rooted in social-ecological systems thinking (Bruckmeier, 2016) within an interdisciplinary human-centred perspective crossing environmental, social, economic, political and cultural contexts and sectors (Gómez-Baggethum and Barton, 2013). Both require the adoption of participatory approaches incorporating equity and shared values (Reed et al., 2013); both involve a change in values and thinking from negative associations of policies associated with protection, control and restraint towards more holistic, proactive and development-led visions and interventions (Scott et al., 2013).

The journeys and experiences of two environmental mainstreaming case studies are now used to help illuminate the mainstreaming continuum revealing its wider applicability and transferability. Table 36.2 summarises these two case studies.

Birmingham Green Living Spaces Plan (GLSP)

Birmingham has an ambition to become a leading global green city. Against this backdrop, it established a Green Commission, a cross cabinet-level body, who collectively agreed a new green vision. As part of the development of the City's Local Plan the need for a Green Infrastructure Strategy was identified and approved. This opportunity was exploited to take advantage of the latest science emerging from the UK National Ecosystem Assessment (UKNEA, 2011) and provisions within the Natural Environment White Paper and the emerging National Planning Policy Framework (NPPF) (DCLG, 2012) (paragraph 109 recognising the value of ecosystem services).

One of the barriers to adopting an ecosystem approach at a city scale had been the level of understanding required. Therefore, a series of studies were undertaken, applying the ecosystem services methodology to six dominant urban issues and displaying these as GIS maps of the city. These six chosen topics were aesthetics and mobility, flood risk, local climate, education, recreation and biodiversity. These were depicted in both supply and demand maps showing areas of need and overprovision. These six maps were then super-imposed into a single multi-layered challenge map for Birmingham (Figure 36.3).

These maps can then be overlaid onto the street plan and reduced to district or neighbourhood scale for more detailed interpretation and used as evidence maps by non-specialists including community groups and elected members.

As part of the mainstreaming process in the GLSP, Birmingham established a cross disciplinary working group, who brought together their evidences, policies and delivery plans. This dialogue generated seven cross cutting principles that could form the backbone of the green infrastructure policy. These seven principles have then been locked into the statutory Birmingham Local Plan (now approved). The seven chosen principles are: (1) An Adapted City; (2) The City's Blue Network; (3) A Healthy City; (4) The City's Productive Landscapes; (5) The City's Greenways; (6) The City's Ecosystem; and (7) The City's Green Living Spaces.

Table 36.1 Mapping spatial planning and ecosystem approach paradigms

Spatial planning principles	*Ecosystem approach principles*
'shaping economic, social, cultural, and ecological dimensions of society through "place making" . . . positive, integrated and resource-based contexts' Allmendinger and Haughton (2010, p. 83).	'strategy for the integrated management of land, water and living resources that promotes conservation and sustainable use in an equitable way' (CBD, 2010, p. 12).
The Governance Principle (e.g. authority, legitimacy, institutions, power, decision making)	1 The objectives of management of land, water and living resources are a matter of societal choice. 3 Ecosystem managers should consider the effects (actual or potential) of their activities on adjacent and other ecosystems. 9 Management must recognise the change is inevitable.
The Subsidiarity Principle (e.g. delegation to lowest level, shared responsibility, devolution)	2 Management should be decentralised to the lowest appropriate level.
The Participation Principle (e.g. consultation, inclusion, equity, deliberation)	11 The ecosystem approach should consider all forms of relevant information, including scientific and indigenous and local knowledge, innovations and practices. 12 The ecosystem approach should involve all relevant sectors of society and scientific disciplines.
The Integration Principle (e.g. holistic, multiple scales and sectors, joined up)	3 Ecosystem managers should consider the effects (actual or potential) of their activities on adjacent and other ecosystems. 5 Conservation of ecosystem structure and functioning, in order to maintain ecosystem services, should be a priority target of the ecosystem approach. 7 The ecosystem approach should be undertaken at the appropriate spatial and temporal scales. 8 Recognising the varying temporal scales and lag effects that characterise ecosystem processes, objectives for ecosystem management should be set for the long-term. 10 The ecosystem approach should seek the appropriate balance between, and integration of, conservation and use of biological diversity.
The Proportionality Principle (e.g. deliverable viability, pragmatism, best available information)	4 Recognising potential gains from management, there is usually a need to understand and manage the ecosystem in an economic context. 9 Management must recognise the change is inevitable.
The Precautionary Principle (e.g. adaptive management, limits, uncertainty, risk)	6 Ecosystem must be managed within the limits of their functioning. 8 Recognising the varying temporal scales and lag effects that characterise ecosystem processes, objectives for ecosystem management should be set for the long-term. 10 The ecosystem approach should seek the appropriate balance between, and integration of, conservation and use of biological diversity.

Source: Scott et al. (2018, p. 233)

Table 36.2 Case studies of environmental mainstreaming

Case study	*Environmental planning challenge*	*Key ESP principles*	*Approach to mainstreaming*	*Hooks /bridges*	*Key tools used to help mainstreaming*
Birmingham City Council non statutory Green Living Spaces Plan 2014	What is the value of green infrastructure (GI) to the residents and businesses of the city? How can the council embed this information to improve its policies, plans and investment opportunities?	Governance Integration Proportionality Precautionary	ES assessment of GI. Created green commission at Cabinet level. Used ES data sets to create demand and supply maps showing areas requiring ES investment.	H Climate change national performance indicators H NPPF paragraph 109[2] B Green Infrastructure B Multiple Benefits B Risk	Green Commission Ecosystem Services Mapping (Demand and Supply) Public engagement workshops
South Downs National park	How can the EcA be used within the park authority in general and a park local plan in particular to improve policy and decision making?	Governance Integration Proportionality Precautionary	Developing an ES policy as one of 4 core polices pervading across all local plan policies	H UKNEA H NEWP H NPPF paragraph 109 B Park Management Plan Infrastructure B Multiple Benefits	Strong and effective leadership at officer and board level Mapping Ecosystem Services Statutory planning process. Board, staff and public engagement workshops Technical advice for householders/ developers to optimise policy outcomes.

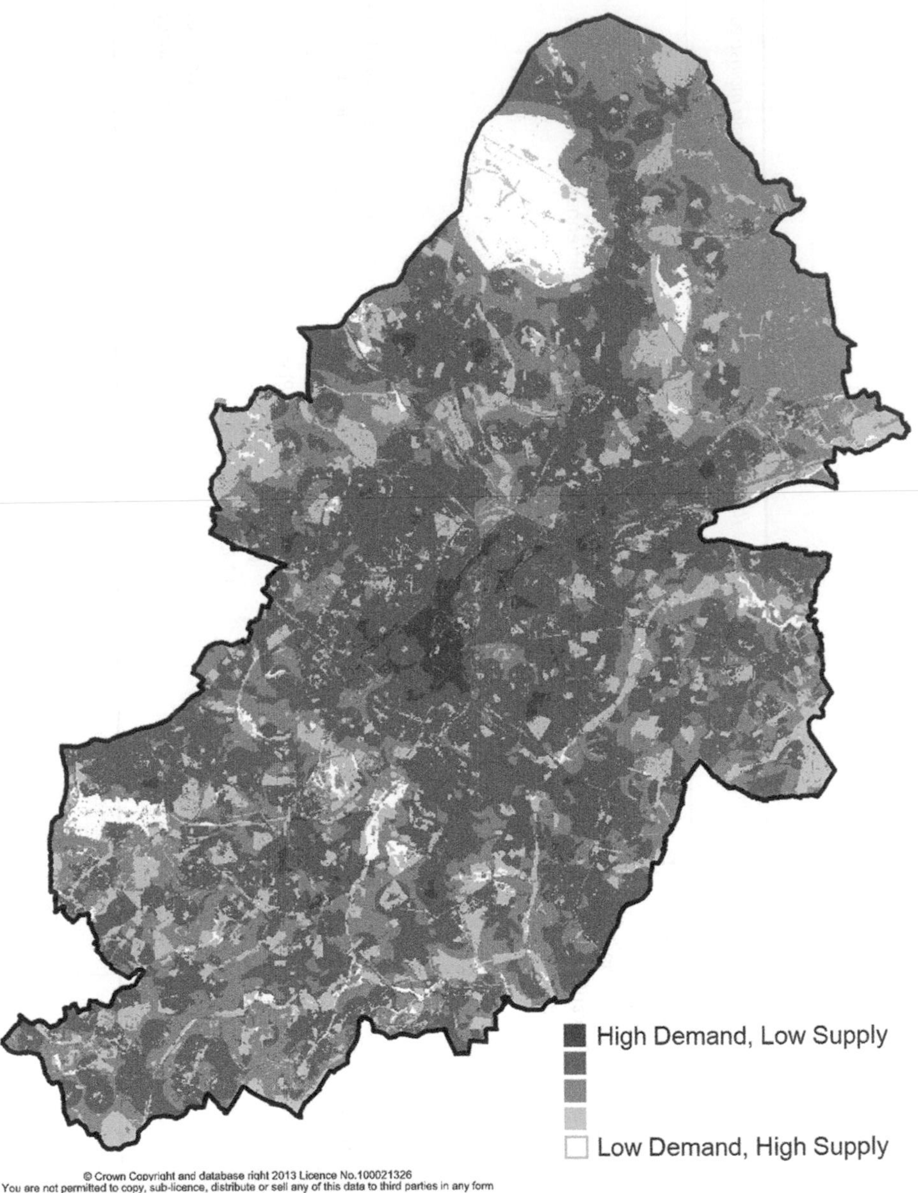

Figure 36.3 Multi-layered challenge map for Birmingham, UK

Source: Birmingham City Council (2013, p. 26)

Additional OS copyright notice:

Plans contained within this document are based upon Ordnance Survey material with the permission of Ordnance Survey on behalf of the Controller of Her Majesty's Stationery Office.

What is the added value of this process to environmental planning?

- Green Commission endorsement of the Ecosystem Services Framework has brought added value to driving forward the city's green vision and appetite for change;
- Supply and demand maps offer a tangible and understandable output for action in particular areas based on evidence which can help secure investment;
- Applying the Ecosystem Approach brings together a wider range of stakeholders and potential budget-holders/investors into a shared dialogue that would not normally happen.

What are the lessons learnt?

- Strong and effective leadership to drive the change agenda through despite obstacles;
- Linking the work to national government policy (hooks) and other departmental interests and concerns (for example, risk and health) is key to making progress;
- Demonstrate who benefits and who loses in easy visualisations;
- Lock-in the proposed changes to existing and future city policy and spatial planning policy;
- Recruit local and political champions across different communities of interest to maximise legacy;
- Updating the data and ongoing communication to members to also inform any new layers of governance that may emerge.

South Downs National Park Local Plan

South Downs National Park Authority (SDNPA) was created in 2011 with statutory responsibilities for the protection of the natural beauty and the promotion of informal recreation. As a new national park, it positioned itself as an innovator in environmental planning and delivery seeking to mainstream the EcA into its plans and policy processes. The UKNEA (2011); NEWP (HM Government, 2011b) and NPPF (DCLG, 2012) were seen as key hooks to facilitate this.

Its first statutory park management plan (SDNPA, 2013) set out the framework for the protection of the park and its special qualities using the Ecosystems Services Framework (ESF). The draft local plan built on this plan providing the statutory planning policy framework and area plans for deciding planning applications within the park boundary.

Initially there was a targeted strategy of consultation and awareness-raising amongst its members, partnership board and 15 planning districts through a number of meetings and workshop events. Here the NPPF paragraph 109 hook on 'recognising the value of ecosystem services' (DCLG, 2012) helped secure the involvement of the entire planning team (strategic and development management) with strong leadership and enthusiasm from the director of planning. This created a bridge to communicate and work jointly with other section leads across the SDNPA enabling the park to secure resources for mapping key ecosystem services (ECOSERV) as an evidence base to feed subsequent policy development. Crucially at that time, they also engaged with several academics and researchers to maximise knowledge exchange within their method. The cumulative social learning resulted in draft policy (SD2, Ecosystem Services (SDNPA, 2015)) which sits as one of only four higher-level policies that all other policies in the plan are subservient to.

Draft Core Policy SD2 Ecosystem Services (submission to Local Plan inquiry)

Proposals that deliver sustainable development and comply with other relevant policies will be permitted provided that they do not have an unacceptable adverse impact on the natural environment and its ability to contribute goods and services. Proposals will be expected to:

a Provide more and better joined up natural habitats;
b Conserve water resources;
c Sustainably manage land and water environments;
d Improve the National Park's resilience to, and mitigation of, climate change;
e Increase the ability to store carbon through new planting or other means;
f Conserve and improve soils;
g Reduce pollution;
h Mitigate the risk of flooding;
i Improve opportunities for peoples' health and wellbeing;
j Stimulate sustainable economic activity;
k Deliver high-quality sustainable design.

Development proposals must be supported by a statement that sets out how the development proposal impacts, both negatively and positively on ecosystem services (SDNPA, 2015).

The policy provides a negotiating tool for planners to have a dialogue about ES outcomes. Note how the ES language is 'translated' into plain English concepts in categories a–k which are accessible to planning applicants and wider publics. This has been supplemented with bespoke technical advice to developers and householders showing how positive environmental outcomes can be secured. This helps deliver the innovative requirement for all developments to produce an assessment of the impact on the ecosystem services, a key prerequisite for changing behaviours.

Under the NPPF (DCLG, 2012) and Localism Act 2011 (HM Government, 2011a), the park is required to undertake a Duty to Cooperate (DTC) across cross boundary strategic issues to ensure that ES are protected and enhanced. Their interim statement on these issues (SDNPA, 2015, p. 4.2) is below:

- Conserving and enhancing the natural beauty of the area;
- Conserving and enhancing the region's biodiversity;
- The delivery of new homes, including affordable homes and pitches for Travellers;
- The promotion of sustainable tourism;
- Development of the rural economy;
- Improving the efficiency of transport networks by enhancing the proportion of travel by sustainable modes and promoting policies which reduce the need to travel.

This statutory obligation to cooperate with the SDNPA also helps them engage with other planners across 15 neighbouring authorities on environmental topics providing the initial traction to what are likely to be challenging discussions.

What is the added value of this process to environmental planning?

- Securing the involvement of all planning staff and board members helps mainstreaming processes develop from the outset. There is a new paradigm which all staff buy into;
- The positive framing of Policy SD2 with bespoke guidance for households and developers enables beneficial ES outcomes to be discussed and negotiated from all planning applications across multiple scales and stakeholders;

- Building a pyramid of plans (park management plan – local plan – neighbourhood plans and green infrastructure plan) based on the same paradigm overcomes silos within the SDNPAs and its residents.

Lessons learnt

- Importance of securing political and officer leadership from the start;
- Value of working with outside academics brings credibility and rigour to processes;
- Need both regulatory and incentive approaches to maximise mainstreaming potential; regulation helps establish initial traction but nudging and negotiation processes are critical;
- Up-front investment in participation with residents, landowners and neighbouring local planning authorities is key to successful mainstreaming efforts.

Conclusion

This chapter has exposed the mainstreaming challenge within environmental planning signposting various ways of addressing it effectively. At present, we are experiencing significant policy disintegration which inhibits and obfuscates mainstreaming processes leaving them trapped largely within environmental sectors and stakeholders. Mainstreaming is a dynamic and evolutionary diffusion process constrained by capacities, capabilities and micro-politics within a given setting. We have identified key drivers that influence success: the need for political support; the interplay between statutory and informal procedures; effective leadership; safe social learning spaces; and a willingness to experiment by stepping outside usual comfort zones and developing new funding and investment tools. Crucially, the language and process of environmental mainstreaming needs to be collectively owned and positively shaped by those engaged with it rather than relying on one department or staff member for its success. This requires a significant culture and behaviour change.

By working across different paradigms, synergies can be identified within which to position mainstreaming efforts. Here the concepts of hooks and bridges provide the key translation devices to help the transition from environmental silos to more effective mainstreaming, enabling key actors and gatekeepers to accept, use and ultimately legitimise environmental concepts within their own policy and practice vocabularies and priorities, thereby creating the traction for further exploration and development of the idea within an adoption process as exemplified by the Birmingham and South Downs examples. In such pioneering endeavours this chapter has hopefully given a stronger theoretical and practical basis for mainstreaming together with a call for more collective social learning from both successes and failures to provide improved opportunity spaces for policy and decision making in the future.

Notes

1 This chapter is based on work carried out as part of the following grant: Mainstreaming Green Infrastructure in Planning Policy and Decision Making NE/R00398X/1.

2 NPPF (DCLG, 2012) paragraph 109 'The planning system should contribute to and enhance the natural and local environment by: . . . recognising the wider benefits of ecosystem services'.

References

Adams, D., Scott, A. J. and Hardman, M. (2013). 'Guerrilla warfare in the planning system: revolutionary progress towards sustainability?' *Geografiska Annaler, Series B: Human Geography*, 95(4): 375–87.

Allmendinger, P. and Haughton, G. (2010). 'Spatial planning, devolution, and new planning spaces'. *Environmental Planning C: Politics and Space*, 28(5): 803–18.

Bagstad, K. J., Semmens, D. J., Waage, S. and Winthrop, R. (2013). 'A comparative assessment of decision-support tools for ecosystem services quantification and valuation'. *Ecosystem Services*, 5: 27–39.

Birmingham City Council (2013). *Green living spaces plan*. Birmingham: Birmingham City Council.

Bruckmeier, K. (2016). *Social-ecological transformation: reconnecting society and nature*. London: Palgrave Macmillan.

Convention on Biological Diversity (CBD) (2010). *Ecosystem Approach*. Available at: www.cbd.int/ecosystem/

Cowell, R. and Lennon, M. (2014). 'The utilisation of environmental knowledge in land-use planning: drawing lessons for an ecosystem services'. *Environment and Planning C: Government and Policy*, 32(2): 263–82.

Department for Communities and Local Government (DCLG) (2012). *The national planning policy framework*. London: DCLG.

Deloitte (2017). *At what price? The economic, social and icon value of the Great Barrier Reef*. Brisbane: Deloitte Access Economics. Available at: https://www2.deloitte.com/content/dam/Deloitte/au/Documents/Economics/deloitte-au-economics-great-barrier-reef-230617.pdf

EEA (2005). *Environmental policy integration in Europe: state of play and an evaluation framework*. Technical Report No. 2/2005. Copenhagen: European Environment Agency.

ESPA (Ecosystem Services Poverty Alleviation) (2018). *ESPA research programme*. Available at: www.espa.ac.uk

Gómez-Baggethun, E. and Barton, D. N. (2013). 'Classifying and valuing ecosystem services for urban planning'. *Ecological Economics*, 86: 235–45.

Hanley, N. and Barbier, E. B. (2013). *Pricing nature. Cost-benefit analysis environmental policy*. Cheltenham: Edward Elgar.

HM Government (2011a). *The Localism Act*. Available at: www.legislation.gov.uk/ukpga/2011/20/contents/enacted

HM Government (2011b). *The natural choice: securing the value of nature*. CM 8082. London: HM Government. Available at: www.gov.uk/government/uploads/system/uploads/attachment_data/file/228842/8082.pdf

IPBES (Intergovernmental Science-Policy Platform on Biodiversity and Ecosystem Services) (2018). *Work programme*. Available at: www.ipbes.net

Lockwood, M. (2010). 'Good governance for terrestrial protected areas: a framework, principles and performance outcomes'. *Journal of Environmental Management*, 91(3): 745–66.

Luck, G. W., Lavorel, S., McIntyre, S. and Lumb, K. (2012). 'Improving the application of vertebrate trait-based frameworks to the study of ecosystem services'. *Journal of Animal Ecology*, 81(5): 1065–76.

Mackrodt, U. and Helbrecht, I. (2013). 'Performative Bürgerbeteiligung als neue Form kooperativer Freiraumplanung' [Performative participation – a new cooperative planning instrument for urban public spaces]. *disP – The Planning Review*, 49(4): 14–24.

Millennium Ecosystem Assessment (MEA) (2003). *Ecosystems and human well-being. A framework for assessment*. Washington, DC: Island Press.

Posner, S., Getz, C. and Ricketts, T. (2016). 'Evaluating the impact of ecosystem service assessments on decision-makers'. *Environmental Science & Policy*, 64: 30–7.

Reed, M. S., Hubacek, K., Bonn, A., Burt, T. P., Holden, J., Stringer, L. C., Beharry-Borg, N., Buckmaster, S., Chapman, D., Chapman, P., Clay, G. D., Cornell, S., Dougill, A. J., Evely, A., Fraser, E. D. G., Jin, N., Irvine, B., Kirkby, M., Kunin, W., Prell, C., Quinn, C. H., Slee, W., Stagl, S., Termansen, M., Thorp, S. and Worrall, F. (2013). 'Anticipating and managing future trade-offs and complementarities between ecosystem services'. *Ecology and Society*, 18(1): 5. http://dx.doi.org/10.5751/ES-04924-180105

Reed, M. S., Allen, K., Attlee, A., Dougill, A. J., Evans, K., Kenter, J., McNab, D., Stead, S. M., Twyman, C., Scott, A. J., Smyth, M. A., Stringer, L. C. and Whittingham, M. J. (2017). 'A place-based approach to payments for ecosystem services'. *Global Environmental Change*, 43: 92–106.

Rogers, E. M. (2003). *Diffusion of innovations* (5th edn). London: Simon and Schuster.

Santos, R., Schroter-Schlaack, C., Antunes, P., Ring, I. and Clemente, P. (2015). 'Reviewing the role of habitat banking and tradable development rights in the conservation policy mix'. *Environmental Conservation*, 42(4): 294–305.

Schröter, M., Albert, C., Marques, A., Tobon, W., Lavorel, S., Maes, J., Brown, C., Klotz, S. and Bonn, A. (2016). 'National ecosystem assessments in Europe: a review'. *Bioscience*, 66(10): 813–28.

Scott, A. J. (2001). 'Contesting sustainable development: a case study of Brithdir Mawr'. *Environment Policy and Planning*, 3: 273–87.

Scott, A. J., Carter, C. E., Larkham, P., Reed, M., Morton, N., Waters, R., Adams, D., Collier, D., Crean, C., Curzon, R., Forster, R., Gibbs, P., Grayson, N., Hardman, M., Hearle, A., Jarvis, D., Kennet, M. Leach, K., Middleton, M., Schiessel, N., Stonyer, B. and Coles, R. (2013). 'Disintegrated development at the rural urban fringe: re-connecting spatial planning theory and practice'. *Progress in Planning*, 83: 1–52.

Scott, A. J., Carter, C., Hardman, M., Grayson, N. and Slaney, T. (2018). 'Mainstreaming ecosystem science in spatial planning practice: exploiting a hybrid opportunity space'. *Land Use Policy*, 70: 232–46.

South Downs National Park Authority (SDNPA) (2013). *Partnership management plan.* Available at: www.southdowns.gov.uk/national-park-authority/our-work/key-documents/partnership-management-plan/

South Downs National Park Authority (SDNPA) (2015). *Local plan.* Available at: www.southdowns.gov.uk/planning/planning-policy/national-park-local-plan/

TEEB (2010). *The economics of ecosystems and biodiversity: mainstreaming the economics of nature: a synthesis of the approach, conclusions and recommendations of TEEB.* Geneva: UNEP TEEB.

UK National Ecosystem Assessment (UKNEA) (2011). *Synthesis of the key findings.* Cambridge: UNEP-WCMC.

UK National Ecosystem Assessment Follow-On (UKNEAFO) (2014). *Synthesis of the key findings.* Cambridge: UNEP-WCMC.

United Nations Economic Commission for Europe (UNECE) (2008). *Spatial planning: key instrument for development and effective governance with special reference to countries in transition.* Geneva: UN.

Vivid Economics (2017). *Natural capital accounts for public green space in London.* London: Vivid Economics. Available at: www.vivideconomics.com/publications/natural-capital-accounts-for-public-green-space-in-london

Wolch, J. R., Byrne, J. and Newell, J. P. (2014). 'Urban green space, public health, and environmental justice: the challenge of making cities "just green enough"'. *Landscape and Urban Planning*, 125: 234–44.

World Bank (2010). *Environmental valuation and greening the national accounts: challenges and initial practical steps.* Washington, DC: World Bank. Available at: http://siteresources.worldbank.org/EXTEEI/Resources/GreeningNationalAccountsDec19.pdf

WWF (2016). *Living planet report 2016. Risk and resilience in a new era.* Gland, Switzerland: WWF International.

Index

Note: page numbers in *italics* indicate figures and in **bold** indicate tables on the corresponding pages.

Printed and bound by CPI Group (UK) Ltd, Croydon, CR0 4YY
17/10/2024
01775694-0005